SYMMETRIES AND NUCLEAR STRUCTURE

NUCLEAR SCIENCE RESEARCH CONFERENCE SERIES

A series of monographs edited by Alexander Zucker, Oak Ridge National Laboratory

VOLUME 1 **GIANT MULTIPOLE RESONANCES**
Edited by Fred E. Bertrand

VOLUME 2 **CONTINUUM SPECTRA OF HEAVY ION REACTIONS**
Edited by T. Tamura, J. B. Natowitz and D. H. Youngblood

VOLUME 3 **LASERS IN NUCLEAR PHYSICS**
Edited by C. E. Bemis, Jr. and H. K. Carter

VOLUME 4 **HIGH ANGULAR MOMENTUM PROPERTIES OF NUCLEI**
Edited by Noah R. Johnson

VOLUME 5 **HEAVY IONS AND NUCLEAR STRUCTURE**
Edited by B. Sikora and Z. Wilhelmi

VOLUME 6 **NUCLEAR PHYSICS WITH HEAVY IONS**
Prospects at Energies Below 20 MeV/amu
Edited by Peter Braun-Munzinger

VOLUME 7 **INSTRUMENTATION FOR HEAVY ION NUCLEAR RESEARCH**
Edited by Dan Shapira

VOLUME 8 **COLLECTIVE MOTION AND GIANT RESONANCES**
Edited by Z. Wilhelmi and M. Kicińska-Habior

VOLUME 9 **HEAVY IONS IN NUCLEAR PHYSICS**
Edited by Z. Wilhelmi and M. Kicińska-Habior

VOLUME 10 **NUCLEAR PHYSICS: RECENT TRENDS AND DEVELOPMENTS**
Edited by Z. Wilhelmi, G. Szeflińska, and M. Kicińska-Habior

VOLUME 11 **NUCLEAR FISSION AND HEAVY-ION-INDUCED REACTIONS**
Edited by Wolf-Udo Schröder

VOLUME 12 **COHERENT EFFECTS IN HIGHLY EXCITED NUCLEI**
Edited by Z. Wilhelmi and G. Szeflińska

VOLUME 13 **SYMMETRIES AND NUCLEAR STRUCTURE**
Edited by R. A. Meyer and V. Paar

Additional volumes in preparation

This book is part of a series. The publisher will accept continuation orders which may be cancelled at any time and which provide for the automatic billing and shipping of each title in the series upon publication. Please write for details.

SYMMETRIES AND NUCLEAR STRUCTURE

Proceedings from the International
Symposium on Symmetries and Nuclear Structure
held at the Inter-University Centre, Dubrovnik, Yugoslavia,
5–14 June 1986

Edited by

R. A. Meyer
Lawrence Livermore Laboratory, USA

and

V. Paar
University of Zagreb, Yugoslavia

harwood academic publishers
chur • london • paris • new york • melbourne

Harwood Academic Publishers

Post Office Box 197
London WC2E 9PX
England

58, rue Lhomond
75005 Paris
France

Post Office Box 786
Cooper Station
New York, New York 10276
United States of America

Camberwell Business Center
Private Bag 30
Camberwell, Victoria 3124
Australia

Library of Congress Cataloging-in-Publication Data

International Symposium on Symmetries and Nuclear
Structure (1986 : Dubrovnik, Croatia)
Symmetries and nuclear structure.

(Nuclear science research conference series,
ISSN 0250-4375 ; v. 13)
Includes index.
1. Symmetry (Physics)—Congresses. 2. Nuclear
structure—Congresses. I. Meyer, R. A. (Richard
Albert), 1927- . II. Paar, Vladimir. III. Title.
IV. Series.
QC793.3.S916 1986 530.1'42 87-8764
ISBN 3-7186-0400-0

Printed in the United States of America.

CONTENTS*

---

*Only the first author is given here. For a complete listing of all authors see the last section titled "AUTHOR INDEX".

PREFACE TO THE SERIES

Nuclear science is now in full flower. From obscure origins in a few university laboratories it has grown to be one of the richest fields of scientific endeavor in the history of science. It has spun off two virtually unlimited sources of energy for mankind, fission and fusion. It is used broadly in many applied areas that range from medicine to agriculture to mining to metallurgy, and withal it continues to be a happy and productive basic science. New approaches, new particles, new energy regimes, and above all new ideas continue to provide astonishing theoretical insights and frequently unexpected experimental results.

Burgeoning growth is a healthy sign, but it carries with it a concomitant information explosion: as the science expands, the number of scientists increases, and specialized areas appear where once there existed but one unified field. To be sure, the technical literature performs its task adequately, but if a practitioner in a narrow field of nuclear science is to be aware of everything that he needs to know, he must read a dozen or more journals. Frequently a year or more elapses before a physicist's results appear in the journal literature. Because of such formidable information barriers, rapid communication in science is now principally oral rather than written. With this realization, specialized scientific conferences become a prime arena of scientific exchange and scientific publication. Spectacular results are now nearly always first communicated at conferences where they also receive their first baptism under fire. Sometimes new ideas may die at the same conference where they were first revealed. Conferences provide the stage where experiment and theory are joined, where results of different groups are compared, and new syntheses are evolved.

A conference will frequently broaden its program to include a paper or two on ancillary but essential developments. For example, a meeting on heavy-ion nuclear reactions might well include a session on new accelerator ideas or ion source developments. Finally, a conference ends with a summary by a noted practitioner in the field, and this summary can have great and lasting value for the reader.

It is for all these reasons that we have undertaken to publish a conference series in nuclear science, NUCLEAR SCIENCE RESEARCH. We expect this series, while it specializes in the areas of nuclear physics, to cover topics both basic and applied. We expect each conference to have its specialized nature so well defined that the proceedings will serve as a reference volume of lasting importance, and we expect that the resource provided to the working physicist by this series will enable those who were not participants to get both the data and the flavor of the conference. Speed of publication is essential for such proceedings. We are told that with modern publishing methods the book can appear 3 to 4 months after receipt of the last manuscript. It is our intent that the series NUCLEAR SCIENCE RESEARCH publish proceedings of important conferences in the field, and our hope that it will provide a record of advances in a way that is most useful to the working scientist.

Alexander Zucker

PREFACE

This book is organized into nine chapters, which consist of manuscripts of invited presentations at the international symposium on Symmetries and Nuclear Structure (SANS). The first two chapters cover theoretical and experimental studies of symmetry in nuclei. The next two chapters concern investigations into aspects of nuclear properties with Chapter 4 consisting of investigations of the octupole and dipole modes of nuclei. Chapters 5 and 6 address the use of heavy ions and high energy to investigate aspects of nuclear structure, respectively. Chapters 7 and 8 contain individual theoretical studies and new experimental techniques, respectively. The last chapter has been included at the request of many participants who expessed a desire to have a permanent copy of the material distributed in association with the SANS formal concert.

The SANS symposium was held at the Inter-University Centre (IUC) in Dubrovnik, Yugoslavia from 5 to 14 June 1986. The total attendance of approximately 200 scientists from approximately 50 countries certainly attests to the high and renewed interest in the area of nuclear structure that exists today. In fact, it was because of this high interest that the original conference bifurcated into the SANS symposium and a parallel conference on Nuclear Structure, Reactions, and Symmetries (NSRS).

The symposium benefitted greatly from being located at the IUC in Dubrovnik. Founded in 1970 by Prof. Ivan Supek, the IUC serves as a location for both conferences and small study groups, which can utilize its large auditorium and smaller independent seminar and course rooms. Mrs. Berta Dragicevic serves officially as executive secretary of the IUC, but in fact does the work of an executive director of the facility. Her efforts on behalf of the SANS symposium are greatly appreciated.

A special effort was made to include graduate students who are presently engaged in thesis work in nuclear science. A separate session of student presentations took place and prizes were given for those who were voted to be most outstanding in content and presentation. The names of the students who were awarded these prizes, named in honor of Prof. Gaja Alaga of the University of Zagreb, are given in the Alaga Award announcement page.

No conference can take place without sponsors and financial contributors. On a separate page we present a list of those organizations to which we are grateful for their support of the SANS symposium. Prof. Daeg S.

Brenner of Clark University served as co-finance officer with Prof. Ruth P. Yaffe of San Jose State University. The symposium is indebted to their efforts in obtaining support and administering the grants. On a separate page we list the many individuals who worked on-site and who volunteered their time and arranged for their own overseas travel support. We wish to thank Mrs. Mary Keith for her help in producing the manuscript of the proceedings.

We are indebted to Dr. Slobodan Brant of the University of Zagreb, who acted as on-site secretariat and contributed many hours every day assuring the smooth operation of the conference. We also want to thank Mrs. Ruth Randig Meyer, whose efforts in providing a positive social environment for both the participants and their companions led to the title of ''conference house mother''.

Last, but hardly least, we want to thank the administration of the Lawrence Livermore National Laboratory, particularly Drs. C. Gatrousis and G. L. Struble, for their support of activities in behalf of the conference, and to the Institut fur Kernphysik II, KFA-Julich GmbH and its director Prof. Dr. O. W. B. Schult for the hospitality and support received during part of the post-conference editing of these proceedings.

R. A. Meyer
V. Paar

The International Symposium
on
SYMMETRIES AND NUCLEAR STRUCTURE
and
its participants
are
indebted to the following organizations
for their role as:

Sponsors:

INTERNATIONAL UNION OF PURE AND APPLIED PHYSICS

LAWRENCE LIVERMORE NATIONAL LABORATORY

Providing fellowships for attendance
and staff activities

U.S. National Science Foundation

U.S. Council for International Exchange of Scholars

(Fulbright Fellowships)

Bundesministerium fur Forschungs und Technologie

Kernforschungsanlage Julich GmbH

Pan American World Airways, Inc.

THE SYMMETRIES AND NUCLEAR STRUCTURE SYMPOSIUM
and its participants
wish to give special recognition to the following
who acted as:

On-site secretariat:
Dr. SLOBODAN BRANT
Prirodoslovno-matematicki fakultet
Zavod za Teorijsku Fiziku
University of Zagreb
41000 Zagreb, Yugoslavia

On-Site Staff:
Ms. L. Reic, Institut Rudjer Boskovic, Zagreb
Mrs. Z. Paar, Samobar, Yugoslavia
Ms. Joyce Plis, LLNL, Livermore, California
Ms. Jackie Mooney, BNL, Upton, New York
Mrs. Ruth Randig Meyer, Livermore, California
Dr. V. Lopac, University of Zagreb
Dr. N. Kusnezov, Lockheed Res. Co., Palo Alto, CA
Mr. C. B. Houser, Los Altos, California
R. & M. Husta, University of Zagreb
Ms. M. Galekovic, University of Zagreb
Ms. V. Cicin, University of Zagreb
Mr. D. Sunko, University of Zagreb
Mr. D. Vretenar, University of Zagreb
Mr. D. Vorkapic, University of Zagreb

STUDENT SUPPORT
U.S. National Science Foundation

Ariel Barfield, University of Arizona
John Engel, Yale University
Kevin Fowler, Carnegie-Mellon University
Paul Fuchs, Oregon State University
Dimitri Kusnezov, Princeton University
John David Robertson, University of Maryland
Craig Stone, University of Maryland
Mark Stoyer, University of California, Berkeley

Symposium Sponsored
Lauren Wood, University of Melbourne, Australia
Dario Vretenar, University of Zagreb
Marie Harder, University of Sussex, U.K.
Raoul Davie, University of Oxford, U.K.
Johnathan Copnell, University of Sussex, U.K.
Stephen Collins, University of Sussex, U.K.

GAJA ALAGA AWARD WINNERS

The international symposium Symmetries and Nuclear Structure along with the international conference on Nuclear Structure, Reactions, and Symmetries established the Alaga Award. This award was named in honor of Prof. Gaja Alaga of the University of Zagreb, who has contributed much to the field of nuclear structure and to science education in Yugoslavia. The Award was given to the graduate students who had the most outstanding thesis content and presentation of their work at the conference. Because of the high quality of graduate student presentations and content, two main awards and two secondary awards were made. These were given to:

MARIE HARDER
(Advisor: Prof. D. Hamilton)
Physics Division, University of Sussex
Brighton BN1 9QH, U.K.

DENIS SUNKO
(Advisor: Prof. V. Paar)
Zavodza Teorijsku Fiziku, University of Zagreb
41000 Zagreb, Yugoslavia

LAUREN WOOD
(Advisor: Prof. I. Morrison)
School of Physics, University of Melbourne
Parkville, Victoria, 3052, Australia

MARK STOYER
(Advisor: Prof. J. O. Rasmussen)
Chemistry Department, University of California
Berkeley, California, U.S.A.

Chapter 1

THEORETICAL STUDIES OF NUCLEAR SYMMETRY

Moderator:

A.F. BARFIELD
University of Arizona
Tuscon, Arizona 85712, U.S.A.

SU(6)-SYMMETRY IN THE INTERACTING BOSON MODEL:
A CONVENIENT TOOL OR PHYSICAL REQUIREMENT?

R.V.Jolos and S.P.Ivanova
Joint Institute for Nuclear Research, Dubna, USSR

1.

It is known that the square root factor $\sqrt{1-\frac{1}{N_{max}}\sum_{\mu} d^{+}_{2\mu}\, d_{2\mu}}$ appearing in the collective Hamiltonian based on SU(6)-symmetry (IBM-model) is the reflection of the Pauli principle acting among the fermions. Starting with the microscopic nuclear model, we have shown that the square root factor is the result only of the approximations made in calculations of the normalization factors of the manyfermion states. And only in this simplest approximation we get the result corresponding to the SU(6)-dynamical symmetry model that gives us the possibility to introduce the S-boson.

Our aim is to develop a method of constructing the collective quadrupole Hamiltonian based on the microscopical nuclear model and to show that the possibility to express the collective Hamiltonian in terms of the SU(6) symmetry group generators is the result of an approximation made in calculations of the normalization factors of the manyfermion states belonging to the collective subspace of the fermion state space.

2.

It is well known that the quadrupole branch of the nuclear excitations is separated well enough from the other degrees of freedom in many even-even nuclei: spherical, transitional and deformed ones. The experimental data give us enough arguments to describe the low-lying states in even-even nuclei based on the Hamiltonian containing only the collective quadrupole variables. But how to construct such a Hamiltonian?

In the phenomenological nuclear model the components of the quadrupole deformation tensor are chosen as collective variables. In the microscopical approach the structure of the collective quadrupole mode depends on the properties of the concrete nucleus: single-particle energies and wave functions, energy gap and the quadrupole-quadrupole interaction constant. This structure is determined by the solution of dynamical equations in the Tamm-Dancoff or random phase approximation. For instance, in the Tamm-Dancoff approximation, the operator $A^{+}_{2\mu}$ generating the lowest 2^{+} excitation has the following structure:

$$A^+_{2\mu} = \frac{1}{\sqrt{2}} \sum_{ss'} \psi_{ss'} (\alpha^+_s \alpha^+_{s'})_{2\mu} ,$$

$$(\alpha^+_s \alpha^+_{s'})_{2\mu} = \sum_{mm'} C^{2\mu}_{j_s m j_{s'} m'} \alpha^+_{sm} \alpha^+_{s'm'} ,$$

$$\sum_{ss'} \psi^2_{ss'} = 1, \quad \psi_{ss'} \sim \langle s || r^2 Y_2 || s' \rangle (E_s + E_{s'} - \omega).$$

Here α^+_{sm} is the quasiparticle creation operator in the single-particle state with quantum numbers s, m; E_s is the quasiparticle energy; ω is the lowest solution of the secular equation for quadrupole excitations in the Tamm-Dancoff approximation. Using the operators $A^+_{2\mu}$ we can construct the basis of the collective fermion states in the same way as the boson basis:

$\|0\rangle$	- quasiparticle vacuum	$\leftrightarrow$	$\|0)$	- phonon vacuum
$A^+_{2\mu} \|0\rangle$		$\leftrightarrow$	$d^+_{2\mu} \|0)$	- one-phonon state
$\frac{1}{\sqrt{2}}(A^+_2 A^+_2)_{IM} \|0\rangle$		$\leftrightarrow$	$\frac{1}{\sqrt{2}}(d^+_2 d^+_2)_{IM} \|0)$	
$(A^+_2 \ldots A^+_2)_{N\nu\Omega IM} \|0\rangle \equiv \|N\nu\Omega IM$		$\leftrightarrow$	$(d^+_2 \ldots d^+_2)_{N\nu\Omega IM} \|0)$	

The states $|0\rangle, A^+_{2\mu}|0\rangle, \frac{1}{\sqrt{2}}(A^+_2 A^+_2)_{IM}|0\rangle, \ldots (A^+_2 \ldots A^+_2)_{N\nu\Omega IM}|0\rangle$ form the basis in the collective subspace inside the complete fermion space. It is our ansatz.

So the problem of construction of the collective quadrupole Hamiltonian is reduced to calculations of the matrix elements of the microscopical Hamiltonian between the collective fermion states. To this end let us firstly investigate more carefully the collective fermion basis. The amplitudes $\psi_{ss'}$ are determined so that $\langle 0|A_{2\mu} A^+_{2\mu}|0\rangle = 1$.

All other collective fermion states determined so that the corresponding multi-phonon states

$$\frac{1}{\sqrt{2}}(d^+_2 d^+_2)_{IM} |0), \ldots (d^+_2 \ldots d^+_2)_{N\nu\Omega IM} |0)$$

normalized to unity have the norm smaller than 1 due to the Pauli principle acting among quasiparticles.

For instance:

$$<0|(A_2 A_2)_{IM} \frac{1}{\sqrt{2}} \cdot \frac{1}{\sqrt{2}}(A_2^+ A_2^+)_{IM} |0> =$$

$$= 1 - 50 \sum_{1,2,3,4} \psi_{13} \psi_{12} \psi_{34} \psi_{24} \begin{Bmatrix} j_1 & j_3 & 2 \\ j_2 & j_4 & 2 \\ 2 & 2 & I \end{Bmatrix}$$

With increasing number of quasiparticles in the collective fermion state the norm of this state becomes smaller and smaller. So one of our tasks is to normalize the collective fermion states, i.e., to calculate scalar products:

$$N_{N\nu\Omega IM} \equiv <0|(A_2 \dots A_2)_{N\nu\Omega IM} (A_2^+ \dots A_2^+)_{N\nu\Omega IM} |0>. \quad (1)$$

3.

The following step is to calculate the matrix elements of the microscopic Hamiltonian between the collective fermion states. Since the microscopic Hamiltonian is a combination of terms linear and quadratic in two-quasiparticle operators:

$$\alpha_s^+ \alpha_{s'}^+, \quad \alpha_s^+ \alpha_{s'}, \quad \alpha_{s'} \alpha_s \quad (2)$$

it is enough for our purposes to calculate the matrix elements of the operators (2).

The results are the following:

$$<N\nu\Omega IM|\alpha_s^+ \alpha_{s'}^+|N-1, \nu'\Omega'I'M'> =$$

$$= \frac{1}{2\sqrt{2N}}(N-1, \nu'\Omega'I'; 2|\}N\nu\Omega I) \sum C^{2\mu}_{j_s m_s j_{s'} m_{s'}} C^{IM}_{2\mu I'M'} \frac{\partial}{\partial \psi_{ss'}} N_{N\nu\Omega I} , \quad (3)$$

$$<N\nu\Omega IM|\alpha_s^+ \alpha_{s'}|N\nu'\Omega'I'M'> = \frac{\sqrt{N}}{4} \sum (N-1, \nu''\Omega''I''; 2|\}N\nu\Omega I) \times$$

$$\times (N-1, \nu''\Omega''I''; 2|\}N\nu'\Omega'I') \times \quad (4)$$

$$\times 5\sqrt{(2I+1)(2I'+1)} \begin{Bmatrix} \ell & j_s & j_{s'} \\ j_\eta & 2 & 2 \end{Bmatrix} \begin{Bmatrix} \ell & I & I' \\ I'' & 2 & 2 \end{Bmatrix}$$

$$\times(-1)^{M+m_s+I''+j_\eta} C^{\ell m}_{IM\,I'-M'} C^{\ell m}_{j_s m_s j_{s'} -m_{s'}} \left(\psi_{\eta s'} \frac{\partial}{\partial\psi_{\eta s}} N_{Nv\Omega I} + \right.$$

$$\left. + \psi_{\eta s} \frac{\partial}{\partial\psi_{\eta s'}} N_{Nv'\Omega' I'}\right).$$

The approximation made in these calculations is the following:

We take into account only two-quasiparticle operators coupled to angular momenta equal to 2, i.e.,

$$a_s^+ a_{s'}^+ = \sum_{\ell m} C^{\ell m}_{j_s m_s j_{s'} m_{s'}} (a^+ a^+)_{\ell m} \approx C^{2\mu}_{j_s m_s j_{s'} m_{s'}} (a^+ a^+)_{2\mu} .$$

Expressions (3) and (4) can be written in another form from which it is more easy to find the boson representation of the fermion operators

$$\langle Nv\Omega IM | a_s^+ a_{s'}^+ | N-1, v'\Omega' I' M'\rangle =$$

$$= \frac{1}{2\sqrt{2N}} (Nv\Omega IM | d^+_{2\mu} | N-1, v'\Omega' I' M') C^{2\mu}_{j_s m_s j_{s'} m_{s'}} \frac{\partial}{\partial\psi_{ss'}} N_{Nv\Omega I} , \tag{5}$$

$$\langle Nv\Omega IM | a_s^+ a_{s'} | Nv'\Omega' I' M'\rangle =$$

$$= \frac{1}{2\sqrt{N}} \Sigma C^{\ell m}_{j_s m_s j_{s'} - m_{s'}} (-1)^{\ell+\mu+j_\eta+m_s} (Nv\Omega IM | (d_2^+ d_2)_{\ell m} | Nv'\Omega' I' M') \times$$

$$\times \begin{Bmatrix} \ell & j_s & j_{s'} \\ j_\eta & 2 & 2 \end{Bmatrix} \left(\psi_{\eta s'} \frac{\partial}{\partial\psi_{\eta s}} N_{Nv\Omega I} + \psi_{\eta s} \frac{\partial}{\partial\psi_{\eta s'}} N_{Nv'\Omega' I'}\right). \tag{6}$$

4.

From the consideration presented above, it is seen that we can solve the task to calculate the matrix elements of the microscopic Hamiltonian between the collective fermion states if we get an expression for $N_{Nv\Omega I}$. Let us try to do this. For brevity we get an equation for $N_{Nv\Omega I}$ in the case of states with the maximum angular momentum I (at a given number of phonons), i.e., I = 2N, v = N and $N_{Nv\Omega I}$ reduce to N_N. In this case:

$$N_N = N\Sigma\,\psi^2_{ss'}\,N_{N-1} + \frac{N(N-1)}{2}\Sigma\,C^{IM}_{2\mu\,I'M'}\,C^{IM}_{2\nu\,\tilde{I}\tilde{M}}\,C^{\tilde{I}\tilde{M}}_{2\nu'\,I''M''}\,\times$$

$$\times \langle N-1|[[A_{2\mu}\,,\,A^+_{2\nu}],A^+_{2\nu'}]|N-2\rangle, \tag{7}$$

$$(I = 2N,\ I' = 2N-2,\ I'' = 2N-4,\ \tilde{I} = 2N-2).$$

Here

$$[[A_{2\mu}\,,\,A^+_{2\nu}],\,A^+_{2\nu'}] =$$

$$= -2\sqrt{2}\,\Sigma\,\psi_{st}\psi_{s't}\psi_{s\eta}\,C^{2\mu}_{j_s m_s j_t m_t}\,C^{2\nu}_{j_{s'} m_{s'} j_t m_t}\,C^{2\nu'}_{j_s m_s j_\eta m_\eta}\,\alpha^+_{s'}\,\alpha^+_\eta\ .$$

If, as above, we shall keep only two-quasiparticle operators $\alpha^+_{s'}$, α^+_η coupled to angular momenta equal to 2,

$$[[A_{2\mu}\,,\,A^+_{2\nu}],\,A^+_{2\nu'}] \approx$$

$$\approx -2\sqrt{2}\,\Sigma\,\psi_{st}\psi_{s't}\psi_{s\eta}\,C^{2\mu}_{j_s m_s j_t m_t}\,C^{2\nu}_{j_{s'} m_{s'} j_t m_t}\,C^{2\nu'}_{j_s m_s j_\eta m_\eta}\,C^{2\mu'}_{j_{s'} m_{s'} j_\eta m_\eta}\,(\alpha^+_{s'}\,\alpha^+_\eta)_{2\mu'} =$$

$$= -2\Sigma\,C^{2\mu}_{j_s m_s j_t m_t}\,C^{2\nu}_{j_{s'} m_{s'} j_t m_t}\,C^{2\nu'}_{j_s m_s j_\eta m_\eta}\,C^{2\mu'}_{j_{s'} m_{s'} j_\eta m_\eta}\,\frac{\partial}{\partial\psi_{s'\eta}}\,A^+_{2\mu'}\ .$$

If we put this result in (7), we get:

$$N_N = N\sum_{ss'}\psi^2_{ss'}N_{N-1} + \frac{1}{2}N\Sigma(-1)^{j_s - j_{s'}}\times$$

$$\times \langle (j_t\, j_{s'})2;(j_s\, j_\eta)2;4|(j_t\, j_s)2;(j_{s'}\, j_\eta)2;4\rangle\times \tag{8}$$

$$\times\,\psi_{st}\,\psi_{s't}\,\psi_{s\eta}\,\frac{\partial}{\partial\psi_{s'\eta}}\,N_{N-1}\ .$$

Let us consider the approximate solution of this equation. We suppose that the normalization factor N_N depends only on traces of the products of matrices containing $\psi_{ss'}$ in the degree not higher than 2. For instance: we suppose, calculating derivatives, that

$$\Sigma(-)^{j_s - j_{s'}} \langle (j_t j_{s'})2;(j_s j_\eta)2;4|(j_t j_s)2;(j_{s'} j_\eta)2;4\rangle \psi_{st}\psi_{s't}\psi_{s\eta}\psi_{s'\eta} =$$

$$= \frac{1}{2\Omega}(\sum_{ss'} \psi^2_{ss'})^2 \qquad (\Omega \text{ - constant value})$$

Strictly speaking, it means that all amplitudes are equal and the number of the single-particle states in summation $\sum_{ss'}$ is finite.

In this case we can use the ansatz $N_N = N!(\sum_s \psi^2_{ss'})^N \bar{N}_N$ and eq.(8) reduces to

$$\bar{N}_N = (1 - \frac{N-1}{\Omega})\bar{N}_{N-1}, \tag{9}$$

that is the well-known result of the SU(6) dynamical symmetry group approximation with $\Omega = N_{max}$. It means that

$$A^+_{2\mu} \to d^+_{2\mu}\sqrt{1 - \frac{\sum_\nu d^+_{2\nu} d_{2\nu}}{\Omega}}. \tag{10}$$

Substituting (10) into (1) we get (9). The expression (10) means that we may introduce the S-boson.

Now let us investigate the equation for N_N more carefully. For simplicity, we consider the case of monopole bosons. In this case eq.(8) takes the form

$$N_N = N\Sigma\,\psi^2_{ss'} N_{N-1} - \frac{N}{2}\Sigma\,\psi_{st}\psi_{s't}\psi_{s\eta}\frac{\partial}{\partial\psi_{s'\eta}}N_{N-1} \equiv$$

$$\equiv -N\,\mathrm{Sp}\,\psi^2 \cdot N_{N-1} + \frac{N}{2}\Sigma\,(\psi^3)_{s\nu}\frac{\partial}{\partial\psi_{s\nu}}N_{N-1}\,.$$

It is convenient to introduce: $\bar{N}_N \equiv \frac{1}{N!}N_N$ and the generator function $\bar{N}(\lambda) \equiv \sum_{N=0}^{\infty}\frac{\lambda^N}{N!}\bar{N}_N$.

For $\bar{N}(\lambda)$ we get the following result $\bar{N}(\lambda) = e^{\frac{1}{2}\mathrm{Sp}\ln(1-2\lambda\psi^2)}$. Consider the expressions for

$\mathrm{Sp}\,\psi^{2N}$:

$$\mathrm{Sp}\,\psi^2 = \sum_{1,2} \psi_{12}\psi_{21} \equiv \sum_1 \chi_{11}$$

$$\mathrm{Sp}\,\psi^4 = \sum_{1,2,3,4} \psi_{12}\psi_{23}\psi_{34}\psi_{41} \approx \sum_{1,2,4} \psi_{12}\psi_{21}\psi_{14}\psi_{41} = \sum_1 \chi_{11}^2$$

$$\cdots\cdots\cdots$$

$$\mathrm{Sp}\,\psi^{2N} \approx \sum_1 \chi_{11}^N .$$

Here we take into account only the coherent terms in expressions for $\mathrm{Sp}\,\psi^{2N}$. In such an approximation

$$\bar{N}(\lambda) = e^{-\frac{1}{2}\sum_1 \ln(1-2\lambda\chi_{11})} . \tag{11}$$

A. If $\chi_{11} = -\frac{1}{2\Omega}$ and the number of single-particle states in summation is restricted to 2Ω, then

$$\bar{N}(\lambda) = e^{\frac{1}{2}2\Omega\ln(1+\frac{\lambda}{\Omega})} = e^{\ln(1+\frac{\lambda}{\Omega})^{\Omega}} = (1+\frac{\lambda}{\Omega})^{\Omega} =$$

$$= \sum_{N=1} \frac{\lambda^N}{N!} 1\cdot(1-\frac{1}{\Omega})(1-\frac{2}{\Omega})\ldots(1-\frac{N-1}{\Omega}),$$

$$\bar{N}_N = (1-\frac{N-1}{\Omega})\bar{N}_{N-1} .$$

It is the result of the SU(6) dynamical symmetry group approximation.

B. Expression (11) can be written in another form:

$$N(\lambda) = e^{-\frac{1}{2}\sum_1 \int_0^\infty d\epsilon \ln(1-2\lambda\chi(\epsilon))\cdot\delta(\epsilon-\epsilon_1)} = e^{-\frac{1}{2}\int_0^\infty d\epsilon \ln(1-2\lambda\chi(\epsilon))\cdot g(\epsilon)} ,$$

where $g(\epsilon) \equiv \sum_1 \delta(\epsilon-\epsilon_1)$.

If we put $\chi(\epsilon) = -\chi_0 \frac{1}{(1+\epsilon/\Delta)^2}$ and $g(\epsilon) = \mathrm{const}$, then

$$\bar{N}(\lambda) = e^{\frac{1}{\chi_0}(\sqrt{1+2\lambda\chi_0}-1)} .$$

We find numerically the coefficients $\bar{N}_N$ for $\chi_0 = 0.1$ and compare these results with the results of the SU(6) approximation fixing N_{max} so that the values of N_2 coincide in both approximations:

	$\chi(\epsilon) = -\frac{1}{10}\frac{1}{1+(\epsilon/\Delta)^2}$	SU(6)
$\bar{N}_1$	1	1
$\bar{N}_2$	0.9	0.9
$\bar{N}_3$	0.73	0.72
$\bar{N}_4$	0.535	0.504
$\bar{N}_5$	0.356	0.302
$\bar{N}_6$	0.215	0.151
$\bar{N}_7$	0.119	0.060
$\bar{N}_8$	0.060	0.018
$\bar{N}_9$	0.028	0.004
$\bar{N}_{10}$	0.012	0.0004
$\bar{N}_{11}$	0.0050	0
$\bar{N}_{12}$	0.0019	
$\bar{N}_{13}$	0.0007	
. . . .		

The results on the first column can be approximated by the recursive relation $\bar{N}_N = \frac{\bar{N}_{N-1}}{(1+\frac{N-1}{\Omega})^2}$ (compare it with $\bar{N}_N = (1-\frac{N-1}{\Omega})N_{N-1}$).

SYMMARY

1. The problem of constructing the collective quadrupole Hamiltonian is reduced to calculations of scalar products of the collective fermion state.

2. The expressions for scalar products of the collective fermion states are obtained. It is shown that in the general case they do not reduce to the SU(6) result and the subspace of the collective fermion states is unrestricted in contrast with the IBM.

THE SCHWINGER (=IBM), HOLSTEIN-PRIMAKOFF (=TQM) AND DYSON (=FQM) REALIZATIONS OF THE QUADRUPOLE COLLECTIVE ALGEBRA IN THEIR INTERRELATION

Georgi Kyrchev

Institute of Nuclear Research and Nuclear Energy,
Bulgarian Academy of Sciences, Sofia 1784
BULGARIA

ABSTRACT

SU(6) collective models (Interacting Boson Model-IBM, Truncated Quadrupole Phonon Model-TQM and Finite Quadrupole Phonon Model-FQM) are seen as resulting from Schwinger, Holstein-Primakoff and Dyson realizations, respectively, of the SU(6)-Quadrupole Collective Algebra (QCA). An examination of the interrelation between these realizations of QCA enables a deductive inference of SU(6) collective models, as well as their unification.

1. INTRODUCTION

Two new collective models based on SU(6) dynamical symmetry were introduced in the last decade: TQM[1)], a particular quadrupole phonon model, and IBM[2)], a particular boson model in terms of s- and d-bosons. Extensive successful developments and applications have drawn attention to IBM in recent years. Thus, understandably, there had been much interest in comparing IBM with the other collective models.

Much attention has been (and is being) paid to the study of the relationship between IBM and the geometrical model of Bohr and Mottelson (BMM) as representations of the "new"- and "old phenomenology", resp.[3-7] In refs.[6,7] was explicated the complexity of this relation. The cause seems to lie in the fact that the dynamical symmetry groups underlying BMM and IBM differ (cf. ref.[7]). A more direct way to connect IBM to the "old phenomenology" is to relate it to TQM, a particular BMM, based on SU(6) dynamical symmetry[1]. In this case QCA (which is isomorfic to SU(6) and plays a central role in the sequel) underlies both IBM and TQM[8].

IBM Hamiltonian and Quadrupole operator have been shown to arise in Schwinger realization (SR) of QCA.[13] The other possible bosonization of QCA is Holstein-Primakoff realization (HPR) which leads to TQM[1]. There is also a third realization of QCA, namely Dyson realization (DR) which corresponds to a new (not yet sufficiently studied) SU(6) collective model referred to as FQM[15].

Thus there is SU(6) algebra-QCA, such that the models under consideration - IBM, TQM and FQM - correspond to different realizations of this algebra. Following refs. [9,10] we assign fundamental importance to QCA and investigate its representations in their interrelation rather than taking each one separately as is usually done. Such an approach permits[8]: i) the establishment of a unitary equivalence of IBM and TQM in the physical subspace, ii) a deductive inference of IBM, TQM and FQM on the same basis, iii) a unification of SU(6) collective models.

2. THE QUADRUPOLE COLLECTIVE ALGEBRA AND THE SU(6) COLLECTIVE HAMILTONIAN

Since the introduction of TQM and IBM the question of TQM-IBM relationship has been in issue[2,11-14,8]. If this issue is approached fundamentally, a set of interrelated problems occurs which require detailed mathematical elaboration. The links, labelled by (i) - (x) in Fig. 1 schematically represent the interrelation.

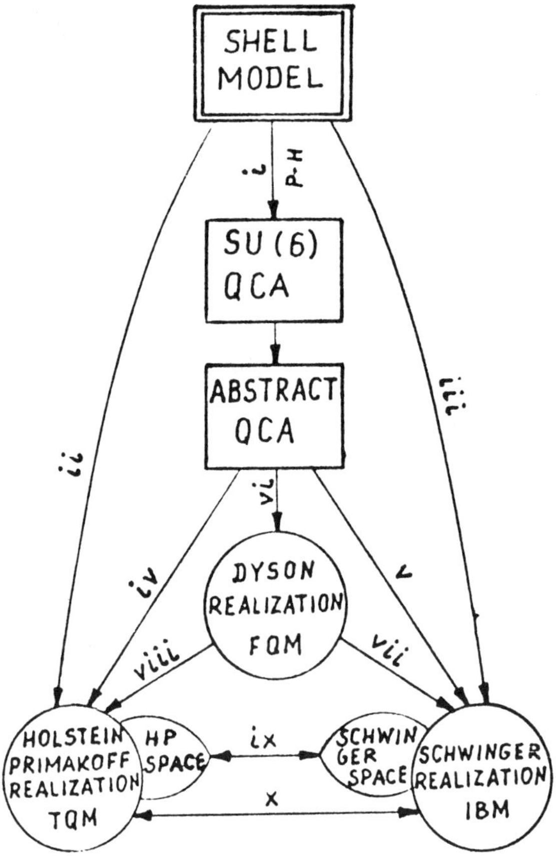

Fig.1. Full scope of boson realizations of QCA

The use of adequate mathematical formalism enables deductive inference of SU(6) collective models and their unification between the above mentioned problems. They are: (i) Microscopic derivation of QCA in p-h channel[1], (ii)-(iii) Direct microscopic derivations of H_{TQM} and H_{IBM}, (iv) Introduction of HPR[1], (v)-(vi) Explicit constructions of SR[13] and DR[15], (vii) Transformation of SR into DR[14], (viii) Transformation of DR into HPR[8], (ix) Determination of the relation between HP and Schwinger spaces[8], (x) Explicit construction of the operator which intertwins SR and HPR[8].

Links (iv)-(x) present a full scope of boson realizations of QCA.

QCA is closed by the quadrupole collective coordinates $\hat{q}_{2\mu}$ $(\mu=0,\pm1,\pm2)$, conjugated moments $\hat{p}_{2\mu}$ and their commutators. The explicit form of the commutation relations satisfied by QCA generators $\{\hat{g}_\lambda\}$ $(\lambda=1,2,...,35)$ is given in ref.[1]. The SU(6) collective Hamiltonian is SU(3) - invariant built from QCA generators[1]:

$$H_{QCA}=ie\sum_{\mu}[\hat{q}_{2\mu},\hat{p}_{2\mu}]-\frac{1}{4}\sum_{LM}\frac{1}{2}\,(1+(-)^L)d_L([\hat{q}_2,\hat{p}_2])_{(LM)}(-)^M \tag{1}$$

$$([\hat{q}_2,\hat{p}_2])_{(L-M)} + s\sum_{\mu}(-)^{\mu}\hat{q}_{2\mu}\hat{q}_{2-\mu} + v\sum_{\mu}(-)^{\mu}\hat{p}_{2\mu}\hat{p}_{2-\mu} +$$

$$+ \frac{1}{2} iw \sum_{\mu} \hat{q}_{2\mu}(-)^{\mu}$$

$$([\hat{q}_2,\hat{p}_2])_{(2-\mu)} - \frac{1}{4}\sum_{LM}\frac{1}{2}(1-(-)^L)d_L([\hat{q}_2,\hat{q}_2])_{(LM)}(-)^M$$

$$([\hat{q}_2,\hat{q}_2])_{(L-M)}$$

3. SR, HPR AND DR BOSON REALIZATIONS OF QCA

SR of QCA has been explicitly constructed in ref.[13] and reads:

$$\hat{q}^{SR}_{\mu} = sd^{+}_{\mu} + d_{\bar{\mu}}s^{+}$$
$$\hat{p}^{SR}_{\mu} = i\,(sd^{+}_{\bar{\mu}} - s^{+}d_{\mu})$$
$$i\,[\hat{q}_{\mu},\hat{p}_{\mu'}]^{SR} = d^{+}_{\mu}d_{\mu'} + d^{+}_{\bar{\mu}'}d_{\bar{\mu}} - 2\delta_{\mu\mu'}s^{+}s$$
$$[\hat{q}_{\mu},\hat{q}_{\mu'}]^{SR} = d^{+}_{\mu}d_{\bar{\mu}'} - d^{+}_{\mu'}d_{\bar{\mu}} \qquad (2)$$

Here $d^{+}_{\bar{\mu}} = (-)^{\mu}d^{+}_{-\mu}$, $\mu=0,\pm1,\pm2$.

Carrier spaces in which QCA generators (2) operate are spanned by the set of monomials[8]:

$$| [N] n_s \{n_\nu\}>_{SR} = \prod_{\nu=-2}^{2} (n_\nu !)^{-\frac{1}{2}} (d_\nu^+)^{n_\nu} (n_s !)^{-\frac{1}{2}} (s^+)^{n_s} |0> \qquad (3)$$

subject to the constraint

$$n_s + \sum_\nu n_\nu = N = \text{fixed} . \qquad (4)$$

In Eq. (3) symbol $|0>$ stands for the vacuum of s- and d-bosons. The explicit form of HPR of QCA generators is [1,8]:

$$\hat{q}_\mu^{HPR} = (N-\hat{N})^{\frac{1}{2}} b_{\bar{\mu}} + b_\mu^+ (N-\hat{N})^{\frac{1}{2}}$$

$$\hat{p}_\mu^{HPR} = i(b_{\bar{\mu}}^+ (N-\hat{N})^{\frac{1}{2}} - (N-\hat{N})^{\frac{1}{2}} b_\mu) \qquad (5)$$

$$i [\hat{q}_\mu, \hat{p}_{\mu'}]^{HPR} = b_\mu^+ b_{\mu'} + b_{\bar{\mu}'}^+ b_{\bar{\mu}} - 2\delta_{\mu\mu'} (N-\hat{N})$$

$$[\hat{q}_\mu, \hat{q}_{\mu'}]^{HPR} = b_\mu^+ b_{\bar{\mu}'} - b_{\mu'}^+ b_{\bar{\mu}} ,$$

where b_μ^+ are creation operators of quadrupole bosons, $\hat{N} = \sum_\mu b_\mu^+ b_\mu$ .The set of monomials

$$| [\hat{N}] \{n_\nu\} >_{HPR} = \prod_{\nu=-2}^{2} (n_\nu !)^{-\frac{1}{2}} (b_\nu^+)^{n_\nu} |\tilde{0}> \qquad (6)$$

subject to the constraint

$$n_b = \sum_\nu n_\nu \leq N \tag{7}$$

forms a basis for the totally symmetrix IR of SU(6) algebra. In (6) $|\hat{0}>$ is the vacuum of b-bosons, i.e. $b_\mu|\hat{0}> = 0$. It is natural to name "physical" the basic states (6), (7) and "unphysical" - the vectors (6) with $n_b > N$, because the generators $\hat{q}^{HPR}$ and $\hat{p}^{HPR}$ are meaningful if condition (7) is fulfilled.

Detailed derivation of DR in the case of QCA is presented in refs.[8,15]. Only the final result is given here:

$$\begin{aligned}
\hat{q}^{DR}_\mu &= b^+_\mu (N-\hat{N}) + b_{\bar{\mu}} \\
\hat{p}^{DR}_\mu &= i\,[\,b^+_{\bar{\mu}} - (N-\hat{N})\,b_\mu\,] \\
i\,[\hat{q}_\mu, \hat{p}_{\mu'}]^{DR} &= b^+_\mu b_{\mu'} + b^+_{\bar{\mu}'} b_{\bar{\mu}} - 2\delta_{\mu\mu'}(N-\hat{N}) \\
[\hat{q}_\mu, \hat{q}_{\mu'}]^{DR} &= b^+_\mu b_{\bar{\mu}'} - b^+_{\mu'} b_{\bar{\mu}}
\end{aligned} \tag{8}$$

Obviously DR is finite, but nonhermitian.

4. UNITARY EQUIVALENCE OF IBM AND TQM

We introduce an operator which maps SR carrier space (3) into carrier space (6):

$$P_{N_o} = \sum_{n_\nu} |[N_o]\{n_\nu\}\rangle_{HPR\ SR}\langle \{n_\nu\}n_s\ [N_o]| \qquad (9)$$

It has been shown [8] that P_N intertwins SR and HPR, i.e. for any QCA generator there is:

$$P_{N_o} g_\lambda^{SR} = g_\lambda^{HPR} P_{N_o}\ , \qquad \lambda = 1,2,...,35 \qquad (10)$$

Let $\{M\}$ denotes the set of monomials of the type (3),(4). It follows from (9) that for any $N' < N_o$ there holds: $P_{N_o}\ |[N']n_s\ \{n_\nu\}\rangle_{SR} = 0$, which means that P_{N_o} has a kernel. So it is natural to introduce the relevant factor space $\{M\}/$ Ker P_{N_o} and the operator $P^f_{N_o}$ defined as:

$$P^f_{N_o}\ (\ |m\rangle + \text{Ker}\ P_{N_o}\ ) = P_{N_o}\ |m\rangle \qquad (11)$$

where $|m\rangle \varepsilon \{M\}$. Definition (11) warrants that $P^f_{N_o}$ has a well defined inverse. From the conjunction of (9)-(11) and (3)-(7) the following precise statement has been inferred for any $|\ m\rangle\ \varepsilon\{M\}$:

$$P^f_{N_o} g^{f\ SR}\ (|m\rangle + \text{Ker}\ P_{N_o}) = g^{HPR}/_{phys} P^f_{N_o}\ (|m\rangle + + \text{Ker}\ P_{N_o}) \qquad (12)$$

Relation (12) symbolically expresses the fact that $\{g_\lambda^{f\ SR}\}$ - the SR induced in the factor space $\{M\}/\mathrm{Ker}\ P_{N_p}$ - is isomorphic to $\{g_\lambda^{HPR}|_{phys}\}$ - the HPR restricted to the physical subspace specified by (6),(7). It can be shown that $P^f_{N_o}$ preserves the norm in the factor space and $(P^f_{N_o})^+$ does the same in the physical subspace of HPR. From the isomorfism of $\{g_\lambda^{f\ SR}\}$ and $\{g_\lambda^{HPR}|_{phys}\}$ Eq. (12), and the isometry of $P^f_{N_o}$ it follows that these two realizations of QCA are factually unitarily equivalent, i.e.:

$$g_\lambda^{f\ SR} = (P^f_{N_o})^{-1}\, g_\lambda^{HPR}\,|_{phys}\, P^f_{N_o} \tag{13}$$

Thus the unitary equivalence of SR (=IBM) and HPR (=TQM) is rigorously proved and the domain where the equivalence is valid - the physical subspace (6), (7) - is explicitly identified.

5. DEDUCTIVE INFERENCE AND UNIFICATION OF SU(6) COLLECTIVE MODELS

A body of facts[8, 11-14] concerning the relationship between IBM and TQM represents a good foothould for axiomatization of SU(6) collective models. We can derive them taking as a starting point H_{QCA} given by (1). Indeed: if QCA generators in SR (cf.(2)) are substituted for their resp. counterparts in (1), the well-known IBM Hamiltonian is obtained[2], the replacement $\{g_\lambda\} \to \{g_\lambda^{HPR}\}$ (cf.(5)) leads to the TQM Hamiltonian[1]; if DR for QCA generators (cf.(5)) is taken, H_{FQM} is derived[8,15]. Thus SU(6) collective models can all be deductively inferred from the same source H_{QCA} (cf.(1)). On the other hand

$$H_{IBM} = (P^f_{N_o})^{-1} H_{TQM} \Big|_{phys} P^f_{N_o} \tag{14}$$

directly follows from (13). Thus in the physical subspace defined by (6),(7) IBM and TQM are <u>indistinguishable</u>: they simply represent two unitarily equivalent forms of an unified SU(6) collective model determined by H_{QCA}.

The treatment of SR (=IBM), HPR (=TQM) and DR (=FQM) in terms of their interrelation brings to a close the issue of IBM - TQM relationship.

I am indepted to V.Paar for useful discussions.

REFERENCES

1. Jolos R.V., Dönau F. and Janssen D., Theor.Mat. Phys.20,112(1974); Janssen D., Jolos R.V. and Dönau F., Nucl.Phys. A224,93(1974)

2. Arima A. and Iachello F., PR.Lett.35,1069(1975); Ann.Phys. 99,253(1976),201(1978),ibid.123,468 (1979).

3. Moshinsky M., Nucl.Phys. A338,156(1980).

4. Ginocchio J.N. and Kirson M.W., Nucl.Phys.A350, 31(1980).

5. Dieperink A.E.L., Scholten O. and Iachello F., Phys.Rev.Lett.44,1747(1980).

6. Kirson M.W., Ann.Phys. 143,448(1982).

7. Gilmore R. and Feng D.H., Phys.Rev. C26,776 (1982).

8. Kyrchev G. and Paar V., Ann.of Phys., 170(1986) 257

9. Wightman A.S., Phys.Rev. 10,860(1956).

10. Haag R. and Kastler D., J.Math.Phys. 5,848 (1964).

11. Blaizot J.P. and Marshalek E.R., Nucl.Phys. A309,422(1978).

12. Paar V. e.a., Nucl.Phys. A378, 41(1982).

13. Kyrchev G., Nucl.Phys. A349, 416 (1980).

14. Klein A. and Vallieres M., Phys.Rev.Lett. 46,586 (1981).

15. Kyrchev G. and Paar V., Fizika 161,291 (1984).

SOME GROUP THEORETICAL ASPECTS OF THE INTERACTING BOSON-FERMION MODEL

Yu.F. Smirnov and V.N. Tolstoy
Institute of Nuclear Physics, Moscow State University;
Moscow, 119899, USSR

ABSTRACT

The Dyson-type boson realizations (for Hamiltonians constructed from the generators of semisimple Lie algebras) found on the base of Verma's module theory is shown to be obtained using the generalized coherent states taken in a factorized form.
The supersymmetric $U(6/12) \supset U(6) \otimes U(1/2)$ scheme is considered. The fractional parentage supercoefficients (FPSC) are introduced. The application of these coefficients to calculation of the IBFM Hamiltonian matrix in a $U(6) \supset U(3)$, $U(1/2) \supset U_B(1) \otimes SU_F(2)$ basis is discussed.
The explicit form of the extremal projection operators for the arbitrary finitedimensional contragredient Lie superalgebra is given. The projectors allow one to calculate the Clebsch-Gordan coefficients for superalgebras and to solve other problems of the superalgebra representation theory.

1. DYSON-TYPE BOSON REALIZATIONS OF REPRESENTATIONS FOR SEMISIMPLE LIE ALGEBRAS

1.1 Introduction

In connection with the successful description of collective excitations of spherical, deformed and transition nuclei by the interacting boson model (IBM)[1,2] the problem of the microscopical foundation of IBM is given much attention in the literature. In order to solve this problem it is necessary to obtain the exact mapping of the nuclear (many fermion) Hamiltonian H onto some boson Hamiltonian H_B. Most recent papers are concerned with the Hamiltonian H having some symmetry properties. Usually it means that H is constructed from the generators J_i of a Lie algebra G (see for example the generalized seniority scheme[3]), SO(8) model[4]), two j-shell model with SO(5) symmetry[5]) etc.). Therefore the physical problem of $H \rightarrow H_B$ mapping is reduced in the group theoretical language to obtaining the exact (or approximate) boson realization of the G irreducible representation D^λ to which the wave functions Ψ_m^λ of many particle fermion system belong. If the boson realization of generators J_i is known, the problem of construction of the boson image H_B for the initial Hamiltonian $H=f(J_i)$ can be solved easily.

There are various types of boson realizations for semisimple Lie algebras: the boson representation of Schwinger, the boson realizations of Dyson and Holstein-Primakoff. In what follows we discuss only the boson realization of Dyson-type. Among various papers concerned with this question, it is worthwhile to mention two papers[6,7]) where the universal approaches to the construction of Dyson realizations for semisimple Lie algebras were suggested. The generalized coherent state (GCS) technique is used in Ref.[6]) where an algorithm of construction of the single definite Dyson-type boson realization for algebra G is given. The method of Ref.[7]) employs the properties of the elementary representation of algebra G (Verma module theory) and permits a set of equivalent boson realizations for the same algebra G to be obtained.

The construction of boson realization for IBM is now an actual problem and, moreover, it is intensively used in the development of the symplectic collective models[8]). It is therefore rather important in our opinion to consider the interrelations of the methods described in Refs.[6]) and [7]) at greater length. The deep understanding of the connection between the two approaches would enable us to select most convenient form of boson realizations in correspondence with specifical properties of the physical system under consideration. The best way to show more clearly the common and different points of the two methods is to derive the results of both the methods in the frame of a single formalism. In this connection we shall rederive the results of Ref.[7] using the GCS technique similarly to Ref.[6]. It should be noted that the literature on boson realizations is rather rich. The earlier papers can be found in Refs. of the papers[6,7] or in the recent survey paper[8]. Some other approaches to the bosonization problem are also being developed (see, for example, Ref.[9]).

1.2 Boson Realizations and the Theory of Generalized Coherent States

In Ref.[6] the boson realizations of Lie algebra generators (H_i, $i = 1,2,\dots,r$, E_α, $E_{-\alpha}$, $\alpha > 0$, where r is a rank of an algebra G, α's are its positive roots) were obtained by applying these generators to the Perelomov type GCS's (see for example[10]):

$$\Psi_c^\lambda = \exp\left(\sum_\alpha z_\alpha E_{-\alpha}\right)\Psi^\lambda \tag{1}$$

where z_α is a complex parameter. It is assumed that the raising (E_α), lowering ($E_{-\alpha}$) and diagonal (H_i) generators are taken in the standard Carthan-Weyl form and satisfy the conditions

$$[H_i, E_{\pm\alpha}] = \pm\alpha(H_i)E_{\pm\alpha}\,, \qquad [E_\alpha, E_{-\alpha}] = H_\alpha \tag{2}$$

In Eq.(1) Ψ^λ is a highest weight vector having the usual properties

$$H_i\Psi^\lambda = \lambda_i\Psi^\lambda\,, \qquad E_\alpha\Psi^\lambda = 0 \quad \text{for } \alpha > 0. \tag{3}$$

The general idea of obtaining the boson images for generators is the following. Let us consider an expression $E_{-\beta}\Psi^\lambda$. If $E_{-\beta}$ commutates with all lowering generators $E_{-\alpha}$ this expression can be rewritten as $E_{-\beta}\Psi_c^\lambda = (\partial/\partial z_\beta)\Psi_c^\lambda$. In

Ref.[6] the following interpretation of this equation is given: the image of the generator E_β in the GCS representation coincides with $\partial/\partial z_\beta$:

$$\rho(E_\beta) = \frac{\partial}{\partial z_\beta}$$

In the general case a more complicate expression can be found $\rho(E_\beta) = (\partial/\partial z_\beta + \ldots)$ where omitted terms are some combinations of operators z_α and $\partial/\partial z_\alpha$. These terms are present in $\rho(E_\beta)$ if $E_{-\beta}$ and the exponent in Eq.1 do not commutate. As for the generators H_i and $E_{-\beta}$ $(\beta > 0)$ the procedure of obtaining their expressions in the GCS representation consists in the following: 1) first we transpose H_i (or E_α) with the exponent factor in Eq.(1) by using the well-known identity

$$A \exp B = (\exp B) \sum_{k=0}^{\infty} (1/k!) \hat{A}^k B \qquad (4)$$

(here $\hat{A}B = [A,B]$) and commutation rules (2); 2) now the properties (3) must be taken into account; 3) then it is necessary to represent the obtained combination of lowering generators as a result of applying of some polynomial construction from operators z_α and $\partial/\partial z_\alpha$ to the exponent factor in Eq.(1). Thus the explicit expressions for the generators E_β, H_i and $E_{-\beta}$ in terms of z_α and $\partial/\partial z_\alpha$ operators can be obtained. And at last, the substitution of z_α , $\partial/\partial z_\alpha$ operators by boson annihilation and creation operators must be done:

$$z_\alpha \to b_\alpha^+ , \quad \partial/\partial z_\alpha \to b_\alpha$$

The resulting expressions coincide with the boson images of generators in some definite Dyson-type boson realization.

In what follows it will be shown that the results of Ref.[7] can be obtained in the same scheme if another factorized form of the GCS

$$\Phi_c^\lambda = \left[\prod_{\alpha>0} \exp(z_\alpha E_{-\alpha}) \right] \Psi^\lambda \qquad (5)$$

is used instead of the symmetrical form (1). The preliminary version of this approach was given in the report[11]. It should be noted that individual factors in the r.h.s. of (5) are in the so-called "normal ordering". It means (see Ref.[12]) that the factor corresponding to the arbitrary composite root $\beta = \alpha + \gamma$ stands between the factors corresponding to the roots α and γ (i.e. these three factors are going in the order ... $\exp(z_\alpha E_{-\alpha})$...$\exp(z_\beta E_{-\beta})$...$\exp(z_\gamma E_{-\gamma})$ ... or in the inverse order ... $\exp(z_\gamma E_{-\gamma})$ ... $\exp(z_\beta E_{-\beta})$ ... $\exp(z_\alpha E_{-\alpha})$...). In a given root system the normal ordering of roots is not unique and a number of variants of normal ordering is possible. Each variant generates the GCS (5) and the corresponding Dyson-type boson realization.

Let us illustrate this fact with an example of gl(3,C) algebra. Its generators A_{ik} $(i,k = 1,2,3)$ satisfy the commutation rules

$$[A_{ik}, A_{pq}] = \delta_{kp} A_{iq} - \delta_{iq} A_{pk}$$

Two variants of normal orderings of lowering generators are possible: A_{32}, A_{31}, A_{21} and A_{21}, A_{31}, A_{32}. As a result two boson realizations for gl(3,C) generators can be obtained. In particular, if the GCS is taken in the form

$$\Phi_c^\lambda = \exp(z_{23} A_{32}) \exp(z_{13} A_{31}) \exp(z_{12} A_{21}) \Psi[f]$$

where $[f] = [f_1 f_2 f_3]$ is the highest weight of the gl(3,C) irreducible representation, then we obtain the expression

$$\rho_B^{[f]}(A_{23}) = b_{23}\,, \qquad \rho_B^{[f]}(A_{13}) = b_{13}$$

Applying Eq.(4) for A_{12} generator and taking account of the equation

$$\frac{\partial}{\partial z_{12}}\Phi_c^\lambda=(A_{21}+z_{23}A_{31})\Phi_c^\lambda=(A_{21}+z_{23}\frac{\partial}{\partial z_{13}})\Phi_c^\lambda$$

we may conclude that

$$\rho_B^{[f]}(A_{12})=b_{12}-b_{23}^+b_{13}$$

In analogy with this result we can obtain the boson realizations for the rest gl(3,C) generators

$$\begin{aligned}
&\rho_B^{[f]}(A_{11})=f_1-b_{12}^+b_{12}-b_{13}^+b_{13}\,,\\
&\rho_B^{[f]}(A_{22})=f_2+b_{12}^+b_{12}-b_{23}^+b_{23}\,,\\
&\rho_B^{[f]}(A_{33})=f_3+b_{23}^+b_{23}+b_{13}^+b_{13}\,,\\
&\rho_B^{[f]}(A_{21})=b_{12}^+(f_1-f_2-b_{12}^+b_{12})-b_{13}^+b_{23}\,,\\
&\rho_B^{[f]}(A_{32})=b_{23}^+(f_2-f_3-b_{23}^+b_{23}-b_{13}^+b_{13}+b_{12}^+b_{12})+b_{13}^+b_{12}\,,\\
&\rho_B^{[f]}(A_{31})=b_{13}^+(f_1-f_3-b_{23}^+b_{23}-b_{13}^+b_{13}-b_{12}^+b_{12})+b_{23}^+b_{12}^+(f_1-f_2-b_{12}^+b_{12})
\end{aligned}\qquad(6)$$

In the general case of any Lie algebra G at arbitrary fixed normal ordering of factors in the r.h.s. of the Eq.(5) the Dyson-type boson realization for diagonal and rasing generators has a form

$$\rho_B^\lambda(H_i)=\lambda_i-\sum_{\ell=1}^{N}\alpha_\ell(H_i)b_\ell^+b_\ell\,,\quad i=1,2,\ldots,r,$$

$$\rho_B^\lambda(E_{\alpha_p})=b_p+\sum_{\ell=2}^{p-1}\Big[\prod_{i=1}^{\ell-1}\frac{(b_i^+)^{K_i^\ell}}{(K_i^\ell)!}\Big(\prod_{q_i=0}^{K_i^\ell-1}N_{-\alpha_p-K_1^\ell\alpha_1-\ldots-q_i\alpha_i,-\alpha_i}\Big)\Big]b_\ell$$

$$\text{for }\forall\,\alpha_p>0.\qquad(7)$$

As for the lowering generators their boson expressions are rather cumbersome and therefore we have restricted ourselves only to $\rho_B^\lambda(E_{-\alpha_s})$ for simple roots $\alpha_s>0$:

$$\rho_B^\lambda(E_{-\alpha_s}) = b_s^+\Big[(\alpha_s,\lambda) - \frac{1}{2}(\alpha_s,\alpha_s)b_s^+b_s - \sum_{l=s+1}^{N}(\alpha_s,\alpha_l)\,b_l^+ b_l\Big] +$$
$$+ \sum_{l=1}^{s-1}\Big[\prod_{i=l+1}^{s-1}\frac{(b_i^+)^{K_i^l}}{K_i^l!}\Big(\prod_{q_i=0}^{K_i^l-1} N_{-\alpha_i,\,\alpha_s - K_{s-1}^l\alpha_{s-1} - \ldots - q_i\alpha_i}\Big)\Big]b_l \qquad (8)$$
$$+ \sum_{s+1}^{N}\Big[\prod_{i=s+1}^{l-1}\frac{(b_i^+)^{K_i^l}}{K_i^l!}\Big(\prod_{q_i=0}^{K_i^l-1} N_{\alpha_s - K_{s+1}^l\alpha_{s+1} - \ldots - q_i\alpha_i,\,-\alpha_i}\Big)\Big]b_l$$

Here K_i^l are nonnegative integers satisfying the equation $\alpha_l - \alpha_p = K_1^l\alpha_1 + \ldots + K_{l-1}^l\alpha_{l-1}$,

$$\alpha_l + \alpha_s = K_{l+1}^l\alpha_{l+1} + \ldots + K_{s-1}^l\alpha_{s-1} \quad \text{for } l<s,$$
$$\alpha_l + \alpha_s = K_{s+1}^l\alpha_{s+1} + \ldots + K_{l-1}^l\alpha_{l-1} \quad \text{for } l>s$$

provided that there exist all intermediate roots:

$$\alpha_p+\alpha_1,\ \alpha_p+2\alpha_1,\ \ldots,\ \alpha_p+K_1^l\alpha_1,\ \alpha_p+K_1^l\alpha_1+\alpha_2,\ \ldots,$$
$$\alpha_p+K_1^l\alpha_1+K_2^l\alpha_2,\ \ldots,\ \alpha_p+K_1^l\alpha_1+\ldots+(K_{l-1}^l-1)\alpha_{l-1} \text{ and so on.}$$

The boson realization of lowering generators corresponding to composite roots can be found from (8) by using the commutation relations

$$[\rho_B^\lambda(E_{-\alpha}),\rho_B^\lambda(E_{-\beta})] = N_{-\alpha,-\beta}\,\rho_B^\lambda(E_{-\alpha-\beta}).$$

The results (7), (8) can be easily extended onto the boson-fermion realizations of Lie superalgebra representations[7)]

2. SUPERSYMMETRICAL SCHEME U(6/12) $\supset$ U(6) $\otimes$ U(1/2) IN THE INTERACTING BOSON-FERMION MODEL

2.1 Introduction

The supersymmetry idea was introduced in nuclear theory in Refs.[13,14)]. It was suggested to describe the spectra of energy levels of nuclei in the Pt-region by using the irreducible representation of the U(6/4) superalgebra. Later the investigations concerned with the interacting boson and fermion model (IBFM) employed another superalgebras U(6/M) (for example U(6/12) in Ref.[15,16]). The accuracy of these supersymmetric schemes in the description of energy level spectra, reduced probabilities B(E2), one and two nucleon spectroscopic factors is rather hopeful (about 10-30% as a rule) which explains increased interest of theoreticians and experimentalists to the problem of supersymmetry in nuclei. The survey of physical applications of supersymmetric models and their experimental status can be found in Refs.[17,18].

In this section we consider the variant of the supersymmetric scheme U(6/12) corresponding to the reduction of this superalgebra to the following chain of subalgebras:

$$U(6/12) \begin{array}{l} \nearrow U_{BF}(6) \supset SU_{BF}(3) \supset SO_L(3) \searrow \\ \searrow U(1/2) \supset U_B(1) \otimes U_F(1) \otimes SU_S(2) \nearrow \end{array} SU_J(2) \qquad (9)$$

It should be noted that we are not interested in the case of a "good" dynamical supersymmetry that was given main attention in the previous papers[13-18]. Exact supersymmetry is probably very seldom. Therefore we shall attempt to construct the formalism convenient for analysis of the violated supersymmetry U(6/12). Namely, we shall use the wave functions Φ_K, corresponding to the superalgebra chain (9), as a basis of expansion of the IBFM wave function

$$\Psi = \sum_K C_K \Phi_K$$

As usual, the energy eigenvalues and the corresponding C_k coefficients can be found by diagonalization of the IBFM Hamiltonian matrix $H_{kk'}$ calculated in this basis. Here "k" stands for all quantum numbers with respect to subalgebras (9). The set of quantum numbers will be discussed below. Also we shall consider the calculation of the fractional parentage supercoefficients (FPSC). These coefficients enable one to find the matrix of the IBFM Hamiltonian and matrix elements of other operators (quadruple momentum, single and two-particle transfer etc.). Thus, a rather complete and effective method of IBFM calculations will be formulated, (at least for the nuclei with unpaired nucleons occupying the shells with j = 1/2, 3/2, 5/2). In Ref.[15,16] and similar papers, dedicated to the same U(6/12) scheme, the superalgebra U(6/12) was reduced onto the subgroup chain containing only usual Lie algebras. Instead of this, the subgroup chain (9) contains the superalgebra U(1/2). The advantage of our version of U(6/12) scheme is a possibility of using effectively the SU(3) mathematical formalism developed by us earlier in connection with the development of IBM in the SU(3) basis[20].

2.2 State Classification in the $U(6/12) \supset U(6) \otimes U(1/2)$ Scheme

In this scheme the low lying levels of a nucleus are interpreted as states of a boson-fermion system containing a number N_B of s,d-bosons and a number N_F of fermions occupying single particle orbitals with j = 1/2, 3/2, 5/2. In the spirit of pseudo-SU(3) model it is possible to consider these single-particle wave functions as states $\tilde{s}_{1/2}$, $\tilde{d}_{3/2}$, $\tilde{d}_{5/2}$ of some fermion with orbital quasimomentum $\tilde{l} = 0$ or 2 and spin $\tilde{s} = 1/2$ which are split by a "spin-orbit interaction". As a result, bosons and fermions have 6 single particle orbital states. In the space of the boson orbital states there act the algebra $U_B(6)$. The algebra $U_F(6)$ is determined in the space of the fermion quasiorbital states $\tilde{s},\tilde{d}$. The presence of the $U_{BF}(6)$ algebra in the chain (9) means that we consider the s,d boson states and the $\tilde{s},\tilde{d}$ fermion states as identical. In other words, in the first approximation (in the case of an exact U(6/12) symmetry) we have 6 single particle orbitals s and d which can be occupied by N particles. Each of these particles can be either a boson B with vanishing spin s=0 or a fermion F with s = 1/2. This unified superparticle will be called below a superon σ . Therefore our problem is a classification of states in the N-superon configuration σ^N. According to the subalgebra chain (9) these states can be enumerated by the following quantum numbers

$$|\sigma^N [f] \alpha (\lambda\mu) \varkappa L, N_F S : J M\rangle \qquad (10)$$

Here the Young scheme $[f] = [f_1 f_2 \cdots f_6]$ defines simultaneously the irreducible representations of $U_{BF}(6)$ and U(1/2) algebras. The integers f_i, i=1,2,...6 satisfy the conditions $f_1 \geq 2 \geq f_2 \geq \ldots \geq f_6$, $f_1 + \ldots + f_6 = N$. The allowed values of the SU(3) symmetry symbol $(\lambda\mu)$ at a given $U_{BF}(6)$ representation [f] can be found by usual shell model methods[19-21,22]. The values of the orbital angular momentum L obtained after the $SU(3) \supset SO_L(3)$ reduction satisfy Elliott's rules[19,21,22]. It should be noted that L is an orbital quasimomentum if $N_F \neq 0$. Therefore, the IBFM Hamiltonian is nondiagonal with respect to this quantum number (i.e. L and S are non-conserving quantum numbers at $N_F \neq 0$; only total angular momentum $\vec{J} = \vec{L} + \vec{S}$ is an integral of motion). The number of bosons in a state (10) is equal to $N_B = N - N_F$, $N_F = 0,1,2,\ldots,\min(N,12)$. If $[f] = [f_1 2^n 1^m]$ the allowed values of N_F and S are given by formulae:

$$N_F = \nu_1 + \nu_2 \,, \quad S = (\nu_1 - \nu_2)/2 \,,$$

$$n+m+1 \geq \nu_1 \geq n+m \,, \quad n+1 \geq \nu_2 \geq n \,, \quad \nu_1 \geq \nu_2 \,.$$

2.3 Supercoefficients of the Fractional Parentage

The wave functions (10) must be supersymmetrized, i.e. they are symmetrical with respect to boson-boson and boson-fermion permutation and antisymmetrical with respect to fermion-fermion permutation[23]. The construction of such wave functions and the corresponding calculations can be realized by using the fractional parentage supercoefficients (FPSC) that are determined by the equation

$$|\sigma^N[f]\alpha(\lambda\mu)N_F S\rangle = \\ = \sum |\sigma^{N-1}[f']\alpha'(\lambda'\mu')N_F' S', \sigma_N : (\lambda\mu) S\rangle \\ \langle N[f]\alpha(\lambda\mu)N_F S|N-1[f']\alpha'(\lambda'\mu')N_F' S'\rangle \tag{11}$$

Thus single particle FPSC's $\langle N/N-1\rangle$ enable us to express the supersymmetric wave function of N superons in terms of supersymmetric wave functions of N-1 superons coupled with the wave function of the last N-th superon. In r.h.s. of Eq.(11) the vector coupling of spins $\vec{S}' + \vec{S}_N = \vec{S}$ is assumed. It is also assumed the SU(3) irreducible representation ($\lambda'\mu'$) is coupled with the irreducible representation (20) for N-th superon to obtain the total SU(3) representation ($\lambda\mu$). The indices of the summation in (11) are: $[f'], \alpha', (\lambda'\mu'), N_F', S'$. In analogy with this result the two particle FPSC's can be introduced

$$|\sigma^N[f]\alpha(\lambda\mu)N_F S\rangle = \\ = \sum |\sigma^{N-2}[f']\alpha'(\lambda'\mu')N_F' S', \sigma^2[f^\circ](\lambda^\circ\mu^\circ)N_F^\circ S^\circ : (\lambda\mu) S\rangle \\ \langle N[f]\alpha(\lambda\mu)N_F S|N-2[f']\alpha'(\lambda'\mu')N_F' S', 2[f^\circ](\lambda^\circ\mu^\circ)N_F^\circ S^\circ\rangle \tag{12}$$

Here we imply $\vec{S}' + \vec{S}^\circ = \vec{S}$, $(\lambda'\mu') \otimes (\lambda^\circ\mu^\circ) \to (\lambda\mu)$. The summation over $[f']\alpha'(\lambda'\mu')N_F' S', [f^\circ](\lambda^\circ\mu^\circ)N_F^\circ S^\circ$ must be done. The pair of separated superons N-1 and N is supersymmetrical, if

$$[f^\circ];(\lambda^\circ\mu^\circ);S^\circ = \begin{cases} [2];(40),(02);0 & \text{at } N_F^\circ = 0, N_B^\circ = 2 \\ \left.\begin{array}{l}[2];(40),(02);0 \\ [11];(21);1\end{array}\right\} & \text{at } N_F^\circ = 2, N_B^\circ = 0. \end{cases} \tag{13}$$

The wave function $|2\,[f^\circ]\,(\lambda^\circ\mu^\circ)\; N_F^\circ = 1\;\; S^\circ = \frac{1}{2}>$ must be symmetrical with respect to permutation, i.e.

$$|\sigma^2\,[f^\circ]\,(\lambda^\circ\mu^\circ)\,N_F^\circ=1\;S^\circ=\tfrac{1}{2}> = \frac{1}{\sqrt{2}}\Big[\,|B_{N-1}F_N\,(\lambda^\circ\mu^\circ)\;S^\circ=\tfrac{1}{2}>+ + (-1)^{\lambda^\circ+\mu^\circ}\,|F_{N-1}B_N\,(\lambda^\circ\mu^\circ)\;S^\circ=\tfrac{1}{2}>\Big] \quad (14)$$

where B_i and F_i are the boson and fermion single particle states of i-th superon. The interrelations of two-particle and one-particle FPSC is of the form

$$<N[f]\,\alpha(\lambda\mu)N_F S\,|\,N-2\,[f']\,\alpha'(\lambda'\mu')N_F' S',\;2\,[f^\circ](\lambda^\circ\mu^\circ)N_F^\circ S^\circ> =$$
$$= \sum <N[f](\lambda\mu)N_F S\,|\,N-1\,[f'']\,\alpha''(\lambda''\mu'')N_F'' S''>$$
$$<N-1\,[f'']\,\alpha''(\lambda''\mu'')\,N_F'' S''\,|\,N-2\,[f']\,\alpha'(\lambda'\mu')\,N_F' S'> \quad (15)$$
$$U((\lambda'\mu')(20)\,(\lambda\mu)(20);(\lambda''\mu'')(\lambda^\circ\mu^\circ))\cdot U(s' s_{N-1} S\, s_N\,;\,S''\,S^\circ)$$

at $N_F{}^\circ = 2$ ($S_{N-1} = S_N = 1/2$) or $N_F^\circ = 0$ ($S_{N-1} = S_N = 0$) (the summation over all double primed indices must be done). At N_F=1 the corresponding formula has a form

$$<N[f]\,\alpha\,(\lambda\mu)N_F S\,|\,N-2\,[f']\,\alpha'(\lambda'\mu')\,N_F-1\,S',\;2[f^\circ](\lambda^\circ\mu^\circ)N_F^\circ=1\;S^\circ=\tfrac{1}{2}>=$$
$$=\frac{1}{\sqrt{2}}\sum\Big[<N[f]\alpha\,(\lambda\mu)\,N_F S\,|\,N-1\,[f'']\,\alpha''\,(\lambda''\mu'')\,N_F-1\,S'>$$
$$<N-1\,[f'']\,\alpha''\,(\lambda''\mu'')\,N_F-1\,S'\,|\,N-2\,[f']\,\alpha'(\lambda'\mu')\,N_F-1\,S'> + \quad (16)$$
$$+(-1)^{\lambda^\circ+\mu^\circ}<N[f]\,\alpha\,(\lambda\mu)\,N_F\,S\,|\,N-1\,[f'']\,\alpha''(\lambda''\mu'')\,N_F\,S>$$
$$<N-1\,[f'']\,\alpha''\,(\lambda''\mu'')\,N_F S\,|\,N-2\,[f']\,\alpha'(\lambda'\mu')\,N_F-1\,S'>\Big]$$
$$U((\lambda'\mu')(20)(\lambda\mu)(20);\,(\lambda''\mu'')(\lambda^\circ\mu^\circ)).$$

From Eqs. (15), (16) the set of Racah equations for FPSCs can be easily obtained. For this purpose it is necessary to set the r.h.s. of Eq.(15) equal to zero at the following combinations of quantum numbers: $N_F^0 = 2, (\lambda^0\mu^0)S^0 = (40)1, (02)1$ and $(21)0$; $N_F^0 = 0, (\lambda^0\mu^0)S^0 = (21)0$. Substituting the sign "+" by "-" in the r.h.s. of Eq.(16) and considering the resulting as vanishing one we obtain additional equations. The set of Racah equations is reciprocal and enables one to calculate the FPSC $\langle N/N-1\rangle$ if the coefficients $\langle N-1/N-2\rangle$ are known. The initial point of this calculation is the fact: $\langle 2\, N_F/N'_F\rangle = 1$ at $N_F = 0,2$ and $\langle 2\, N_F = 1/N_F\rangle = (-1)^{N_F}/\sqrt{2}$.

3.4 The IBFM Hamiltonian Matrix in the U(6/12) Scheme

In the general case the IBFM Hamiltonian has a form

$$H_{IBFM} = H_0 + H_{int} ,$$

where $H_0 = \sum_{i=0}^{N} h_i$ is a superon single-particle energy, $H_{int} = \sum V_{ik}$ is the superon interaction energy. Using standard FPC technique, we obtain[19)]

$$\langle \sigma^N [f] \alpha (\lambda\mu) \varkappa L N_F S : J | H_0 | \sigma^N [f'] \alpha' (\lambda'\mu') \varkappa' L' N_F S' : J \rangle =$$

$$= N \sum \langle N [f] \alpha (\lambda\mu) N_F S | N-1 [f''] \alpha'' (\lambda''\mu'') N_F'' S'' \rangle$$

$$\langle N [f'] \alpha' (\lambda'\mu') N_F S' | N-1 [f''] \alpha'' (\lambda''\mu'') N_F'' S'' \rangle$$

$$\langle (\lambda''\mu'') \varkappa'' L''; (20) \ell | (\lambda\mu) \varkappa L \rangle \langle (\lambda''\mu'') \varkappa'' L''; (20) \ell | (\lambda'\mu') \varkappa' L' \rangle$$

$$\varepsilon_{\ell S_N j} \begin{pmatrix} L'' & S'' & J'' \\ \ell & S_N & j \\ L & S & J \end{pmatrix} \begin{pmatrix} L'' & S'' & J'' \\ \ell & S_N & j \\ L & S & J \end{pmatrix} \tag{17}$$

where $\varepsilon_{\ell S_N j}$ is a single particle energy of a boson ($S_N = 0$) or a fermion ($S_N = 1/2$) occupying an orbital with angular momentum j. In round brackets the transformation coefficients for the transition from LS to JJ coupling are given. The summation over double primed indices is assumed. The matrix of H_{int} can be calculated by the formulae

$$\langle \sigma^N [f] \alpha (\lambda\mu) \chi L N_F S : \gamma | H_{int} | \sigma^N [f'] \alpha' (\lambda'\mu') \chi' L' N_F S' ; \gamma \rangle =$$

$$= \frac{1}{2} N(N-1) \sum \langle N [f] \alpha (\lambda\mu) N_F S | N-2 [f''] \alpha'' (\lambda''\mu'') N_F'' S'', 2 [f^\circ] (\lambda^\circ\mu^\circ) N_F^\circ S^\circ \rangle$$

$$\langle N [f'] \alpha' (\lambda'\mu') N_F S' | N-2 [f''] \alpha'' (\lambda''\mu'') N_F'' S'', 2 [\tilde{f}^\circ] (\tilde{\lambda}^\circ\tilde{\mu}^\circ) N_F^\circ \tilde{S}^\circ \rangle$$

$$\langle (\lambda''\mu'') \chi'' L'', (\lambda^\circ\mu^\circ) \chi^\circ L^\circ | (\lambda\mu) \chi L \rangle \langle (\lambda''\mu'') \chi'' L'', (\tilde{\lambda}^\circ\tilde{\mu}^\circ) \tilde{\chi}^\circ \tilde{L}^\circ | (\lambda'\mu') \chi' L' \rangle$$

$$\langle N_F^\circ (\lambda^\circ\mu^\circ) \chi^\circ L^\circ S^\circ \gamma^\circ | V_{N-1N} | N_F^\circ (\tilde{\lambda}^\circ\tilde{\mu}^\circ) \tilde{\chi}^\circ \tilde{L}^\circ \tilde{S}^\circ \gamma \rangle$$

$$\begin{pmatrix} L'' & L^\circ & L \\ S'' & S^\circ & S \\ \gamma'' & \gamma^\circ & \gamma \end{pmatrix} \begin{pmatrix} L'' & \tilde{L}^\circ & L' \\ S'' & \tilde{S}^\circ & S' \\ \gamma'' & \gamma^\circ & \gamma \end{pmatrix} \qquad (18)$$

(the summation of $[f'']$, α'', $(\lambda''\mu'')$, L'', S'', γ'', L°, S°, γ°, $\tilde{L}^\circ$, $\tilde{S}^\circ$, $[f^\circ]$, $(\lambda^\circ\mu^\circ)$, $[\tilde{f}^\circ]$, $(\tilde{\lambda}^\circ\tilde{\mu}^\circ)$ must be done). The two-body matrix elements of V_{N-1N} in the r.h.s. of Eq.(18) are the fitting parameters in the IBFM characterizing the constants of boson-boson, boson-fermion and fermion-fermion interactions. Similarly we can derive the formulae for the calculations of the reduced probabilities B(E2) and one- and two-particle transfer spectroscopic factors. Thus the complete and effective mathematical formalism for description of the superparticle systems can be formulated. It would be interesting to apply it to the nuclei with violated supersymmetry.

3. EXTREMAL PROJECTORS FOR LIE SUPERALGEBRAS

3.1 Introduction

It is well known that the extremal projectors for usual semisimple Lie algebras[24,25] is a very effective tool of investigation of various problems in the Lie algebra representation theory[26,27]. In particular, they allow one to elaborate the elements of Wigner-Racah algebra (Clebsch-Gordan coefficients, 6j-symbols etc., see for example[28]) that are necessary for physical calculations. The extremal projectors for Lie superalgebras play the same important role (the first example, of the extremal projector for OSP(2.1) superalgebra was considered in Ref.[29]). The problem of construction of extremal projection operators for all contragredient superalgebras has been solved in Ref.[30]. Below we give some information on the structure of these projectors for superalgebras gl(n/m), osp(2n/M), D(2,1;α) F(4) and G(3).

3.2 Extremal Projectors

Contragredient Lie superalgebras have the following properties common to usual semisimple Lie algebras. In both the cases the root subspace corresponding to the vanishing root coincides with Cartan subalgebra; the nonvanishing roots are distributed into pairs α , $-\alpha$ and the root subspaces corresponding to the nonvanishing roots are one-dimensional. Thus in both cases there is the Cartan-Weyl basis $E_{\pm\alpha}$, H_i satisfying conditions

$$
\begin{aligned}
&[H_i, H_j] = 0, \\
&[H_i, E_\alpha] = \alpha(H_i) E_\alpha, \\
&[E_\alpha, E_\beta] = N_{\alpha,\beta}\, E_{\alpha+\beta}
\end{aligned}
\tag{19}
$$

($N_{\alpha+\beta} \neq 0$ if $\alpha+\beta$ is a root). In the case of Lie superalgebras the brackets [,] mean a supercommutator.

Let Δ_+^0 (Δ_+^1 be a system of even (odd) positive roots an arbitrary contragredient Lie superalgebra G. The extremal projector P = P(g) is given by the formula[30]

$$
P = \prod_{\alpha > 0} P_\alpha \tag{20}
$$

where

$$P_\alpha = \sum_{k=0}^{\infty} \frac{(-1)^k}{k!} \varphi_{\alpha,k}^{-1} E_{-\alpha}^k E_\alpha^k \qquad \text{at} \quad \alpha \in \Delta_+^0 \tag{21}$$

$$P_\alpha = 1 - \varphi_{\alpha,1}^{-1} E_{-\alpha} E_\alpha \qquad \text{at} \quad \alpha \in \Delta_+^1 \tag{22}$$

$$\varphi_{\alpha,k} = \prod_{i=1}^{k} (\varphi_\alpha + i) \tag{23}$$

$$\varphi_\alpha = H_\alpha + \rho(H_\alpha) + \delta_\alpha \qquad \text{at} \quad \alpha \in \Delta_+^0 \tag{24}$$

$$\varphi_\alpha = H_\alpha + \rho(H_\alpha) + \frac{1}{2}\alpha(H_\alpha) \qquad \text{at} \quad \alpha \in \Delta_+^1 \tag{25}$$

$\delta_\alpha = 1/2$ if $\alpha/2 \in \Delta_+^1$ and $\delta_\alpha = 0$ if $\alpha/2 \notin \Delta_+^1$, ρ is a difference between a halfsum of even roots and a halfsum of odd roots. In Eq.(20) the factors P_α ($\alpha \in \Delta_+ \equiv \Delta_+^0 \cup \Delta_+^1$) are arranged according to the normal order of positive root system Δ_+. The explicit form of the extremal projector does not depend on the choice of the normal order. It satisfies the conditions

$$E_\alpha P = P E_{-\alpha} = 0 \qquad \text{for all} \quad \alpha \in \Delta_+ \tag{26}$$

and

$$P^2 = P \tag{27}$$

For a more detailed description of extremal projectors and their applications see Refs.[30,31].

The authors are thankful to I.F. Istomina and A.V. Fursa for the assistance with the preparation of this paper.

---

x) It is assumed that $\alpha(H_\alpha) = 2$ for all roots $\alpha \in \Delta_+^0$ $H_\alpha = [E_\alpha, E_{-\alpha}]$.

REFERENCES

1) Janssen, B., Dzholos, R.V. and Donau, F., Nucl. Phys. A224, 93 (1974).
2) Arima, A. and Iachello, F., Ann.Phys. 99, 253(1976); 111, 201 (1978); 123, 468 (1979).
3) Otsuka T., Arima, A. and Iachello, F., Nucl. Phys. A309, 1(1978).
4) Ginocchio J.N., Ann. Phys. 126, 234 (1980).
5) Geyer H.B. and Hahne, W., Phys. Lett. B97, 173(1980).
6) Dobaczewsky, J., Nucl. Phys. A369, 237 (1981).
7) Smirnov Yu.F., Tolstoy, V.N., Sakharuk, A.A., Havliček, M. and Burdik, C., in:"Group Theor. Methods in Physics", Moscow, Nauka, vol.2, 517 (1983).
8) Elliott, J.P., Rep. Prog. Phys. 48, 171 (1985).
9) Takada, K. and Tamura, T., Phys. Rev. C31, No.5, 1948 (1985).
10) Perelomov, A.M., Uspekhi Fiz. Nauk 123, 23(1977).
11) Smirnov, Yu.F., Tolstoy, V.N., Burdik, C., Havlicek, M. and Navratil, O., Nuclear spectroscopy and atomic nuclei structure', Rept. thesis of 36th Conference, Nauka, Leningrad, 192(1986).
12) Asherova, R.M., Smirnov Yu.F. and Tolstoy V.N., Matem. zametki 26, 15 (1979).
13) Iachello, F., Phys. Rev. Lett. 44, 772(1980).
14) Balantekin, A.B., Bars, I. and Iachello, F., Nucl. Phys. A370, 284(1981).
15) Balantekin, A.B., Bars, I., Bijker, R. and Iachello, F., Phys. Rev. C27, 1761 (1983).
16) Van Isacker, P., Frank, A. and Sun, B.Z. Ann. Phys. 157, 183 (1984).
17) Bars, I., in:"Proc. Workshop on Bosons in Nuclei", Singapore, 155 (1984).
18) Cizewski, J.A., ibid., 175 (1984).

19) Guseva, I.S., Smirnov, Yu.F. and Kharitonov, Yu.I., Izv. Akad. Nauk SSSR, ser.fiz. 49, 37 (1985).
20) Guseva, I.S., Smirnov Yu.F., Tolstoy, V.N. and Kharitonov, Yu.I., Preprints of the Leningrad Institute of Nuclear Physics, No.678 (1981); No.837(1983); No.999 (1984).
21) Elliott, J.P., Proc. Roy, Soc. A245, 128, 562(1958).
22) Neudatchin, V.G. and Smirnov, Yu.F. "Nucleon associations in light nuclei", Moscow, Nauka, 1968.
23) Chen, J.Q., Gan, M.J., Chi, V.J., Vallieres, M. and Feng, D.H., Nucl. Phys. A419, 77 (1984).
24) Smirnov, Yu.F. "Clustering Phenomena in Nuclei" (Vienna: IAEA) 153(1969).
25) Asherova, R.M., Smirnov, Yu.F. and Tolstoy V.N., Sov. Journ. Teor. and Math. Fizika 8, 255 (1971).
26) Asherova R.M., Smirnov, Yu.F. and Tolstoy, V.N., Teor. and Math. Fizika 15, 107 (1973).
27) Zhelobenko, D.P., Dokl. Akad. Nauk SSSR, 273, 785, 1301 (1983).
28) Pluhar Z., Smirnov, Yu.F. and Tolstoy, V.N., J. Phys. A19, 21 (1986).
29) Berezin F.A., Tolstoy, V.N., Commun. Math. Phys. 78, 409 (1981).
30) Tolstoy, V.N., Uspekhi Mathem. Nauk 40, 225 (1985); in: "Group Theor. Methods in Physics", Moscow, Nauka v.2, 46(1986).
31) Tolstoy, V.N., Istomina, I.F. and Smirnov, Yu.F., in: "Group Theor. Methods in Physics", Moscow, Nauka, 2, 55(1986).

THE RESTRICTED DYNAMICS MODELS AND BOSON REALIZATIONS

V.Vanagas
Institute of Physics, Academy of Sciences of the Lithuanian SSR, 232600 Vilnius, USSR

ABSTRACT

Conceptions and some applications of various approaches to the many-body problem in nuclear physics based on the idea of the restricted dynamics are discussed, including the boson-type realizations for the nuclear collective motion and gluon field states.

1. DEFINITIONS AND TYPES OF THE RESTRICTED HAMILTONIANS

We start with the Schrodinger equation $H\,\Psi = E\,\Psi$ and simplify it to $H_o\Psi_o = E_o\Psi_o$, where H_o is obtained by means of the special decomposition $H = H_o + H'$. We use the term "special" because H_o acts within some subspace R_o of the space R, i.e. H_o is " H restricted to $_o$". The idea of the restricted dynamics, proposed and developed in [1], gives a method for constructing nuclear models.

There are two main kinds of the restricted Hamiltonians - the collective H_{coll} and the anti-collective H_{Acoll}. In general, the restricted dynamics Hamiltonian H_o consists of both terms, i.e.

$$H_o = H_{coll} + H_{Acoll}. \tag{1}$$

Our aim is to discuss nuclear models defined by the Schrodinger equation

$$(H_{coll} + H_{Acoll})\ \Psi_o = E_o \Psi_o\ . \tag{2}$$

In particular, if interested in collective effects, we take only H_{coll}.

There are various types of both collective and anticollective Hamiltonians with the restricted dynamics beginning with sophisticated and ending with simplest ones. Simplified versions of the general microscopic collective Hamiltonian H_{coll}, introduced in [2], are shown in Fig.1, where the subscripts used read as follows: the first letters M or ME in subscripts mean "Microscopic"

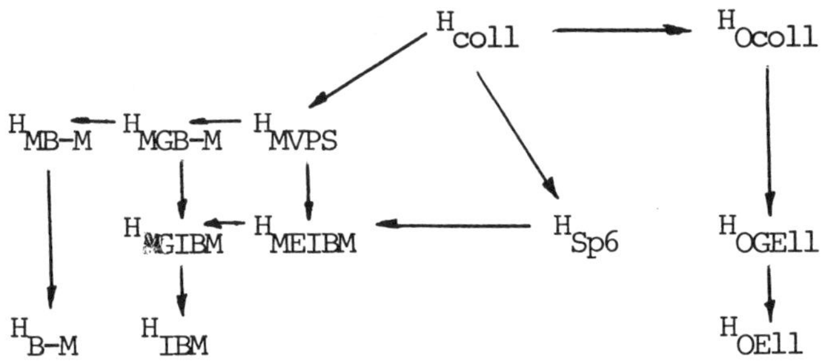

Fig.1. Simplified versions of the general microscopic collective Hamiltonian

or "Microscopic Extension", O indicates the collective Hamiltonians with strictly restricted dynamics, G stands for "Generalized"; Ell, B-M, IBM, VPS and Sp6 correspondigly denote the Hamiltonians of Elliott, Bohr-Mottelson, Interacting Boson Model, Violated Permutation Symmetry and Symplectic Collective.

In Fig.1 three main branches are shown, beginning with H_{coll}. Two of them end with Bohr-Mottelson and IBM Hamiltonians, while the third one - with Elliott Hamiltonian. From Fig.1 we see, that the Microscopic Generalized Bohr-Mottelson Hamiltonian H_{MGB-M} is the last "cross-road" on the way from H_{coll} to both H_{B-M} and

H_{IBM}. Using the symplectic type decomposition [3], H_{MGB-M} can be decomposed in terms of boson-type interactions. This decomposition provides the microscopic derivation of H_{IBM} and establishes its relationship with H_{B-M} (see for details [1,4,5]). More general Hamiltonian H_{MVPS} gives an analogue of H_{IBM} extended to odd nuclei. Detailed comments of Fig.1 and a similar scheme for the anti-collective Hamiltonians with additional references will be described in [6].

2. SOME FEATURES OF THE RESTRICTED DYNAMICS MODELS

The microscopic collective Hamiltonian H_{coll} was obtained by averaging, roughly speaking, H over particle indices. The relationship between the microscopic and phenomenological approaches can be established using the following bilinear and threelinear forms of Jacobi variables ρ_i^s :

$$q^{ss'} = q^{ss'}(\rho\beta\gamma,\phi'\Theta\phi'') = \sum_{i=1}^{A-1} \rho_i^s \rho_i^{s'}$$

$$q^{ss's''} = q^{ss's''}(\rho\beta\gamma,\phi'\Theta\phi'';\Theta_1,\ldots,\Theta_{3A-9}) =$$

$$= \sum_{i=1}^{A-1} \rho_i^s \rho_i^{s'} \rho_i^{s''} . \qquad (3)$$

The set of $q^{ss'}$ variables is equivalent to β, γ, three Euler angles $\phi'\theta\ \phi''$ and the global radius ρ. The operator H_{MGB-M} depends on $q^{ss'}$ which makes it possible to introduce of the "deep collectivity" conception and to propose a new interpretation of the IB model [5]. The $q^{ss's''}$ variables can be linked with the octupole degrees of freedom; from the microscopic point of view they depend on both collective $\rho, \beta, \gamma, \phi', \theta, \phi''$ and internal $\theta_1, \ldots, \theta_{3A-1}$ microscopic variables.

Every term of the nuclear Hamiltonian H contains its collective part, thus H_{coll} consists of kinetic H^K_{coll}, Coulomb H^e_{coll}, central H^c_{coll}, vectorial H^v_{coll} and tensorial H^t_{coll} terms, i.e.

$$H_{coll} = H^K_{coll} + H^e_{coll} + H^c_{coll} + H^v_{coll} + H^t_{coll}. \tag{4}$$

The general microscopic collective Hamiltonian H_{coll} is defined as the O_{A-1}-scalar part of H. From this definition, proposed in [2], it follows, that H_{coll} possesses additional integrals of motion: the eigenvalues E_{coll} of H_{coll} depend on space symmetry λ, isospin T of nucleus and the new important integral of motion ω, labelling the O_{A-1}-irreducible representation $\omega = (\omega_1\omega_2\omega_3)$. Thus we conclude, that $E_{coll} = E_{coll}(J^{\pi};\omega\lambda T)$. In both even and odd nuclei this spectrum consists of ω -bands, additionally labelled by λT as well as by the total momentum L and parity π . Depending on the nucleon-nucleon potentials used, ω-bands possess the infinite number of discrete, continuum or both discrete-continuum states.

3. NEW COLLECTIVE EFFECTS IN EVEN AND ODD NUCLEI

In the lowest states of the even-even nuclei the last two terms in (4) do not contribute. As a consequence $E_{coll} = E_{coll}(L^{\pi};\omega\lambda\ ST)$, where L and S give the total space angular momentum and spin of the nucleus. For this reason H_{coll} possesses for even-even nuclei the "classical" type spectrum enriched by the following new microscopic features: within the ω-band there are λST-bands, and (depending both on ω and the potentials used) within the λST-band there are traditional K-bands as well as (in the case of the discrete spectrum) - $(\lambda\mu)$-bands. Both K - and - $(\lambda\mu)$ characteristics are approximate and can be destroyed by the sophisticated nucleon-nucleon interaction potentials.

Vectorial H^{v}_{coll} and tensorial H^{t}_{coll} terms of H_{coll} contribute even in the case of the low-lying states of odd and odd-odd nuclei, destroying L and S and coupling them to J. In the simplest cases of low-lying states of odd nuclei with $\lambda = [4...432...2]$ or $\lambda = [4...42...21]$ when $S = 1/2$, thus H^{t}_{coll} gives zero, the interaction H^{v}_{coll} couples L and 1/2 to $J = L + 1/2$. As a result we have the ω -bands of a pure collective spectrum with half integers J. The origin of this collective spectrum essentially differs from the spectrum obtained using a well known picture of the odd nucleus with a deformed core and extra particles moving in Nilsson orbits.

Anticollective effects emerge, when the dynamic equations are, roughly speaking, averaged over the Carthesian indices, which stress the many-particle structure of the nucleus. There are a few types of the anticollective Hamiltonians (see for details [1,6]). The spectrum of the momentum-independent anticollective Hamiltonian $H_{MIAcoll}$ consists of the group of levels, labelled by λST as well as by the other characteristics (e.g. $(\lambda\mu)$), depending on the basis functions used.

Both the collective and anticollective effects arise in the equivalent fermion system. This property has been checked by means of the "computer experiment"[7]. It lies at the bottom of the Schrodinger equation (2), which can be solved in the strictly restricted dynamics approximation. From such a simplified analysis it follows [8], that ω-bands, developed by H_{coll}, are built up on the levels of H_{Acoll}, belonging to definite states of the internal degrees of freedom. In a general case the Schrodinger equation (2) produces a very rich spectrum in both even and odd nuclei.

4. "DEEP COLLECTIVITY" INTERACTING BOSON MODEL AND ITS ANALOGUE FOR ODD NUCLEI

Let us present H in the form of

$$H = H_{MGB-M}(q^{ss'}) + H' , \tag{5}$$

with the first term, depending only on $q^{ss'}$.

The decomposition (5) in the matrix representation has been studied in [1] making use of the fact, that H_{MGB-M_K} acts only in the space with $\omega = (0)$, thus in H_{MGB-M} only H^K_{coll}, H^e_{coll} and H^c_{coll} terms of H_{coll} contribute. The products of the creation and annihilation operators for $g^{ss'}$ variables give Sp(12,R) algebra, used in [9], containing U(6) subalgebra. This is the reason, why the Hamiltonian H_{MGB-M} covers both Bohr-Mottelson and IBM Hamiltonians, which explains the fragment in Fig.1, beginning with H_{MGB-M}. Few simplest terms of H_{MGB-M} obtained using the symplectic-type decomposition[3] give the IBM Hamiltonian H_{IBM} introduced on the grounds of "deep collectivity" conception, without referring to active fermions coupled to s and d bosons.

We conclude, that H_{coll}, restricted to the space $R^{(\omega)}$ with $\omega = (0)$, reduces to H_{MGB-M}. This provides the first, simplest example of the Microscopic Violated Permutation Symmetry Hamiltonian H_{MVPS}. If ω is replaced by the O_{A-1}-irreducible representation (1), then H_{coll} acts in $R^{(1)}$. In this case the Schrodinger equation for

H_{coll} gives the "deep collectivity" IBM analogue in the odd nuclei (the second example of H_{MVPS}). The chain of Hamiltonians beginning with H_{Sp6}, shown in Fig.1, is obtained from H_{coll} using the symplectic-type decompositions [3].

5. COLLECTIVE EFFECTS IN CONSTANT NON-ABELIAN GLUON FIELD MODEL

In[10] the functional Schrodinger equation for the gluon field has been simplified to the usual Schrodinger equation. The explicit expression for H_{Gluon} has been obtained at first for SU(2) two- and three-dimentional case[11] and later on - for the general SU(N) field in τ_o-dimensions [12]. Strong microscopic features in H_{Gluon} have been discovered[12]. In particular it was shown, that H_{Gluon} for SU(2) model in τ_o-dimension can be presented in the form of

$$H_{Gluon} = H^k_{coll} + \frac{1}{96} \lambda(\lambda+1)(\lambda+2)\left(\frac{\tau_o^{-1}}{\tau_o}\rho^4-\beta^2\right), \tag{6}$$

where ρ and β -collective variables, employed in the nuclear theory. Using both analytical and numerical methods the spectrum of H_{Gluon} has been studied [13] and the following features of the constant non-abelian field states have been discovered.

The first term of the "deep collectivity" IBM-type decomposition for H_{Gluon}, presented in the form of Casimir operators of the unitary U_3 and U_{τ_o} groups describe the gross-structure of the H_{Gluon} spectrum. Next IBM-type terms give splitting of $U_3 \times U_{\tau_o}$ states into ω levels. Finally, the rest terms of H_{Gluon} composed from the $Sp(6\tau_o,R)$ infinitesimal operators are responsible for the compressing effects of the ω levels. As a result H_{Gluon} possesses the collective-type spectrum with inversed semi-fine ω level structure [13].

REFERENCES

1. Vanagas,V., "The Microscopic Nuclear Theory within the Framework of the Restricted Dynamics", Lecture Notes, University of Toronto, 1977.

2. Vanagas,V., Sov.J.Nucl.Phys., 23,950-959 (1976).

3. Vanagas,V., "The Symplectic Models of Nucleus". In "Group Theoretical Methods in Physics",Proc. 1982 Zvenigorod Int.Sem., ed. M.Markov and V. Man´ko. Moscow, Nauka, vol.2, 144-164 (1983); Engl.transl., Chur,Paris,London: Harwood Academic publ. Vol.1, 259-282 (1985).

4. Vanagas,V., "The Microscopic Theory of the Collective Motion in Nuclei", Lecture Notes in "Group Theory and its Applications in Physics - 1980. Latin American School in Physics, Mexico City". N.Y.American Institute of Physics, AIP Conference Proceedings, No 71, 220-293, 1981.

5. Vanagas,V., Bulg.J.Phys., 9, No 3, 231-245 (1982).

6. Vanagas,V., "The Restricted Dynamics Nuclear Models: Conceptions and Applications "Lecture Notes, Poiana Brasov,Romania, September 9-13, 1986(to be published).

7. Katkevičius,O. and Vanagas,V., Liet.Fiz.Rinkinys, 25, No 2, 10-18(1985); No 3, 3-11 (1985).

8. Vanagas,V., Sabaliauskas,L.and Eriksonas,K., Izv.Acad.Nauk. USSR,Ser.Fiz.47,963-968(1983).

9. Vanagas,V., Nadjakov,E. and Raychev,P.,"Group-Theoretical Approach to Nuclear Vibration-Rotation States". Preprint IC/75/40, Trieste (1975); Bulg.J.Phys., 11, No 6, 558-569 (1975).

10. Simonov,Y.A., Phys.Lett., 136B, 105 (1984).

11. Badalyan,A., Sov. J.Nucl.Phys., 39, 947-955 (1984).

12. Vanagas,V., Sov.J.Nucl.Phys., 41, 1474-1484 (1985).

13. Vanagas,V. and Katkevičius, O., "The Structure and Spectrum of the Constant SU(2) Non-Abelian Gluon Field Hamiltonian" (to be published).

MICROSCOPIC BOSON OPERATORS FROM THE DYSON MAPPING AND A SENIORITY DICTATED SIMILARITY TRANSFORMATION

HENDRIK B GEYER
Institute of Theoretical Nuclear Physics
University of Stellenbosch, Stellenbosch 7600,
South Africa

Abstract We show that seniority boson images of bifer= mion operators can indeed be constructed as proposed by Geyer and Lee, namely by first using the Dyson mapping, followed by a similarity transformation which restores in the boson space the association between seniority and boson number. The original results of Otsuka, Arima and Iachello are thereby recovered and can at the same time be generalised to include g-bosons or even bosons with $J>4$ as well as any higher order boson terms. Further= more the technique produces seniority boson images *valid for any number of d- or g-bosons* - a result which cannot be obtained within the framework of the usual Marumori- or OAI-method for the complete boson image of an opera= tor.

INTRODUCTION

In the construction of a microscopically founded interacting boson model of nuclei two vital stages are encountered irre= spective of the particular approach taken, namely the trans= cription of the fermion problem to a boson space and the identification of the physically relevant (collective) bosons which facilitate a truncated description.

Although phenomenology obviously guides this attempted construction, the resulting boson model need of course not be restricted to the structure of any presently existing version of the successful phenomenological interacting boson model (IBM)[1-2]. In fact, from microscopic considerations one hopes to identify and understand those situations for which the existing IBM is applicable as well as those for which modifications should and could be made.

The order of first mapping and then truncating can of course also be reversed, although it seems that some of the advantages associated with the possibility of performing manipulations in the boson space are then lost. This aspect, however, requires further investigation. At present we prefer to stick to our original order and have therefore to choose a particular mapping procedure. As far as this aspect is concerned the Dyson boson mapping (DBM), characterized by finite boson images, has been demonstrated[3-12] to be a very efficient tool for formal investigations as well as ideally suited for the construction of boson models directly linked to fermion microscopy. Due to the non-unitarity of the DBM, the equivalent boson picture obtained from a direct application of this mapping is characterized by a Hamiltonian which is non-hermitian when represented in an ideal boson basis. This, however, is no reason for concern[4-5] since one is dealing with an exact boson realization of some Lie algebra. In fact, the very nature of this pair boson Dyson Hamiltonian has served as the basis for economic and elegant further developments[8-12]. Here we show that the ideas of ref. 12 can be generalized to the case of a single-j shell model resulting in a description in terms of seniority bosons.

Direct application of the Dyson mapping leads to a description in terms of pair bosons, as does of course any boson mapping based on the preservation of the complete bifermion operator algebra. In general this situation, however, presents an unfavourable point of departure when considering a truncated description in terms of the physically relevant bosons, such as s- and d-bosons in the IBM. This point was emphasized by Geyer and Lee[8] and illustrated by numerical examples in a single-j shell model.

For this model Otsuka, Arima and Iachello[13] had previous=

ly shown that, at least in the SU(5) vibrational limit of IBM, the physical s- and d-bosons are to be associated with a seniority description. In the OAI-scheme boson operators are accordingly constructed by requiring equality between boson matrix elements and the corresponding fermion matrix elements calculated in a seniority basis. This Marumori-type method has, however, limitations once the seniority basis includes states with an arbitrary number of d- or g-bosons.

After a brief resumé of the Dyson mapping we proceed to show that this problem is circumvented by the introduction of a similarity transformation characterized by the fact that it restores in the boson space the association between seniority and twice the number of non-s bosons.

THE DYSON MAPPING

The generalized DBM[3-7] is defined by the mapping

$$b^{\alpha\beta} \to R^{\alpha\beta} \equiv B^{\alpha\beta} - B^{\alpha\theta} B^{\beta\lambda} B_{\theta\lambda} ,$$
$$b_{\alpha\beta} \to R_{\alpha\beta} \equiv B_{\alpha\beta} , \; b^{\alpha}_{\beta} \to R^{\alpha}_{\beta} \equiv B^{\alpha\theta} B_{\beta\theta} \tag{1}$$

between bifermion operators b and ideal boson operators B. (See e.g. ref. 10 for notation). For a single-j shell it is convenient to introduce spherical boson operators

$$B^{JM} \equiv 2^{-\frac{1}{2}} \sum \langle jmjm' | JM \rangle B^{jmjm'} ; \; B_{JM} = (B^{JM})^{\dagger} . \tag{2}$$

From the basic mapping and standard angular momentum coupling it is now straightforward to obtain the Dyson images of the spherical fermion operators

$$A^{JM} \equiv 2^{-\frac{1}{2}} \left[c^j \times c^j \right]^J_M ; \; U^J_M \equiv \left[c^j \times \tilde{c}_j \right]^J_M \tag{3}$$

in terms of the spherical pair boson operators (2), namely

$$(A^{JM})_D = B^{JM} - 2 \sum \hat{J}_1 \hat{J}_2 \hat{J}_3 \hat{L} \begin{Bmatrix} j & J_1 & j \\ j & J_2 & j \\ J_3 & L & J \end{Bmatrix} \left[\left[B^{J_1} \times B^{J_2} \right]^L \times \tilde{B}_{J_3} \right]^J_M , \tag{4}$$

$$(A_{JM})_D = B_{JM} \quad , \tag{5}$$

$$(U^J_M)_D = -2 \sum \hat{J}_1 \hat{J}_2 \begin{Bmatrix} J_1 & J_2 & J \\ j & j & j \end{Bmatrix} \left[B^{J_1} \times \tilde{B}_{J_2} \right]^J_M \tag{6}$$

Note that the pair boson image $(A^{00})_D$ contains terms such as $d^\dagger . d^\dagger s$ which changes the number of non-s bosons, in contrast to the fermion operator A^{00} which does not change the senio= rity. (A notation where $s \equiv B_{00}$, $d^\mu \equiv d^\dagger_\mu \equiv B^{2\mu}$, etc. is used).

Of particular importance for further developments is the structure of the Dyson image for the pairing Hamiltonian $H = -G\Omega A^{00} A_{00}$ where $2\Omega = 2j + 1$. From expressions (4-6) one finds ($\sum'$ denotes a summation over non-zero J)

$$H_D = -G\left[\Omega n_s - n_s(n_s - 1) - 2n_s \sum{}' n_{2J}\right] + G\sum{}' B^J . B^J ss$$
$$-G\sqrt{2\Omega} \sum{}' \hat{J}_1 \hat{J}_2 \hat{J}_3 \begin{Bmatrix} J_1 & J_2 & J_3 \\ j & j & j \end{Bmatrix} \left[\left[B^{J_1} \times B^{J_2}\right]^{J_3} \times \tilde{B}_{J_3}\right]^0_0 s. \tag{7}$$

CONSTRUCTION OF THE SIMILARITY TRANSFORMATION

We note that H_D, when represented in an ideal boson basis, such as the one defined by the SU(5)$\supset$O(5)$\supset$SO(3) chain, is of triangular form. (In ref. 10 it has been proved that the use of an ideal boson basis in conjunction with a mapped operator, such as H_D, is always permissible). This means that $H_D = H_o + W$ has the same spectrum as its diagonal part H_o and eigenstates $|\psi\rangle$ and $|\phi\rangle$ of H_D and H_o, respectively, are related by

$$|\psi\rangle = \sum_{k=o}^{\infty} \left(\frac{1}{E - H_o} W\right)^k |\phi\rangle \equiv Z^{-1}|\phi\rangle \tag{8}$$

for a spesific energy E. If we now operate on the equation

$$(H_o + W)\, Z^{-1} Z|\psi\rangle = EZ^{-1}Z|\psi\rangle \tag{9}$$

from the left by Z, the resulting expression is

$$Z(H_o + W)Z^{-1}|\phi\rangle = E|\phi\rangle = H_o|\phi\rangle \ , \tag{10}$$

which explicitly shows that the similarity transformation Z transforms away the term W in H_D. If Z also commutes with $(A_{00})_D = s$, it is clear that the same Z also transforms away those terms in $(A^{00})_D$, which change the number of non-s bosons. Explicitly this means that one gets the structure envisaged previously, namely

$$(A^{00})_{sen} \equiv Z(A^{00})_D\, Z^{-1} = s^\dagger(1 - \frac{n_s}{\Omega} - \frac{2}{\Omega}\sum{}' n_{2J}), \qquad (11)$$

from which it becomes possible to directly associate senio= rity and boson number through $\nu = 2\sum{}' n_{2J}$.

We note that the choice $j=3/2$ allows only s- and d-bosons in the expression for $(A^{00})_D$. In particular the term W in H_D consists of only one contribution which contains the combination $d^\dagger \cdot d^\dagger ss$. Due to this simplification a closed form for Z^{-1} can be obtained. First write Z^{-1} as

$$Z^{-1} \equiv \sum_{k=o}^{\infty} \left(\frac{1}{E-H_o} W \right)^k \equiv \sum_{k=o}^{\infty} \left(\frac{1}{\hat{H}_0 - H_0} W \right)^k \wedge \qquad (12)$$

Here a positional operator as introduced by Kim and Vincent[11] appears - the position of the hat namely indicates where in an expression a number dependent hat operator is to be evalu= ated. In this notation Z^{-1} can then be re-written as

$$Z^{-1} = \vee\; e^{-\frac{1}{2}d^\dagger \cdot d^\dagger ss} \frac{(\Omega - 1 - n_d - \check{n}_d)!!}{(\Omega - 1 - 2n_d)!!} \qquad (13)$$

In general one now finds for any Θ from expressions (4-6) the seniority boson image from

$$\Theta_{sen} = Z\, \Theta_D\, Z^{-1} . \qquad (14)$$

For $\Theta = A^{00}$ one confirms that $(A^{00})_{sen}$ is given by the expression (11). Furthermore $(A_{00})_{sen} = s$, since Z indeed commutes with s. In fact, the seniority images we obtain are given, except for $(A_{2\mu})_{sen}$ which is incorrectly given there, as eqs. (15) of ref. 14 where, however, these images were obtained in OAI-fashion from a comparison of matrix elements.[15] For general $j>3/2$ we note that in the non-diagonal part W of H_D in eq. (7), there now appear two classes of terms with the structure $B^J \cdot B^J ss$ and $[B^{J_1} B^{J_2} \tilde{B}_{J_3}]^0_0 s$. While a closed form for the corresponding transformation Z now seems unattainable, we can still find approximate seniority boson images by truncating the sum over k in eq. (12). Furthermore, it is clear from the structure of Z^{-1} that the inclusion of higher values of k will only introduce higher order multi-boson contributions (characterized by having more and more s-boson annihilation operators) to these seniority images, without changing the structure of lower order contributions. We now adopt the lowest order approximation to Z^{-1} which retains only the $k = 0$ and $k = 1$ terms. In this approximation, together with a restriction to s- and d-boson only, one finds e.g.

$$(U^2_\mu)_{sen} \cong \sqrt{\frac{2}{\Omega}} \left[s^\dagger \tilde{d}_\mu + d^\mu s \left(\frac{\Omega - N - n_d}{\Omega - 1 - 2n_d}\right) \right] - 10 \begin{Bmatrix} 2 & 2 & 2 \\ j & j & j \end{Bmatrix} \left(\frac{\Omega - 2N}{\Omega - 2n_d}\right) [d^\dagger \times \tilde{d}]^2_\mu \, , \tag{15}$$

Expression (15) is not the OAI-result yet, since $s^\dagger \tilde{d}_\mu$ and $d^\mu s$ have different coefficients. One has to bear in mind, however, that for a given fermion operator Θ the Dyson scheme prescribes the association[5]

$$\langle\psi|\Theta|\psi'\rangle = \left[{}_L(\psi|\Theta_D|\psi')_R \; {}_L(\psi'|(\Theta^\dagger)_D|\psi)^*_R \right]^{\frac{1}{2}} \tag{16}$$

between the fermion matrix element and its boson equivalent - a direct result of the non-unitarity of the Dyson mapping. For any seniority boson image, such as $(U^2_\mu)_{sen}$ in eq. (15), of which the different boson terms connect different pairs of boson states of good seniority (good n_d) one can accordingly write down a hermitian equivalent boson operator $\bar{\Theta}$ for which the boson matrix element is calculated in the usual way, namely

$$\langle\psi|\Theta|\psi'\rangle = (\psi|\bar{\Theta}|\psi') \ . \tag{17}$$

In the present case the hermitian equivalent $\bar{U}^2_\mu$ of $(U^2_\mu)_{sen}$ is given by

$$\bar{U}^2_\mu = \sqrt{\frac{2}{\Omega}}\left[d^\mu s\sqrt{\frac{\Omega - N - n_d}{\Omega - 1 - 2n_d}} + \sqrt{\frac{\Omega - N - n_d}{\Omega - 1 - 2n_d}}\ s^\dagger \tilde{d}_\mu\right] - 10\begin{Bmatrix}2 & 2 & 2\\ j & j & j\end{Bmatrix}\frac{\Omega - 2N}{\Omega - 2n_d}\left[d^\dagger \times \tilde{d}\right]^2_\mu \ . \tag{18}$$

The result (18) is the same as the one originally given in ref. 16 and differs from the OAI-result[13], rederived by Li[9], which is only valid for the case $n_d = 1$.

As in ref. (17) we can now go beyond a seniority map= ping in terms of s- and d-bosons by including g-bosons or even bosons with J>4 in the seniority boson images, as well as retaining higher order many-body boson operators. For $(U^2_\mu)_{sen}$ this means that we also take into account terms of the type $[d^\dagger \times \tilde{g}]^2_\mu$, $[g^\dagger \times \tilde{d}]^2_\mu$, $[d^\dagger d^\dagger \tilde{d}]^2_\mu$ s and $s^\dagger[d^\dagger \tilde{d}\tilde{d}]^2_\mu$ etc. As already stated, the inclusion of such terms does not change the structure of the terms $s^\dagger\tilde{d}_\mu$, $d^\mu s$ and $[d^\dagger \times \tilde{d}]^2_\mu$ already calculated. In contrast to the commutator method of ref. 17 where the inclusion of such higher order terms requires a much increased effort to handle all the unknown functions introduced, the calculation of boson number depen= dent coefficients of terms like $[d^\dagger \times \tilde{g}]^2_\mu$ requires no more effort than that of e.g. $d^\mu s$.

For the terms $[d^\dagger \times \tilde{g}]^2_\mu$ and $[g^\dagger \times \tilde{d}]^2_\mu$ we again find the same results as in ref. 17. On the other hand we find no term of the form $s^\dagger[d^\dagger\tilde{d}\tilde{d}]^2_\mu$. This only means that we cannot write down a hermitian equivalent $\bar{U}^2_\mu$ which contains the con= jugate pair $s^\dagger[d^\dagger\tilde{d}\tilde{d}]^2_\mu$ and $[d^\dagger d^\dagger\tilde{d}]^2_\mu$ s. However, as long as we observe the proper prescription (16) to calculate senio= rity boson matrix elements, these matrix elements will con=

tain all information that can be carried by a seniority boson image of a chosen order.

CONCLUSIONS

In terms of the OAI-prescription we have succeeded through the similarity transformation to obtain the seniority boson images of operators *valid for states with an arbitrary number of d-bosons*, *g-bosons* or higher j-bosons. This clearly goes beyond what is normally considered in the OAI framework, namely to find matrix element equivalence only for some low seniority states containing one or two d- or g-bosons. In fact the OAI method cannot be used to determine seniority images for general numbers of $J \geqq 2$ bosons as soon as these images contain more than one term which connects the same set of basis states, since one would then attempt to find the (number dependent) coefficients of all these terms from one equation which equates the fermion and boson matrix element. This difficulty is similar to the one which appears when one attempts to construct a hermitian equivalent operator when the seniority image contains more than one term which can connect the same set of basis states. There, however, we can of course simply utilize the non-hermitian boson image in the proper way as given in eq. (16).

REFERENCES

1. A. Arima and F. Iachello, Adv. Nucl. Phys. 13. 139 (1984).
2. F. Iachello, ed., Interacting Bose-Fermi Systems in Nuclei (Plenum, New York, 1981).
3. D. Janssen et. al., Nucl. Phys. A172, 145 (1971).
4. P. Ring and P. Schuck, Phys. Rev. C16, 801 (1977).
5. F.J.W. Hahne, Phys. Rev. C23, 2305 (1981).
6. H.B. Geyer and F.J.W. Hahne, Phys. Lett. 97B, 173 (1980).
7. H.B. Geyer and F.J.W. Hahne, Nucl. Phys. A363, 45 (1981).

8. H.B. Geyer and S.Y. Lee, Phys. Rev. C26, 642 (1982).

9. C.T. Li, Phys, Lett. 120B, 251 (1983).

10. H.B. Geyer, C.A. Engelbrecht and F.J.W. Hahne, Phys. Rev. C33, 1041 (1986).

11. G.K. Kim and C.M. Vincent, Univ. of Pittsburgh preprint, 1985.

12. H.B. Geyer, Phys. Lett. B, submitted (1986).

13. T. Otsuka, A. Arima and F. Iachello, Nucl. Phys. A309, 1 (1978).

14. A. Arima, N. Yoshida and J.N. Ginocchio, Phys. Lett. 101B, 209 (1981).

15. J.N. Ginocchio, private communication.

16. T. Otsuka, A. Arima, F. Iachello and I. Talmi, Phys. Lett. 76B, 139 (1978).

17. D. Bonatsos, A. Klein and C.T. Li, Nucl. Phys. A425, 521 (1984).

SOME NEW ASPECTS OF THE SDG INTERACTING BOSON MODEL

H.C. WU* and A.E.L.DIEPERINK
Kernfysisch Versneller Instituut, 9747 AA Groningen, The Netherlands
* Permanent address: Suzhou University, Suzhou, China

Abstract The intrinsic states and electromagnetic M1 and E4 properties are discussed for the SU(3) limit of the sdg IBA Model.

1. INTRODUCTION

In the original version of the IBA model[1] it is assumed that low-lying quadrupole collective states can be described in terms of L=0 (s) and L=2 (d) bosons. Since the model was introduced it has been suggested that the L=4 (g) boson also play a role. Indeed some phenomenological analyses[2-4] indicate that the incorporation of this additional boson leads to improved predictions, especially for deformed nuclei.

In this contribution we first discuss the derivation of analytic expressions for intrinsic states in the SU(3) limit of the sdg IBA model. By using these intrinsic states, results are derived for magnetic dipole properties and E4 transition strength.

2. INTRINSIC STATES

Intrinsic states have been shown to be a useful concept in the geometric model, the Elliott SU(3) fermion model[5] and the IBA model. For the SU(3) limit of the IBA model, a reasonable choice is to define intrinsic states as the eigenstates of the z-component quadrupole operator $Q_0^{(2)}$, as was used in the Elliott model.

In the sdg IBA model the symmetry group is U(15), and the generators of the SU(3) algebra, $L_\mu^{(1)}$ and $Q_\mu^{(2)}$, are given by ref. 3. In the version of sdg IBA in which one distinguishes between the neutron and proton degrees of freedom the symmetry group is $U^{(\pi)}(15)\times U^{(\nu)}(15)$. In this case the appropriate group chain is

$$\begin{array}{ccccccccc} U^{(\pi)}(15)\otimes U^{(\nu)}(15) & \supset & U^{(\pi+\nu)}(15) & \supset & SU^{(\pi+\nu)}(3) & \supset & SO^{(\pi+\nu)}(3) \\ [N_\pi] \quad [N_\nu] & & [N,F] & & (\lambda,\mu) & & L \end{array} \tag{2.1}$$

and the SU(3) generators are

$$L_\mu^{(1)} = L_{\pi,\mu}^{(1)} + L_{\nu,\mu}^{(1)} \quad \text{and} \quad Q_\mu^{(2)} = Q_{\pi,\mu}^{(2)} + Q_{\nu,\mu}^{(2)} . \tag{2.2}$$

The eight generators $\{Q_\mu^{(2)}, L_\mu^{(1)}\}$ of the group SU(3) (in the SO(3) basis) can be rewritten in the SU(2)×U(1) basis. The three operators $\{\frac{1}{2}L_0^{(1)}, \mp\sqrt{\frac{2}{3}}Q_{\pm 2}^{(2)}\}$ generate the SU(2) group and provide the labels Λ and K which are the eigenvalues of the SU(2) Casimir invariant and L_0, respectively. Furthermore the operator $\sqrt{8}\,Q_0^{(2)}$ with the eigenvalue ε is the sole generator of U(1). The socalled ladder operators are

$$\xi_\pm = \sqrt{8}\,Q_{\pm 1}^{(2)} + \sqrt{3}\,L_{\pm 1}^{(1)} \quad \text{and} \quad \eta_\pm = \sqrt{8}\,Q_{\pm 1}^{(2)} - \sqrt{3}\,L_{\pm 1}^{(1)} . \tag{2.3}$$

Within a given SU(3) irrep we choose the SU(2)×U(1) highest-weight state (h.w.s.) as the intrinsic state. For the single-boson case (N=1) the SU(2)×U(1) eigenstates are obtained by an SO(3) to SU(2) transformation[6]. For the general case with N>1 a direct approach to construct the h.w.s. is the diagonalization of the relevant Casimir operator in the SU(2)×U(1) basis. A more elegant method[6] is the use of ladder-operators starting with the h.w.s. of the lowest SU(3) irrep (corresponding to the g.s.b.) and by using a successive orthogonality procedure.

The leading SU(3) irreps (λ,μ) contained in the totally symmetric irrep [N] of the U(15) are

$$[N]=(4N,0)+(4N-4,2)+(4N-6,3)+(4N-8,4)_1+(4N-8,4)_2+\cdots \quad (2.4)$$

To construct the intrinsic state we begin with the g.s.b. $(\lambda,\mu)=(4N,0)$. The h.w.s. for the g.s.b. is given by $|gsb\rangle=\frac{1}{\sqrt{N!}}(\omega_0^+)^N|0\rangle \equiv |N\rangle$. By acting twice with the η_+ operator and through an orthogonality procedure, one obtains the intrinsic states for β and γ bands

$$|\beta\rangle=\frac{1}{\sqrt{2N-1}}\,u_0^+\,|N-2\rangle \quad \text{and} \quad |\gamma\rangle=\frac{1}{\sqrt{2N-1}}\,u_{+2}^+\,|N-2\rangle \quad (2.5)$$

where the quadrupole phonon creation operators u_0^+, u_{+2}^+ are[6]

$$u_0^+=\sqrt{2}\,\sigma_0^+\omega_0^+-\sqrt{\frac{3}{2}}\,\tau_{+1}^+\tau_{-1}^+ \quad \text{and} \quad u_{+2}^+=\sqrt{2}\,\sigma_{+2}^+\omega_0^+-\frac{\sqrt{3}}{2}(\tau_{+1}^+)^2 . \quad (2.6)$$

This procedure can be repeated for the construction of the intrinsic states of other bands. As for the doubly-degenerate irreps (4N-8,4) the socalled 'SU(3) seniority' operator[7] is used to distinguish these degenerate irreps. All the intrinsic states of $\mu\leqslant 4$ for the sdg IBA -1 and $\mu\leqslant 3$, $F=F_{Max}$ and $F_{Max}-1$ for the sdg IBA-2 are constructed. These states are used to calculate the corresponding isoscalar factor(I.S.F.).

It is interesting to consider the large N limit of the present results. As an example we consider the two (4N-8,4) K=4 intrinsic states. For $N\to\infty$ one finds

$$\begin{aligned} |g_4\rangle &= |(4N-8,4)_1\,K=4\rangle \rightarrow g_{+4}^+\,|N-1\rangle \\ |\gamma\gamma\rangle &= |(4N-8,4)_2\,K=4\rangle \rightarrow (u_{+2}^+)^2|N-4\rangle \sim (\sigma_{+2}^+)^2|N-2\rangle . \end{aligned} \quad (2.7)$$

Therefore in the large N limit one may interpret the intrinsic state $|\gamma\gamma\rangle$ as a double γ-phonon excitation, whereas the $|g_4\rangle$band represents a g-boson (hexadecapole) excitation from the g.s.b.

3. M1 FRAGMENTATION

One of the most interesting recent developments in the nuclear spectroscopy has been the prediction[8,9] and experimental confirmation[10] of the collective magnetic dipole transitions in deformed nuclei. In the SU(3) limit of the sd IBA-2 model, the M1 strength is predicted to be concentrated into one level. More realistic calculations in the framework of HFB plus particle-hole excitations in the RPA[11] predict a fragmentation of the M1 strength over many levels. The recent results of (e,e') and (γ,γ') experiments[12] seem to indicate the existence of such a fragmentation.

We investigate the M1 fragmentation in a group-theoretical way[13] in the SU(3) limit of the sdg IBA-2 model. The allowed SU(3) irreps can be obtained from the branching rule:

$$[N_\pi]\otimes[N_\nu] = \sum^{|N_\pi-N_\nu|}[N-i,\, i] \tag{3.1}$$

where

$$\begin{aligned}[N] &= (4N,0)+(4N-4,2)+(4N-6,3)+(4N-8,4)^2+\cdots \\ [N-1,1] &= (4N-2,1)+(4N-4,2)+(4N-6,3)^2+\cdots\end{aligned} \tag{3.2}$$

One sees that in the irreps with $\mu\leqslant 3$, there are four K=1 bands: the symmetric one with $\mu=3$, a mixed symmetry ($F=F_{Max}-1$) band with $\mu=1$ and two bands which belong to the doubly-degenerate mixed-symmetry SU(3) irreps (4N-6,3). A schematic spectrum for 1^+ states is shown in Fig. 1.

In the sdg IBA-2 scheme, there are four one-body rank-one operators, which are linear combinations of the generators

$$L^{(1)}_{d,\rho}=\sqrt{10}\,(d^+_\rho\tilde{d}_\rho)^{(1)} \quad \text{and} \quad L^{(1)}_{g,\rho}=\sqrt{60}\,(g^+_\rho\tilde{g}_\rho)^{(1)} \quad (\rho=\pi \text{ or } \nu) \tag{3.3}$$

While one of these combinations is the angular momentum, the

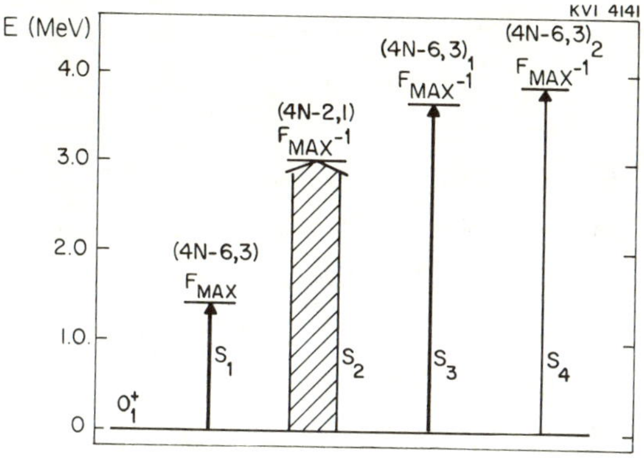

FIGURE 1. The lowest 1^+ levels in the SU(3) limit of the sdg IBA-2 (the width of the arrows represents the M1 strength)

others can be chosen in several ways. For the physics it appears most convenient to construct these operators such that they are orthogonal to the angular momentum in the g.s.b. in the large-N limit:

$$\Lambda_\rho^{(1)} = \frac{1}{2\sqrt{5}}(4L_{d,\rho}^{(1)} - 3L_{g,\rho}^{(1)}) \quad \text{and} \quad L_a^{(1)} = \frac{1}{N}(N_\nu L_\pi^{(1)} - N_\pi L_\nu^{(1)}). \tag{3.4}$$

The most general one-body M1 operator can thus be expressed as

$$T(M1) = \sqrt{\frac{3}{4\pi}}\left(g_R L^{(1)} + g_a L_a^{(1)} + \sum_\rho h_\rho \Lambda_\rho^{(1)}\right) \tag{3.5}$$

where g_R, g_a, and h_ρ are the gyromagnetic ratios. The M1 strengths from the g.s.b. to the four 1^+ states are given by (for N_π, $N_\nu \gg 1$)

$$S_2 = \frac{3}{4\pi}\frac{8N_\pi N_\nu}{N}g_a^2 \quad ; \quad S_1 = \frac{3}{4\pi}\frac{24}{5}\frac{(N_\pi h_\pi + N_\nu h_\nu)^2}{N}$$
$$S_3 = \frac{3}{4\pi}\frac{48}{5}\frac{N_\pi N_\nu}{N}(h_\pi - h_\nu)^2 ; \quad S_4 = \frac{3}{4\pi}\frac{36}{5}\frac{N_\pi N_\nu}{N^2}(h_\pi - h_\nu)^2 \rightarrow 0 . \tag{3.6}$$

The M1 strengths S_i (i=1-4) depend on g_R, g_a and h_ρ, which can be calculated through a mapping procedure[14]. Typical values for ^{156}Gd are

$$g_R = 0.672\,\mu_N\ ,\quad g_a = 1.226\,\mu_N\ ,\quad h_\pi = -0.076\,\mu_N\ ,\quad h_\nu = -0.019\,\mu_N$$

Whereas the main contribution to $g_{d,\rho}$ and $g_{g,\rho}$ (and thus also g_R, g_a) comes from orbital motion, the h_ρ, (being the difference between $g_{d,\rho}$ and $g_{g,\rho}$) can be associated with the spin degree of freedom. Therefore the transition strengths S_1, S_3 and S_4 originate from intrinsic degrees of freedom, while S_2 is due to the out-of-phase motion of the protons and neutrons.

4. E4 TRANSITION

Several experiments have been carried out to investigate the need for the incorporation of the g boson. Among these, the measurement of E4 strength, which cannot be explained very well in sd IBA, seems most relevant[15]. In the sdg IBA model, there are eleven L=4 states (with $\mu \leqslant 4$), all of which can be connected by a general E4 operator to the g.s. We again employ the intrinsic states to obtain an estimate of the E4 strength using a tentative E4 operator

$$T(E4) = \alpha\,(g^\dagger s + s^\dagger \tilde{g})^{(4)} \tag{4.1}$$

The calculated E4 strengths (for the large N limit) are shown in Fig. 2.

We note that one predicts that B(E4;g.s.-γ)>>B(E4; g.s.-β). The strength of the transitions from g.s. to 4^+_{9-11} is rather weak, because these transitions correspond to double-boson excitations. If the 'SU(3) seniority' is broken, configuration mixing will increase those E4 strengths for 4^+_{9-11} considerably.

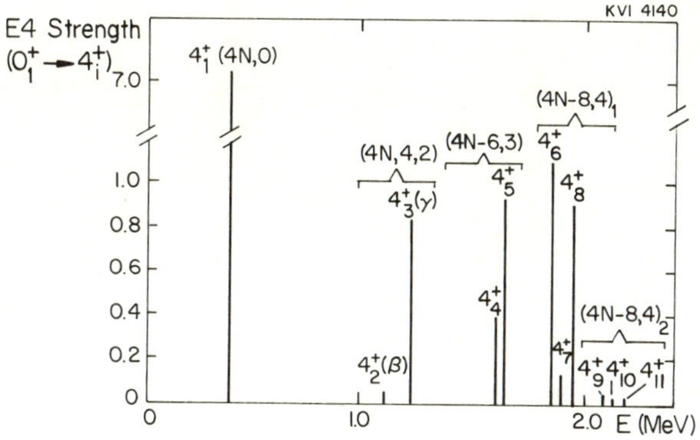

FIGURE 2. Relative E4 strengths in the SU(3) limit of the sdg IBA-2

REFERENCES

1. A.Arima and F.Iachello, Ann. Phys. (N.Y.) 99(1976)253; 111(1978)201; 123(1979)468.
2. R.D.Ratna Raju, Phys. Rev. C23(1981)518.
3. H.C.Wu, Phys. Lett. 110B(1982)1.
4. N.Yoshinaga, Y.Akiyama and A.Arima, Preprint.
5. J.P.Elliott, Proc. Roy. Soc. 245(1958)562.
6. H.C.Wu, A.E.L.Dieperink and S.Pittel, Phys. Rev., to be published.
7. Y.Akiyama, Nucl. Phys. A433(1985)369.
8. F.Iachello, Nucl. Phys. A358(1981)89c.
9. A.E.L.Dieperink, Prog. Part. Nucl. Phys. 9(1983)121.
10. D.Bohle et al., Phys. Lett. 137B(1984)27; 148B(1984)260.
11. R.R.Hilton et al., to be published.
12. D.Bohle et al., to be published.
13. H.C. Wu, A.E.L. Dieperink and O. Scholten, Preprint.
14. O.Scholten, Phys. Rev. C28(1983)1783.
15. P.B.Goldhoorn et al., Phys. Lett. 103B(1981)291.

INTEGRITY BASES FOR DYNAMICAL GROUP CHAINS IN THE IBA MODEL

JORIS VAN DER JEUGT*
Seminarie voor Wiskundige Natuurkunde, Rijksuniversiteit Gent, Krijgslaan 281-S9, B9000 Gent, Belgium.

Abstract In this paper we solve the question of which higher-order operators can be added to the Hamiltonian in one of the three dynamical symmetries for the interacting boson model. By means of generating function techniques, we construct integrity bases for O(3) scalar operators in the enveloping algebra of G, where G is one of the physical subgroups SU(3), O(6) or SU(5) of SU(6).

INTRODUCTION AND THE WELL-KNOWN SU(3) INTEGRITY BASES

In the original interacting boson model (IBM), initially introduced by Arima and Iachello [1], a dynamical symmetry arises whenever the Hamiltonian H can be written in terms of invariants only of maximal subgroups $G \subset SU(6)$. In these papers, the Hamiltonian H was an expression up to second order in the SU(6) generators (i.e. only two-boson interaction terms were considered). As a consequence, the operator form of H was expressed purely in terms of Casimir invariants of a chain of subgroups. Such Hamiltonians gave already rise to a very good approximation for the energy levels of several nuclei. Recently, Vanden Berghe et al.[2] introduced higher-order terms which conserve the dynamical symmetry in the Hamiltonian. Such terms can be interpreted as symmetry-conserving higher-order interaction terms. For the corresponding operators, one must require rotational invariance. Hence, these operators must belong to the set of O(3) scalars in the enveloping algebra of G, where G is one of the subgroups in the IBM-symmetry, namely G=SU(5), SU(3) or O(6). Such a set of operators (in fact, they form a ring without unity) can be studied by means of constructing an integrity basis [3]. For the SU(3) symmetry, this integrity basis is determined by the generating function [3,4]

$$(1+U^6)/(1-U^3)(1-U^4). \qquad (1)$$

When expanded in the form $\sum_u n_u U^u$, the generating function (1) yields the number (n_u) of O(3) scalar operators of degree u in the enveloping algebra of SU(3). This generating function (GF) shows that, besides the Casimir operators of SU(3) and the O(3) Casimir operator L^2 (for which the corresponding denominators are already left out of (1)), there are still two functionally independent operators of third and fourth degree in the SU(3) generators. Moreover, there is an operator of degree 6 whose square can be expressed in terms of these and of Casimirs. The functionally independent operators, sometimes denoted by Ω and Λ, were added to the Hamiltonian in the SU(3) dynamical chain [2], and they gave rise to a much better approximation of the energy spectrum. Also, they removed the degeneracy which exists for members of the β and γ bands by using the original Hamiltonian.

For the other two dynamical symmetries, namely $SU(6)\supset O(6)\supset O(5)\supset O(3)$ and $SU(6)\supset SU(5)\supset O(5)\supset O(3)$, the relevant integrity bases of O(3) scalars are so far unknown. However, if one wishes to preserve only the O(5) symmetry in both chains, the corresponding GF can be found in Giroux et al.[5], eq. (3.7) :

$$(1+U^9)/(1-U^4)(1-U^6). \tag{2}$$

Hence, acting on states of symmetric O(5) representations, there are two functionally independent O(3) scalars in the O(5) enveloping algebra, namely one of fourth and one of sixth order. In the notation of Iachello [6], the fourth order operator can be constructed by coupling the $G^{(3)}(dd)$ tensor together with three $G^{(1)}(dd)$ tensors to a tensor of rank 0.

THE O(6) CHAIN

It is the aim of this paper to obtain the structure of the integrity basis for O(3) scalars in the O(6) enveloping algebra, including the far-reaching restriction that all operators act only on totally symmetric representations of O(6). Since the Lie algebra of O(6) is the same as of SU(4), and the symmetric O(6) irreducible representations (irreps) $(\sigma,0,0)$ correspond to the representations $(0,\sigma,0)$ of SU(4), we must first construct the generating function for λ-tensors in the degenerate enveloping algebra acting on $(0,\sigma,0)$-irreps. From the work of Giroux et al. [5] (eq. (4.14)), it follows that this GF is given by

$$[(1-U^2\Lambda_2^2)(1-U\Lambda_1\Lambda_3)]^{-1} . \tag{3}$$

Herein, Λ_1 carries the Dynkin label λ_i of SU(4) irreps as exponent. Now, we have to find the GF for O(3) scalars contained in SU(4) irreps of the form (v,w,v). This is performed in two steps. First, we have constructed the branching rule GF for SU(4) irreps (v,w,v) into O(5) irreps (p,q) :

$$[(1-VQ)(1-VP^2)(1-W)(1-WQ)]^{-1} . \tag{4}$$

For example, when such a GF is expanded in the form $\sum_{v,w} \sum_{p,q} (c_{v,w,p,q} P^p Q^q) V^v W^w$, the integer number $c_{v,w,p,q}$ yields the multiplicity of the O(5) irrep (p,q) in the decomposition of the SU(4) irrep (v,w,v). Then, the GF for O(3) scalars contained in O(5) irreps (p,q) is given by eq.(3.5) of Giroux et al.[5] :

$$(1+P^6Q^3)\,[(1-P^4)(1-Q^3)(1-P^4Q^2)\,]^{-1} \; . \tag{5}$$

Replacing (P,Q) by (P^{-1},Q^{-1}) in (4), multiplying by (5) and retaining the coefficient in P^0Q^0 gives

$$\begin{aligned}&[V^2W^2/(1-V^2)(1-VW) + (1+V^2W+VW^2)/(1-V^3)\\ &\quad + (V^4+V^5W^2+V^3W)/(1-V^2)(1-V^3)]\,[(1-W)(1-V^2)(1-W^3)\,]^{-1}.\end{aligned} \tag{6}$$

This is the GF for the number of O(3) scalars contained in an SU(4) irrep (v,w,v). Now, according to (3), we only have to multiply (6) by $[(1-U^2W^{-2})(1-UV^{-1})\,]^{-1}$ and retain the part in V^0W^0. Omitting a factor $(1-U^2)^{-1}$ corresponding to the O(3) second degree Casimir, this gives rise to

$$\frac{1 + 3U^4 + U^5 + 3U^6 + U^{10}}{(1-U^2)(1-U^3)^2(1-U^4)} \; . \tag{7}$$

Eq.(7) is the equivalence of (1), but for the O(6) chain. Hence, disregarding the O(6) and O(3) Casimir operators, there are four functionally independent O(3) scalar operators in the enveloping algebra of O(6) for symmetric $(\sigma,0,0)$ irreps : one of second degree, two of third degree, and one of fourth degree. The second degree operator is the O(5) Casimir. With $\ell_\mu = G_\mu^{(1)}(dd)$, $p_\mu = G_\mu^{(2)}(ds) + G_\mu^{(2)}(sd)$ and $q_\mu = G_\mu^{(3)}(dd)$, the third degree operators can be written as $[\,p\times[\ell\times\ell\,]^{(2)}\,]^{(0)}$ and $[\,[\,p\times q\,]^{(1)}\times\ell\,]^{(0)}$. The operator of degree four is the O(3) scalar in O(5), namely

$[q \times [\ell \times \ell \times \ell]^{(3)}]^{(0)}$. Besides these four functionally independent operators, the integrity bases of O(3) scalars in the degnerate enveloping algebra of O(6) still contains three other fourth degree operators, one fifth degree operator, and three operators of degree six.

In view of this result, the first operator in the O(6) chain that should be added to the Hamiltonian H is $[p \times [\ell \times \ell]^{(2)}]^{(0)}$. For this purpose, Vanthournout et al.[7] have made a general analysis of matrix elements of the p-tensor in the physical basis $|N,\sigma,\tau,\nu_\Delta,\ell,m>$.

THE SU(5) CHAIN

Omitting most of the details, we now give a similar construction of the GF for the O(3) scalars in the degenerate enveloping algebra of SU(5), i.e. when acting on states of symmetric irreps $(n_d,0,0,0)$ of SU(5). First, the GF for λ-tensors in the degenerate enveloping algebra is given by[8]

$$(1 - U\Lambda_1\Lambda_4)^{-1} . \tag{8}$$

Hence, we have to find the GF for O(3) scalars contained in SU(5) irreps of the form (v,0,0,v). Just as in the previous situation, we first construct the branching rule GF for SU(5) irreps (v,0,0,v) into O(5) irreps (p,q) :

$$\frac{1+V^2Q^2}{(1-VP^2)(1-VQ^2)(1-V^2)(1-V^2Q^2)} . \tag{9}$$

For O(5) irreps (p,q), the O(3) scalar contents is given by (5). Replacing (P,Q) by (P^{-1},Q^{-1}) in (9), multiplying by (5) and retaining the coefficient in P^0Q^0 gives the GF for the number of O(3) scalars contained in an SU(5) irrep (v,0,0,v).

Then, according to (8), we only need to multiply this latter expression by $(1-UV^{-1})^{-1}$ and retain the part in V^0. Omitting a factor $(1-U^2)^{-1}$ corresponding to the O(3) second degree Casimir, one obtains :

$$\frac{1 + 3U^4 + 2U^5 + 3U^6 + U^{10}}{(1-U^2)(1-U^3)^2(1-U^4)} . \tag{10}$$

This expression is very close to GF (7). In fact, it shows that the integrity basis of O(3) scalars in the degenerate enveloping algebra of SU(5) contains four functionally independent scalars, three operators of fourth degree, two operators of fifth degree and three operators of degree six. The four functionally independent operators have the same degrees as in the case of O(6). The second degree operator is again the O(5) Casimir. With $T^{(k)}_{\kappa} = G^{(k)}_{\kappa}(dd)$ $(k = 1,2,3,4)$, the third degree operators can be taken as $[T^{(2)} \times [T^{(1)} \times T^{(1)}]^{(2)}]^{(0)}$ and $[[T^{(3)} \times T^{(2)}]^{(1)} \times T^{(1)}]^{(0)}$. It follows from (10) that any other third degree O(3) scalar, for instance $[[T^{(4)} \times T^{(3)}]^{(1)} \times T^{(1)}]^{(0)}$, is indistinguishable from a linear combination of the first two when acting on states of totally symmetric representations $(n_d,0,0,0)$. Finally, the set of functionally independent operators also contain one fourth degree operator, for which one can take the previously mentioned O(3) scalar in O(5).

The GF's (7) and (10) solve the question of symmetry-conserving higher-order interaction terms in the O(6) and SU(5) limit of IBM completely. For the O(6) limit, the matrix elements needed have already been determined [7], and soon we hope to report on their actual influence in the Hamiltonian. As it was the case in the SU(3) limit [2], we expect that the introduction of those higher-order terms will give rise to a much better agreement with experimental data.

REFERENCES

1. A. Arima and F. Iachello, Ann. Phys., NY 99, 253-317 (1976); Ann. Phys., NY 111, 201-238 (1978); Ann. Phys., NY 123, 468-492 (1979).
2. G. Vanden Berghe, H. De Meyer and P. Van Isacker, Phys. Rev. C32 (3), 1049-1056 (1985).
3. R. Gilmore and J.P. Draayer, J. Math. Phys. 26, 3053-3067 (1985).
4. B.R. Judd, W. Miller, Jr., J. Patera and P. Winternitz, J. Math. Phys. 15, 1787-1799 (1974).
5. Y. Giroux, M. Couture and R.T. Sharp, J. Phys. A : Math. Gen. 17, 715-725 (1984).
6. F. Iachello, in Nuclear Structure, ed. K. Abrahams, K. Allaart and A.E.L. Dieperink (New York : Plenum), p. 53-89 (1980).
7. J. Vanthournout, J. Van der Jeugt, H. De Meyer and G. Vanden Berghe, J. Math. Phys. (1986) (submitted).
8. R.C. King, J. Patera and R.T. Sharp, J. Phys. A : Math. Gen. 15, 1143-1958 (1982).

★ Senior Research Assistant N.F.W.O. (Belgium)

A BOSE-FERMI MAPPING IN THE SU(3) LIMIT

L.D. WOOD and I. MORRISON
School of Physics, University of Melbourne, Parkville, Victoria, Australia 3052.

Abstract We present a method of mapping bose-fermi systems in the SU(3) limit of the IBM, using sdg bosons with spin and isospin.

The phenomenological success of the IBM has prompted many attempts at explaining the parameterisation microscopically. Perhaps the best known of these is the O.A.I. (Otsuka-Arima-Iachello) map[1] in which the boson states are mapped by coupled fermion pairs

$$|j^{2n}\ ,\ \nu=0\ ,\ J=0>_F \rightarrow |(s^+)^n\ ,\ J=0>_B$$

$$|j^{2n}\ ,\ \nu=2\ ,\ J=2>_F \rightarrow |(s^+)^{n-1}\ d^+\ ,\ J=2>_B$$

and the boson operators are then obtained by equivalence.

To map odd-even systems one couples a quasiparticle onto the even-even core, where by definition the quasiparticle operator commutes with the boson operators. Then the real fermion operator will be some (infinite) function of the boson and quasiparticle operators which is truncated to a usable length in the mapping procedure.

This procedure has been carried out in the SU(5) limit with the seniority conserving mapping (SCM)[2] where one takes a spherical pairing (BCS) limit for the ground states of neighbouring nuclei and adds a quadrupole perturbation to reach the 2^+ state of the central even-even nucleus.

If we interchange the relative importance of pairing and quadrupole field effects, we can, in the SU(3) limit, use a quadrupole field to map rotational bands analogously with the pairing field causing band mixing.

The Elliott fermion SU(3) model[3] is well known and works well in the sd shell, while the pseudo-SU(3) scheme based on it[4] has had successes in heavier nuclei. At this stage in our work we treat only the sd-shell for simplicity. The SU(3) group is used to label the oscillator states, and representations of this group can be decomposed into R(3) orbital angular momentum representations. The spin and isospin labels can be treated as SU(2) group labels. For fermions in the sd shell we can thus label states by

$$\begin{array}{cccccccc} [f] & \supset SU(3) & \supset R(3) & \times & [g] & \supset SU(2) & \times & SU(2) \\ SU(6) & \alpha(\lambda\mu) & K_L L & & SU(4) & S & & T \end{array}$$

We can use a similar scheme for bosons. For simplicity we treat a boson as two maximally spatially aligned fermion SU(3) quanta (i.e. $(\lambda\mu) = (40)$ or sdg bosons), and use the IBM-4 developed by Elliott and Evans[5] with bosons with T=1,S=0 or T=0,S=1, i.e. we can incorporate odd-odd systems. So we label the boson states by

$$\begin{array}{ccccccc} [f] & \alpha(\lambda\mu) & K_L L & [g]\bar{\beta} & (\overline{\lambda\mu\nu})\beta S\ T \\ SU(15) \supset & SU(3) \supset & R(3) \times & SU(6) \supset & SU(4) \supset SU(2) \times SU(2) \end{array}$$

Note for fermions [g] is contragredient to [f] while for bosons [g] is cogredient to [f].

We then map even-even and odd-odd states by $(40)^n$ and couple in an orthogonal quasiparticle to map odd-even systems. If we denote the real fermion operator by 'c', the boson by 'b' and the orthogonal quasiparticle operator by 'a' then we can write a first-order mapping as

$$c^{+(20)}_{\ell sjmtm_t} \rightarrow u a^{+(20)}_{\ell sjmtm_t} + v \left(b^{+(40)}_{S_b T_b} \times \tilde{a}^{(02)}_{S_a T_a} \right)^{(20)}_{\ell sjmtm_t} \qquad (1)$$

where

$$\tilde{x}^{(02)}_{\ell - m_\ell s - m_s t - m_t} = (-)^{\ell - m_\ell + s - m_s + t - m_t} x^{(20)}_{\ell m_\ell s m_s t m_t}$$

and by equal matrix elements in boson and fermion space obtain the mass scaling u, v. We assume strong coupling and impose the same $(\lambda\mu)$ distribution on the boson system as exists in the fermion system. Note that this requirement, together with spin-isospin requirements, may require a non-maximally symmetric SU(15) representation, but this does help to impose some form of Pauli blocking in that we ignore spurious $(\lambda\mu)$ states where these would be forbidden in the fermion case by the Pauli principle.

The u and v parameters are given in terms of recoupling coefficients and coefficients of fractional parentage

$$\langle [f'] \alpha'(\lambda'\mu') S'T' \,|||\, c^{+(20)}_{\frac{1}{2}\frac{1}{2}} \,|||\, [f] \alpha(\lambda\mu) ST \rangle_F$$

$$= u \, \delta f_{a,0} \, \delta f_{a',1} \, \delta f_{b}, f_{b'} \; +$$

$$v \; 2\sqrt{6} \sum_{\substack{S_b T_b \\ S_b'' T_b''}} \hat{S}\hat{T} \, (-)^{S_b + S' + T_b + T'} \begin{Bmatrix} T & \frac{1}{2} & T' \\ T'' & T_b & \frac{1}{2} \end{Bmatrix} \begin{Bmatrix} S & \frac{1}{2} & S' \\ S'' & S_b & \frac{1}{2} \end{Bmatrix}$$

$$\times \begin{bmatrix} (\lambda_b \mu_b) & (20) & (\lambda\mu) \\ (40) & (02) & (20) \\ (\lambda'\mu') & (00) & (\lambda'\mu') \end{bmatrix}$$

$$\langle [f_b'] \alpha_b' (\lambda'\mu') S'T' \,|||\, b^{+(40)}_{S_b'' T_b''} \,|||\, [f_b] \alpha_b (\lambda_b \mu_b) S_b T_b \rangle$$

Where b refers to boson and a to quasiparticle and ×( ) is a 9-(λμ) coefficient[6].

In analogy with the SCM map, the mass scaling (u,v) is given in terms of recoupling and parentage coefficients of the appropriate groups:- Sp(2j+1), O(3) for the SCM case and SU(3), O(3), SU(2) for the rotational case discussed here.

One problem is with (λμ) multiplicities. The (λμ) multiplicity in the boson $[f_b]$ is often much larger than in the corresponding fermion [f] and the problem of assigning the correct equivalence has not been solved in the general case, although Akiyama's operator[7] does give us a method under some circumstances.

As no tables of SU(6) ⊃ SU(4) and SU(15) ⊃ SU(3) cfp's exist, we have generated these for some states of interest. This is relatively easy under the standard IBA prescription of including [N] (4N,0) and other Pauli violating representations but is more complex if we restrict [f] (λμ) to minimise Pauli violation. One benefit of this approach in highlighting the transformation (Lie group) properties of the single particle map could be to help resolve the current problems of particle transfer operators in supersymmetry schemes.

In conclusion, we have developed a bose-fermi map in the SU(3) limit, where we map (λμ) rotational bands. There are still some problems and unanswered questions, but the general method looks promising.

REFERENCES

1. T. Otsuka, A. Arima and F. Iachello, Nucl. Phys. A309, 1 (1978).
2. O. Scholten and A.E.L. Dieperink, in Interacting Bose-Fermi Systems in Nuclei, ed. F. Iachello (Plenum, 1981).

3. J.P. Elliott, Proc. Roy. Soc. A245, 128, 562 (1958).
4. R.D. Ratna Raju, J.P. Draayer and K.T. Hecht, Nucl. Phys. A202, 433 (1973).
5. J.P. Elliott and J.A. Evans, Phys. Lett. 101B, 216 (1981).
6. K.T. Hecht and D. Braunschweig, Nucl. Phys. A244, 365 (1975).
7. Y. Akiyama, Nucl. Phys. A433, 369 (1985).

ROTATIONAL BANDS IN AN EXTENDED IBM-2 MODEL

KEVIN FOWLER AND RAYMOND A. SORENSEN
Carnegie Mellon, Pittsburgh, Pennsylvania 15213, USA

Abstract The IBM-2 model, expanded to include proton and neutron bosons of spins 4,6,...ħ, is used to investigate rotational bands in deformed nuclei. Excited bandheads are generated from the ground state configuration using a Tamm Dankov approximation. The energies of these excitations and the rotational bands built upon them are calculated using angular momentum projection. Results for the beta, gamma and magnetic 1^+ vibrations are discussed.

INTRODUCTION

The success of the IBM and IBM-2 descriptions of low lying collective nuclear spectra has lead to extensions of the model in order to encompass a broader range of collective nuclear states. Extending the basis to include proton and neutron bosons of spins 4,6,...ħ has been useful in obtaining a better physical description of high spin states. Such an extension prevents the use of group theoretical techniques to obtain an exact solution, thus approximations such as the Hartree-Bose(HB) method have been developed to treat these models. Pittel and coworkers[1] have generated collective vibrational states in an s-d-g IBM-2 using a Tamm Dankoff(TD) approximation after the HB calculation.

In this paper we describe a similiar study of vibrational states using a different hamiltonian. We take a furthur step by using angular momentum projection to generate the rotational bands built upon the vibrations.

FORMALISM

Our model hamiltonian contains the following terms

$$H = \Sigma_{\rho=\pi,\nu} \Sigma_j \epsilon_j^\rho n_j^\rho + \kappa Q^\pi \cdot Q^\nu - \lambda F^2 .$$

This differs from that of Ref.1 in the last term. Those authors use an n-p hexadecapole interaction in the sdg space. Due to large K=1 matrix elements of this interaction the symmetric and antisymmetric 1^+ states separate in energy, placing the antisymmetric state at a physically reasonable energy. The K=0,2,.. states do not experience such a large symmetry splitting.

We have chosen instead to use an F-spin symmetrizing interaction in the extended boson space. In the sd subspace this interaction is related to the usual Majorana interaction[2] by

$$F^2 = F_{max}(F_{max} + 1) - M_{\pi\nu} .$$

With this interaction the state with maximum F-spin symmetry is lowered in energy by an amount determined by the interaction strength λ. Thus, all states of mixed symmetry lie near or above the level of the 1^+ mixed symmetry state. Experimental data is availiable[3] for the 1^+ mixed symmetry state, thus allowing a rough fit of the single parameter λ. As little is known about the mixed symmetry $0^+, 2^+, \ldots$ states in well deformed nuclei we avoid the more complicated parameterizations which affect the relative energies of these states.

Having chosen the size of the boson basis and assigned

the hamiltonian parameters we use the HB method to calculate the condensate wave function

$$|N_\pi N_\nu\rangle = \frac{(\Gamma^\dagger_{\pi o})^{N_\pi} (\Gamma^\dagger_{\nu o})^{N_\nu}}{[N_\pi ! N_\nu !]^{1/2}} |0\rangle \quad ; \quad \Gamma^\dagger_{\rho\alpha} = \Sigma_{jm}\, \eta^\alpha_{\rho jm}\, \gamma^\dagger_{\rho jm} \ ,$$

where the $\gamma^\dagger_{\rho jm}$ are the basis boson creation operators and the $\eta_{\rho jm}$ are the coefficients determined in the HB calculation. The vibrational excitations which serve as rotational bandheads are then calculated in the TD approximation using as a basis the excitations

$$|\rho\alpha; N_\pi N_\nu\rangle = \frac{\Gamma^\dagger_{\rho\alpha}\, \Gamma_{\rho o}}{\sqrt{N_\rho}} |N_\pi N_\nu\rangle \ .$$

We next improve upon this calculation by projecting these states to good angular momentum. Angular momentum projection[4] is quite feasible within this model due to the simple form of the wave functions. In the ground state condensate the bosons are restricted to occupy a single level while the TD states replace a condensate boson with an excited boson. This structure of the wave function causes the matrix elements needed for the projection to collapse to simple product forms. For instance, the matrix element needed for the normalization integral of the condensate is given by

$$\langle N_\pi N_\nu | R(\Omega) | N_\pi N_\nu\rangle = (f^{oo}_\pi)^{N_\pi}\, (f^{oo}_\nu)^{N_\nu} \quad ;$$

$$f^{\alpha\beta}_\rho = \Sigma_{jmm'}\, \eta^\alpha_{\rho jm}\, \mathcal{D}^j_{m'm}\, \eta^\beta_{\rho jm'} \ .$$

This is the simplest matrix element; the others are not much more complicated. With projection possible we can generate reliable estimates for the energies of the rotational bands by projecting various angular momentum states from the same intrinsic excitation.

RESULTS

Results of such a calculation are shown in Figs.1,2 using the hamiltonian parameters of Table 1. The α-parameters are used to generalize the quadrupole coefficients away from their exact SU(3) limit[1]. We have included only up to g- bosons for this example but can extend the basis up to 12ħ- bosons if necessary. In Fig.1 we show the low-lying bands symmetric in neutron and proton degrees of freedom, projected up to spin 6ħ. The compression of the I=5,6ħ levels of the γ-band is due mostly to the limited boson basis; however, staggering effects in this band are common in these calculations. The β-band, while clearly rotational has a smaller moment of inertia than the ground state(gs) band. This also is a common feature of the TD calculation and at odds with the experimental data.

TABLE 1 Hamiltonian input parameters.

j	ε^{π}	ε^{ν}	α^{π}	α^{ν}
0	0.0	0.0	1.0	1.0
2	0.7	0.7	0.9	0.9
4	1.2	1.2	0.8	0.8

$N_{\pi} = N_{\nu} = 6 \quad \kappa = -0.028 \quad \lambda = 0.15$

Fig.2 shows the low-lying bands antisymmetric in proton and neutron degrees of freedom. The 1^+ band shows an interesting staggering effect, this result has been found to persist to varying degrees for other parameter sets. In some cases the order of the levels become inverted with the higher even spin levels dropping below the lower odd spin levels, indicating a departure from rotational behavior. Suprisingly, the β_a- and γ_a-bands have much smoother energy spacings than their symmetric partners. The β_a-band has a higher moment of inertia than the gs-band and the γ_a-band shows no level compression like the γ-band. If these bands occurred low in energy they would be perfectly acceptable candidates for the experimental β- and γ-bands.

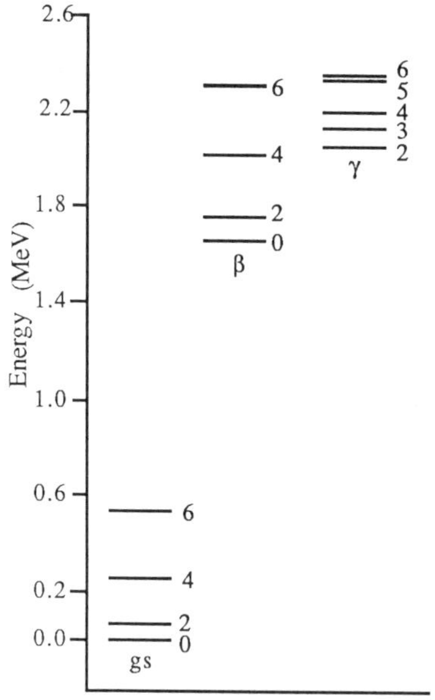

FIGURE 1. Ground state, β and γ bands up to $I=6\hbar$

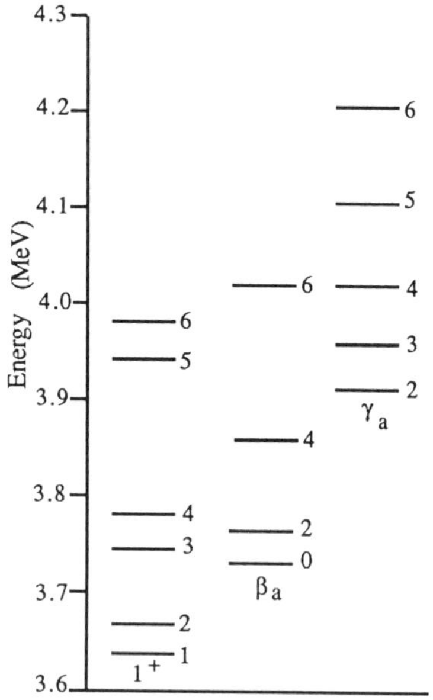

FIGURE 2. Mixed symmetry 1^+, β_a and γ_a bands up to $I=6\hbar$

SUMMARY

Using this approximate treatment of a boson system we can in principle generate the entire low energy excitation spectrum accessible to the chosen basis. This allows comparison to a much broader collection of experimental data than possible with the HB or TD treatment alone. In turn, this should allow a detailed analysis of the parameters used for a given hamiltonian, thus restricting their physically acceptable values. Such an understanding of these boson model parameters is required if we want to continue to take advantage of the simplicity of the boson model space to investigate nuclear structure. Clearly, methods of experimentally determining the F-spin symmetry character of excited states are an immediate need.

REFERENCES

1. S.Pittel, J.Dukelsky, R.P.J.Perazzo, H.M.Sofia, Phys. Lett. 144B, 145 (1984)
2. P.VanIsacker, K.Heyde, J.Jolie, M.Waroquier, J,Moreau, O.Scholten, Phys. Lett. 144B, 1 (1984)
3. D.Bohle, A.Richter, W.Steffen, A.E.L.Dieperink, N.LoIudice, F.Palumbo, O.Scholten, Phys. Lett. 137B, 27 (1984)
4. R.A.Sorensen and K.Fowler, "Extended IBM-2 for High Spin States", these Proceedings

A REVIEW OF THE UNBROKEN SU(3) SYMMETRY AND THE SYSTEMATICS OF THE IBM COEFFICIENTS

J.B. GUPTA
Ramjas College, Delhi Univ., Delhi, India

Abstract The predictions of the exact SU(3) limit of IBM-1 are reviewed for the Dy - Pt region. A linear dependence of the coefficient α of the energy expression and of the $B(E2; 0_1 \rightarrow 2_1)$ on N_pN_n product is derived. No such relation for the β-parameter is apparent. Improved one-parameter symmetry breaking measure is proposed.

INTRODUCTION

In the SU(3) dynamical subgroup of the U(6) symmetry of Interacting Boson Model[1] (IBM) one obtains an analytical solution of H_{IBM}. Deviations are treated by perturbed symmetries. In most studies, the emphasis has been on determining the parameters of H to reproduce the energy spectra of a nucleus or a group of nuclei. Here we review the SU(3) symmetry for a whole region of deformed rare earths Dy - Pt, N = 90 - 104 and search for a correlation of input parameters to boson number N or N_p, N_n. This can provide a link with the shell model.

THE VALIDITY OF SU(3) SYMMETRY

Arima and Iachello identified the following four main tests for the SU(3) symmetry of axially deformed nuclei :-

1. The energies of the (2N,0) g-band (and of excited bands) have the L(L+1) pattern.
2. The β- and γ-bands in the (2N-4,2) representation are degenerate.
3. The rotational bands are predicted to a cut off at L = 2N, a consequence of compact algebra.
4. Strong β-γ band coupling is predicted as opposed to weak coupling assumed in Bohr-Mottelson (BM) model[2] applications.

The deformed nuclei in this region do exhibit the L(L+1) spectrum as also described in BM model. Also there exist a few nuclei with nearly degenerate β- and γ-bands, e.g. ^{156}Gd, ^{170}Er, ^{172}Yb, ^{178}W and ^{234}U. Most other nuclei exhibit broken degeneracy, often by large amounts. The condition of band cut off arises from the explicit dependence of the B(E2;L+2 → L) values on the total boson number N. Only a slow fall off was observed in actual nuclei. It is now recognized[3-5] that this unique feature of IBM cannot be tested at all since other degrees of freedom such as the core excitation[3,4] or the L = 4 g-bosons emerges at such spins. The latter gives rise to the SU(15) group which leads[5] to a reduced (x 1/2) fall off of B(E2) values and a cut off at double the previous value.

The fourth condition of a strong β-γ band coupling implies a large β-γ/g B(E2) ratio. Only a few β-γ interband transitions have been observed in this region, e.g. in ^{156}Er and ^{164}Dy $B(E2;2_\beta \rightarrow 2_\gamma)/B(E2;2_\beta \rightarrow 2_g)$ ratio is 38 and 15 respectively, showing strong β-γ coupling.

Warner et al[6] showed that in ^{168}Er, the band mixing parameter $Z_{\beta\gamma}$ is smaller than Z_β, Z_γ and thus has small effect on β-g, γ-g transition strength but the same $Z_{\beta\gamma}$ induced large $\beta \rightarrow \gamma$ transition strength . Using the Dynamic PPQ model[7] based on the BM model, we predicted[8] large E2 matrix element (= 0.93) for $2_\gamma \rightarrow 2_\beta$ compared to only 0.47 and 0.40 for $2_\gamma \rightarrow 2_g$ and $2_\beta \rightarrow 2_g$ in ^{156}Gd. Thus the strong β-γ coupling also arises in BM model, especially if $2_\beta, 2_\gamma$ levels are close.

THE SYSTEMATICS OF SU(3) PARAMETERS

Here we present further tests of the validity of the SU(3) symmetry. For the quadrupole Hamiltonian H = - K Q.Q, the analytical solution is,

$$E = (3/4)K\ L(L+1) - K\ C(p,q) \tag{1}$$

This gives

$$E(2_2^+) - E(2_1^+) = (4/3)(2N-1)E(2_1^+) \tag{2}$$

For ^{156}Gd, ^{170}Er and ^{234}U, the good SU(3) nuclei[1], total boson number N comes out to be 5.0, 4.6 and 8.2 respectively instead of the required 12, 17 and 13. This points to the total inadequacy of Eq.(1). If one adds the L.L term one gets H = - K Q.Q - K'L.L giving

$$E([N],L,p,q) = \alpha L(L+1) - \beta\ C(p,q) \tag{3}$$

Now both α (= $\frac{3}{4}$ K - K') and β are input parameters

$$6\beta(2N-1) = E(2_2^+) - E(2_1^+)$$

$$\text{and} \quad 6\alpha = E(2_1^+) \tag{4}$$

The totally symmetric 2_1^+ states form : F-spin (= N/2) multiplets.[9] For each constant N, the parameter α was found[10] to be fairly constant and varied with N. If we plot $1/\alpha$ against $(N_p.N_n)^{1/2}$, a fairly linear relationship is exhibited. In this we have achieved the folding of the seperate curves of the moment of inertia Θ (= $1/\alpha$) versus A (p.74 of Ref.2) into a single curve.

A similar folding of the $B(E2, 2_1 \rightarrow 0_1)$ versus the neutron number curves for same $N = N_p + N_n$ or for same Z is obtained in a N_pN_n plot (Fig.2). The fast rising portion corresponds to the SU(5) to SU(3) transition.

THE β-COEFFICIENT OF CASIMIR INVARIANT

The value of β from Eq. (4) does not show a systematic variation with N or N_p, N_n etc. Even for two nuclei of same N, β varies sharply, e.g. in ^{170}Er and ^{174}Yb, with N = 16, β is 4.0 and 7.5 respectively. Further, β- , γ-bands being non-degenerate, same β-value will not fit both the bands. For transition towards O(6); the pairing term will break the degeneracy raising the 0_2^+ state.

$$H = - K\, Q.Q - K'L.L + K''\, P.P \qquad (5)$$

Casten and Warner[11] suggested K''/K as a measure of asymmetry, which was related to

$$R_B = \frac{E(0_\beta)}{E(2_\gamma) - E(2_g)} - 1 \qquad (6)$$

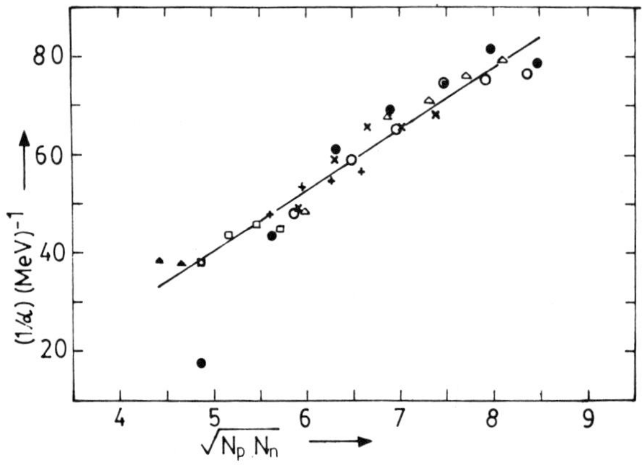

Fig. 1 $1/\alpha = a + b\sqrt{N_pN_n}$, a= -25, b= 38/3.

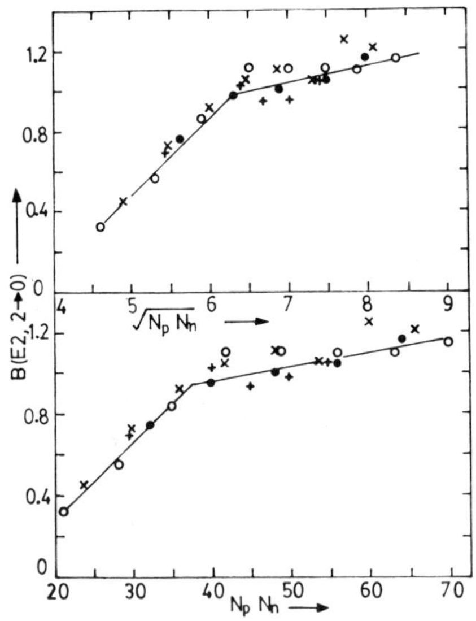

Fig. 2 Plot of $B(E2;\ 2_1^+ - 0_1^+)$.

For the Dy - Os nuclei, R_B does not show[10] any systematic dependence on N or N_p, N_n etc. We define a modified expression for the symmetry breaking term

$$R_B' = \frac{E(2_\beta) - E(2_\gamma)}{E(0_\beta)} . \tag{7}$$

Here all numbers relate to the same multiplet. R_B' shows much less variation with N, but still there is no systematic trend.

CONCLUSION

Review of SU(3) conditions shows that the O(3) symmetry is well represented by the boson subspace of valence nucleons, as reflected in the N_pN_n dependence of moment of inertia $\theta = 1/\alpha$ and $B(E2, 2_1 \rightarrow 0_1)$ values. However, the intrinsic vibration represented through C(p,q) term requires arbitrary values of β. The complexity of nuclear structure remains hidden in the parameter β or K'' of P.P symmetry breaking term.

The work is supported in part by the Department of Atomic Energy, Government of India.

REFERENCES

1. A. Arima and F. Iachello, Ann. Phys.111,201 (1978).
2. A. Bohr and B.R. Mottelson, Nuclear Structure, Vol. II (Benjamin, N.Y. 1975).
3. A. Gelberg and A. Zemel, Phys. Rev. C22,937 (1980).
4. N. Yoshida, A. Arima and T. Otsuka, Phys. Lett. 114B, 86(1982).
5. Wu Hua-Chuan, Phys. Lett. 110B, 1(1982).
6. D.D. Warner, R.F. Casten and W.F. Davidson, Phys. Rev. C24, 1713(1981).
7. K. Kumar and M. Baranger, Nucl. Phys.A122, 273(1968).
8. J.B. Gupta, K. Kumar and J.H. Hamilton,Phys. Rev. C16, 427(1977).
9. P. Von Brentano, A. Gelberg, H. Harter and P. Sala, J. Phys. G11, 631(1985).
10. J.B. Gupta, Phys. Rev. C33, 1505(1986).
11. R.F. Casten and D.D. Warner, Phys.Rev.Lett. 48, 666(1982).

THE INTERACTING VECTOR BOSON MODEL AND THE COLLECTIVE MOTIONS IN NUCLEI

A. Georgieva, P.Raychev, R.Roussev
Institute of Nuclear Research and Nuclear Energy Bulgarian Academy of Science, Sofia, Bulgaria

In this communication we shall briefly describe the model of collective motions in nuclei, which is based on the assumption that the collective excitations may be'built up' of two kinds of vector bosons, called p- and n-bosons. We call this model an Interacting Vector Boson Model (IVBM). The main features of IVBM are given in [1-4].

To avoid confusion, it must be said that the names p- and n-bosons have nothing to do with the real protons and neutrons (or at least the connection between them and the real particles in nuclei is now unclear) and they simply label two different kinds of excitations, which 'built up' the real, observable collective motions in nuclei.

The hypothesis of the existanse of two

kinds of vector bosons may seem strange to some extend so some additional explanations must be given.

The first one may be regarded as a supplement to the 'historical overview' part of review article [5] devoted to the symplectic models of nuclei. In [6] we suggested that if the main collective properties of the deformed even-even nuclei may be described by the angular momentum

$$L_{\mu\nu} = \sum_{\alpha=1}^{N} (x_{\mu\alpha} p_{\nu\alpha} - x_{\nu\alpha} p_{\mu\alpha}); \; L_{\nu} = \varepsilon_{\nu\mu\lambda} L_{\mu\lambda}$$

and the quadrupole momentum

$$Q_{\mu\nu} = \sum_{\alpha} (x_{\mu\alpha} x_{\nu\alpha} - \frac{1}{3} \delta_{\mu\nu} (x_{\alpha} \cdot x_{\alpha}))$$

where $x_{\mu\alpha}$ and $p_{\mu\alpha}$ ($\mu = 1,2,3$; $\alpha = 1,\ldots,N$) are the coordinates and momenta of the particles in the nucleus, then the symplectic group Sp(6,R) (which in [6] was denoted as Sp(3,3)) arises in a natural way. As a matter of fact, if we introduce the boson creation and annihilation operators

$$u^{+}_{\mu\nu} = \frac{1}{\sqrt{2}} (x_{\mu\alpha} - ip_{\mu\alpha}); \; u_{\mu\alpha} = \frac{1}{\sqrt{2}} (x_{\mu\alpha} + ip_{\mu\alpha})$$

the quadrupole operator splits into three parts

$$Q_{\mu\nu} = Q^{(0)}_{\mu\nu} + Q^{(+)}_{\mu\nu} + Q^{(-)}_{\mu\nu}$$

where in $Q^{(+)}_{\mu\nu}$ ($Q^{(-)}_{\mu\nu}$) two creation (annihilation) operators apear; $Q^{(0)}_{\mu\nu}$ contain only terms of type $\sum_{\alpha} u^{+}_{\mu\alpha} u_{\nu\alpha}$ and is called the 'truncated' quadrupole operator, which together with L_{μ}, closes the algebra of SU(3). So in [6] it was shown that in such a way the symplectic group Sp (G,R) and its main chains of subgroups

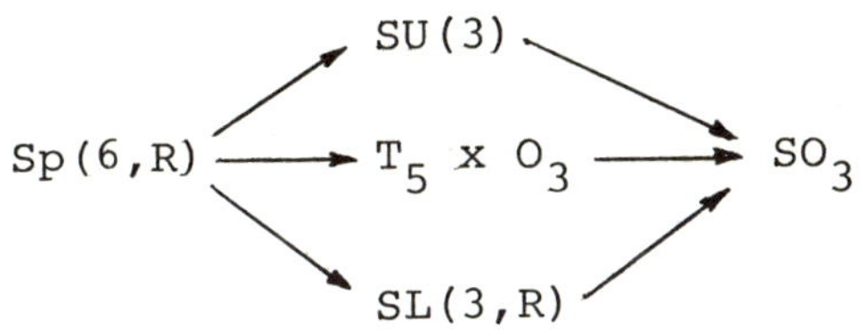

arises as a dynamical (i.e. spectrum generating) group for the deformed even-even nuclei. (T_5 is the five-dimensional abelian group, generated by $Q_{\mu\nu}$).

Of course now all this is well known.However it is worth recolling it in order to underline, as it was clear from the beginning, that a description of collective modes by means of the symplectic model can be accomplished through the use of N different vector bosons associated with the coordinates and momenta of particles (and not with the 'real' collective modes, such as phonons in nuclei).

The conclusion that N different vector-bosons are needed for group theoretical description of collective modes in nuclei finds a strong support in the Vanagas approach to nuclear structure. He pointed out [7,8], that a quite general nuclear Hamiltonian with an arbitrary nucleon-nucleon potential can be decomposed into parts, each of which can be expressed in terms of generators of a 'maximal' (or full) symplectic group for a given nucleus, namely $Sp(6N,R)$. Without going into detail it must be said, that in this case we are practically dealing again with N kinds of bosons constructed from coordinates and momenta of particles in nucleus.

If this is the case, the question arises why in IVBM we are dealing with only two kinds of vector bosons?

To answer the question we must mention that in [9] we investigate the particular chain $SU(3) \supset O(3)$ applied to the deformed heavy even-even nuclei. The basic assumption made in [9] was that the ground and gama-vibrational bands in this nuclei can be united in a single multiplet of SU(3) of the type (λ ,2) where λ is even. The interaction Hamiltonian breaks down the SU(3) symmetry to O(3) and so the spectrum of nuclei is generated. The energy spectrum of the ground- and gama-vibrational bands and the B(E2)-transition rations within and between the bands as well are calculated analitically in this basis. We come to the conclusion that a good agreement with the experimental data can be attained if the symmetry breaking term includes all four basic O(3)-scalars in the enveloping algebra of SU(3). In practice this means that Hint must include three- and four-body interaction terms such as Q.Q.Q and L.Q.Q.L. What is most important for our purposes however, is that four basic O(3)-scalars taking part in the Hamiltonian can be built up of only two kinds of vector bosons. Obviously with the help of these bosons we can also built up the eigenstates of the Hamiltonian.

This bring us to the conclusion that these vector bosons can be used as a 'language' which allow us to describe the collective Hamiltonian and the collective states in nuclei. In other words, it is only the group theoretical properties of the Hamiltonian that are important for the description of collective modes (and not the way in which we describe them). The vector bosons are only some 'auxiliary material' which is used to construct the Hamiltonians and the states of nucleus with well defined group-theoretical properties and there is no need to give them an interpretation as real excitations (for example as a pair of nucleons with angular momentum $L=1$).

This interpretation may be extended also to the standard IBM with its s- and d-bosons. If this is the case our vector boson model has some advantage over IBM because with the help of the vector bosons we can built up the interaction terms and states with more general group-theoretical properties.

Most physicists however prefer to work with more 'tangible' quantities like the real excitations, because they are more appealing to our intuition. For this reason it is preferable to consider these vector bosons as elementary excitations which build up the real ones (i.e. to consider them as 'excitation quarks'). We can also hope to find out the deeper microscopic sense of the vector bosons, but this is a problem which must be discussed in another paper.

And so we assume that the creation operators of these vector bosons, denoted by p and n, transform like O(3)-vectors and that they also transform according to two independent fundamental representations (1,0) of the group SU(3) containing SO(3) as a subgroup. (The annihilation operators p and n transform according to (0,1).) Further we introduce an additional quantum number T, called 'pseuvdospin', which distinguishes p-bosons from n-bosons; its projection take the value T=+1/2 for the p-bosons and -1/2 for the n-bosons. In such a way the two vector bosons can be united in one fundamental representation of U(6) of the type [1,0,0,0,0,0] . They can be also labeled by the quantum numbers T=1/2, L=1 and M=+1,0,-1 according to the chain

$$\begin{array}{ccccccc} U(6) & \rightarrow & U(3) & \times & U(2) & \rightarrow & O(3) & \rightarrow & O(2) \\ 1 & & [1,0] & & T & & L & & M \end{array}$$

Hence one can introduce the notations

$$u_m^+(\alpha = \tfrac{1}{2}) = p_m^+, \quad u_m^+(\alpha = -\tfrac{1}{2}) = n_m^+$$

Finally, in order to explain the negative parity states we shall postulate that under the space reflections p transforms as a pseudovector, while n - as a vector, i.e.

$$P: p \to -p, \quad ; \quad n \to n$$

Using $u_m^+(\alpha)$ and $u_n(\alpha)$ one can introduce the bilinear combinations

$$F_M^L(\alpha,\beta) = \sum_{m,n} C_{1m1n}^{LM} u_m^+(\alpha) u_n^+(\beta); \quad G_M^L(\alpha,\beta) =$$

$$= [F_M^L(\alpha,\beta)]^+ \quad ; \quad A_M^L(\alpha,\beta) = \sum_{m,n} C_{1m1n}^{LM} u_m^+(\alpha) u_n(\beta)$$

It can be easily seen that these operators (which transform like O(3) tensors with angular momenta L and projection M) also satisfy the commutational relations given in [1] (formulae (4.1, a-d)) from which we can conclude that they close the algebra of the Sp (12,R).

The most general one- and two-body Hamiltonian of the nucleus can be constructed in terms of these operators and must be an O(3) scalar. Evidently, the eigenstates of the Hamiltonian can be expanded in series of the basis vectors of UIR of Sp(12,R). In such a way the energy spectrum of the system is generated, i.e. Sp(12, R) is a dynamical group of the system under consideration.

Further we may add the condition that the Hamiltonian must conserve the number of bosons. Under this condition the Hamiltonian should contain the generators of U(6) group only. The Hamiltonian truncated in such a way can be writen in the form

$$H = \sum_{\alpha\beta} h_{\sigma}(\alpha,\beta)\, A^{0}(\alpha,\beta) + \\ + \sum_{L\alpha\beta\gamma\delta} V^{L}(\alpha\beta\gamma\delta)\,(-1)^{M}\, A^{L}_{M}(\alpha\gamma) A^{L}_{-M}(\beta\delta)$$

where $V^{L}(\alpha\beta\gamma\delta)$ are phenomenological constants related to two body interaction. It is well known (Moshincky,Quesne), that by means of six creation operators one can construct only two types of UIR of Sp(12,R) - even and odd.Further, under the reduction Sp(12,R) $\supset$ U(6) the UIR of Sp(12,R) splits into a countless number of symmetric UIR of U(6) of the type $[N,0,0,0,0,0]$

where N=0,2,4... for the even UIR of Sp(12,R) and N=1,3,5... for the odd one.

The consequence of U(6)-structure of the Hamiltonian is that in the basis of Sp(12,R) its matrix elements vanish between the UIR of U(6) with different N, i.e. in this basis the matrix of H is block-diagonal, which significantly simplifies the mathematical considerations. This of course does not diminish the role of Sp(12,R) because the complite spectrum of the system lies in the Sp(12,R) basis and only part of its states are in each particular UIR of U(6).

For symplicity we shall limit ourselve with U(6)-truncated Hamiltonian, which is physicaly meaningfull at least for the low-lying states of nuclei. The arising U(6) algebra of IVBM has a rich structure and contains the following chains of subgroups

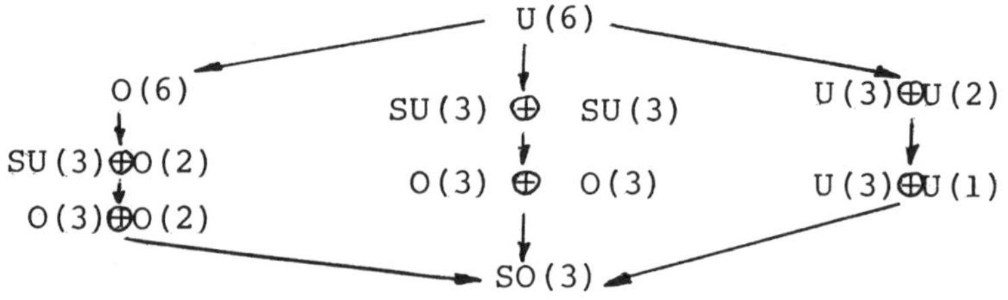

Of particular interest are the limiting cases when the Hamiltonian contains only the Casimir operators of one of these chains. The chain

$$U(6) \supset SU(3) \times U(2) \supset SO(3)$$

it describes the rotational motions in the deformed even-even nuclei. The operators of the angular momenta and the truncated quadrupole momentum are

$$L_M = -\sqrt{2}\,(A^1_M(p,p) + A^1_M(n,n))$$

and

$$\mathcal{Q}_M = \sqrt{6}\,(A^2_M(p,p) + A^2_M(n,n))$$

respectively. The generators of 'pseudospin' connected with U(2) are given by

$$T_{+1} = \sqrt{\frac{3}{2}}\,A^0(p,n) \qquad T_{-1} = \sqrt{\frac{3}{2}}\,A^0(n,p)$$

$$T_0 = -\sqrt{\frac{3}{2}}\,(A^0(p,p) - A^0(n,n))$$

The remaining operators $A^L_M(\alpha,\beta)$ (L=1,2) can be treated as vector and quadrupole transition operators between the states with different numbers of p- and n-bosons. The states of the nuclei can be labeled by the quantum numbers of the subgroup along the chain, namely N (for U(6)), T (for U(2)) and L,M (for SO(3) $\supset$ SO(2)). It should be mentioned that the groups SU(3) and U(2) are mutually complementary, and hence the quantum numbers (λ,μ) of SU(3) are related to N and T via

$$N = \lambda + 2\mu \quad , \quad T = \frac{\lambda}{2}$$

The energy spectrum corresponding to the chain under consideration is calculated in [2,3] and is given by

$$E = a N + \alpha_6 N(N+5) + \alpha_3(\lambda^2 + \mu^2 + \lambda\mu + 3\lambda + 3\mu) + \beta_3 L(L+1) + \alpha_1 \frac{\lambda^2}{4}$$

Here α_n (n=6,3,1) are the coefficients before the quadratic invariant operators of corresponding U(n) groups and β_3 is the coefficient before SO(3) Casimir operator.

The theoretically predicted energy spectrum for the deformed heavy even-even nuclei which is generated by the chain U(6)$\supset$SU(3) $\otimes$ U(2)$\supset$SO(3) is shown on fig. 1. However not all of these states are observed experimentally. The comparison with the experimental data shows that in the region of rare-earth nuclei one can find only the SU(3) multiplets with $\lambda > \mu$. So the ground band belongs to the multiplet $(\lambda,0)$ and is characterised by quantum numbers $K^{+} = 0^{+}$. The other bands (in sequence of increasing energy) belong to multiplets $(\lambda - 2,1)$, $(\lambda - 4,2)$ ect., and have quantum numbers $K^{\pi} = 0^{+}$, $K^{\pi} = 1^{-}$ ('superband'), $K^{\pi} = 2^{+}$ (β- and γ- band) and so on. A typical experimental situation and a comparison with the theoretical predictions are shown in fig. 2 for the case of ^{170}Er.

Contrary to this, in the region of actinides the multiplets with $\mu > \lambda$ are observed.An explanation of this fact can be given by the assumption that in rare-earth nuclei the number of p-bosons prevails over the number of n-bosons, whereas the situation in the actinides is opposite - the number of n-bosons prevails. (Of course this explanation badly needs a microscopical motivation, but in framework of IVBM we can stop at this point). The consequence of the fact that actinides $\mu > \lambda$ is that the states of the ground band belong to the $(0,\mu)$ representation and have quantum numbers $K^{\pi} = 0^{+}$. Then the bands with negative parity (and with $K^{\pi} = 0^{-}$ and 2^{-} respectively) (usually referred to as 'octupole' bands) appear as two members of multiplet $(2, \mu-1)$, whereas: β- and γ-bands appear in multiplet $(4, \mu-2)$. A typical experimental situation and comparison with the theory is shown in fig.3 for the case of ^{236}U.

In such a way in the framework of IVBM a unified description of the positive and negative parity states can by attained, which is the advantage of this model. The model also allowes to obtain the analitic results for some of the electromagnetic transitions whithin and between the bands, which are in good agreement with the experimental data [2,3].

At the end a few words must be said about the remaining chains. The chain $U(6) \supset SU(3) \oplus SU(3) \supset SO(3)$ contains a big variety of near lying states and is connected with the description of giant resonance. On the other hand the chain $U(6) \supset SO(6) \supset SO(3)$ causes a strong mixing of the rotational SU(3) multiplets and is presumably appropriate for the description of transitional nuclei.

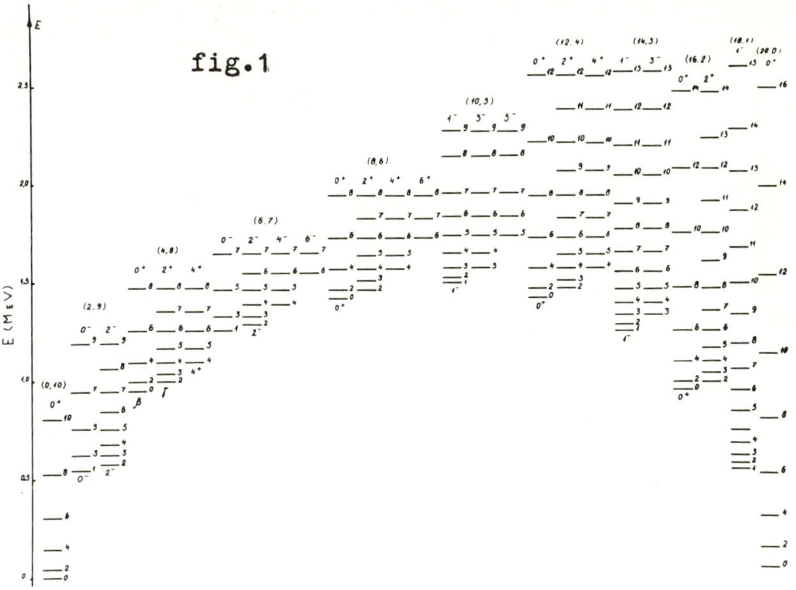

fig.1

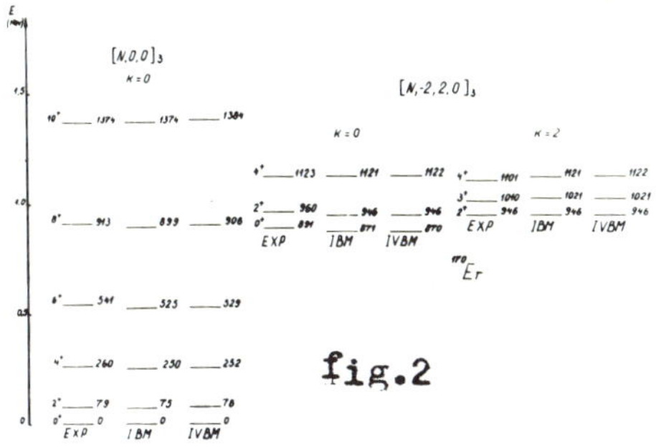

fig.2

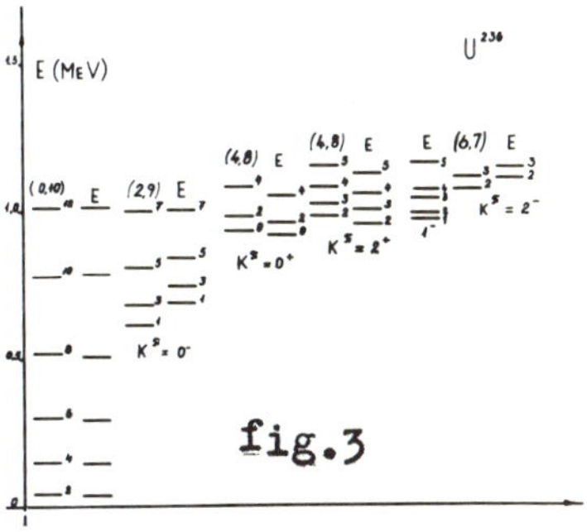

fig.3

REFERENCES

1. A.Georgieva, P.Raychev, R.Roussev
   I. of Phys. G. 8 (1982) 1377

2. A.Georgieva, P.Raychev, R.Roussev
   I. Phys. G. 9 (1983) 521

3. A.Georgieva, P.Raychev, R.Roussev
   Bulg. I.Phys. 12 (1985) 147

4. S.Alisanskos, A.Georgieva, P.Raychev, R.Roussev
   Bulg. I.Phys. 11 (1984) 2

5. D.I.Rowe, Microscopic Theory of Nuclear
   Collective Model, Rep. of Progress in Phys., 1986

6. Raychev P. Sov., I.Nucl.Phys. 16 (1972) 1171
   Raychev P. C.R.Bulg.Acad.Sci. 25 (1972) 1503

7. V.Vanagas Lecture Notes; University of Toronto,
   Toronto (1977)

8. V.Vanagas;

9. P.P.Raychev, R.P.Roussev Sov. I.Nucl. Phys.
   27 (1978) 792

Chapter 2

SYMMETRY PROPERTIES
OF
NUCLEI

Moderator:

U. KAUP

Physik Department

Technische Universität München

James-Frank-Strasse

D-8046 Garching, F.R. Germany

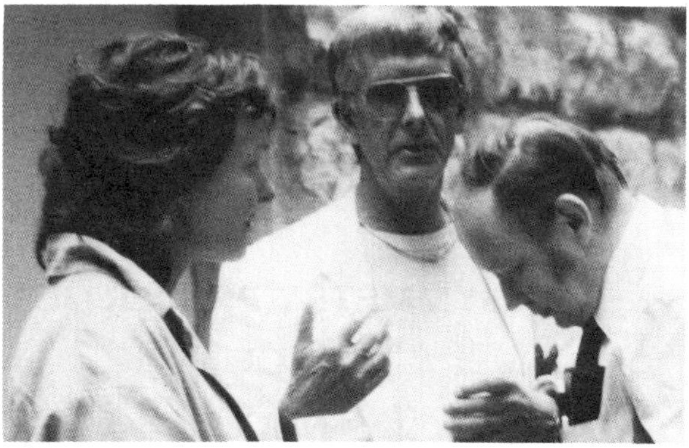

THE IDENTIFICATION OF MIXED-SYMMETRY STATES IN THE A ≈ 50 MASS REGION

S.P. Collins*, S.A. Hamada*, W.D. Hamilton*, F. Hoyler+ and S.J. Robinson+

* Physics Division, University of Sussex, Brighton, BN1 9QH, U.K.

+ Institute von Laue-Langevin, 38042 Grenoble cedex, France.

Abstract Multipole mixing ratios have been measured in ^{48}Ti, ^{54}Cr and ^{58}Fe. B(E2) and B(M1) values have been deduced for these and neighbouring nuclei and we have used them to identify states of mixed proton-neutron symmetry. These have been discussed within the framework of IBA-2.

INTRODUCTION

Measurements of multipole mixing ratios provide information about the relative magnitude and sign of E2 and M1 electromagnetic transition matrix elements. Combined with lifetime measurements the absolute magnitudes of the matrix elements can be deduced. Of particular interest are strong M1 transitions which are forbidden in collective models where the proton and neutron wavefunctions are not treated separately. We have measured multipole mixing ratios in ^{48}Ti, ^{54}Cr and ^{58}Fe by angular correlation of prompt gamma rays following thermal neutron capture. We show that there are strong M1 transitions linking low-lying states in these nuclei and that they are in reasonable agreement with those predicted for a transition from a mixed-symmetry state to a symmetric state in the SU(5) (vibrational) limit of IBA-2.

THE GAMMA-GAMMA ANGULAR CORRELATION EXPERIMENTS

Excited states in ^{48}Ti, ^{54}Cr and ^{58}Fe were populated by the capture of thermal neutrons from the high flux reactor at the I.L.L., Grenoble. The experimental set-up at the I.L.L. is described in detail in refs.[1,2]. The targets used in these experiments were 0.7 gramme-barn of enriched (79%) ^{47}Ti in TiO_2 form, 1 gramme-barn of natural chromium metal, and 0.5 gramme-barn of ^{57}Fe metal, enriched to 95%.

Coincidences were measured between one fixed Ge(Li) detector and a second Ge(Li) detector which was moved through five angles between 90^0 and 180^0. After correction for accidentals and source mis-centering coincidence count rates were fitted to the normal correlation function

$$\omega(\Theta) = 1 + A_2B_2Q_2P_2(\cos\Theta) + A_4B_4Q_4P_4(\cos\Theta) \qquad (1)$$

where the detector solid angle correction factors Q_k were taken from[3].

Fig. 1 shows some corrected coincidence count rates at each angle and the fitted correlation function.

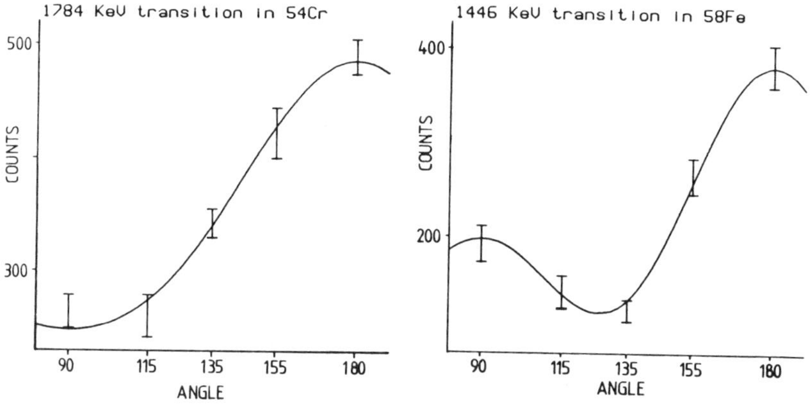

Fig 1a Fig 1b

Fig. 1. Normalized coincidence count rates and fitted correlation function for a) $2^+_2 \rightarrow 2^+_1 \rightarrow 0^+_1$ cascade in ^{54}Cr and b) $0^+_2 \rightarrow 2^+_1 \rightarrow 0^+_1$ cascade in ^{58}Fe, indicating the correct functioning of the experiment.

The results of these experiments will be published subsequently.

SYSTEMATICS OF 2^+ STATES IN THE A ≈ 50 MASS REGION

The systematics of low-lying 2^+ states in some even-even nuclei in the mass 50 region have been plotted out in figs. 2,3 and 4. Combining the present results with previously measured multipole mixing ratios[4-11] and lifetimes[7,9-15], B(M1) and B(E2) values have been calculated. The most striking feature of the decays of all these nuclei is the strong M1 transitions from 2^+ states to the first 2^+ state.

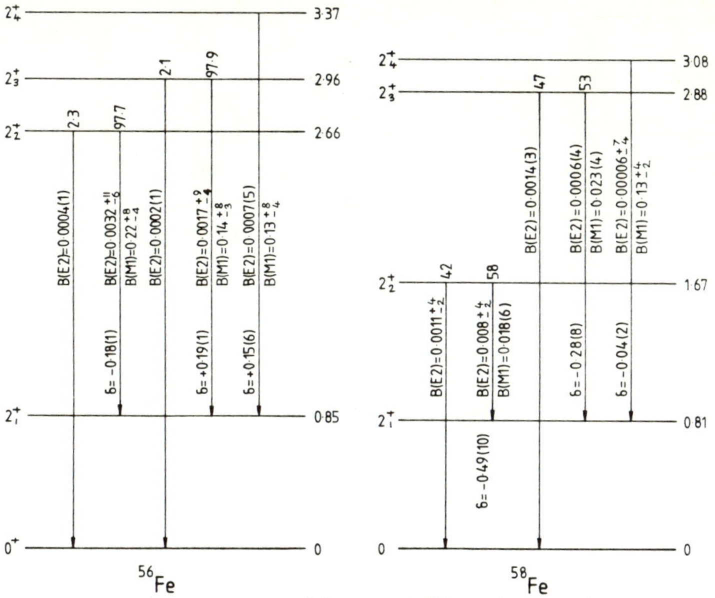

Fig 2. 2^+ states in ^{56}Fe and ^{58}Fe (Z = 26)

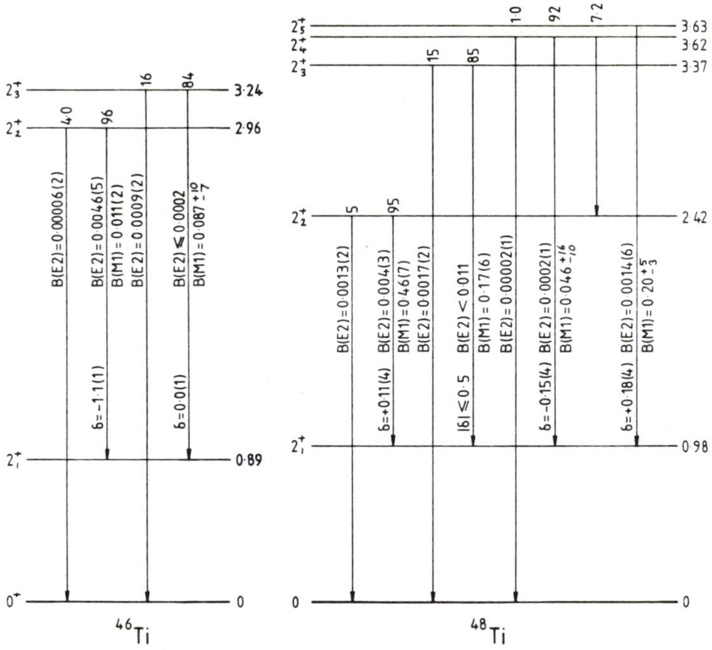

Fig. 3. 2^+ states in ^{46}Ti and ^{48}Ti (Z = 22)

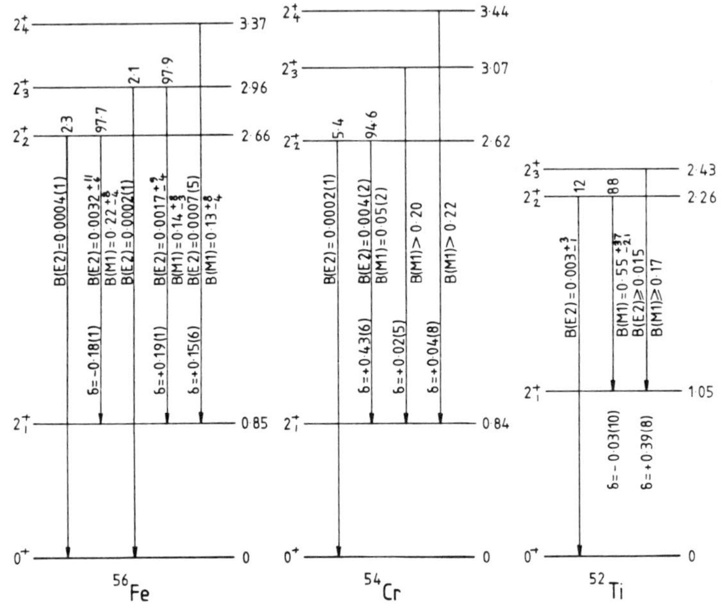

Fig 4. 2^+ states in ^{56}Fe, ^{54}Cr and ^{52}Ti (N = 30)

DISCUSSION

In the IBA-2 model strong M1 transitions are predicted between states of mixed proton-neutron symmetry and states of full symmetry. In the SU(5) (vibrational) limit the lowest lying mixed symmetry state is expected to be a 2^+. Levels in ^{140}Ba, ^{142}Ce and ^{144}Nd have been shown[16] to have decay properties consistent with mixed-to full-symmetry transitions. In ^{56}Fe two 2^+ levels close in energy have been interpreted[4] as being admixtures of mixed-symmetry and symmetric states. Such admixtures can cause a fragmentation of the M1 strength over several states.

In the SU(5) limit of IBA-2 the B(M1) for a transition from the one d-boson mixed symmetry state to the one d-boson symmetric state can be written

$$B(M1;\ 2^+_{ms} \rightarrow 2^+_{s}) = \frac{9}{2\pi}\,(g_\nu - g_\pi)^2\,\frac{N_\nu N_\pi}{N^2} \tag{2}$$

and the magnetic dipole moment of the first 2^+ state is given[17] by

$$\mu(2^+_1) = \frac{2}{N}\,(N_\pi g_\pi + N_\nu g_\nu) \tag{3}$$

where $N_\nu(N_\pi)$ are the number of neutron (proton) bosons available and g_ν and g_π are the boson g-factors. It has been shown[18] that in heavy nuclei g_ν and g_π are close to 0 and $1\mu_N$ respectively. We do not have good enough statistics in magnetic moment measurements to deduce accurately the g-factors, so we shall adopt these values. We can, however, check the predicted magnetic moments against experimental[19,4,20,7,11] values and this has been done in table 2.

Table 2. Experimental and predicted moments. N_ν and N_π have been counted from the nearest closed shell.

NUCLEUS	EXPERIMENTAL $\mu(2_1^+)/\mu_N$	N_π	N_ν	PREDICTED $\mu(2_1^+)/\mu_N$
^{58}Fe	0.92 (20)	1	2	0.7
^{54}Cr	1.12 (20)	2	1	1.3
^{48}Ti	0.86 (38)	1	1	1.0

The predicted and experimental values are shown to be in good agreement. Using these g-factors B(M1) values have been calculated.

Table 3 shows total B(M1) strengths for all observed low-lying 2^+ states showing large M1 transitions to the first 2^+ and the IBA-2 predictions.

The experimental B(M1) values are generally close to the IBA-2 predictions. Differences between experimental and theoretical values can be explained by:-

(i) A poor choice of g-factors, on which the B(M1) values depend critically.

(ii) Some of the M1 strength being distributed over unobserved states.

(iii) Departure from the SU(5) limit.

It can be seen that the predicted B(M1) values do not change much with boson number.

In ^{56}Fe a 4^+ state at 3.12 MeV and a 3^+ state at 3.45 MeV both show strong M1 transitions to the first 4^+ state at 2.09 MeV. The experimental B(M1) values are:-

$$B(M1;\ 4_2^+ \rightarrow 4_1^+) = 0.43^{+0.43}_{-0.14}\ \mu_N^2$$

and $$B(M1;\ 3_1^+ \rightarrow 4_1^+) \geqslant 0.11\ \mu_N^2$$

Table 3. Experimental and predicted B(M1) values.

NUCLEUS	N_π	N_ν	TOTAL $B(M1;2^+\rightarrow 2_1^+)/\mu_N^2$	IBA-2 $B(M1;2^+\rightarrow 2_1^+)/\mu_N^2$
^{56}Fe	1	1	0.5 (1)	0.36
^{58}Fe	1	2	0.17 (3)	0.32
^{54}Cr	2	1	>0.4	0.32
^{52}Ti	1	1	>0.5	0.36
^{46}Ti	1	2	0.10 (2)	0.32
^{48}Ti	1	1	0.9 (2)	0.36

Although in general in the SU(5) limit there are expected to be two d-boson mixed-symmetry states with spin 0,1,2,3,4, in ^{56}Fe we only have one boson of each kind. This would mean that the only two d-boson mixed-symmetry states allowed would have spin 1 and 3 . This raises the question of how bosons should be counted. It may be more appropriate to count boson particles from the nearest lowest closed shell. This would give for ^{56}Fe $N_\pi = 3$ and $N_\nu = 1$. In the SU(5) limit the B(M1) from the 4^+ two d-boson mixed-symmetry state to the 4^+ two d-boson symmetric state is given by

$$B(M1;\ 4^+_{ms} \rightarrow 4^+_s) = \frac{15}{2\pi}\,(g_\nu - g_\pi)^2\,\frac{N_\nu N_\pi}{N^2}\left[\frac{N-2}{N-1}\right] \qquad (4)$$

Taking $N_\pi = 3$, $N_\nu = 1$ and using the same g-factors as used previously this gives

$$B(M1;\ 4^{+}_{ms} \rightarrow 4^{+}_{s}) = 0.30\ \mu^{2}_{N}$$

which is in excellent agreement with the experimental value. This choice of boson numbers could also account for the four low-lying 2^{+} states in ^{56}Fe, otherwise only 3 would be allowed with two bosons.

We gratefully acknowledge useful discussions on the interacting boson model with Professor J.P. Elliott.

REFERENCES

1. Hungerford, P. and Kerr, S.A., I.L.L. Internal report 81 KE12T (1981).
2. Hamada, S., Contribution to this conference.
3. Camp, D.C. and Van Lehn, A.L., Nucl. Instrum. Methods 76, 192 (1969).
4. Eid, S.A., Hamilton, W.D. and Elliott, J.P., Phys. Lett. 166B, 267 (1986).
5. Krane, K.S., At. Data Nucl. Data Tables 22, 269 (1978).
6. Bradley, D.I., Stone, N.J., Rikovska, J., Novakova, D. and Ferencei, J., J. Phys. G 12, 115 (1986).
7. Peker, L,K., Nucl. Data Sheets 42, 457 (1984).
8. Verheul, H. and Auble, R.L., Nucl. Data Sheets 23, 455 (1978).
9. Pronko, J.G., Bardin, T.T., Becker, J.A., Fisher, T.R., McDonald, R.E. and Poletti, A.R., Phys. Rev. C9, 1430 (1974).
10. Auble, R.L., Nucl. Data Sheets 24, 1 (1978).
11. Alburger, D.E., Nucl. Data Sheets 45, 557 (1985).
12. Auble, R.L., Nucl. Data Sheets 20, 253 (1977)
13. Kosyak, Y.G., Kaipov, D.K., Chekushina, L.V., Izv. Akad. Nauk SSR Ser. Fiz. 47, 2118 (1983).
14. Stuchbery, A.E., Morrison, I., Kennedy, D.L. and Bolotin, H.H., Nucl. Phys. A337, 1 (1980)
15. Rahman, M., Nottrodt, H.P. and Rauch, F., Nucl. Phys. A 401, 253 (1983).
16. Hamilton, W.D., Irbäck, A. and Elliott, J.P., Phys. Rev. Lett. 53, 2469 (1984)
17. Van Isaker, P., Heyde, D., Jolie, J. and Sevrin, A., Ann. Phys. (N.Y.) To be published.
18. Sambataro, M., Schoten O., Dieperink, A.E.L. and Piccatto, G., Nucl. Phys. A 423, 333 (1984).
19. Auble, R.L., Nuclear Data Sheets 28, 103 (1979).
20. Fahlander, C., Johansson, K., Karlsson, E. and Possnert, G., Nucl. Phys. A 291 (1977).

TRIAXIALITY IN pf-SHELL?

J.RIKOVSKA and N.J.STONE

Clarendon Laboratory,Parks Road, Oxford OX1 PU, U.K.

INTRODUCTION

The even-even iron nuclei (Z=26,N=30,32,34 and 36) are close to the shell closure at Z=N=28. However, definite deviation from the expected spherical shape, at least for some states, has been observed[1-14].Experimental data suggests prolate collective excitations of both ^{56}Fe and ^{58}Fe which could be regarded as those of a rigid rotor, with the posibility of γ-vibrations around a spherical shape. The spectroscopic information on ^{60}Fe and ^{62}Fe is much less complete. The existing data points to a possible vibrational character for the low-lying excitations of these nuclei.

Many attempts to understand the structure of even Fe nuclei in a nuclear model framework have been made. Shell-model calculations[2,3], pairing-plus-quadrupole model[6,15-16] and Hartree-Fock-Bogoliubov calculation[17] show reasonable agreement with qualitative trends of experimental data. In detail, there are numerous problems with quantitative agreement,especially with magnetic properties. The general picture drawn by the experimental data as well as the model calculations shows a considerable contribution of collective degrees of freedom to at least low-lying states in even Fe nuclei. Going from ^{54}Fe, the long range neutron-proton interaction breaks the pairing forces, lowers the energy of the 2_1^+ state and causes deformation and collectivity of some states. Approaching the N=40 subshell closure, the equilibrium deformation decreases and the collective excitations resemble vibrations about a spherical shape. However, these collective states coexists with majority of other excitations, with pronounced single-particle character.

All the low-lying states in $^{56-62}Fe$ are summarized in Fig.1. No special selection has been made, except for ^{56}Fe, where 6_2^+ was taken instead of 6_1^+ which is much less oollective[4,16]. The 4_2^+ state in ^{56}Fe did not correspond to the expected energy of the 4^+ member of a quasi-vibrational band, so the 4_3^+ state was tentatively chosen to complete the sequence of the 2_2^+ and 3_1^+ states. The most striking feature of this picture, suggesting a transition from rotational-like spectrum in ^{56}Fe, to vibration-like in ^{62}Fe is the sudden decrease in energy of the quasi-gamma vibrational band in ^{58}Fe as compared to ^{56}Fe. The 2_2^+ state in ^{60}Fe is again much higher and the gamma-band structure has disappeared. Another characteristic feature is the neutron number dependence of the energy of the 0_2^+ state (from ~3 MeV for ^{56}Fe to ~1.8 MeV for ^{62}Fe). The 2_3^+ states show similar behaviour although not so dramatically. The energies of the 2_1^+ and 4_1^+ states are almost constant for $^{56-62}Fe$ and these states are clearly of different nature from the 0_2^+ and 2_3^+ excitations. The neutron number dependence of the 0_2^+ and 2_3^+ states recalls strongly the characteristic behaviour of intruder excitations across a major shell gap. However, the large M1 components in the $2_3^+ - 2_1^+$ transition in ^{56}Fe and ^{58}Fe suggests[18] the possibility of a strong single particle component in the 2_3^+ states. Recently, another tentative interpretation of the 2_2^+ and 2_3^+ states in ^{56}Fe has been suggested in terms of mixed proton-neutron symmetry states.[19]

The present calculation aimed at extension of the region of lighter nuclei to which the IBM hass been recently applied (Sr(Z=38)[20], Kr(Z=36)[21], Se(Z=34)[22], Zn(Z=30)[23]). Straihtforward application of the IBM model to Fe(Z=26) in the usual way, counting the number of bosons from the nearest closed shells (N=Z=28) leads to severe truncation of the boson space, certainly inadequate for calculation of ^{56}Fe and ^{58}Fe. However, there is a possibility to take $^{48}_{20}Ca_{28}$ core which was done in some shell model calculations. There is no mechanism in the IBM which can distinguish between boson-hole and boson-particle states[24,25]. Our choice introduces 3 proton particle bosons instead of 1 proton-hole boson in the f7/2 shell, leaving all the neutrons in the pfg shell.

RESULTS AND DISCUSSION

The calculation was performed using standard IMB2 Hamiltonian. Explicit quadrupole-quadrupole interaction between like bosons has been included to allow for inclusion of deviations from axial symmetry of the nuclear shape for ^{58}Fe.

The resulting parameters are given in Table 1 and the comparison between calculated energy levels and experiment is seen in Fig.1. The only way to achieve agreement with experiment for ^{58}Fe was to change the sign of the parameter χ_ν from the large negative value in ^{56}Fe, which is normal at the beginning of a major shell, to a large positive value. In the higher mass isotopes χ_ν is again negative.

Table 1 Parameters of the IBM2 Hamiltonian

A	ε	κ	χ_ν	$c_{0\nu}$	$c_{2\nu}$	$c_{4\nu}$	QN	QP
56	1.32	-0.36	-1.3				0	0
58	1.38	-0.28	+1.1	-0.35	+0.40	+0.35	+0.14	-0.14
60	1.51	-0.23	-0.8	-0.65	-0.50	+0.35	0	0
62	1.56	-0.20	-0.6	-0.65	-0.60	+0.35	0	0

All parameters are in MeV except of $\chi_\nu(\chi_\pi)$ (dimensionless) χ_π = -1.2 and $c_{L\pi}$ = 0.9, 0.6 and 0.1 for L=0,2,4 respectively. The Majorana force parameters were fitted to $\xi_1 = \xi_3 = -2\ \xi_2$. ξ_2= 0.13, 0.08, 0.05 and 0.02 for A=56,58,60 and 62, respectively.

This sign change we take to indicate a change in symmetry in going from ^{56}Fe, which is close to the SU3 limit ($\chi_\nu = \chi_\pi = \sqrt{7}$) towards a triaxial shape in ^{58}Fe [26,27] close to the SU3* limit ($\chi_\nu = -\chi_\pi = \sqrt{7}$). SU3* symmetry corresponds, in a geometrical model, to an asymmetric triaxial rotor[28] . This geometrical picture gave the first model explanation for the observed decrease the levels of the gamma-vibrational band in terms of increasing deviation from axial symmetry. Such a shape is believed to have its origin in different deformations of the neutron and proton matter with maximum overlap of the two shapes. The most likely situation for observation of the symmetry to arise is when there are comparable numbers of valence protons and neutrons occupying states in different major oscillator shells, as in the case for ^{58}Fe.

Calculation was carried out with the parameters close to those required to fit ^{58}Fe for all the other isotopes but not sensible agreement with experiment was achieved. It seems clear that for the even iron isotopes, rather than a smoothly varying sequence of IBM parameters there is

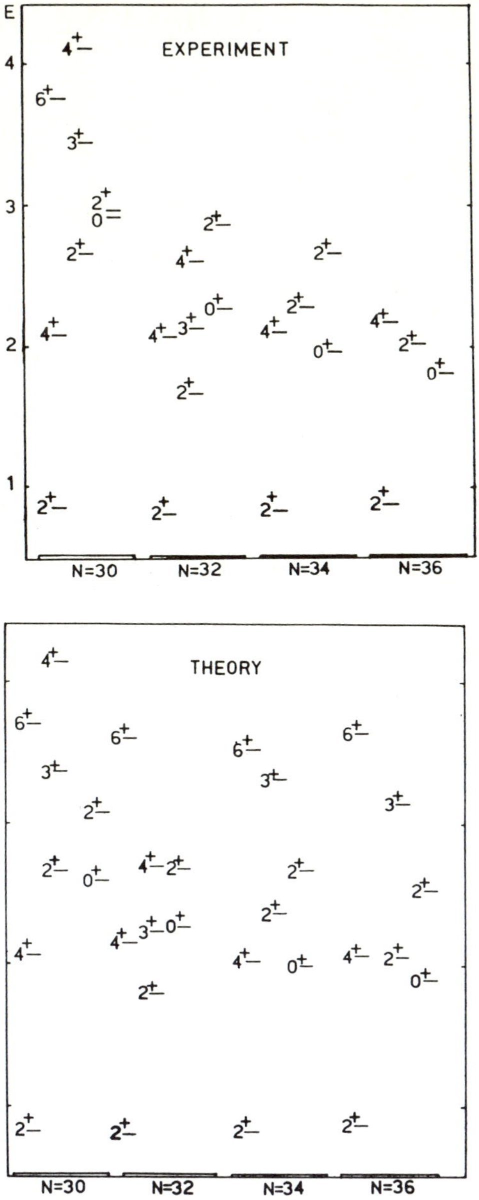

Fig. 1 Experimental and calculated energy levels in $^{56-62}Fe$ isotopes.

TABLE 2 ELECTROMAGNETIC PROPERTIES OF $^{56-60}Fe$ ISOTOPES[a)]

	A = 56		A = 58		A = 60	
	exp	calc	exp	cal	exp	calc
$B(E2)[e^2 fm^4]$ [b)]						
$2_1^+ \to 0_1^+$	204(11)	212	247(7)	211	186(40)	204
$4_1^+ \to 2_1^+$	295(80)	299	540(310)		190^{+63}_{-38}	320
$6_1^+ \to 4_1^+$	250(43)	279	205(23)			
$\frac{B(E2, 2_2^+ - 0_1^+)}{B(E2, 2_2^+ - 2_1^+)}$	0.11(3)	0.10	0.11(4)	0.06	0.06(5)	0.03
$Q_{2_1^+}[e\ fm^2]$ [b)]	- 19(8)	- 26	- 27.3(5)	- 24	-	- 25
$\mu_{2_1^+}$[n.m.][c)]	0.60(10)	0.69	0.63(12)	0.065	-	0.52
$\delta(E2/M1)$						
$2_2^+ \to 2_1^+$	- 0.185(15)	- 0.15	- 0.56(3)	- 0.28	$4.3^{+3.8}_{-1.5}$ or $-(0.27^{+0.13}_{-0.11})$	+ 1.2
$2_3^+ \to 2_1^+$	+ 0.27(1)	+ 0.43	$- 0.33^{+0.18}_{-0.11}$	+ 0.05	-	- 0.022

Experimental data taken from Ref. 1-14, 18.

a) No experimental data on ^{62}Fe

b) Effective boson charges $e_\pi = 6\ e\ fm^2$ ($^{56,60}Fe$), $e_\nu = 6, 3\ e\ fm^2$ for $^{56,60}Fe$ respectively, $e_\pi = 7\ e\ fm^2$, $e_\nu = 2.5\ e\ fm^2$ for ^{58}Fe.

c) Effective g-factors $g_\pi = 1$, $g_\nu = 0$ for $^{56-60}Fe$.

an abrupt change in one shape parameter which reflects a specific physical development.

It is well known that agreement with energy level systematics is inadequate to determine a uniquely best set of IBM parameters. The calculated wave functions were used to evaluate electromagnetic properties of gamma transitions in the iron isotopes (see Table 2). The good over all agreement lends strong support to the validity of the chosen parameters.

To conclude, the IBM2 model succesfully reproduces the observed behavious of the collective states with I 6 below 4 MeV in even Fe isotopes. To achieve agreement with experiment, the model requires a change of symmetry SU3 to SU3* in going from ^{56}Fe to ^{58}Fe, reflecting a change from axial to triaxial rotor form. More experimental data are needed to make detailed examination and definitive conclusion on these symmetries.

References

1. M.J. Le Vine, E.K. Warburton and D. Schwalm, Phys. Rev. C23, (1981) 244.
2. J.B. McGrory and S. Raman, Phys. Rev. C20 (1978) 83.
3. R.B.M. Mooy and P.W.M. Glaudemans, Z. Phys. A312 (1983) 59.
4. N. Bendjaballah, J. Delaunay, A. Jaffrin, T. Nomura and K. Ogawa, Nucl. Phys. A284 (1977) 513.
5. N. Bendjaballah, J. Delaunay, T. Nomura and H.J. Kim, Phys. Rev. Lett. 36 (1976) 1536.
6. G.K. Hubler, H.W. Kugel and D.E. Murnick, Phys. Rev. C9 (1974) 1954.
7. M. Niwano, T. Ishimatsu, R. Asano, T. Suchiro and M. Tanaka, Nucl. Phys. A377 (1982) 148.
8. S. Cavallaro, J. Delaunay, R. Ballini, T. Nomura, N. Bendjaballah and C. Tosello, Nucl. Phys. A293 (1977) 125.
9. C.W. Towsley, thesis, University of Rochester (1974) unpublished.
10. D.C. Kocker and R.L. Auble, Nucl. Data Sheets 19 (1976) 445.
11. H.H. Bolotin, A.E. Stuchbery, K. Amos and I. Morrison, Nucl. Phys. A311 (1978) 75.
12. E.B. Norman, C.N. Davids, M.J. Murphy and R.C. Pardo, Phys. Rev. C17 (1978) 2176.
13. E.B. Norman, C.N. Davids and C.E. Moss, Phys. Rev. C18 (1978) 102.

14. E. Runte et al, Nucl Phys. A399 (1983) 163.
15. V. Paar, Nucl. Phys. A185 (1972) 544.
16. M. Girod and P.G. Reinhard, Nucl. Phys. A384 (1982) 179.
17. J.K. Parikh, Phys. Rev. C5 (1972) 153, and Phys. Rev. C10 (1974) 2568.
18. J. Lange, K. Kumar and J.H. Hamilton, Rev. Mod. Phys. 54 (1982) 119.
19. S.A.A. Eid, W.D. Hamilton and J.P. Elliott, Phys. Lett. 166B (1986) 267.
20. K. Heyde, J. Moreau and M. Waroquier, Phys. Rev. C29 (1984) 1859.
21. U. Kaup and A. Gelberg, Z. Phys. A293 (1979) 311.
22. U. Kaup, C. Mönkemeyer and P.V. Brentano, Z. Phys. A310 (1983) 129.
23. C.H. Druce, J.D. McCullen, P.D. Duval and B.R. Barrett, J. Phys. G. 8 (1982) 1565.
24. S. Pittel, P.D. Duval and B.R. Barrett, Phys. Rev. C25 (1982) 2834.
25. S. Pittel, P.D. Duval and B.R. Barrett, Annals of Phys. 144 (1982) 168.
26. A.E.L. Dieperink and R. Bijker, Phys. Lett. 116B (1982) 77.
27. A.E.L. Dieperink and I. Talmi, Phys. Lett. 131B (1983) 1.
28. A.S. Davydov and G.F. Filippov, Nucl. Phys. 8 (1958) 237.

IBM-2 CALCULATION AND NEW RESULTS IN EVEN-EVEN Sr NUCLEI

J. RIKOVSKA*, R.F. DAVIE†, A. KUCHARSKA† and D. SINCLAIR†
*Clarendon Laboratory, Parks Road, Oxford OX1 3PU, U.K.
†Nuclear Physics Laboratory, Keble Road, Oxford OX1 3RH, U.K.

Abstract Energy levels and transition probabilities in $^{78-86}$Sr isotopes are studied[1] in the framework of IBM-2. The reduction of the number of proton bosons, reflecting the effect of Z = 40 subshell closure, is important for understanding of experimental B(E2) values.

INTRODUCTION

Nuclei in the region around N~Z~40 have been of considerable theoretical and experimental interest in recent years. Nuclear shapes in this region vary rapidly as a function of particle number and there is a strong evidence for shape coexistence.[1] The subshell closure effects (see Ref.2 for example) that occur for both protons and neutrons at particle numbers between 36 and 42 for deformed nuclei, in addition to the spherical subshell closure at particle number 40, make the understanding of nuclear phenomena in this region and their dependence on neutron and proton number rather complicated. Recent experiments[3-5] show a smooth variation in the shape of the light Sr nuclei from a strongly deformed prolate shape for ^{78}Sr to a soft near spherical shape for ^{84}Sr. This observation is in agreement with the theoretical calculations of Ragnarsson and Sheline[6], and Bucurescu et al.[7]

There have been several attempts to reproduce more detailed properties of nuclei in this region. Various forms of the interacting boson model have been applied to the Kr, Se and Sr isotopes (see Ref. 8 and references therein). These investigations have been fairly successful in reproducing positive parity energy levels up to higher spins, and transition rates and negative parity states in some cases. Sambataro and Molnar[9] studied the shape coexistence in the neutron rich Mo nuclei (A = 96-104) using the IBM-2 model with configuration mixing based on the interaction between "normal" and "intruder" configurations, the latter having its origin in 2p-2h proton excitation across the subshell gap at Z = 40. In contrast to the interacting boson descriptions, Nazarewicz et al[2] have used the "cranking" model to describe properties of high spin states in N~Z~40 nuclei as a function of both particle number and spin, with particular reference to the light Sr nuclei.

Recent experimental studies of the light Sr isotopes have provided new information on B(E2) transition probabilities ($^{80-84}$Sr)[10,11], excited state g-factors (^{82,84}Sr)[11] and isotope shifts ($^{80-88}$Sr).[12,13] The availability of this new experimental data, together with discrepancies between experiment and theory for previous IBM calculations, stimulated a revised IBM-2 calculation of the even-even Sr isotopes in the region of A = 78-86.

THE MODEL

The interacting boson Hamiltonian was used in the form

$$H = \varepsilon(n_{d_\pi} + n_{d_\nu}) + \kappa Q^{(2)}Q^{(2)} + V_{\pi\pi} + V_{\nu\nu} + M_{\pi\nu},$$

where $n_{d_{\pi(\nu)}}$ is the boson number operator for protons (neutrons), $Q^{(2)} = Q_\pi^{(2)} + Q_\nu^{(2)}$ is the quadrupole operator, $V_{\pi\pi(\nu\nu)}$ is the residual interaction between like bosons and $M_{\pi\nu}$ is the Majorana operator. Our choice of the quadrupole-quadrupole interaction included, in addition to the proton-neutron interaction, a quadrupole interaction between like bosons which is thought to be important for nuclei with triaxial and more complex equilibrium shapes.[13] As a consequence, our parameter κ is only half of the value normally used for the interaction strength $Q_\pi^{(2)}Q_\nu^{(2)}$. This more complex form of the quadrupole-quadrupole interaction introduces no additional parameters. All the other parameters of the above Hamiltonian which were used in the present calculation have the standard meaning (see e.g. Ref. 15).

Besides calculating both energy levels and reduced E2 transition probabilities and their ratios, g-factors and isotope shifts were calculated and compared with experiment to find optimum values of the parameters for the IBM Hamiltonian. The reduced E2 transition probability is calculated as

$$B(E2, I_i \rightarrow I_f) = \left(e_\pi \langle I_i || Q_\pi^{(2)} || I_f \rangle + e_\nu \langle I_i || Q_\nu^{(2)} || I_f \rangle \right)^2 / (2I_i + 1)$$

where $e_{\pi(\nu)}$ is an effective boson charge.

RESULTS AND DISCUSSION

For the light Sr nuclei (Z = 38, N = 40-48) there are several possible ways of choosing the number of proton and neutron bosons. If we use the full proton shell (28-50) then we have $N_\pi = 5$, while if the subshell closure at Z = 40 is considered dominant then both $N_\pi = 1$ configurations and $N_\pi = 3$ states, corresponding to 2p-2h excitations across the subshell gap, will predominate. In principle, the closure at N = 40 should also be taken into account. However, this closure is thought to be less pronounced than the corresponding proton subshell closure.[1,8] Therefore, the neutron subshell closure was not included explicitly into the calculation by restricting the neutron boson number; however, it was reflected implicitly in the derived dependence of the neutron parameter χ_ν on neutron number.

We have carried out calculations for one, three and five proton bosons to investigate the effects of the proton subshell gap at Z = 40. Checks have been made on convergence of the calculation process in the spaces with fewer bosons. In Fig. 1 we compare the resulting energy levels for $N_\pi = 1$ and $N_\pi = 3$ with the existing experimental data. The

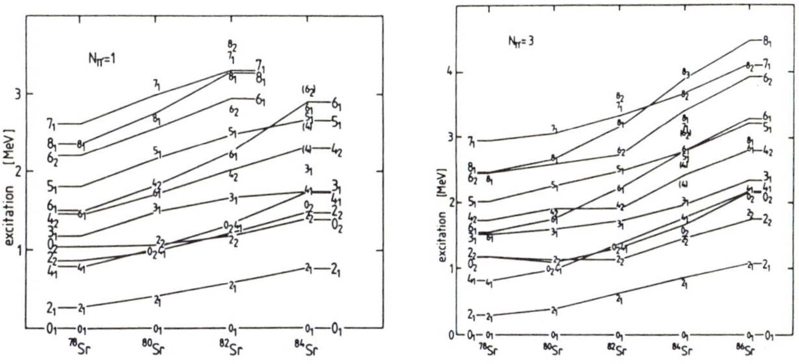

FIGURE 1 IBM-2 fit to energy levels for $N_\pi = 1$ and 3. Experimental points are represented by corresponding I^π values.

parameters used, which are summarised in Table 1, vary

Table 1 Parameters of the IBM-2 Hamiltonian

A	N_π	ε	κ	χ_ν	C_0^ν	C_2^ν	C_4^ν	ξ_2
78	5	0.84	-0.066	-0.30	-0.35	-0.15	0.02	0.005
	3	0.69	-0.082					0.015
	1	0.40	-0.090					0.015
80	5	0.87	-0.060	-0.20	-0.35	-0.15	0.02	0.005
	3	0.72	-0.080					0.007
	1	0.52	-0.082					0.120
82	5	0.89	-0.042	-0.10	-0.35	-0.15	0.08	0.015
	3	0.81	-0.080					0.020
	1	0.60	-0.084					0.150
84	5	0.995	-0.032	+0.10	-0.55	-0.15	0.30	0.050
	3	0.93	-0.080					0.070
	1	0.76	-0.084					0.200
86	5	1.15	-0.015	+0.45	-0.55	-0.15	0.30	0.100
	3	1.13	-0.080					0.100

All parameters are in MeV except $\chi_{\nu,\pi}$ (dimensionless).

smoothly with neutron number. The parameter χ_π was kept constant ($\chi_\pi = -1.3$) through the whole sequence (despite the change in N_π), parameters of $V_{\pi\pi}$ were $C_L^\pi = -0.3$, 0.2 and 0.02 (L = 0,2,4), respectively, for $N_\pi = 5$, and were set equal to zero otherwise. Nonzero values of these parameters improve the fit of energy levels somewhat, but their influence on electromagnetic properties is not significant. In previous calculations the neutron number dependence of the parameter χ_ν,which depends upon the number of neutrons relative to the closed shell, was taken relative to the whole 28-50 neutron shell. In contrast, the present values of χ_ν also reflect the weak neutron subshell closure at N = 40. The Majorana force parameter ξ_2 only was used to achieve the best fit, with $\xi_1 = \xi_3 = 0$ for all cases except N = 84-86 $N_\pi = 3$ where $\xi_3 = 0.08$. These values differ from those used by Heyde et.al.[8]; we have also carried out calculations using this previous set of parameters, in order to compare them with the results of the present study.

As can be seen from Fig. 1 and Ref. 8, it is possible to obtain a good fit to the experimental energy levels independent of the number of proton bosons used in the calculation. However, the predictions for the B(E2) values for these various treatments of the Z = 40 subshell closure are quite different. In Tables 2 and 3 we summarise both results of

TABLE 2 B(E2)-values for g.s.b. and q.g.v. band in light Sr Isotopes [e^2b^2]

A	$2_1^+ \to 0_1^+$	$4_1^+-2_1^+$	$6_1^+-4_1^+$	$8_1^+-6_1^+$	$2_2^+-0_1^+$	$4_2^+-2_2^+$	$6_2^+-4_2^+$	$8_2^+-6_2^+$
78 exp	0.22(3)	0.34(3)	-	-	-	-	-	-
a)	0.36	0.54	0.60	0.62	0.015	0.25	0.38	0.42
b)	0.40	0.60	0.69	0.69	0.010	0.20	0.40	0.46
c)	0.27	0.40	0.44	0.43	0.013	0.14	0.24	0.26
d)	0.17	0.24	0.24	0.21	0.010	0.10	0.13	0.11
80 exp	0.17(1)	0.23(3)	0.16(8)	0.37(4)	-	-	-	-
a)	0.22	0.42	0.47	0.48	0.064	0.22	0.31	0.32
b)	0.28	0.45	0.52	0.53	0.006	0.17	0.30	0.34
c)	0.19	0.29	0.32	0.31	0.007	0.12	0.17	0.18
d)	0.12	0.16	0.16	0.13	0.003	0.075	0.082	0.054
82 exp	0.100(5)	0.23(5)	0.11(2)	-	0.0008(3)	0.08(2)	0.15(8)	0.20(10)
a)	0.21	0.31	0.35	0.35	0.0002	0.18	0.23	0.22
b)	0.18	0.30	0.36	0.38	0.0010	0.12	0.20	0.23
c)	0.11	0.18	0.21	0.20	0.0010	0.072	0.11	0.10
d)	0.08	0.103	0.10	0.06	0.0004	0.042	0.033	-
84 exp	0.057(6)	0.037(5)	0.045(a)	0.009(1)*	-	-	-	-
a)	0.15	0.22	0.25	0.24	0.003	0.12	0.13	0.14
b)	0.13	0.22	0.26	0.27	0.0002	0.08	0.13	0.15
c)	0.08	0.13	0.14	0.1	0.0001	0.044	0.052	0.007
d)	0.048	0.060	0.043	-	0.0000	0.001	-	-
86 exp	-	-	-	-	-	-	-	-
a)	0.103	0.15	0.18	0.16	0.01	0.08	0.11	0.11
b)	0.09	0.16	0.18	0.18	0.0000	0.062	0.09	0.10
c)	0.07	0.10	0.10	0.064	0.0004	0.03	0.03	-
d)	-	-	-	-	-	-	-	-

a) Heyde et al ref. 8, b) $N_\pi = 5$, c) $N_\pi = 3$, d) $N_\pi = 1$. Experiment data from Refs. 3,10,11,18,19.

* (for$(\nu_{g_{9/2}})^2$ $8_2^+ - 6_1^+$ B(E2) = 0.027(1))

TABLE 3 B(E2) branching ratios for some transitions in light Sr isotopes

A	$\frac{2_2^+-2_1^+}{2_2^+-0_1^+}$	$\frac{3_1^+-2_1^+}{3_1^+-2_2^+}$	$\frac{4_2^+-4_1^+}{4_2^+-2_1^+}$	$\frac{4_2^+-2_2^+}{4_2^+-2_1^+}$
78 exp	-	-	-	-
a)	9.32	0.084	0.360	115
b)	2.03	0.040	0.139	140
c)	5.15	0.069	0.217	49
d)	6.12	0.077	0.351	75
80 exp	>86(9)	>0.007(2)	0.94(5)	>120(40
a)	25	0.055	0.53	117
b)	13	0.029	0.13	167
c)	15	0.058	0.44	85
d)	43	0.042	0.73	308
82 exp	160(25)	0.018(3)	0.10(2)	-
a)	1344	0.002	0.85	232
b)	31	0.010	0.12	150
c)	16	0.034	0.24	44
d)	215	0.016	0.95	356

A	$\frac{2_2^+-2_1^+}{2_2^+-0_1^+}$	$\frac{3_1^+-2_1^+}{3_1^+-2_2^+}$	$\frac{4_2^+-4_1^+}{4_2^+-2_1^+}$	$\frac{4_2^+-2_2^+}{4_2^+-2_1^+}$
84 exp	390(2)	-	-	-
a)	70	0.018	0.848	1070
b)	134	0.0024	0.143	242
c)	166	0.0030	0.289	70
d)	405	0.0029	-	-
86 exp	1000(80)	-	-	-
a)	11.6	0.1030	0.784	42
b	2360	0.0001	0.124	1080
c)	2090	0.0090	0.254	700
d)	-	-	-	-

a) Heyde et al., ref. 8, b) N_π - 5, c) N_π - 3, d) N_π - 1

Experimental data taken from ref. 10,11 and herein

our calculations and the experimental data on the B(E2) values for the ground state (g.s.) bands and the quasi-vibration (q.g.v.) band, in addition to some B(E2) ratios for the light Sr isotopes. The values of effective boson charges were $e_\pi = e_\nu = 0.12$ eb. Different values of e_π and e_ν, for example $e_\pi = 1.5$ eb and $e_\nu = 0.5$ eb used in Hartree-Fock calculations in this region,[16] were considered; it was found that the qualitative results of the study were independent of these alternative choices of the effective charges.

Examination of Tables 2 and 3 shows that both sets of parameters for $N_\pi = 5$ yield the poorest agreement with experiment. Reduction in number of proton bosons considerably improves the agreement between theory and experiment. For the g.s. band, the configuration with $N_\pi = 3$ gives good results for N = 40 and 42, particularly for the B(E2) values which are accurately known in some cases, while for N = 44 and 46 the data favours $N_\pi = 1$. Unfortunately, there is very little experimental data on the q.g.v. bands. However, the limited amount of experimental data on these bands also favours $N_\pi = 3$ for N = 40 and 42, and $N_\pi = 1$ for N = 44 and 46, in good agreement with the results for the g.s. band.

Preliminary results on g-factors and mean square radii show good agreement with experiment. Detailed analysis will be published elsewhere.

The difficulties in reproducing experimental data using $N_\pi = 5$, i.e. the full proton shell, indicates the importance of the subshell closure around Z = 40 in light even-even Sr nuclei. All the results presented can be understood in terms of interplay between a normal (N_π = 1) and intruder (N_π = 3) configurations. The assumption that the collective intruder configuration is responsible for increasing collectively for N < 44 is strengthened by the fact the excited protons are in the $g_{9/2}$ orbital which is the same orbital as that of the valence neutrons (holes), which increases their interaction.

Detailed configuration mixing evaluation is now in progress. The N_π = 3 configuration is dominant in $^{78-82}$Sr while there is a transition to vibration-like structure with N_π = 1 in $^{84\ 86}$Sr. Full results will be published soon.

1. J.H. Hamilton, P.G. Hansen and E.F. Zganjar, Reports on Progress in Physics, 48 631 (1982).
2. W. Nazarewicz, J. Dudek, R. Bengtsson, T. Bengtsson and I. Ragnarsson, Nucl. Phys., A435, 397 (1985).
3. C.J. Lister, B.J. Varley, H.G. Price and J.W. Olness, Phys. Rev. Lett., 49, 308 (1982).
4. P.S. Haskins et al., Phys. Rev., C32, 1897 (1985).
5. C.A. Fields, F.W.N. De Boer, E. Sugarbaker and P.M. Walker, Nucl. Phys., A361, 352 (1981).
6. I. Ragnarsson and R.K. Sheline, Phys. Scr., 29, 385 (1984).
7. D. Bucurescu, G. Constantinescu and M. Ivascu, Rev. Roum. Phys., 24, 971 (1979).
8. K. Heyde, J. Moreau and M. Waroquier, Phys. Rev., C29, 1859 (1984).
9. M. Sambataro and G. Molnar, Nucl. Phys., A376, 201 (1982).
10. R.F. Davie et.al., submitted to Nucl. Phys. and contribution to this conference.
11. A. Kucharska, Ph.D. Thesis, Oxford, 1985.
12. F. Buchinger, R. Corriveau, E.B. Ramsay, D. Berdichovsky and D.W.L. Spring., Phys. Rev., C32, 2058 (1985).
13. D.E. Eastham et. al., Daresbury Laboratory preprint DL/NUC/P234E submitted to J. Phys. G.
14. A.E.L. Dieperink and R. Bijker, Phys. Lett., 116B, 77 (1982).
15. A. Arima and F. Iachello, Annu. Rev. Nucl. Sci., 31, 75 (1981).
16. D.P. Ahalpara, A. Abzouzi and K.H. Bhatt, Nucl. Phys., A445 (1985).

E0 AND E2 DECAY OF 0^+ STATES IN ^{78}Se, ^{124}Te, ^{172}Yb and ^{200}Hg

Subber[†] A.R.H., Hamilton[†] W.D., Schreckenbach[$] K., Colvin[$] G. and Kumar[*] K.

† School of Mathematical and Physical Sciences
University of Sussex, Brighton BN1 9QH, U.K.

$ Institute von Laue-Langevin, BP 156
38042 Grenoble Cedex, France.

* Physics Department, Tennessee Technological University
Cookville,
TN 38505, U.S.A.

Abstract E0 transition in the isotopes ^{78}Se, ^{124}Te, ^{172}Yb and ^{200}Hg have been investigated by means of conversion-electron and gamma-ray spectroscopy. Values or upper limits for P(E0) and the ratio X(E0/E2) of E0 to E2 reduced transition probabilities have been derived for $0^+ \rightarrow 0^+$ transitions and for $\Delta I = 0$, $I \neq 0$ transitions. The experimental results are discussed in framework of the DDM and IBM.

INTRODUCTION

The study of excited 0^+ states in even-even nuclei can provide important facts on the nuclear structure in a given region of the nuclear chart. A variety of theoretical descriptions, for the existance of the first excited 0^+ state, have been suggested during the last three decades. In the vibrational model[1)] the low lying excited 0^+ level is considered as a member of the two phonon triplet. Configuration mixing of two particles in the shell model produces excited 0^+ levels[2)]. In the collective models[3,4)] the excited 0^+ states are described as the band head of β-vibrational bands. While in the γ-soft model, the first excited 0^+ level is assumed to be a member of three phonon triplet which is pushed down due to the γ-softness of the nucleus[5)]. Pairing-vibration[6)], and excitation of a pair of neutrons[7)] or protons[8)] are also suggested as leading to possible configurations for excited 0^+ states.

Internal conversion is the only decay mode between 0^+

states, but E0 transitions can also occur in other cases where the levels have the same spin and parity.

A convenient measure of the strength of E0 matrix elements is given Rasmussen's[9] dimensionless ratio which may be written as

$$X(E0/E2) = B(E0;\ I_i - I_f)/B(E2;\ I_i - I_f' \qquad (1)$$

where $I_f = I_f'$ for $I_i \neq 0$ and $I_f = 0$, $I_f' = 2$ for $I_i = 0$. This ratio may be calculated using γ-ray and conversion electron intensities and the relation

$$X(E0/E2) = 2.54\ 10^9\ A^{4/3}\ E\ /\Omega_\gamma^5\ \alpha(E2)\ \frac{I_e(E0)}{I_e(E2)} \qquad (2)$$

The atomic mass number is denoted by A, E_γ is the transition energy measured in MeV, $I_e(E0)$ and $I_e(E2)$ are the measured electron intensities and Ω_K is an electronic factor[10] depending on Z and E_γ.

In the cases where Coloumb excitation is possible for the 0^+ state the X-value and absolute B(E2) can be used to derive absolute value for the monopole matrix element $\rho(E0)$.

EXPERIMENTS AND RESULTS

The group of experiments were carried out at the High Flux Reactor at the ILL. The target in each case was a separated isotope. Internal conversion spectra were recorded with the high resolution electron spectrometer BILL[11]. The sources were made by sedimentation onto a 1mg cm^{-2} thick alumimium foil. The area of the source in each case was 30 x 100 mm^2. The thickness of the target is partly determined by the thermal neutron capture cross-section of the separated isotope.

The $^{77}Se(n,e^-)$ ^{78}Se Measurement

Several 0^+ levels in ^{78}Se decay scheme[12] were identified from $\gamma\gamma$ angular correlation experiment. We searched for the decay of these 0^+ states via E0 transitions to the low lying 0^+ levels but we were unable to identify clearly any $0^+ \rightarrow 0^+$ transition. An upper limit for their electron intensities were carefully constructed from the (n,e^-) spectra. Internal conversion coefficient for transitions from other levels in the scheme were derived. Fig. 12 shows the decay of excited 0^+ levels in ^{78}Se with the ratios of reduced transition probabilities normalized to $B(E2;\ 0_1^+ - 2_1^+)$ as unity.

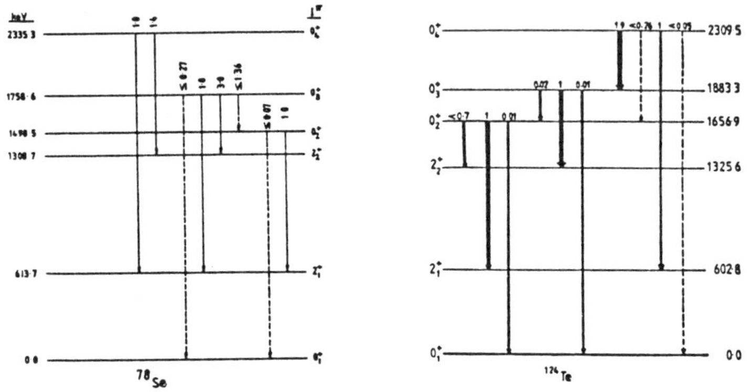

Fig-(1-4) Part of the decay scheme of ^{78}Se, ^{124}Te, ^{172}Yb and ^{200}Hg containing the E0 and E2 transitions of interest in the work with their B(EL) branching ratios.

The ^{123}Te(n,e$^-$) ^{124}Te Measurement

Robinson et al[13] established four excited 0^+ levels, below 3 MeV, in ^{124}Te decay scheme. The positions of these states were different from those suggested earlier[14]. To clear up this ambiguity we performed internal conversion measurement using BILL spectrometer. Several E0 transitions were identified and X-values were obtained, Fig. 2. On the basis of these E0 transitions we confirmed the results of Robinson et al and excluded other proposed 0^+ levels at 1157, 1290 and 2020 KeV. We found no evidence for a 2712 KekV (0^+) level suggested by Robinson et al.

The ^{171}Yb(n,e$^-$) ^{172}Yb measurement

The ^{172}Tm decay experiment by Otteson and Helmer[16] assigned the level at 1043 KeV to be 0^+ band head of $K^\pi = 0^+$ band. This assignment was confirmed by Greenwood et al[17] who also identified 0^+ bands. The monopole transitions in ^{172}Yb have been studied by Larysz et al[8] using the thermal neutron capture method. The large value $X(E0/E2;\ 0_3 - 0_1) = 2.81$ was explained as arising from proton pair excitation. In the present experiment we searched mainly for low energy E0

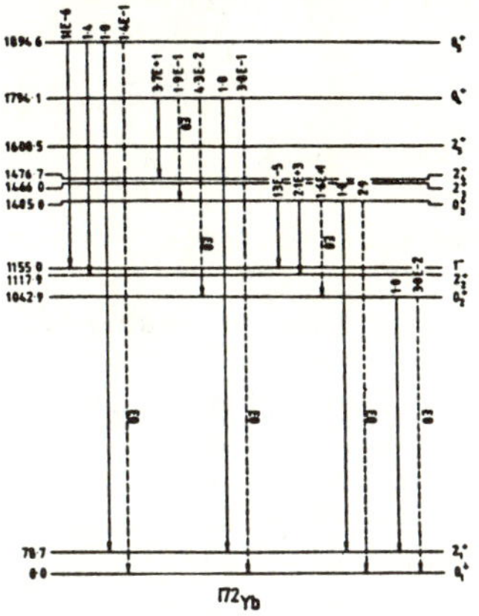

Figure 3.

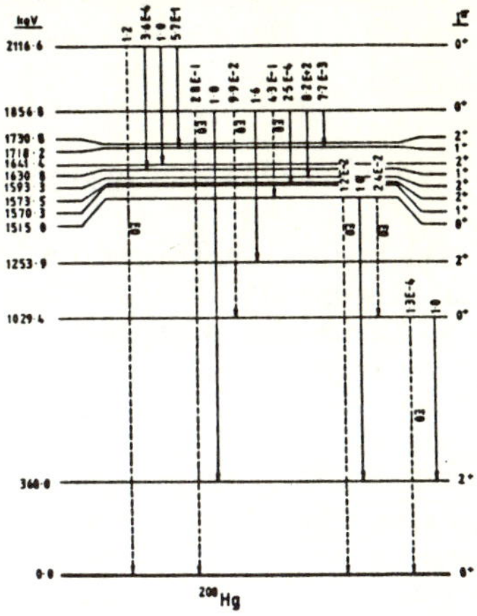

Figure 4.

transitions linking the excited 0^+ states in the scheme. All E0 transitions, previously identified[8)], have been seen in the present measurement plus 362.1 KeV 0_3^+ - 0_2^+, 389.1 KeV 0_4 - 0_3 and 751.2 KeV 0_4^+ - 0_2^+ transitions. Fig. 3 shows the decay of excited 0^+ levels in ^{172}Yb with their branching ratios. The most striking features of the measurement is the observation of the 362.1 KeV 0_3^+ - 0_2^+ transition with the large x-value of 15.6 which leads to the results $\rho(0_3\text{-}0_2)$ = 0.043(5), $\rho(0_3\text{-}0_1)$ = 0.015 (2) and B(E2; $0_3\text{-}2_1$) = 1.4 x 10^{-5} e^2b^2 using the life time measurmeent of ref[7)].

The ^{199}Hg(n,e$^-$) ^{200}Hg Measurement

An extensive thermal and resonance neutron capture study was carried out by Breitig et al[18)] who established a large amount of information about the structure of levels below 3 MeV in ^{200}Hg and confirmed or proposed the positions of the excited 0^+ levels at 1029, 1515, 1857 and 2697 KeV.

It was though, very useful tp provide a more detailed study of the ^{199}Hg(n,e$^-$) reaction by identifying E0 transitions. Special attension is paid, in the present work, to the variation of the strength of the monopole matrix elements in $\Delta I = 0$ transitions. Several $0^+ \rightarrow 0^+$ transitions were observed and confirm the location of the

1029.3 KeV 0_2, 1515.0 KeV 0_3 and 1857.0 KeV 0_4 levels in the scheme. The ratio X(E0/E2) for these transitions was obtained and is shown in Fig. 4.

COMPARISON OF THE RESULTS WITH NUCLEAR MODELS

The Dynamic Deformation Model[21] was used to describe level structure of ^{18}Se, ^{124}Te, ^{172}Yb and ^{200}Hg. The DDM requires no parameter adjustment to fit the experimental results. Low lying levels are reproduced rather well by this model[22,23]. The potential energy surface produced by the DDM for ^{124}Te shows a good example of a γ-unstable nucleus[22]. This prediction is supported by the experimental B(E2) and B(E0). The IBM-2[24] was also applied to these nuclei. Low lying levels were produced closed to the experimental data by adjusting the model parameters. Monopole matrix elements parameters were obtained from the experimental isotope shift[25]. Table 1 contains the experimental X(E0/E2; 0_2-0_1) values together with DDM and IBM-2 predictions.

Data on E0 transitions in ^{124}Te support the γ-unstable predictions of the DDM as proposed by Park et al[22]. The observed E0 transitions are allowed only by the second term of the E0 operator ($\beta^3\cos3\gamma$) with $\Delta\tau = 3$[23]. In the γ-unstable model of Elliott et al[25] the ratio B(E0; 0_3-0_2)/B(E0; 0_3-0_1) is 1.4 in agreement with the experimental value of (2.0). While the absence of E0 transitions in ^{18}Se might be related to the vibrational nature of this nucleus.

E0 transitions in ^{172}Yb provided more detailed information about the nuclear band structure of this nucleus and opportunity to test E0 prediction of the IBM-2.

The E0 transitions in ^{200}Hg show that this nucleus has a similar structure to that of ^{124}Te, this is especially true for the decays of 0_2^+ and 0_3^+ levels.

Table 1: The x-value for $0_2^+ \rightarrow 0_1^+$ transitions in Se, Te, Yb and Hg isotopes.

Isotope	$X(E0/E2; \frac{0_2-0_1}{0_2-2_1})$		
	Exp[(a)]	DDM	IBM-2
^{72}Se	0.058 (20)	0.020	0.058[(b)]
^{74}Se	0.030 (3)	0.078	0.047
^{76}Se	0.031 (7)	0.087	0.31(b)
^{78}Se	$\leqslant$ 0.07	0.05	0.09
^{124}Te	0.012 (3)	0.09	0.031
^{168}Yb	0.11	0.0001	0.003
^{170}Yb	0.0038 (4)	0.012	0.013
^{172}Yb	0.029 (2)	0.06	0.002
^{174}Yb	< 0.051	0.06	0.009
^{200}Hg	1.3 (1) x 10^{-4}	0.07	0.3

(a) Experimental values are from this work unless from refs[19,20)]

(b) Fitted values

REFERENCES

1. Scharff, G.G. and Wenser, J., Phys. Rev 98 212 (1955).
2. Bayman, B.F., Reiner, A.S. and Sheline, R.K., Phys. Rev. 155 1627 (1959).
3. Bohr A. and Mottelson B.R., "Nuclear Structure", (New York: Benjamin) (1975).
4. Davydov, A.S., Nucl. Phys. 24 682 (1961).
5. Meyer, R.A., Grifioen R.D., Lefter, J.G. and Walter, W.B., Phys. Rev. C14 2024 (1976).
6. Bes, D.R. and Broglia, R.A., Nucl. Phys. 80 289 (1966)
7. Andrijtscheff, W., Petkov, P., Protochristow, Ch.,

Kostov, L.K., Hamilton, W.D. and Hoyler, F., "To be published" (1986).

8. Larysz, J.R., Schreckenbach, K., Davidson, W.F., Börner, H.G., Warner, D.D., Casten, R.F. and Gelletly, W., Nucl. Phys. A309 128 (1978).
9. Rasmussen, J.O., Nucl. Phys. 19 85 (1960).
10. Bell, D.A., Aveledo, C.E., Davidson, M.G. and Davison J.P., Can. J. Phys. 48 2542 (1970).
11. Mampe, W., Jeuch, P., Von Egidy,T., Scheckenbach, K., Braumandl, F. and Larysz, J.R., Nucl. Inst. Meth. 128 585 (1978).
12. Subber, A.R.H., Robinson, S.J., Hungerford, P., Hamilton, W.D., Van Isacker, P., Kumar, K., Park, P., Scherchenbach, K. and Golvin, G., "To be published".
13. Robinson S.J., Hamilton, W.D. and Snelling D.M., J. Phys. G9 961 (1983).
14. Lederer, C.M. and Shirley V.S., "Table of Isotopes 7th edition". Pub. Willey, New York (1978).
15. Ottesen, O.H. and Helmer, R.G., Phys. Rev. 164 1485 (1967).
16. Greenwood, R.C., Reich, C.W. and Vegors S.H., Nucl. Phys. A252 260 (1975).
18. Breitig, D., Casten, R.F. and Cole, G.W., Phys. Rev. C9 366 (1974).
19. Lange, J., Kumar, K. and Hamilton, J.H., Rev. Mod. Phys. 54 119 (1982).
20. Passoja A., Vergnes M., Kantele J. and Luontama M., Physica Scripta, Vol. T5 147 (1983).
21. Kumar, K., "Nuclear models and the search for unity in nuclear Physics", University of Bergen Press, Pergen (1984).
22. Park, P., Subber, A.R.H., Hamilton, W.D., Elliott, J.P. and Kumar, K., J. Phys. G11 L251 (1985).
23. Subber, A.R.H., Park, P., Hamilton, W.D., Kumar, K., Schreckenbach, K. and Colvin G., "To be published in J. Phys. G (1986).
24. Arima, A. and Iachello, F., Adv. Nucl. Phys. 13 139 (1984).
25. King, W.H., "Isotope Shifts in Atomic Spectra", Plenum Press, New York (1984).
26. Elliott, J.P., Evans, J.A., Park P., Phys. Lett. 169B 309 (1986).

COLLECTIVE PROPERTIES OF THE ^{99}Rh NUCLEUS

D.BUCURESCU, G.CATA, G.CONSTANTINESCU, D.CUTOIU, M.IVASCU and N.V.ZAMFIR

Central Institute of Physics, Bucharest, P.O.Box MG6, ROMANIA

Abstract Lifetimes of some high-spin states in ^{99}Rh are measured with the RD method in the ^{88}Sr(^{14}N,3nγ) reaction. Excitation energies and electromagnetic properties in this nucleus are discussed in the frame of the IBFA model (for positive parity states) and the U(6/2) supersymmetric scheme (for the negative parity states), respectively.

INTRODUCTION

Recent investigations, both experimental and theoretical, confirm that in the even-even Ru and Pd nuclei, a shape transition takes place, from vibrational to deformed nuclei, when the neutron number increases above 50 [1]. A similar transition is also present in the odd-mass Rh isotopes [2]. The wealth of recent data on excited levels in these nuclei, attracted a number of studies [3-5] on the systematics of the nuclear structure in this region, based on the IBA and IBFA models. However, detailed tests of particular questions raised by these models, such as the realization of certain possible dynamical symmetries or supersymmetries in these nuclei, are missing, since electromagnetic transition probabilities are rather little studied experimentally.

In this paper we report on γ-ray transition probabilities measured for both positive and negative parity states in ^{99}Rh. The results are discussed by comparison with IBFA calculations (for positive parity) and with predictions of the U(6/2) supersymmetric scheme (for negative parity).

EXPERIMENTAL RESULTS

High spin states in ^{99}Rh have been populated with the reaction ^{88}Sr(^{14}N,3nγ), performed at the Bucharest FN tandem Van de Graaff accelerator at 54 MeV. Recoil distance measurements have been performed with the apparaturs described in [7], with targets of natural SrO of about 0.1 mg·cm^{-2}. Figure 1 shows the level scheme relevant to the plunger measurements. The measured lifetimes are given for each level. Prompt side feedings have been assumed for the levels and for the highest transition in the cascade an effective lifetime was determined.

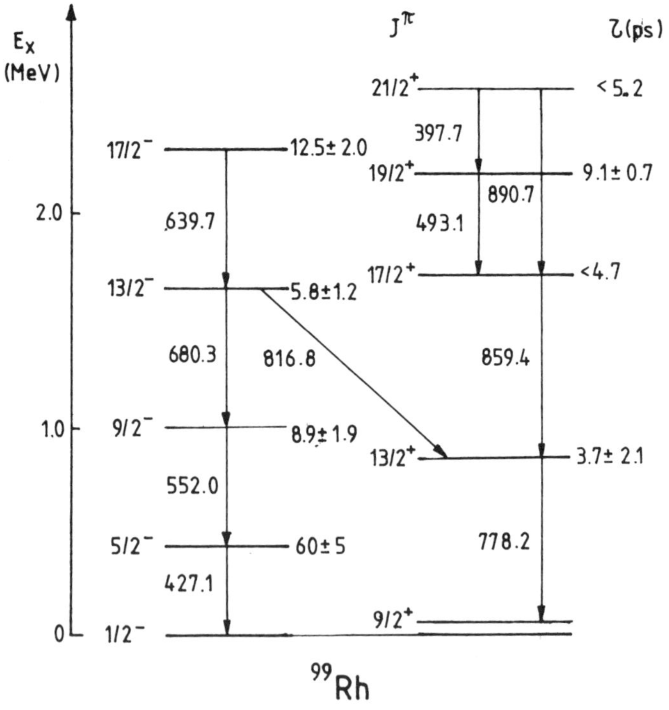

FIGURE 1. Level scheme of ^{99}Rh(ref.[6]).Only the levels relevant to the plunger experiment are shown.

IBFA CALCULATIONS FOR THE POSITIVE-PARITY STATES

In our calculations [3, 4] we treated the Rh isotopes as consisting of the even-even core Ru with a fermion coupled to it, which can occupy either one ($1g_{9/2}$) or two ($1g_{9/2}$ and $2d_{3/2}$) shells. The presence of the second shell generally improves the description. The main results are as follows. For the chain of isotopes ^{97}Rh to ^{107}Rh, the yrast levels up to $21/2^+$ are well reproduced, and also the (j-1) anomaly (inversion of the $7/2^+$ state with the $9/2^+$ one as ground state, in the middle of the chain). The parameters of the boson-fermion interaction terms, found by us on a purely phenomenological basis [3,**4**] are consistent with microscopic estimations. Another result to be emphasized is the importance of the exchange term in the boson-fermion interaction in describing the properties of these Rh isotopes: the positive parity levels could be reasonably described also with just one active shell ($g_{9/2}$) and only the exchange term for the bosonic core-fermion interaction [3], which implies a relatively small role of the quadrupole coupling here. The gamma-ray transition probabilities for the positive parity states agree reasonably well with the theoretical ones [4].

NEGATIVE PARITY STATES AND THE U(6/2) SUPERSYMMETRIC SCHEME

Recently, it has been suggested [8, 9] that the mass region $A \sim 100$ provides experimental evidence for the U(6/2) supersymmetry. The low-lying states in the even-even Ru nuclei are described by the U(5) limit of the IBA, while the negative parity states in the odd-even Rh nuclei are built mainly on the $2p_{1/2}$ shell model orbit. We consider here only the first two members of the $N = 5$ supermultiplet, namely $^{98}_{44}Ru_{54}$ (N=5, M=0) and $^{99}_{45}Rh_{54}$ (N=4, M=1). The experimental energy spectra of these two nuclei are compared with the theoretical spectra in the U(6/2) supersymmetric scheme, in figure 2. The model parameters were obtained by fitting the known experimental levels of ^{98}Ru and ^{99}Rh with n_d = v≤3. A very good representation (ϕ=1%, σ=20 keV, see ref. 9) of these levels has been obtained. In particular the lowering of the levels in ^{99}Rh with respect to the corresponding core states, is nicely reproduced, like in the case of the other Rh isotopes[2, 8].

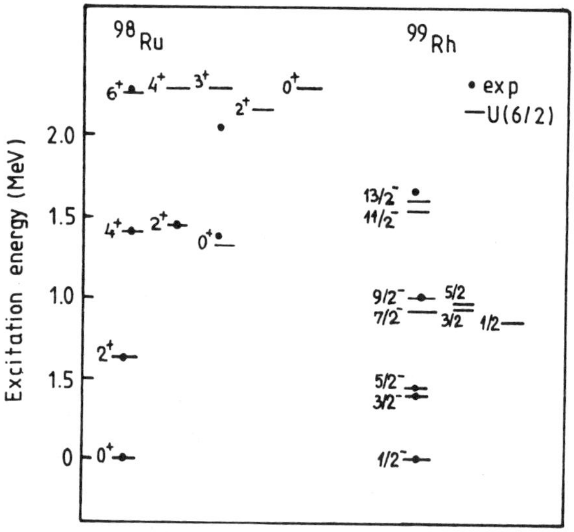

FIGURE 2. Comparison between experimental excitation energies and the U(6/2) supersymmetry predictions in ^{98}Ru-^{99}Rh. The parameters of the calculations (as defined in [8]) are: ε_1=407.1; ε_2=-237.6; α=41.8; β= 5.0; γ_1=-5.2; γ_2=5.0 (all in keV).

The core-particle(weak) coupling model, in its simplest version, predicts that the centroid of the doublets should be close to the energy of the corresponding core states.

In table 1, B(E2) values in ^{98}Ru and ^{99}Rh are compared with the predictions of the U(6/2) scheme (the effective charge: $\tilde{q}_2^2$ = 156 e^2fm^4).

TABLE 1

Nucleus	Transition $J_i^\pi \rightarrow J_f^\pi$	$B(E2;J_i \rightarrow J_f)$ (e^2fm^4) exp	U(6/2)
^{98}Ru	$2_1^+ \rightarrow 0_1^+$	$778 \pm 62^{(a)}$	(778)
	$4_1^+ \rightarrow 2_1^+$	$1078 \pm 122^{(a)}$	1245
	$6_1^+ \rightarrow 4_1^+$		1400
	$2_2^+ \rightarrow 2_1^+$	$1470 \pm 250^{(a)}$	1245
	$2_2^+ \rightarrow 0_1^+$	$13 \pm 2^{(a)}$	0
^{99}Rh	$5/2^- \rightarrow 1/2^-$	938 ± 78	622
	$9/2^- \rightarrow 5/2^-$	1750 ± 370	934
	$13/2^- \rightarrow 9/2^-$	950 ± 200	934

a) ref. [10].

A better judgement of the applicability of the U(6/2) scheme in the present case may be offered by certain ratios of B(E2) values within the supermultiplet. These are shown in table 2, where the predictions of the corresponding geometrical model (vibrational) are also given.

TABLE 2

	Exp.	U(6/2)	Geometrical
$\frac{B(E2;\, 4_1^+ \rightarrow 2_1^+)}{B(E2;\, 2_1^+ \rightarrow 0_1^+)}$	1.39 ± 0.19	1.60	2
$\frac{B(E2;\, 9/2^- \rightarrow 5/2^-)}{B(E2;\, 5/2^- \rightarrow 1/2^-)}$	1.86 ± 0.42	1.50	2
$\frac{B(E2;\, 5/2^- \rightarrow 1/2^-)}{B(E2;\, 2^+ \rightarrow 0^+)}$	1.20 ± 0.14	0.80	1

As it can be seen from this table, while the first two ratios, each one within a certain nucleus, could be considered to favor the algebraic model, the third one, specific to the supersymmetry scheme, indicates a certain degree of its breaking, which was not obvious from the level schemes (figure 2). In other nuclei, also compared with the predictions of the U(6/2) scheme, this ratio was generally found to be lower than 1, fact interpreted as an evidence of a lower collectivity in the odd nucleus due to the number of bosons lower by one unit than in the even-even nucleus[11]. From figure 3 it can be seen that the present data, added to the other existing cases where the $U(6/2) \supset SU(5) \otimes \frac{1}{2}$ scheme has been considered, seem to indicate a systematic departure from this rule when the number N (total number of bosons and fermions) decreases.

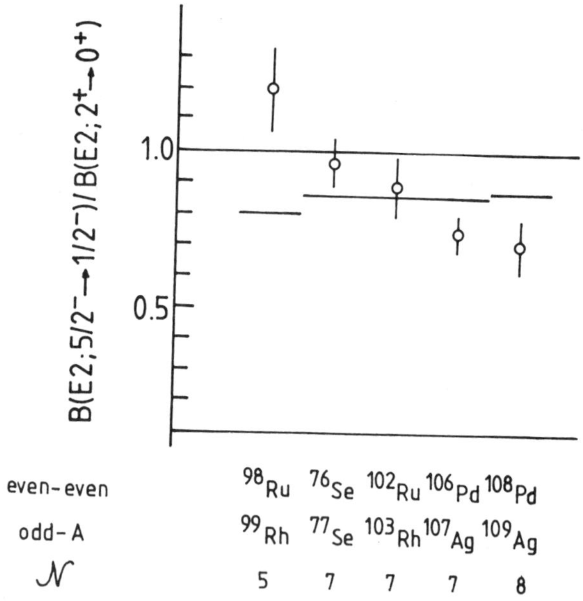

FIGURE 3. Ratio of B(E2)'s in neighbouring nuclei in the supermultiplet, compared with the predictions (lines) of the U(6/2) scheme.

SUMMARY

The properties of the high spin states in the ^{99}Rh have been investigated in the frame of the IBFA model. The positive-parity states were reasonalby described by couplind a fermion in one ($1g_{9/2}$) or two ($1g_{9/2}$ and $2d_{5/2}$) shells to an IBA-1 Ru core. The known negative-parity states are also well described with the U(6/2) supersymmetric scheme; however, their B(E2) properties indicate that ^{99}Rh is a kind of limit of the applicability of this model, where breaking of the supersymmetry appears. Further tests of this would be constituted by the experimental assignment of the rest of low-lying negative parity levels in ^{99}Rh, and by the measurement of B(E2) values of such levels in ^{101}Rh.

REFERENCES

1. J. Stachel, P. Van Isacker and K. Heyde, Phys.Rev. C25, 650 (1982).
2. P. Van Isacker, J. Jolie, K. Heyde, M. Waroquier, J. Moreau and O. Scholten, Phys.Lett. 149B, 26 (1984).
3. D. Bucurescu, G. Cata, G. Constantinescu, D. Cutoiu, M. Ivascu and N.V.Zamfir, Rev.Roum.Phys. 30,181 (1985).
4. D.Bucurescu, G. Cata, G. Constantinescu, D. Cutoiu, M. Ivascu and N.V.Zamfir, Nucl.Phys. A443, 217 (1985).
5. J. Jolie, P. Van Isacker, K. Heyde, J. Moreau, G. Van Landeghem, M. Waroquier and O. Scholten, Nucl.Phys. A438, 15 (1985).
6. G. Kajrys, M. Irshad, S. Landsberger, R. Lecomte, P. Paradis and S. Monaro, Phys.Rev. C26, 138 (1982).
7. D. Bucurescu, G. Constantinescu, M. Ivascu, N.V. Zamfir, and M. Avrigeanu, J.Phys. G7, 399 (1981).
8. R. Bijker and V.K.B. Kota, Ann.Phys. 156, 110 (1984).
9. J. Vervier and R.V.F. Janssens, Phys.Lett. 108B,1 (1982).
10. S. Landsberger, R. Lecomte, R. Paradis and S. Monaro, Phys.Rev. C21, 588 (1980).
11. J. Vervier, Phys. Lett. 133B, 135, (1983).

A HAMILTONIAN WITH BROKEN SUPERSYMMETRY FOR THE XENON ISOTOPES

Peter Reinicke

Physikalisches Institut der Universität Bonn
Nussallee 12, D-5300 Bonn 1, W-Germany

ABSTRACT

A Hamiltonian with broken u(6/2j+1) supersymmetry is proposed that can describe collective excitations in nuclei with high spin, single-j orbitals. Ground-state as well as excitation energies of nine xenon isotopes with both odd and even mass are fairly well described by the so(6) limit of this Hamiltonian with only one set of parameters. Transition elements are discussed shortly, since experimental data are missing.

1. Introduction

A Hamiltonian with broken u(6/2j+1) supersymmetry[1)] is proposed that can be applied to a nuclear region. Its bosonic part is identical with IBA[2)], while the fermionic degrees of freedom are characterized by the seniority quantum number. The physical structure is outlined by an explicit application to the isotopes of xenon where j = 11/2. The supermultiplet scheme of these nuclei is depicted in Fig. 1. Each strip represents a supermultiplet with respect to u(6/12). An asterisk labels the high spin bands of the corresponding nucleus.

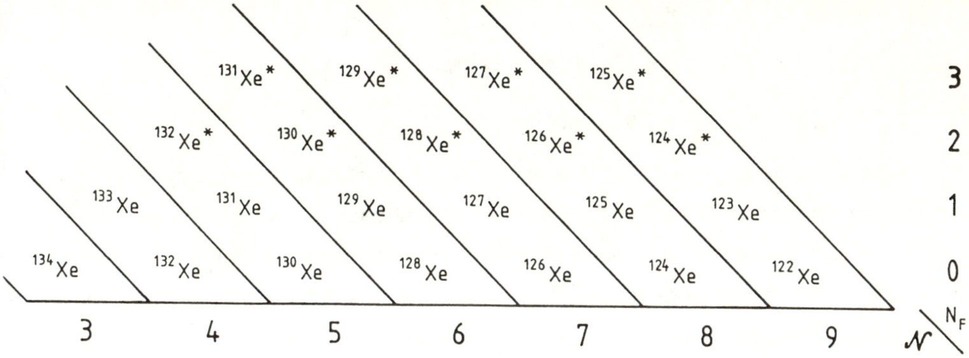

Fig. 1

2. CONSTRUCTION OF THE HAMILTONIAN

Starting from the supersymmetric harmonic oscillator with 6 bosonic and (2j+1) fermionic degrees of freedom,

$$H = \alpha\left(\sum_{\nu=1}^{6} b_\nu^+ b_\nu + \sum_{\rho=1}^{2j+1} a_\rho^+ a_\rho\right) = \alpha(\hat{N}_B + \hat{N}_F) = \alpha\hat{\mathcal{N}}, \quad (1)$$

supermultiplets are defined in terms of the total particle number $\mathcal{N}$, their members, the nuclei, by the quantum numbers N_B and N_F. The degeneracies are partly removed by successive symmetry breaking, through terms which are at most quartic in the creation and annihilation operators and conserve total particle number. In a first step, the symmetry of (1) can be broken without the introduction of interaction terms, namely

$$H = \alpha(\hat{N}_B + \beta\hat{N}_F) \ . \quad (2)$$

This Hamiltonian has an $u_B(6) \times u_F(2j+1)$ symmetry and for $\beta \approx 1$ one can speak of broken supersymmetry.

Then the succeeding breaking steps are achieved by interaction terms. Having in mind that we want to describe single-j orbitals we arrive at the chain

$$u(6/2j+1) \supset u_B(6) \times u_F(2j+1) \supset u_B(6) \times Sp_F(2j+1) \supset u_B(6) \times su_F(2), \quad (3)$$

where now any of the three possible $u_B(6)$ chains of IBA[2] may be inserted. In the so(6) limit which we will focus on the chain continues as follows

$$u_B(6) \times su_F(2) \supset so_B(6) \times su_F(2) \supset so_B(5) \times su_F(2) \supset so_B(3) \times su_F(2) \supset su_{B+F}(2). \tag{4}$$

In principle, the Hamiltonian should contain, for every algebra of the chain, all linearly independent hermitian scalars that conserve total particle number and are at most quartic in the creation and annihilation operators. These invariants are given in ref. 3).

Nevertheless, we will present a simplified ansatz for the Hamiltonian by taking only a sum over Casimir operators

$$H = a_1 + a_2 C_2^{u(6/2j+1)} + a_3 C_1^{u_B(6)} + a_4 C_1^{u_B(6)} + a_5 C_1^{u_F(2j+1)} + a_6 C_2^{u_F(2j+1)} + a_7 C_2^{Sp_F(2j+1)} + a_8 C_2^{so_B(6)} + a_9 C_2^{so_B(5)} + a_{10} C_2^{so_B(3)} + a_{11} C_2^{su_F(2)} + a_{12} C_2^{su_{B+F}(2)} . \tag{5}$$

It should be noticed that the coefficients $a_1,\ldots,a_{12}$ are constants because any dependence on the quantum numbers would mean that higher order terms are covered, too, which is excluded from the beginning. Consequently, there is no reason to restrict the application of H to only one supermultiplet instead of describing a region of nuclei, in our example the isotopes of xenon (see Fig. 1).

3. APPLICATION TO THE XENON REGION

By experimental limitations (for details see ref. 1)), not all constants can be determined independently (e.g., $\nu = N_F$ for all levels). That's why the energy formula has to be rearranged as follows,

$$\begin{aligned} E(N_B,N_F,\sigma,\tau,L_B,L_F,I) = \\ c_0 + c_1 N_B + c_2 N_B^2 + c_3 N_F + c_4 N_F^2 + c_5 N_B N_F \\ + b_1(N_B(N_B+4) - \sigma(\sigma+4)) + b_2 \tau(\tau+3) \\ + b_3(I(I+1) - L_F(L_F+1)) + b_4 L_B(L_B+1) . \end{aligned} \tag{6}$$

The coefficients are separated in two groups, one describing the groundstates, i.e., the negative binding energies, the other describing the excitations. Groundstates (6 parameters for 13 levels) and excited states (4 parameters for about 70 levels) have been fitted separately.

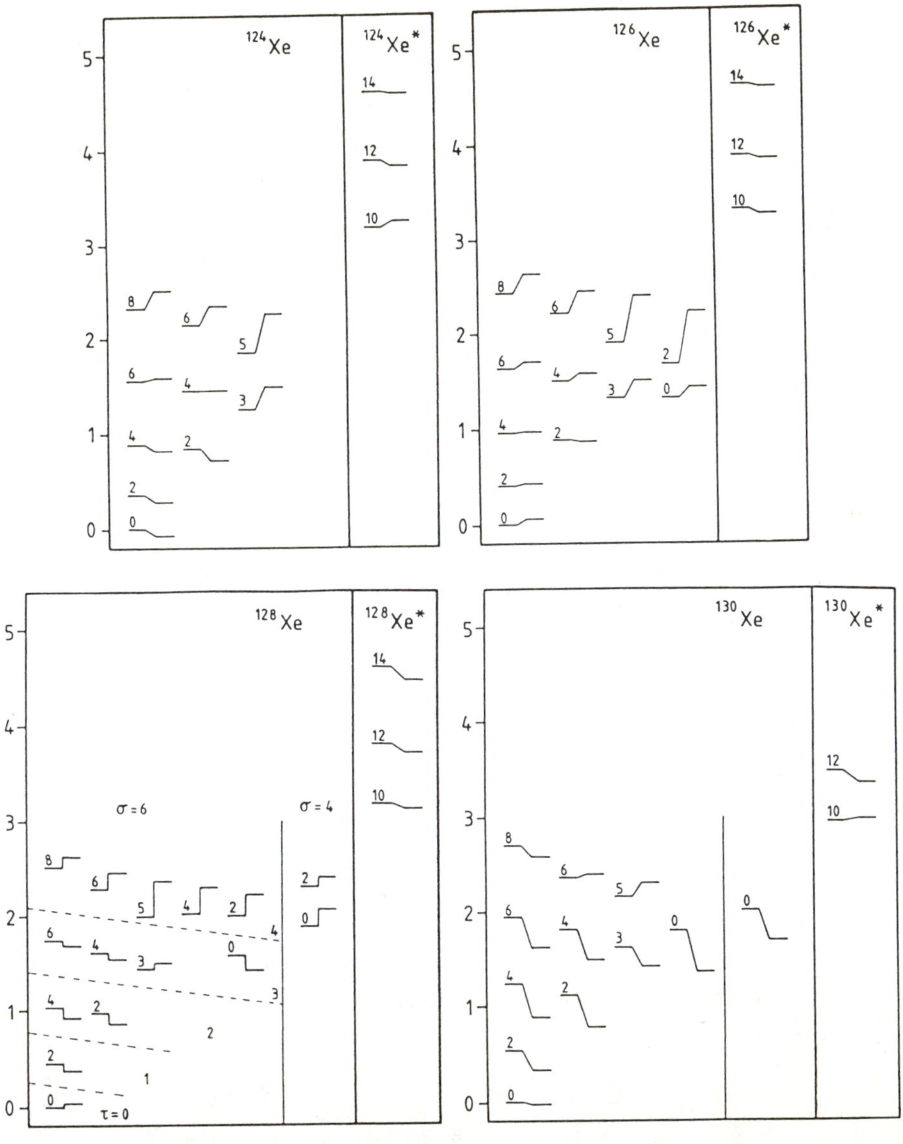

Fig. 2 a

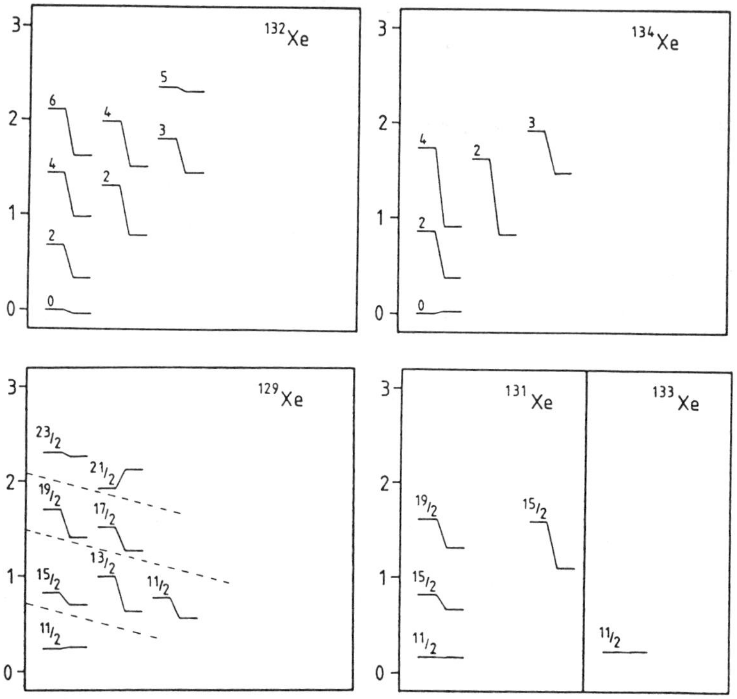

Fig. 2 b

The result is shown in Fig. 2 where energies are given in Mev. Each pair of levels consists of the experimental value on the left and the theoretical one on the right, characterized by the total angular momentum.

From the fit of the groundstates we find $\beta \simeq 0.6$ for the supersymmetry breaking parameter of (2). This means that the splitting ratios are acceptable (cf. fig. 3 of ref. 1)).

4. SOME REMARKS ON MATRIX ELEMENTS

In order to test the wave functions one should compare transition probabilities. In the so(6) limit the electric quadrupole operator is given by[4)]

$$T_\mu^{(E2)} = q_2(d_\mu^+ s + s^+ d_\mu) \ . \qquad (7)$$

Matrix elements for this operator are given in ref. 4).

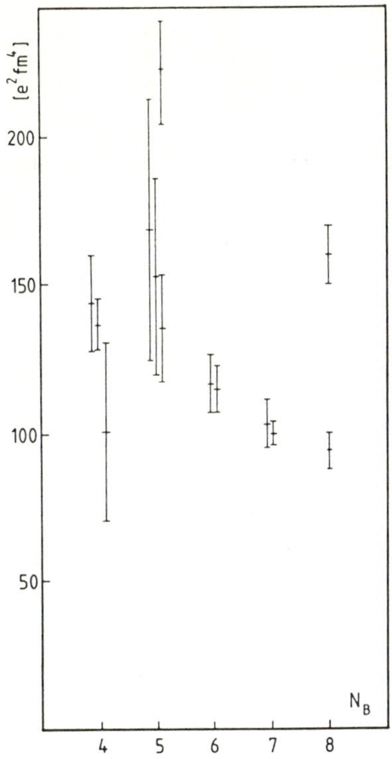

Fig. 3

Fig. 3 depicts experimental data for $B(E2;\ 2_g^+ \rightarrow 0_g^+)\cdot 5/(N_B(N_B+4))$ as a function of N_B. Within the model this quantity should be a constant (= q_2^2). For other E2-transitions (especially for even-odd nuclei and high spin states) data are missing.

Let us now consider nuclear reactions. In view of the fact that there are no data available, we only propose the reaction $^{A}Xe\ (p,\ t)\ ^{A-2}Xe$, which is a good test for the model, since it combines adjacent supermultiplets. Within the model the transition operator for this reaction reads

$$T = \gamma \cdot s^+ \tag{8}$$

(we restrict our considerations to reactions with no angular momentum transfer). The only non vanishing matrix elements are

$$\langle N_B+1,\sigma\pm 1,\tau,(\alpha)|s^+|N_B,\sigma,\tau,(\alpha)\rangle = \pm\, 1/2\left\{(N_B+4\pm x)(x\pm(\tau+2))(x\mp(\tau+1))/(x(x\pm 1))\right\}^{1/2}, \tag{9}$$

where $x=\sigma+2$. This prediction has to be tested experimentally.

REFERENCES

1) M. Baake, P. Reinicke, and A. Gelberg, Phys. Lett. 166B, 10 (1986)

2) A. Arima and F. Iachello, Adv. Nucl. Phys. 13, 139 (1984)

3) M. Baake and P. Reinicke, J. Math. Phys. 27, 1430 (1986)

4) A. Arima and F. Iachello, Ann. Phys. 123,468 (1979)

TRANSFER IN THE Ba REGION AND THE U(6/4) MODEL

M. VERGNES[+], G. BERRIER-RONSIN[+], G. ROTBARD[+],
J. VERNOTTE[+] and R. TAMISIER[+*]
+ Institut de Physique Nucléaire, 91406 Orsay, France
* Institut de Physique de Nantes, 44072 Nantes, France

Abstract Following a recent suggestion that the Ba region is even better than the Pt region for the realization of the O(6) symmetry, we have investigated the possibility to find in this area even-even and odd-A partners reasonnably described by the U(6/4) model. There are indications of a possible realization of the U(6/4) supersymmetry in the ^{134,135,136}Ba and perhaps in the ^{132,133,134}Xe isotopes.

For quite a long time, the textbook example[1] for an O(6) symmetry in nuclei has been ^{196}Pt. Quite naturally, when the interacting boson model was extended to odd-A nuclei, the candidates for the U(6/4) supersymmetry[2], with an O(6) core, were researched and found[2-6] in the neighbourhood of ^{196}Pt. Many tests with proton transfer reactions were also performed for Os, Ir, Pt, Au and Hg nuclei ($76 \leqslant Z \leqslant 80$), with good or reasonnable results for several final odd-A nuclei[6] and appreciable breakings for final even-even nuclei[6]. Because a neutron transfer reaction on the Ba isotopes ($76 \leqslant N \leqslant 80$) should be quite similar to the above mentioned reactions, a few of them were performed at the same stage[7].

It has been suggested recently[8] that the even-even nuclei in the $A \approx 130$ mass region, and particularly the Ba isotopes, were even better examples of the realization of the O(6) symmetry than the Pt nuclei. This can be verified quite easily by simply applying the analytic formula[2] for the energies of the levels :

$$E_{even} = -\frac{1}{4} A \left[\sigma(\sigma + 4)\right] + \frac{1}{6} B \left[\tau(\tau+3)\right] + C \left[L(L+1)\right] \quad (1)$$

where the quantities A, B and C are parameters and σ and τ are the 0(6) quantum numbers. The quality of the agreement can be roughly estimated using the quantity :

$$R_E = \sum_i \left| E_i^{exp.} - E_i^{th.} \right| \Big/ \sum_i E_i^{exp.} \quad (2)$$

The results given in Table I concern the even-even nuclei : $^{132,134,136}Ba$ and $^{130,132,134}Xe$, comparable to Os, Pt and Hg when exchanging N and Z. The method of fit was derived from the one used in Fig. 1 of ref. 8 : the level schemes were compared to the scaled ^{196}Pt level scheme. A gross general agreement was obtained with a common scaling factor of : x 1.8. The individual values of R_E were then deduced from a comparison between the experimental levels and the ones calculated using for ^{196}Pt the parameters of ref. 1.

Although the R_E values for the even-even Xe and Ba nuclei are appreciably larger than the ones previously obtained in the Pt region (shown for comparison in Table I), they can be considered as marginally acceptable for our purpose, in view of the single set of parameters used.

It has always been claimed, and widely accepted, that supersymmetry can only occur in very limited regions of the nuclear chart. It does not appear unreasonnable to look, in the Xe, Ba, "0(6) region" discussed above, for a few odd-A nuclei that could be described by the U(6/4) model, therefore with the same A, B and C parameters, using the energy formula[2,6] corresponding to the U(6/4) SUSY.

TABLE I Fits of the level schemes for even-even and odd-A nuclei (see text)

Nuclei	Num.of excit. levels fitted	R_E(%)	Total number of excit. levels fitted	R_E(%)	Parameters A	$B' = \frac{1}{6}B$	C	Ref
^{132}Ba	6	20.8			333	77	41	a
^{134}Ba	7	13.1						
^{135}Ba	5	7.1	18	14.7	333	77	41	a
^{136}Ba	6	19						
^{130}Xe	7	18.9			333	77	41	a
^{132}Xe	6	13.8						
^{133}Xe	7	11.6	17	14.3	333	77	41	a
^{134}Xe	4	17.6						
^{190}Os	7	11.8	22	19		40	10	b
^{191}Ir	8	22.7						
^{192}Os	7	10.4	15	18.3		40	10	b
^{193}Ir	8	30.8						

a) Analysis done in the present work.
b) Ref. 3.

$$E_{odd} = -\frac{1}{4}A\left[\sigma_1(\sigma_1+4)+\sigma_2(\sigma_2+2)+\sigma_3^2\right]+\frac{1}{6}B\left[\tau_1(\tau_1+3)+\tau_2(\tau_2+1)\right]+CJ(J+1) \quad (3)$$

A quite good agreement is indeed obtained for the nuclei ^{133}Xe and ^{135}Ba, the R_E values shown in Table I being small. The general agreement for $^{134,135,136}Ba$ and $^{132,133,134}Xe$ is not worse than the one shown for accepted supersymmetric U(6/4) partners in the Pt region.

The good agreement obtained for ^{135}Ba implies that the identification of the $3/2_2^+$ and $3/2_3^+$ levels of the U(6/4) model to the $3/2^+$ levels experimentally observed at 855 and 979 keV is correct. Although this identification is strongly suggested by the closeness (see Table II) of the experimental and calculated energies, it should be remarked that there is a lower $3/2^+$ level, experimentally observed at 588 keV, which is considered here as an intruder, not belonging to the U(6/4) space. The suggested identification should clearly be verified. The pick-up reaction $^{136}Ba(d,t)^{135}Ba$ appears as particularly suited for this purpose, because only the $3/2_1^+$ and $3/2_3^+$ levels should be observed, the population of the $3/2_2^+$ level being forbidden[3-6].

Our data[7], obtained at 25.2 MeV using the Orsay tandem accelerator and a split pole spectrometer, have therefore been analyzed using DWBA calculations performed with the parameters already used[9] for the same reaction at a similar energy, on the N = 82 isotones. The spectroscopic factors of interest for the present discussion are presented in Table II.

Simply by looking at column 5, it appears that the only levels strongly populated by $l = 2$ (d3/2) transfer are - in agreement with our suggested identification - the $3/2^+$ levels at 0 and 979 keV. The comparison of columns 4 and 5 shows moreover that there is a rather good agreement with the U(6/4) intensity predictions.

The main data extracted from the analysis of the reaction $^{135}Ba(d,t)^{134}Ba$ are shown in Table III and compared to the predictions of the U(6/4) model. Although the intensity branching between the 0_1^+ and 2_1^+ levels is not very well reproduced, it should be pointed out that the ($l=2$, $d_{3/2}$) transition to the 2_2^+ level is indeed forbidden as predicted.

TABLE II $^{136}Ba(d,t)^{135}Ba$

$E_{exp}(keV)^{a}$	J^{π} a	$E_{SUSY}(keV)^{b}$	C^2S_{SUSY} c l=2, d3/2	C^2S_{exp} l=2	C^2S_{exp} l≠2
0	$3/2^+_1$	0	100^+	100^+	
221	$1/2^+_1$	262			48 l=0
268	$11/2^-_1$				190 l=5
480	$5/2^+_1$	590		≲ 4	
588	$3/2^+_d$			≲ 1	
855	$3/2^+_{2d}$	924	0(F)	3.3	
874	$7/2^+_1$	877			~ 0 (l=4)
979	$3/2^+_{3d}$	999	50	80.9	

+ : normalized to 100
a) : ref. 10
b) : calculated with the parameters of Table I :
A = 333, B′ = 77, C = 41
c) : see ref. 3 to 6. The letter F means : forbidden
d) : see discussion in the text

In summary, it seems that there are indications for a zone of realization of the U(6/4) supersymmetry in the $^{134,135,136}Ba$ region. The same could be true also for $^{132,133,134}Xe$. More work is clearly necessary, however, before a clear conclusion can be drawn.

TABLE III $^{135}Ba(d,t)^{134}Ba$ reaction

$E_{exp}(keV)^{a}$	J^{π} a	$E_{SUSY}(keV)^{b}$	C^2S_{SUSY} c l=2, d3/2	$C^2 S_{exp}$ l=2	C^2S_{exp} l≠2 or j≠ 3/2
0	0^+_1	0	100+	100+	
605	2^+_1	554	180	60	37 l=0
1168	2^+_2	1016	0(F)	< 7	27 l=0
1400	4^+_1	1590			15 l=2, j=5/2

+ : normalized to 100
a) : ref. II
b) : calculated with formula (1) and the parameters A = 333, B′ = 77, C = 41
c) : see ref. 3 - 6. The letter F means : forbidden

REFERENCES

1. J.A. Cizewski, R.F. Casten, G.J. Smith, M.L. Stelts, W.R. Kane, H.G. Borner and W.F. Davidson, Phys. Rev. Letters, 40, 167 (1978)
2. F. Iachello, Phys. Rev. Letters, 44, 772 (1980)
3. F. Iachello, S. Kuyucak, Ann. Phys. (N.Y.) 136, 19 (1981)
   A.B. Balantekin, I. Bars and F. Iachello, Nucl. Phys., A370, 284 (1981) and Phys. Rev. Letters, 47, 19 (1981)
4. J. Wood, Phys. Rev., C24, 1788 (1981)
   J.A. Cizewski, Proc. 2nd seminar on Interacting Bose-Fermi Systems in Nuclei, edited by F. Iachello (Plenum, New-York, 1981) p389
5. J. Vervier, Phys. Letters, 100B, 383 (1981)
6. M. Vergnes, Interacting Boson-Boson and Boson-Fermion Systems, edited by O. Scholten (World Scientific ; Singapore, 1984) p. 91-102 and references therein
7. M. Vergnes, G. Rotbard, G. Berrier-Ronsin, J. Vernotte, R. Tamisier, 1982 (unpublished)
8. R.F. Casten and P. Von Brentano, Phys. Letters, 152B, 22 (1985)

9. G. Berrier, M. Vergnes, G. Rotbard, J. Kalifa, Journal de Physique, 37, 311 (1976)
10. E.A. Henry, Nuclear Data Sheets, 14, 191 (1975)
11. Y.V. Sergeenkov, V.M. Sigalov, Nuclear Data Sheets, 34, 475 (1981)

RAPID SHAPE TRANSITION IN PRASEODYMIUM NUCLEI

EMILIAN DRĂGULESCU, MARIN IVASCU, MANUELA IONESCU-BUJOR, ALEXANDRU IORDACHESCU, COSTEL PETRACHE, DOREL POPESCU, GHEORGHE PASCOVICI, GHEORGHE SEMENESCU, ION GURGU, FLORIN BACIU
Central Institute of Physics, P.O.Box MG-6, R-76900, Bucharest, ROMANIA
and
VLADIMIR PAAR, SLOBODAN BRANT, DARIO VRETENAR
Faculty of Science, University of Zagreb, YU-4100, Zagreb, YUGOSLAVIA

Abstract. By means of in-beam γ-spectroscopy, we have experimentally studied the high spin states in odd-even and odd-odd praseodymium nuclei: $^{133,135,137,139}Pr$ and $^{134,136,138}Pr$. New information on high spin states has been obtained for all isotopes. We observed an abrupt shape transition (O(6) → SU(3)) in sequence of ^{135}Pr to ^{133}Pr nuclei. The calculation of the energy levels in odd-even praseodymium has been performed in the SU(6) particle-vibrational model (PTQM).

1. INTRODUCTION

In recent years there is a considerable interest to study the collective properties of odd-proton nuclei in the Z>50 transition region. This work presents a systematic study undertaken to investigate the nuclear structure of neutron-deficient praseodimium nuclei (Z = 59). These nuclei belong to a region of rapid transition from nearly spherical nuclei to nuclei quite soft towards deformation.[1-9]

The main aim of the present investigation is to perform a systematic study of Pr nuclei below neutron number 82 in order to learn about the shape transition under the influence of high-j orbitals.

A small amount of information has been gathered concerning Pr nuclei. The level structure of these nuclei has been previously studied from β - decay and in-beam γ - ray spectroscopy measurements.[10-15] As far as we know there are no high spin data in the literature concerning the odd-odd Pr nuclei. The experimental data on high spin states are scarce and sometimes not consistent.

We have experimentally studied the odd-even $^{133,135,137,139}Pr$ and odd-odd $^{134,136,138}Pr$ nuclei produced in heavy ion reactions by means of in-beam γ - ray spectroscopy methods. The population of non-yrast states offers the possibility to compare experimental data with the predictions of theoretical models.

2. EXPERIMENTAL PROCEDURE

The high spin states in Pr nuclei were produced by bombarding enriched $^{118-124}Sn$, $^{121,123}Sb$, $^{125-130}Te$, $^{136,138}Ce$ targets with ^{19}F, ^{16}O, ^{14}N, proton beams, respectively. The heavy-ion beams were delivered by Bucharest FN-Tandem coupled to the ALIGATOR postaccelerating system. The ALIGATOR rf postaccelerator uses individually phased spiral-loaded cavities operated at room temperature.[16] In these experiments, different high efficiency and resolution Ge(Li) detectors were used.

The experimental information was deduced from γ - ray singles spectra as well as from excitation functions, γγ(t) coincidences, Xγ(t) coincidences, γt spectra and angular distributions.

3. EXPERIMENTAL RESULTS AND DISCUSSION

The construction of level schemes was obtained from γ - γ coincidence data, intensity balance and informations from γ - ray excitation and population in different reactions. The experimental results extracted from the above measurements, include excitation energies, J^{π} assignements and lifetimes attached to the states.

In the present experiments, we have confirmed the previous results concerning the low-spin states. New information on high spin states is obtained as a result of the higher angular momenta available in these experiments.

3.1. The odd-even Pr nuclei

The first lowest states in odd proton Z=59 Praseodymium nuclei are single quasiparticle states $d_{5/2}$, $g_{7/2}$ and $h_{11/2}$ The energy level variation of these states is shown in Figure 1.

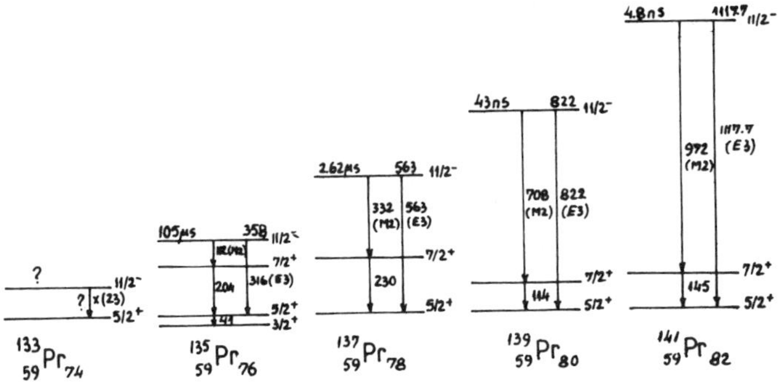

FIGURE 1. Energy systematics of the $5/2_1^+$, $7/2_1^+$ and $11/2_1^-$ states and their transitions in odd-even Pr nuclei

In some of odd-even Pr nuclei the $h_{11/2}$ state is isomeric and decays by M2 and E3 transitions to the $7/2^+$ and $5/2^+$ levels. In ^{137}Pr the 563.0 keV $11/2^-$ state ($T_{1/2}$=2.66 μs) decays by 332 keV M2 to 230 keV $7/2^+$ state and by 563 keV E3 transition to $5/2^+$ state.[23] In Figure 1 the general trends of such a representation show as an increasing deformation of these nuclei with decreasing A.

Several completely new features have been observed in the present study of ^{133,135,137}Pr nuclei, namely (i) a series of $\Delta J = 2$ bands built on $11/2^-_1$, $5/2^+_1$ and $7/2^+_1$ states, (ii) a series of 3-quasiparticle states with different structure of configurations and parities.

The population strength of the positive bands relative to that of $11/2^-_1$ band, in odd-even Pr nuclei, is small for lower mass cases, but increases considerably with increasing mass number. This happens primarly because the excitation energy of $11/2^-_1$ level is increasing, thus becoming less yrast in the higher mass isotopes.

It is interesting to compare the level structure of odd-even Pr nuclei with that of neighbouring even-even Ce nuclei.[17-19] As it results from Figures 2 and 4 the bands built on the $11/2^-_1$, $5/2^+_1$ and $7/2^+_1$ states are interpreted as Coriolis-decoupled bands where an $h_{11/2}$, $d_{5/2}$ and $g_{7/2}$ proton is decoupled from the prolate core. The level correspondence shown in Figure 2 for ^{135}Pr → ^{133}Pr sequence exhibit an almost shape transition from O(6) to SU(3), otherwise supported by the summed γ - γ coincidence spectra for gates sets on the $\Delta J = 2$ transitions in both nuclei (see Figure 3). This shape transition was predicted by Leander and Möller.[4]

All above structures observed in these odd-even Pr nuclei have been successfully explained in the framework of the SU(6) particle-vibtational model (PTQM).[22,23,24] For example, in Figure 5 in the case of ^{135}Pr nucleus earlier calculated by different authors[13,14,20,21] in the framework of triaxial model, it is presented our PTQM calculation concerning the positive-parity states.

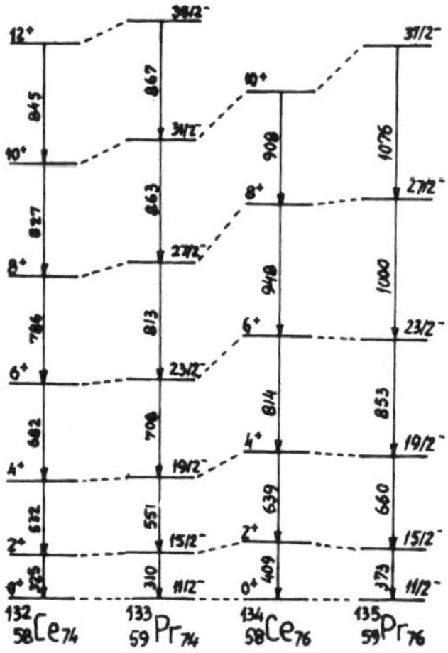

FIGURE 2. Comparison of the $h_{11/2}$ bands in $^{133,135}Pr$ nuclei resulting from the present study with the g.s.b. in $^{132,134}Ce$ nuclei.[17,19]

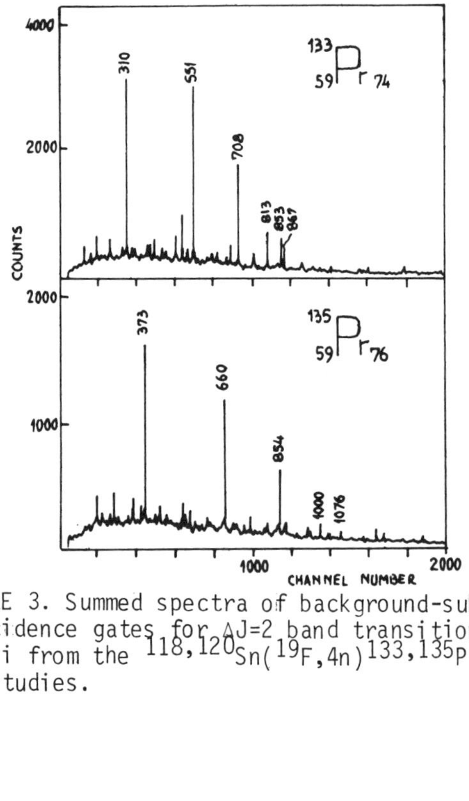

FIGURE 3. Summed spectra of background-substracted γ-γ coincidence gates for ΔJ=2 band transitions in $^{133,135}Pr$ nuclei from the $^{118,120}Sn(^{19}F,4n)^{133,135}Pr$ reactions in our studies.

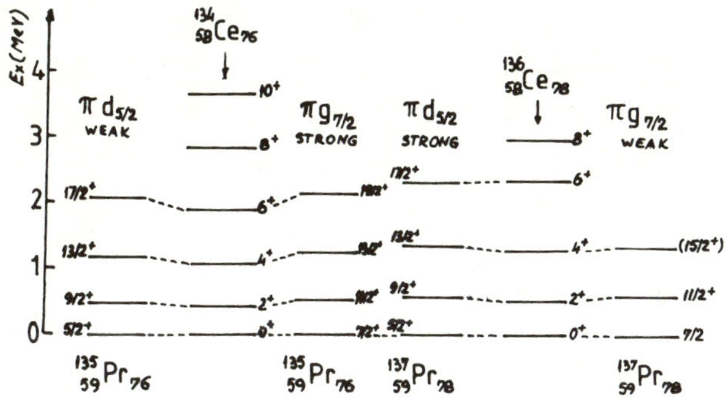

FIGURE 4. Comparison of the $d_{5/2}$ and $g_{7/2}$ bands in ^{135,137}Pr nuclei from the present studies [23,24] with the g.s.b. in ^{134,136}Ce nuclei. [18,19]

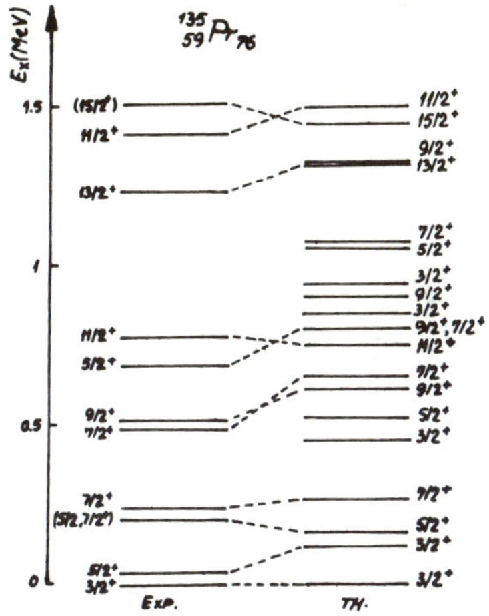

FIGURE 5. Calculated (in PTQM) and positive-parity experimental spectrum of ^{135}Pr resulting from the present study.

A further evicence in favour of one-quasiproton two-quasineutron configurations assigned to 3qp bands in odd-Z nuclei is their decay by $\Delta J=1$ transitions with small mixing ratios δ.($|\delta| < 0.1$). [23] It is demonstrated in ref. [25] that if a pair of neutrons (or protons, respectively) with large alignment i is added to a proton (or neutron) with a relatively large K the M1 transitional moment is strongly enhanced as manifested by the occurence of $\Delta J=1$ transitions with small $|\delta|$. This type of M1 enhancement was first observed by Funke et al.[26] in ^{81}Kr.

3.2. The odd-odd Pr nuclei

From previous studies and our experiments in odd-odd Pr nuclei we have observed that the level schemes of the odd-odd nuclei become much simpler when they are produced by heavy ion reactions: only one or two $\Delta J=1$ bands are evidenced in these experiments. This simplicity is illustrated in Figure 6 for ^{136}Pr. The nuclear structure of odd-odd Pr nuclei can

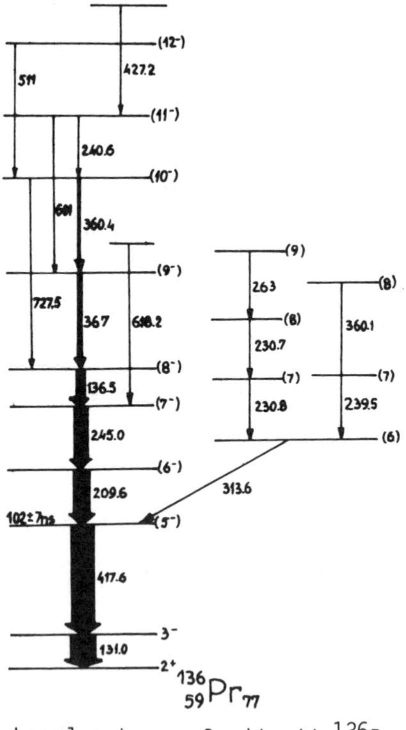

FIGURE 6. Level scheme of odd-odd ^{136}Pr nucleus

be explained by coupling of $d_{3/2}$, $3s_{1/2}$ and $h_{11/2}$ neutron to proton of $d_{5/2}$ or $g_{7/2}$ configurations. Each of these configurations gives rise to a multiplet.

4. CONCLUDING REMARKS

Our experimental studies on Pr nuclei have evidenced new aspects in nuclear structure in this region of nuclei. Experimental data suggest a strong similarity with the nuclei from Pt region. The Pr nuclei (odd-even, and odd-odd) and neighbouring nuclei are good candidates for supersymmetry (SUSY). It appears to be studied the role of triaxiality in this region induced by the competing quasiparticles (protons): $h_{11/2}$, $d_{5/2}$ and $g_{7/2}$. The measurements of quadrupole and dipole moments in Pr nuclei is of great importance.

REFERENCES

1. R. K. Sheline, T. Sikeland, and R. N. Chanda, Phys.Rev. Lett., 7, 446 (1961).
2. D. A. Arseniev, A. Sobiczewski, and V. G. Soloviev, Nucl.Phys., A126, 15 (1969).
3. D. A. Habs, H. Klewe - Nebenius, K. Wisshak, R. Löhken, G. Nowicki, and H. Rebel, Z.Phys., 267, 149 (1974).
4. G. A. Leander and P. Möller, Phys.Lett., 110B, 17 (1982).
5. I. Ragnarson and R. K. Sheline, Physica Scripta, 29, 385 (1984).
6. R. Bengtsson, P. Möller, J. R. Nix, and Jing-Je Zheng, Physica Scripta, 29, 402 (1984).
7. E. Hammaren, K. W. Schmid, F. Grünner, A. Faessler, and B. Fladt, Nucl.Phys., A437, 1 (1985)
8. C. J. Lister et al., Phys.Rev.Lett., 55, 810 (1985).
9. Y. S. Chen, S. Frauendorf, and G. A. Leander, Phys.Rev., C28, 2437 (1983).
10. K. Wisshak, H. Klewe - Nebenius, D. Habs, H. Faust, G. Nowicki, and H. Rebel, Nucl.Phys., A247, 59 (1975).
11. H. Klewe - Nebenius, D. Habs, K. Wisshak, H. Faust, G. Nowicki, S. Goring, H. Rebel, G. Schatz and M. Schwall, Nucl.Phys., A240, 137 (1975).
12. V. Barci, A. Gizon, J. Gizon, and G. Marguier, J.Phys. Soc. of Japan, 49, 1665 (1980).

13. M. Piparinen, M. Kortelahti, A. Pakkanen, T. Komppa, and R. Komm, Nucl. Phys., A342, 57 (1980).
14. M. Kortelahti, R. Juliu, J. Hattula, A. Pakkanen and M. Piparinen, Z.Phys., A321, 417 (1985).
15. C. Erkström, S. Ingelmann, M. Olsmats, and B. Wannberg, Nucl. Phys., A196, 178 (1972).
16. V. Ceauşescu, S. Dobrescu, M. Duma, G. Indreaş, M. Ivaşcu, S. Papureanu, G. Pascovici, and G. Semenescu, Nucl.Instr. and Meth., A244, 295 (1986).
17. A. Kirwan et al., Daresburry Annual Report 1984/1985, 41, (1985).
18. M. Müller - Veggian, Y. Gono, R. M. Lieder, A. Neskakis, and C. Mayer - Böricke, Nucl.Phys., A304, 1 (1978)
19. M. Müller - Veggian, H. Beuscher, D. R. Haenni, R. M. Lieder, and A. Neskakis, Nucl.Phys., A417, 189 (1984).
20. J. Meyer - Ter - Vehn, Nucl.Phys., A249, 111, 141 (1975).
21. H. Toki and A. Faessler, Nucl. Phys., A253, 231 (1975).
22. V. Paar, S. Brant, L. F. Canto, G. Leander, and V. Vouk, Nucl.Phys., A378, 41 (1982).
23. E. Dragulescu, M. Ivaşcu, M.Ionescu-Bujor, A. Iordachescu, C. Petrache, D. Popescu, G. Pascovici, G. Semenescu, I. Gurgu, V. Paar, and S. Brant, to be published.
24. E. Dragulescu, M. Ivascu, I. Gurgu, F. Baciu, C. Petrache, D. Popescu, G. Semenescu, and V. Paar, S. Brant, to be published.
25. F. Donau and S. Frauendorf in Proc. Int. Conf. on high spin properties of Nuclei, Oak Ridge 1982, edited by N. R. Johnson (Harwood Nucl. Sci. Res. Ser., vol. 4), p. 69.
26. L. Funke, F. Donau, J. Döring, P. Kemmitz , E. Will, G. Winter, L. Hildingson, A. Johnson, and T. Lindblad, Phys.Lett. 120B, 301 (1983).

^{168}Er: PARTICLE TRANSFER AND SYMMETRY TESTS

D.G. BURKE[a], J.A. CIZEWSKI[b] and W.F. DAVIDSON[c]

a: McMaster University, Hamilton, Ontario, Canada L8S4K1 and CERN CH-1211, Geneva 23, Switzerland

b: A.W. Wright Nuclear Structure Laboratory Yale University, New Haven, CT 06511 USA

c: National Research Council of Canada Ottawa, Ontario, Canada K1A0R6

Abstract: The successes of the IBA and quasiparticle-phonon model of Soloviev in describing the large body of recently-published single- and two-nucleon transfer data for ^{168}Er are summarized. The Soloviev model is more useful for describing the strengths and for classifying the levels observed in these studies. The IBA is less useful since only 3 of the 18 bands observed up to 2 MeV are within the framework of the model with s and d bosons only. These are the ground state (0_1^+), and the 2_1^+ and 0_2^+ bands. Furthermore, the IBA does not give predictions for the large number of single-nucleon transfer strengths.

Several years ago, a detailed study of the ^{167}Er(n,γ) reaction led to the establishment of 20 rotational bands in ^{168}Er. The unprecedented detail of level energies and radiative decay modes[1] led to a large number of theoretical interpretations based on the traditional collective model[2-4], the shell model[5], the quasiparticle-phonon model of Soloviev[6-8], a microscopic analysis[9], and the interacting boson (IBA) model[10-12]. Of these treatments, most attention has been focussed on the IBA, particularly following the success[10] of this model in explaining the level energies and gamma-ray branching ratios of collective states in ^{168}Er.

Earlier tests of the various interpretations[2-12] have been limited in that comparisons with data have been based almost exclusively on the excitation energies and branching ratios reported in ref. 1. What has been conspicuously missing is a detailed knowledge of the microscopic structure of the ^{168}Er levels, which is of crucial importance in distinguishing among the conflicting interpretations.

Given the widespread interest in the ^{168}Er level scheme, some new measurements have been made[13-17] to provide data that may test the various theoretical predictions. Firstly, further analyses of the original neutron capture data, together with some additional measurements[13-14] of portions of the capture gamma-ray spectrum, have now increased the number of assigned rotational bands to 36. Secondly, an extensive program of single- and two-nucleon transfer reaction studies was initiated to provide additional microscopic nuclear structure information. Measurements of (p,t) and (t,p) reactions[15], (d,p) and (t,d) reactions[16] and (t,α) reactions[17] have now been completed. This work reviews the main findings that are pertinent to tests of nuclear models, in particular the IBA and Soloviev models. To summarize the results, all known rotational bands up to 2 MeV are listed in Table 1. The detailed comparisons of predicted and observed strengths cannot be presented in the space available here, but are available in the published literature[15-17].

All models which have been used to describe ^{168}Er have a ground state (0_1^+) band, and low-lying 2_1^+ and 0_2^+ rotational bands which are similar to the beta and gamma bands in a geometrical picture. Because of the success of the IBA model in predicting the enhanced transitions between the 0_2^+ and 2_1^+ bands, and because of the plethora of bands predicted by this model, it is especially important to investigate further the structures of higher-lying bands to determine whether their properties are consistent with IBA predictions.

The largest body of experimental data which could be analysed in terms of spectroscopic strengths exists for the single-nucleon transfer reactions; for the 18 bands listed in Table 1 there are over 75 (d,p) and (t,α) strengths which have been considered. These results were discussed[16,17] in terms of the quasiparticle-phonon model of Soloviev[6] because it is the only model for which quantitative predictions of the single-nucleon transfer strengths could be obtained. The detailed agreement between the experimental results and the predictions is very good, especially when one considers that the predictions were published before the measurements were made, and thus no optimization of parameters was involved.

Some of the bands listed in Table 1 are predominantly two-quasiparticle configurations which could be identified from the reaction data, and for these the Soloviev predictions are excellent. Others are collective bands for which the microscopic compositions include two-quasiparticle components that can be populated by the single-nucleon transfer reactions. In general, there exists good qualitative agreement between the observed and predicted admixtures for such collective states, and in many cases the agreement is quantitative.

Unfortunately, the IBA has not yet been developed to the stage where predictions of single-nucleon transfer strengths can be made from this model. Although some interesting progress has been reported in this direction using an extended IBA model[18], further development is awaited.

In the framework of the IBA, ^{168}Er has been described[10] as a nucleus near the SU(3) limiting symmetry. In this description a correspondence can be made between the SU(3) bands and the more familiar beta, gamma, and multiphonon excitations of a geometrical description[19]. Labelling the IBA bands in terms of geometrical phonons is convenient for semantic reasons. However, one has to be aware that there is not an exact correspondence in structure between, for example, the 0_2^+ band of the SU(3) limit of the

IBA and a beta vibration of a geometrical model. The significance of this distinction is more important when one is discussing ^{168}Er, which is not a pure SU(3) nucleus.

In the perturbed SU(3) calculations[10] using only s and d bosons for ^{168}Er, the wave functions of states in the 0_3^+, 0_4^+, and 2_2^+ bands have complex structures, but the dominant components in these wave functions[15] have quantum numbers for which direct population in single- or two-nucleon transfer reactions is forbidden. Therefore, the populations can occur only by multistep processes in the reaction, or by the relatively small components in the wave function that can be fed by allowed transitions.

These expectations of the IBA model with s and d bosons only are contradicted by the (t,p) and (t,α) data[15,17]. The (t,p) strengths to the "multiphonon" 0_3^+ and 2_2^+ states are much stronger than the IBA predictions[15] and are also too strong to be attributed to multistep processes in the reaction mechanism. Similarly, the 0_4^+ state at 1833 keV is predicted by the IBA to be predominantly "multiphonon" in character but is found to have significant (t,p) strength[15] and a large (t,α) strength[17]. The (t,α) data show that this band contains a 25% admixture of the {1/2+[411]-1/2+[411]} two-quasiproton configuration. Therefore, the IBA in its simplest form, with s and d bosons only, does not properly describe the 0^+ and 2^+ bands which are higher than the so-called beta and gamma bands.

Recent IBA calculations[12] in which s, d, and g bosons are included have been able to reproduce the observed (t,p) strengths to the 0_3^+, 0_4^+ and 2_2^+ states, and also part of the strength for the $IK^\pi=43^+$ state at 1736 keV. The latter is completely outside the framework of the IBA with s and d bosons only. It is therefore clear that the IBA must be extended to include at least g bosons in order to explain the properties of these states. Since single-nucleon transfer strengths have not yet been calculated it is still not yet known whether the s,d,g boson model could explain the large two-quasiproton admixture in the 0_4^+ band. On the other hand, the Soloviev model predicts this {1/2+[411]-1/2+[411]} admixture to be 30%, in very good agreement with the observed value of 25%. Unfortunately, the (t,p) strengths predicted by the Soloviev model are not yet available. The 0_2^+, 0_3^+, 0_4^+, and 2_2^+ states are all predicted to have single-phonon character in this model, and therefore allowed populations could follow naturally, but it would be of interest to know the predicted strengths.

The successes of the IBA and Soloviev models in reproducing the particle-transfer strengths are summarized in Table 1. The 0_1^+, 0_2^+, and 2_1^+ bands are explained by the IBA with s and d bosons. The 0_3^+, 0_4^+, 2_2^+ bands and the K=3 band at 1736 keV require that g bosons be included in the IBA. The ten negative parity bands listed in Table 1 would obviously require the addition of negative parity bosons. In contrast, the Soloviev model is successful in describing essentially all of the bands in Table 1.

In conclusion, the Soloviev model is more useful for classifying the levels and for providing insight into the detailed structure of ^{168}Er observed in the large body of particle-transfer data. The IBA is less useful because a large fraction of the observed levels are outside the framework of the model with s and bosons only. Another serious shortcoming of the IBA at its present stage of development is that it does not provide microscopic wave functions. It is important that efforts be made to calculate such wave functions from this model, because single-nucleon transfer data have provided insight into the details of the wave functions of collective excitations and it would be illuminating to apply this powerful tool to test the IBA more fully.

Table 1. Particle transfer strengths in ^{168}Er

BAND		IBA WITH S AND D BOSONS (ref. 10,15)			SOLOVIEV MODEL (ref. 6,8)
K^{π}	ENERGY (keV)	(t,p) (p,t) EXPLAINED	MODEL NOT APPLICABLE	DATA CONTRADICT MODEL	(t,α) (d,p),(t,d) EXPLAINED
0_1^+ gsb	0	x			x
2_1^+ γ	821	x			x
4^-	1094		x		x
0_2^+ β	1217	x			x
1^- oct	1358		x		x
0_3^+	1422			x[a)]	x
3^-	1541		x		x
2^- oct	1569		x		x
3^+	1653		x		x
6^-	1773		x		x
0^- oct	1786(I=1)		x		x
3^-	1828		x		(x)
0_4^+	1833			x[a)]	x
2_2^+	1848			x[a)]	x
4^-	1905		x		x
2^+	1930				x
1^-	1936		x		x
3^-	1999		x		x

a) The (t,p) population of these bands has been explained by including g bosons[12].

REFERENCES

1. W.F. Davidson et al., J. Phys. G7(1981)455; J. Phys. G7(1981)843
2. A. Bohr and B. Mottelson, Phys. Scr. 25(1982)28
3. T.S. Dumitrescu and I. Hamamoto, Nucl. Phys. A383(1982)205
4. M. Seiwert et al., Phys. Rev. C30(1984)1779
5. J.P. Draayer and K.J. Weeks, Ann. Phys. 156(1984)41
6. E.P. Grigoriev and V.G. Soloviev, The structure of even deformed nuclei. Nauka Press, Moscow, 1974
7. V.G. Soloviev and N.Yu. Shirikova, Sov. J. Nucl. Phys. 36(1983)799
8. V.G. Soloviev, JETP Lett. 40(1984)1216
9. L. Lin et al., J. Phys. G9(1983)L223
10. D.D. Warner et al., Phys. Rev. Lett. 45(1980)1761; Phys. Rev. C24(1981)1713
11. H.-C. Wu and X.-Q. Zhou, Nucl. Phys. A417(1984)67
12. Y. Akiyama et al., Phys. Lett. 173B(1986)1
13. W.F. Davidson et al., Can. J. Phys. 62(1984)1538, and W.F. Davidson and W.R. Dixon, to be published.
14. W.F. Davidson et al., Phys. Lett. 130B(1983)161
15. D.G. Burke et al., Can. J. Phys. 63(1985)1309
16. D.G. Burke et al., Nucl. Phys. A442(1985)424
17. D.G. Burke et al., Nucl. Phys. A445(1985)70
18. D.D. Warner and A.M. Bruce, Phys. Rev. C30(1984)1066
19. R.F. Casten and D.D. Warner, Phys. Rev. Lett. 48(1982)666

THE TRANSFER REACTIONS EVIDENCE IN THE CASE OF ^{196}Pt

G. ROTBARD, M. VERGNES, G. BERRIER-RONSIN, J. VERNOTTE
Institut de Physique Nucléaire, 91406 Orsay, France

Abstract The nucleus ^{196}Pt, considered as "the best example of an 0(6) nucleus", has been alternatively described recently as a U(5) nucleus. We have reviewed the large available body of transfer reactions data, both one nucleon and two nucleon transfer, and compared the results for the Pt and Hg isotopes. The lightest Hg isotopes appear as rather well described by the U(5) limit. For the heaviest Pt isotopes, no evidence against the 0(6) character has been found and many pieces of data, butressing each other (in particular the g.s. ⟷ g.s two neutron transfer enhancement factor behaviour), are rather in favour of it.

After a long standing status[1] as "the best example of an 0(6) nucleus", ^{196}Pt has recently been alternatively described as a possible U(5) candidate[2]. Comparison of the experimental and theoretical excitation energies and electromagnetic transition rates apparently does not permit[2] to conclusively decide between 0(6), U(5) or even an 0(6) + SU(3) mixture.

Although, as suggested in Ref. 2, the measurement of quantities such as $Q(2_2^+)$ and $Q(4_1^+)$ could be a good step towards a better understanding of ^{196}Pt, one can wonder if the analysis of existing transfer reaction data could not, as previously suggested[3], offer the possibility of a test of the true nature of ^{196}Pt and, more generally, of the Pt isotopes. It is the purpose of the present paper to collect, outline and summarize the results of the analysis of such data.

A single nucleon transfer reaction always links an even-even and an odd-A (odd-proton or odd-neutron) nucleus. It has been suggested that such pairs of nuclei could be described by supersymmetric models and one nucleon transfer reactions have been used[4] as one of the tests of these models, assuming the symmetry of the bosonic core to be known. In the case discussed here, one could consider such reactions as a way to test the nature of the core, assuming the supersymmetric frame to be correct.

In the odd-proton case, the success[5] of the U(6/4) supersymmetry for the ^{192}Os-^{193}Ir pair, including the transfer intensities[4,5] in the ^{192}Os(^{3}He,d)^{193}Ir reaction, could in this way be considered as a confirmation of the O(6) nature of the ^{192}Os bosonic core. A similar success has been shown[6] for the energies and B(E2) values of the ^{196}Pt-^{197}Au pair, but the agreement appears much worse[5,6] for the transfer intensities in the ^{196}Pt(^{3}He,d)^{197}Au reaction and the above confirmation of the O(6) nature cannot therefore be straightforwardly extended to the ^{196}Pt bosonic core.

In the odd-neutron case, the energies and B(E2) values in the nuclei $^{195-197-199}$Pt have been described[7] by the more sophisticated U(6/12) model, also using an O(6) core. It should be noted that a reasonable agreement has been shown[8] also for the transfer intensities in the ^{196}Pt(p,d)^{195}Pt reaction. Although the evidence here is clearly in agreement with an O(6) nature of the bosonic core, the use of such an agreement as a proof would be somewhat unfair, because it is not obvious that a U(5) core could not give a similar agreement. It is for example clear, as far as the above transfer reaction is concerned, that the parameter free ratios[8] $R_{3/2} = C^2S(3/2^-_1)/C^2S(3/2^-_2)$ and $R_{5/2} = C^2S(5/2^-_1)/C^2S(5/2^-_2)$, the experimental values of which are :

$R_{3/2} = 4.65$ and $R_{5/2} = 5.53$, can be reproduced as well (or as badly) with a U(5) core : $R_{3/2} = R_{5/2} = 6$, than with an 0(6) core = $R_{3/2} = R_{5/2} = 3.88$.

The two nucleon transfer reactions link either two even-even or two odd-A nuclei. For even-even targets, formulae have been given[9] for the g.s $\longleftrightarrow$ g.s transfer intensities, showing a different behaviour - as a function of the number N of bosons - for the 3 limits of the interacting boson model. The model also predicts[9] the population of the excited 0^+ levels. In the strict U(5) limit, no excited 0^+ should be populated ; in the other two limits only one 0^+ level should be observed : the 0_2^+ for SU(3), the 0_3^+ for 0(6). Several preliminary tests using these reactions have already been performed[3] but they were not mentionned in Ref. 2 and it is worthwhile to precise and stress the results obtained by mixing together and re-analyzing the available sets of data.

Our (p,t) experiments[10] have been performed at 25MeV, using the Orsay tandem accelerator and a split pole spectrometer. Enhancement factors ε have been obtained through a DWBA analysis with the optical potentials of Ref. 11, slightly modified, using a $(2p_{3/2})^2$ form factor. The targets used were isotopically separated $^{198,196,194,192,190}Pt$, and natural Pt for normalization purposes. These ε values for the g.s $\leftrightarrow$ g.s transitions, as well as ε values taken from Ref. 12 : (t,p) reaction at 17 MeV, have been compared to theory. A first least squares fit was used to determine the best normalization between the 2 sets of data. A second least squares fit (χ^2 given in the figure) was then performed to determine, for each of the 3 IBM-1 curves, the best normalization with the experimental data. The result is shown in Fig. 1. For the purpose of comparison, the ε

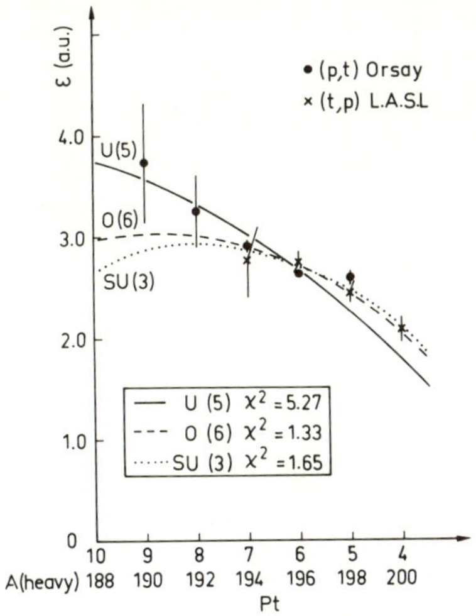

Fig. 1 : **Comparison of the ϵ factors for $Pt_{g.s} \longleftrightarrow Pt_{g.s}$ with the curves calculated for the 3 limits of IBM-1 (see text).**

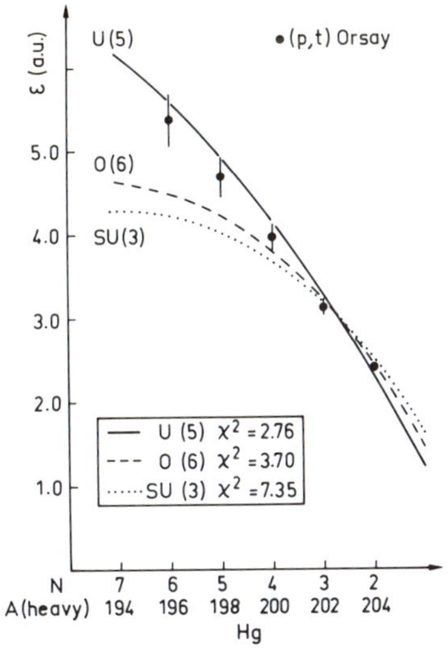

Fig. 2 : **Comparison of the ϵ factors for $Hg_{g.s} \rightarrow Hg_{g.s}$ with the curves calculated for the 3 limits of IBM-1 (see text).**

values for the $^{204,202,200,198,196}Hg$ (p,t) reactions[13], obtained with the same optical potentials and form factor from data measured at the same energy, are compared in Fig. 2 with the predictions for the 3 limits of IBM-1, the normalization being performed as explained just above for Pt.

Although in Fig. 1 the experimental ε values for Pt do not all lie on the curve corresponding to the O(6) limit, a fair description of the results consists to say that the behaviour of the ε values between A = 200 and A = 194 (and perhaps 192) is better reproduced by the O(6), (or O(6) + SU(3)), limit than by the U(5) limit. For A = 192 and particularly 190, the error bars are large but the ε values show a tendency to depart from O(6) towards U(5). The general behaviour of the ε values for the Hg isotopes, shown in Fig. 2, is different and can be said to be quite well reproduced by U(5) for the 4 lightest targets.

The 0_2^+ and 0_3^+ levels have been clearly observed[10] in the Pt(p,t) reactions, the intensity for the population of the 0_3^+ being larger than the one corresponding to the population of the 0_2^+, at least for the heaviest final isotopes $^{194,192,190}Pt$. This is an argument in favour of O(6), but the experimental ratios : $\varepsilon(0_3^+)/\varepsilon(0_1^+)$, are only respectively 6%, 5% and 7%, when they should be 11%, 10% and 9% in a strict O(6) limit.

To summarize our review of the above large body of transfer data, it could be said that we have found no evidence against the O(6) character of ^{196}Pt. Many results are rather in favour of this O(6) nature, the most convincing being the behaviour of the enhancement factor ε as a function of A in the 2 neutron transfer reactions.

REFERENCES

1. J.A. Cizewski, R.F. Casten, G.J. Smith, M.L. Stelts, W.R. Kane, H.G. Börner and W.F. Davidson, Phys. Rev. Letters, 40, 167 (1978)
2. M.P. Fewell, Phys. Letters, 167B, 6 (1986)
3. M. Vergnes, Interacting Bose-Fermi Systems in Nuclei, edited by F. Iachello, (Plenum, New-York, 1981) p. 53 ; J.A. Cizewski, ibid, p. 137
4. M. Vergnes, Proceedings of Nuclear Structure, Reactions and Symmetries International Conference, Dubrovnik, June 5-14, 1986, World Scientific (to be published) and references therein
5. A.B. Balantekin, I. Bars and F. Iachello, Nucl. Phys., A370, 284 (1981)
   F. Iachello and S. Kuyucak, Ann. Phys. (N.Y.), 136, 19 (1981)
6. J. Vervier, Phys. Letters, 100B, 383 (1981)
   J. Vervier, R. Holzmann, R.V.F. Janssens, M. Loiselet and M.A. Van Hove, Phys. Letters, 105B, 343 (1981)
7. D.D. Warner, R.F. Casten, M.L. Stelts, H.G. Börner and G. Barreau, Phys. Rev., C26, 1921 (1982)
   R.F. Casten, D.D. Warner, G.M. Gowdy, N. Rofail and K.P. Lieb, Phys. Rev., C27, 1310 (1983)
8. M. Vergnes, G. Berrier-Ronsin and R. Bijker, Phys. Rev, C28, 360 (1983)
9. A. Arima and F. Iachello, Ann. Phys., (N.Y.), 99, 253 (1976) ; 111, 201 (1978) ; 123, 468 (1979) ; Phys. Rev, C16, 2085 (1977)
10. M. Vergnes, G. Rotbard, J. Kalifa, J. Vernotte, G. Berrier, R. Seltz, H.L. Sharma and N.M. Hintz, Bull. Am. Phys. Soc., II, 21, 8, 976 (1976)
    M. Vergnes, G. Rotbard, J. Kalifa, G. Berrier, J. Vernotte, Y. Deschamps and R. Seltz, Journal de Physique, 17, L-291 (1978) and unpublished data
11. F.G. Perey, Phys. Rev., 131, 745 (1963)
    E.R. Flynn, D.D. Armstrong, J.G. Beery and A.G. Blair, Phys. Rev., 182, 1113 (1969)
12. J.A. Cizewski, E.R. Flynn, R.E. Brown, D.L. Hanson, S.D. Orbesen and J.W. Sunier, Phys. Rev., C23, 1453 (1981)
13. M. Vergnes, G. Berrier-Ronsin, G. Rotbard, J. Vernotte, S. Fortier, J.M. Maison, R. Tamisier, S. Grafeuille, P. Duval and E.R. Flynn, Internal Report Orsay, IPNO-DRE 83-24, and results of more recent analysis.

^{189}Os AND THE U(6/12) SYMMETRY SCHEME

A.M. BRUCE[1,2], D.D. WARNER[3], M.J. CAMPBELL[2] AND W. GELLETLY[2]
[1]Present address: Niels Bohr Institute, 4000 Roskilde, Denmark.
[2]Schuster Laboratory, University of Manchester, Manchester, M139PL, UK.
[3]Brookhaven National Laboratory, Upton, NY 11973, USA.

Abstract This contribution will consider the extension of the U(6/12) symmetry scheme to the region intermediate between the O(6) and SU(3) limits and will compare the predicted level structure with that determined ~~empirically~~ experimentally (ed.) for ^{189}Os.

INTRODUCTION

The extension of the Interacting Boson Model[1] (IBM) description of even-even nuclei to odd-mass nuclei is contained in the Interacting Boson-Fermion Model[2] (IBFM) which considers an odd-mass nucleus in terms of an odd fermion coupled to an even-even core defined within the framework of the IBM. The underlying group structure of the IBFM is of the form $U^B(6) x U^F(m)$ where $U^B(6)$ represents the even-even core and $U^F(m)$ the odd-fermion. The nuclei at the upper end of the N=82-126 neutron shell are well suited to a description of this type. The odd neutrons occupy the $2p_{1/2}$, $2p_{3/2}$ and $1f_{5/2}$ shell model states and this corresponds to an m value of 12. Symmetry schemes corresponding to SU(5), SU(3) and O(6) core symmetries have been deduced[3] for this $U^B(6) x U^F(12)$ framework and comparisons between the level schemes of ^{195}Pt, ref.[4], and ^{185}W, ref.[5], and the O(6) and SU(3) limits respectively suggest that the model has some validity. The question we will consider is whether it is possible to propose an interpretation for the nuclei transitional between these two limiting cases.

THE CONSISTENT-Q-FORMALISM IN ODD-A NUCLEI

One possible answer to this problem has been suggested in terms of the recently proposed Consistent-Q-Formalism in Odd-A nuclei (CQFOA)[6] in which the region transitional between the strict SU(3) and O(6) limits is characterized by one parameter χ and the boson number N. This is a direct extension of the consistent-Q-formalism description of even-even nuclei[7].

In the SU(3) chain decomposition of the U(6/12) scheme, the Hamiltonian which can be used to describe the low-lying structure of ^{185}W is given by:

$$H=aC_{2U^{BF}(6)}+bC_{2SU^{BF}(3)}+cC_{2O^{BF}(3)}+dC_{spin^{BF}(3)} , \qquad (1)$$

where a, b, c, d are constants and $C_{2U^{BF}(6)}$ etc. are the Casimir operators of the subgroups in the chain decomposition. The Casimir operator of the $SU^{BF}(3)$ group contains a quadrupole-quadrupole interaction of the form:

$$Q.Q = (Q_B+Q_F).(Q_B+Q_F) , \qquad (2)$$

where

$$Q_B = (s^+\tilde{d}+d^+s)^{(2)} + \frac{\chi}{\sqrt{5}} (\tilde{d}d^+)^{(2)} , \qquad (3)$$

and

$$Q_F = K_\mu^{(2)}(2,0) + K_\mu^{(2)}(0,2) + \frac{\chi}{\sqrt{5}} K_\mu^{(2)}(2,2) . \qquad (4)$$

Q_B represents the boson part of the interaction and Q_F the fermion part. The $K_\mu^{(2)}(L,L')$ refer to the fermion generators where L and L' denote the pseudo-orbital angular momentum, full details of this are given elsewhere (e.g. ref.[8]). When $\chi=-\sqrt{35}/2$, the strict SU(3) limit is attained and when $\chi=0$ a specific O(6) limit is reached. The Casimirs of the U(6),O(3) and Spin(3) groups contribute only diagonal terms to the Hamiltonian so that for intermediate values of χ, the wavefunctions are specified uniquely by the value of χ and of the boson number N. Then if the E2 operator is defined by:

$$T(E2) = \alpha(Q_B+Q_F) , \tag{5}$$

ratios of B(E2) values can be used for a given N to be a measure of χ. Similarly ratios of single particle structure factors for transitions to levels of the same spin can give a measure of χ.

For nuclei near the end of closed shells, the lowest order transfer operator for the pickup reaction is given by:

$$P_j^+ = \zeta_j a_j^+ , \tag{6}$$

where ζ_j is an independent parameter. The ratio of (d,t) structure factors for states of the same spin is therefore parameter-free, and figure 1 shows the variation of one such ratio as a function of χ. The $K_p=1,2$ notation refers to different bands within the anti-symmetric {[N,1]} representation of states. The calculated ratios for the two spin values are identical because each 3/2, 5/2 pair stems from the same pseudo-L level. For a more detailed explanation of this see e.g. ref.[9]. The dashed lines on figure 1 correspond to the weighted mean of the ratio for the 3/2 states [R(3/2)=0.43(6)] and for the 5/2 states [R(5/2)=0.48(7)] in

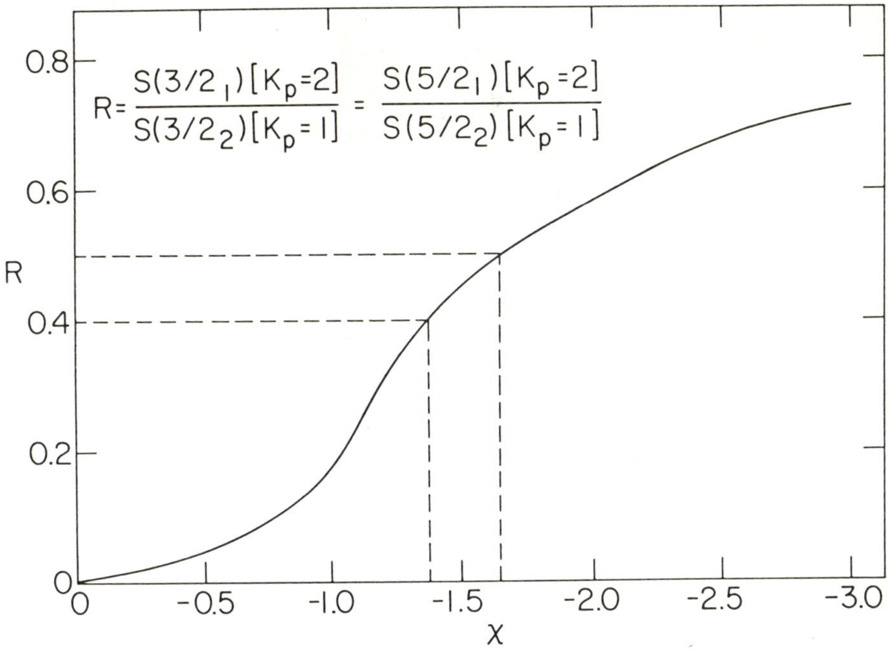

FIGURE 1. The ratio of (d,t) structure factors as a function of χ.

^{189}Os. These two experimental values overlap nicely and define a range of values centered on $\chi=-1.5$. Using this value of χ it is then of interest to make a detailed comparison between the predicted level scheme and that observed for ^{189}Os.

COMPARISON WITH ^{189}Os

In order to make such a comparison meaningful an extensive experimental programme has been carried out to study the structure of ^{189}Os. Initially, experiments were performed at the High Flux Beam Reactor (HFBR) at Brook-

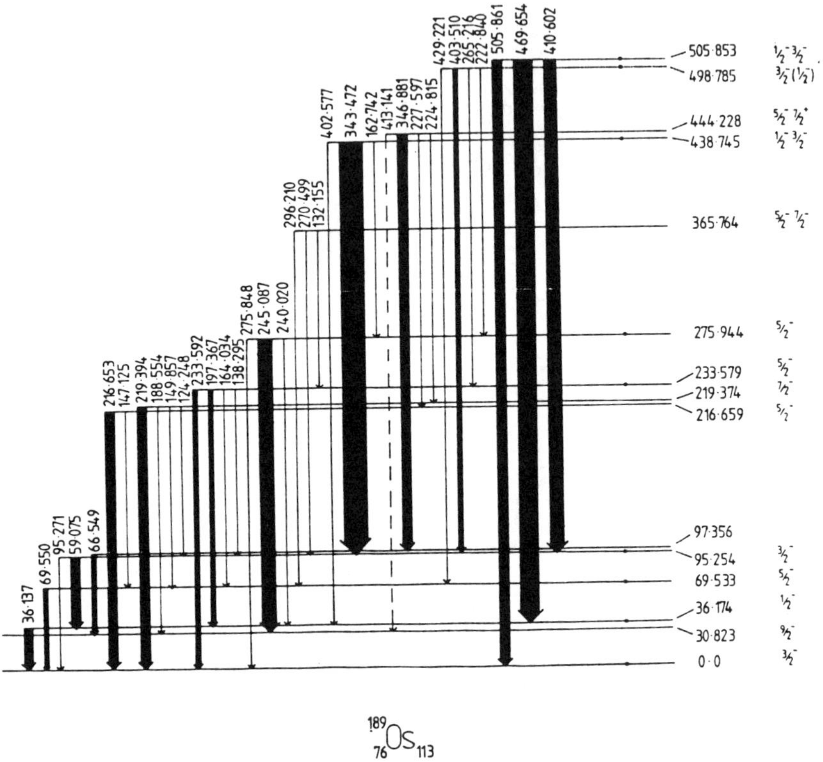

FIGURE 2. The level scheme of ^{189}Os up to 510 keV.

haven National Laboratory with the (n,γ) reaction. Complementary studies involving Coulomb excitation measurements and measurements of the (n,e⁻) reaction were then carried out at the Manchester Van de Graaff accelerator and at the

Institut Laue-Langevin in Grenoble. At the HFBR, the use of the Average Resonance Capture technique ensured that all $1/2^-$, $3/2^-$ levels have been observed up to an excitation energy determined by the sensitivity of the measurement, in this case ~1.5 MeV. In addition, detailed gamma-gamma coincidence and angular correlation studies were carried out following thermal neutron capture and the results enabled the J^π values of other levels to be determined. Figure 2 shows the level scheme of ^{189}Os up to 510 keV, determined from the measurements performed at Brookhaven. Levels which are marked on the right by a dot are the ones which have been observed in the ARC measurement, where in addition to ALL the $1/2^-$,$3/2^-$ levels, some $1/2^+$, $3/2^+$ and $5/2^\pm$ levels were also observed.

The low-lying structure shown in figure 2 is essentially consistent with that previously proposed on the basis of (d,t), (d,p) and (n,γ) studies[10]. One important discrepancy, the full importance of which will become apparent later, involves the 216 keV level which was previously thought to be a $7/2^-$ level and was assigned as the $7/2^-$[503] Nilsson orbit on the basis of the single particle transfer strength observed in the (d,p) and (d,t) reactions[10]. The result of a gamma-gamma angular correlation measurement on the 463-216 keV cascade which originates from the $3/2^-$ level at 679 keV shows that it is less than 0.1% probable that the 216 keV level has $J^\pi = 7/2^-$. Indeed, initial results from the (n,e^-) measurement at the ILL show that the 216 keV transition which depopulates this level is a mixed E2/M1 transition with a δ value[11] of $2.46^{+0.27}_{-0.21}$. Since the 216 keV level is not a $1/2^-$,$3/2^-$ level (because it was not observed in the ARC measurement) the only remaining possibility is that it has $J^\pi=5/2^-$. The transfer strength observed in the (d,t) and (d,p) reactions[10] is inconsistent with this level being assigned as either the $5/2^-[512]$ or the $5/2^-[503]$ Nilsson state and indeed the detailed Nilsson model calculation discussed in ref.[10] predicts that these states should appear at ~1500 keV excitation energy. The simplest and perhaps only logical conclusion of this dilemma is that the observed single particle strength belongs not to the 216 keV level but to the neighbouring 219 keV level which has a J^π value of $7/2^-$ and would then correspond to the $7/2^-[503]$ Nilsson state. This leaves the $5/2^-$ state which since it is not a pure single particle state, must correspond to coupling with some collective excitation in the core.

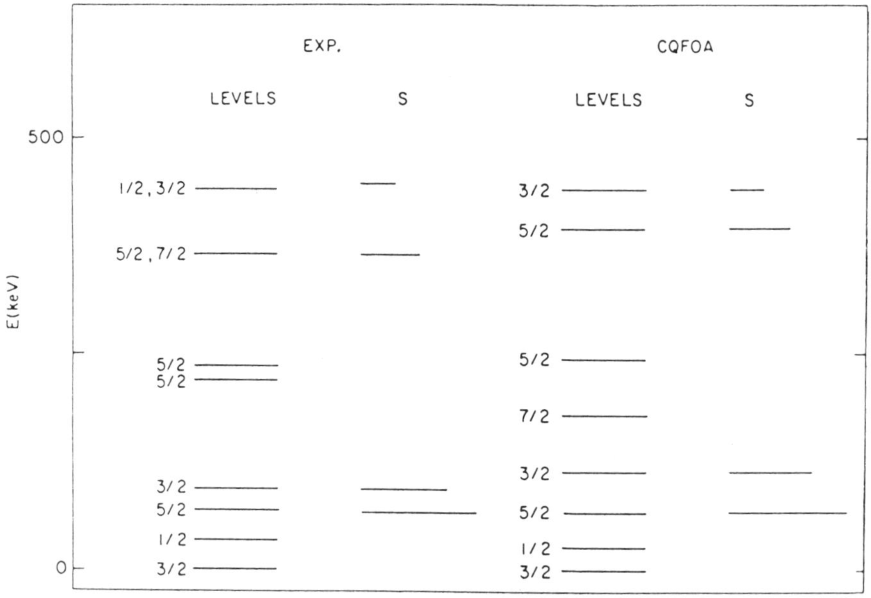

FIGURE 3. Calculated and experimental negative-parity structure in ^{189}Os.

Figure 3 shows the results of a fit to the low-lying level scheme of ^{189}Os using the value of χ=-1.5 deduced from figure 1. On the left of the figure is the experimental energy spectrum up to 475 keV with the level energies being taken from figure 2. The level at 219 keV has been omitted from the figure because it stems from the $h_{9/2}$ shell model state and as such is outside the U(6/12) basis, as are the $9/2^-$ level at 30 keV [which corresponds to the $9/2^-$[505] state] and the $5/2^-$ level at 275 keV (which decays with a measured B(E2) of ~36 s.p.u.[12] to the 30 keV level and has been assigned as the γ-vibrational level built on this state). The levels at 97 and 444 keV have not been included because their J^π values have not yet been uniquely determined, although there is some evidence that they are positive parity states. The single particle structure factors (shown in the column marked S) have been taken from the work of Benson et al.[10]. On the right are shown the level energies and structure factors calculated within the CQFOA using the values of a=-107, b=-15.8, c=18.5 and d=-9 keV in equation 1. Using these values of parameters, no other low spin levels are calculated to lie within the energy range of this figure. It is obvious that in order to be able to make a one-one comparison between empirical and theoretical levels certain assumptions must be made but it is hoped that further analysis will make it possible to distinguish experimentally between the possibilities.

The main problem concerns the 216 keV level which explains why it has been discussed at such length in this contribution. As yet there is no obvious interpretation for this level which since it does not correspond to a pure single particle excitation must involve coupling to excitations in the core. The U(6/12) scheme should automatically include a more complete core description than is included in the Nilsson model, but it is noticeable that the value of χ used in this study is rather different from that which would be used in a description of the core nucleus ^{190}Os (where the value of χ used would be ~0.6). Of course, a much more stringent test of any theoretical model can be made by considering the B(E2) values and analysis of the Coulomb excitation measurement is still progressing. Nevertheless, the initial comparison between the empirical level scheme of ^{189}Os and that predicted by the CQFOA is encouraging and bodes well for further study. Even if the low-lying levels can be adequately explained there is still the question of the higher lying levels which are more numerous and may yet provide a more severe stumbling block for the symmetry scheme.

REFERENCES

1. A. Arima and F. Iachello, Phys. Rev. Letts. 35, 1069 (1975).
2. F. Iachello and O. Scholten, Phys. Rev. Letts. 43, 679 (1979).
3. A.B. Balantekin et al., Phys. Rev. C27, 1764 (1983).
4. D.D. Warner et al., Phys. Rev. C26, 1921 (1982).
5. A.M. Bruce et al., to be published in Nucl. Phys. A.
6. D.D. Warner et al., Phys. Rev. Letts. 54, 1365 (1985).
7. D.D. Warner and R.F. Casten, Phys. Rev. Letts. 48, 1355 (1982).
8. R. Bijker, Ph.D. Thesis, Univ. of Groningen 1984.
9. D.D. Warner, contribution to this conference.
10. D. Benson et al., Phys. Rev. C14, 2095 (1976).
11. W. Gelletly, private communication.
12. S.G. Malmskog et al., Nucl. Phys. A153, 316 (1970).

TEST OF BOSE-FERMI SYMMETRIES IN BETA DECAY

PETR NAVRÁTIL[a] and JAN DOBEŠ[b]
a) MFF, Charles University, Prague
b) Inst. of Nuclear Physics, Řež, Czechoslovakia

Abstract Beta decay of odd nuclei in the Pt/Hg region is studied. The nuclear states are described within the Bose-Fermi symmetry limits of IBFM. Relative transition strengths are well explained.

The interacting boson model (IBM) and the interacting boson-fermion model (IBFM) are confronted with the experiment by comparing energy levels, electromagnetic properties and one- and two-nucleon transfer spectroscopic strengths. Double beta decay with IBM wave functions was also studied[1]. No IBM/IBFM investigation has been till now made for single beta decay characteristics, presumably for two reasons:

i) The agreement of theory and experiment is usually worse for nuclear beta decay than for other characteristics. The fT value often depends on the interplay of several matrix elements. Relatively small variation in the matrix elements can change the resulting fT value considerably.

ii) The beta decay study in even systems needs a knowledge of the structure of odd-odd nuclei, which have been scarcely investigated within IBFM. For odd systems, the pairs of odd-even and even-odd nuclei have not been frequently studied, too.

In the present contribution we investigate beta decay of odd nuclei in the Pt/Hg region. The structure of nuclear states is described within the Bose-Fermi symmetry limits of IBFM. Thus, investigating beta decay we also intend to test the symmetry concept.

The shell model representation of the beta transition operator is written as

$$T^{(K)} = -\hat{K}^{-1} \sum_{j_\pi j_\nu} \xi^{(K)}_{j_\pi j_\nu} (c^+_{j_\pi} \tilde{c}_{j_\nu})^{(K)} \quad . \tag{1}$$

Here, only the total transferred orbital momentum K is explicitely written, $\hat{K}=\sqrt{2K+1}$, $\xi^{(K)}_{j_\pi j_\nu}$ is the single particle (s.p.) matrix element, c^+_j is the single nucleon creation operator and $\tilde{c}_{jm}=(-)^{j-m}c_{j-m}$, indices π and ν refer to protons and neutrons, respectively. The image of the operator (1) in the boson-fermion space is constructed by replacing the shell model operators c^+_j and $\tilde{c}_j$ by the pseudo-particle operators[2]

$$\hat{c}^+_j = u_j a^+_j + v_j \frac{1}{\sqrt{N}} s^+ \tilde{a}_j + \sum_{j'} u_j \bar{\beta}_{j'j} \frac{\sqrt{10}}{\hat{j}} (d^+ \tilde{a}_j)^{(j)} - \sum_{j'} v_j \bar{\beta}_{j'j} \sqrt{\frac{10}{N}} \frac{1}{\hat{j}} s^+ (\tilde{d} a^+_j)^{(j)} \quad . \tag{2}$$

In eq.(2), v_j is occupation probability of the shell model orbit, $u_j^2+v_j^2=1$, coefficients $\bar{\beta}_{j'j}$ are related to the structure of the collective L=2 fermion pair, N is the number of bosons (π or ν) in the system. Keeping only the terms which connect the relevant one-quasi-fermion spaces and are of the first order in the d-boson operators we obtain two types of the beta transition operator in the boson-fermion space. The first type conserves numbers of π and ν bosons separately

$$T^{(K)} = -\hat{K}^{-1} \sum_{j_\pi j_\nu} \xi^{(K)}_{j_\pi j_\nu} \Big\{ u_{j_\pi} u_{j_\nu} (a^+_{j_\pi} \tilde{a}_{j_\nu})^{(K)} - \sum_{j'_\nu} u_{j_\pi} v_{j_\nu} \bar{\beta}_{j'_\nu j_\nu} \sqrt{\frac{10}{N_\nu}} \frac{1}{\hat{j}_\nu} s_\nu (a^+_{j_\pi} (d^+_\nu \tilde{a}_{j'_\nu})^{(j_\nu)})^{(K)} \tag{3}$$

$$- \sum_{j'_\pi} u_{j_\nu} v_{j_\pi} \bar{\beta}_{j'_\pi j_\pi} \sqrt{\frac{10}{N_\pi}} \frac{1}{\hat{j}_\pi} s^+_\pi ((\tilde{d}_\pi a^+_{j'_\pi})^{(j_\pi)} \tilde{a}_{j_\nu})^{(K)} \}$$

and describes transitions between nuclei

$$N_\pi, N_\nu, n_\pi=1 \rightleftarrows N_\pi, N_\nu, n_\nu=1 \qquad (4)$$

The second type changes one ν boson into one π boson together with the conversion of the odd fermion

$$T^{(K)} = \hat{K}^{-1} \sum_{j_\pi j_\nu} \xi^{(K)}_{j_\pi j_\nu} \{ v_{j_\pi} v_{j_\nu} \frac{1}{\sqrt{N_\pi N_\nu}} s^+_\pi s_\nu (\tilde{a}_{j_\pi} a^+_{j_\nu})^{(K)}$$

$$+ \sum_{j'_\pi} u_{j_\pi} v_{j_\nu} \bar{\beta}_{j'_\pi j_\pi} \sqrt{\frac{10}{N_\nu}} \frac{1}{\hat{j}_\pi} s_\nu ((d^+_\pi \tilde{a}_{j'_\pi})^{(j_\pi)} a^+_{j_\nu})^{(K)}$$

$$+ \sum_{j'_\nu} u_{j_\nu} v_{j_\pi} \bar{\beta}_{j'_\nu j_\nu} \sqrt{\frac{10}{N_\pi}} \frac{1}{\hat{j}_\nu} s^+_\pi (\tilde{a}_{j_\pi} (\tilde{d}_\nu a^+_{j'_\nu})^{(j_\nu)})^{(K)} \}$$

and describes processes

$$N_\pi, N_\nu, n_\pi=1 \rightleftarrows N_\pi+1, N_\nu-1, n_\nu=1$$

Only transitions of the first type are discussed here further. The first term of eq.(3) is the usual beta transition quasiparticle operator with the pairing suppression factor $u_{j_\pi} u_{j_\nu}$. The microscopic interpretation of IBFM introduces in eq.(3) the exchange terms in which the change of the even core in the beta decay process is taken into account.

The operator (3) is written in the IBFM-2 formalism with π and ν bosons distinguished. Applying the usual projection method[2] we get the IBFM-1 form by replacing

$$s^+_\rho \tilde{d}_\rho \rightarrow N_\rho / N \; s^+ \tilde{d} \qquad \rho=\pi,\nu \qquad N=N_\pi+N_\nu \quad .$$

There are three cases of beta decay in the Pt/Hg region which are of the first type (4) and in which the structure of the initial and final states is reasonably described within the Bose-Fermi symmetry limits of IBFM: $^{195}Ir \rightarrow {}^{195}Pt$, $^{195}Hg \rightarrow {}^{195}Au$ and $^{197}Hg \rightarrow {}^{197}Au$. All these decays are the first non-unique forbidden transitions with $\Delta J=0,1$ and $\pi_i\pi_f=-1$. For the fT value we have

$$fT = 2\pi^3 \ln 2\, G_\beta^{-2} \left(|M^{(0)}|^2 + |M^{(1)}|^2\right)^{-1}$$

$$M^{(K)} = \langle J_f \| T^{(K)} \| J_i \rangle \qquad K=0,1 \quad .$$

The s.p. matrix elements in eq.(3) are calculated in the impulse approximation with the induced terms neglected[3]. The ratio of the axial to vector coupling constants is $\lambda=-g_A/g_V=1.25$. The radial integrals are calculated with s.p. wave functions of Woods-Saxon potential with the global parameters of Ref.4.

The s.p. levels considered in the wave functions and in the transition operator are $p_{1/2}$, $p_{3/2}$ and $f_{5/2}$ for neutrons and $d_{3/2}$ for protons. Neutron occupancies for $N_\nu=4$ (^{195}Pt and ^{197}Hg) are[5] $v^2_{1/2}=0.16$, $v^2_{3/2}=0.67$ and $v^2_{5/2}=0.46$. For $N_\nu=5$ (^{195}Hg) the values of v's are somewhat decreased from the $N_\nu=4$ case: $v^2_{1/2}=0.10$, $v^2_{3/2}=0.64$ and $v^2_{5/2}=0.44$. The proton parameter for $N_\pi=1$ (Hg isotopes) is in agreement with the graph of Ref.6 - $v^2_{3/2}=0.50$, for Pt is slightly increased as compared to Ref.6 - $v^2_{3/2}=0.46$.

Assuming that the L=2 collective fermion pair exhausts the valence E2 strength the structure coefficients $\bar{\beta}_{j'j}$ are written as

$$\bar{\beta}_{j'j} = \mathcal{N}_\beta^{-1}(u_j v_{j'} + v_j u_{j'})\langle j' \| Y^{(2)} \| j\rangle$$

with the normalization $\sum \bar{\beta}^2_{j'j}=1$.

The nuclear states are described in the Bose-Fermi symmetry limits of IBFM. For ^{195}Au, ^{197}Au and ^{195}Ir the symmetry is[7-9]

$$U^B(6)\otimes U^F(4) \supset SO^B(6)\otimes SU^F(4)\supset \ldots \qquad (5)$$

The ^{195}Pt is a very good example of the symmetry[10]

$$U^B(6)\otimes U^F(12)\supset\ldots\supset U^{BF}(6)\otimes SU^F(2)\supset SO^{BF}(6)\otimes SU^F(2)\ldots \quad (6)$$

For ^{195,197}Hg the symmetry limit is[10,11]

$$U^B(6)\otimes U^F(12)\supset\ldots \quad \ldots\supset U^{BF}(5)\otimes SU^F(2) \quad \ldots \quad (7)$$

Hg isotopes were analysed also within the symmetry (6) with worse results for one-nucleon transfer.

In tables I and II the calculated and experimental log fT values are compared and the quantities $|M|^2=|M^{(0)}|^2+|M^{(1)}|^2$ are given normalized to the strongest transition. The calculated log fT values are greater than experimental ones implying that the calculated matrix elements $|M|^2$ are as much as 10 times smaller than those extracted from experiment. The relative matrix elements are reproduced quite well.

TABLE I ^{195}Ir → ^{195}Pt transitions.

J_f^-	log fT exp[12]	log fT theo	$\|M\|^2$ exp	$\|M\|^2$ theo
$1/2_1$	7.0	7.77	0.16	0.27
$3/2_1$	6.6	7.89	0.40	0.20
$5/2_1$	6.2	7.20	1.	1.
$3/2_2$	-	11.02	-	10^{-4}
$3/2_3$	7.3	8.70	0.08	0.03
$1/2_2$	-	12.06	-	10^{-5}
$5/2_2$	-	9.82	-	0.002

^{195}Ir→^{195}Pt: The unobserved transitions to the $3/2_2^-$ and $1/2_2^-$ states which belong to the [61] ⟨61⟩(11)L=1 irrep of the symmetry limit (6) are theoretically considerably suppressed. The strongest transitions both in theory and experiment lead to the states $3/2_1^-$ and $5/2_1^-$

belonging to the irrep [61]⟨61⟩(1)L=2. The doublet $3/2^-_3$ and $5/2^-_2$ from the irrep [7]⟨7⟩(1)L=2 is weakly populated in agreement with experiment.

TABLE II $^{195,197}Hg \rightarrow {}^{195,197}Au$ transitions.

	J^+_f	log fT exp[12,13]	log fT theo (6)	log fT theo (7)	$\|M\|^2$ exp	$\|M\|^2$ theo (6)	$\|M\|^2$ theo (7)
^{195}Au	$3/2_1$	-	7.72	8.26	-	0.02	0.04
	$1/2_1$	6.4	5.94	6.82	1.	1.	1.
	$3/2_2$	7.9	9.84	-	0.03	10^{-4}	0.
^{197}Au	$3/2_1$	7.20	7.73	8.10	0.03	0.02	0.04
	$1/2_1$	5.71	6.02	6.71	1.	1.	1.
	$3/2_2$	6.61	9.87	-	0.13	10^{-4}	0.

$^{195,197}Hg \rightarrow {}^{195,197}Au$: In table II results are shown using two symmetry limits (6) and (7) for the description of the Hg g.s. The strongest transition is predicted to the $1/2^+_1$ state as observed. The relative strength of the transition to the $3/2^+_1$ level in ^{197}Au is also well explained. There is an appreciable decay to $3/2^+_2$ state. In the limit (7) the transition to the $3/2^+_2$ state is forbidden; it is allowed but still weak in the limit (6). More general form of the transition operator might improve situation. Also the influence of the proton $s_{1/2}$ level in the Au states could be important.

The transition operator is obtained from microscopic considerations so that we have no free parameter to fit the absolute fT values. Only the occupancies are in some cases slightly varied within a reasonable range. The good description of the relative transition strengths seems to support the validity of the Bose-Fermi symmetries in the nuclei investigated.

The discrepancies of absolute values remain to be explained. Here, several points should be considered:

a) The sensitivity of beta decay to nuclear wave functions may be considerable as can be seen from two different symmetry limits studied in Hg→Au decays. Small perturbation of the wave function can change the fT value appreciably.

b) The higher terms in the boson-fermion image of the transition operator might influence results.

c) Details of beta decay theory should be inspected. The decrease of the ratio λ in the nuclear matter coming from the exchange effects has no favourable influence on our results. The core polarization effects[14] usually act in the direction opposite to what would be needed for present calculations.

REFERENCES

1. O.Scholten and Z.R.Yu, Phys.Lett., 161B, 13 (1985).
2. O.Scholten, Thesis, University of Groningen (1980).
3. H.Behrens and W.Bühring, Electron Radial Wave Functions and Nuclear Beta-Decay (Clarendon Press, Oxford, 1982).
4. K.Baer and P.E.Hodgson, J.Phys.G: Nucl.Phys., 4, L 287 (1978).
5. R.Bijker and O.Scholten, Phys.Rev.C, 32, 591 (1985).
6. O.Scholten, Phys.Lett., 127B, 144 (1983).
7. J.L.Wood, in Interacting Bose-Fermi Systems in Nuclei, ed. by F.Iachello (Plenum Press, New York, 1981), p. 381.
8. J.Vervier, Phys.Lett., 100B, 383 (1981).
9. J.A.Cizewski, at NSRS Conference, Dubrovnik 1986.
10. P.Van Isacker et al., Ann.Phys.(N.Y.), 157, 183 (1984).
11. M.Vergnes et al., Phys.Rev.C, 31, 2071 (1985).
12. C.M.Lederer et al., Tables of Isotopes (Willey, New York, 1978).
13. B.Harmatz, Nuclear Data Sheets, 34, 101 (1981).
14. F.Krmpotič et al., Nucl.Phys.A, 342, 497 (1980).

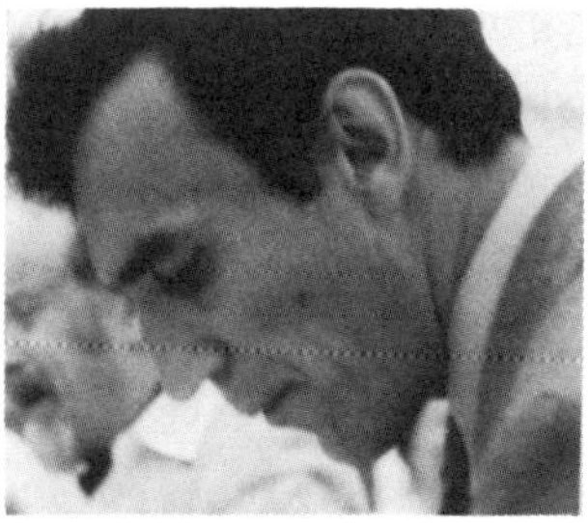

Chapter 3

STUDIES OF INDIVIDUAL NUCLEAR PROPERTIES

Moderator:

W.B. WALTERS
Chemistry Department
University of Maryland
College Park, Maryland 20742, U.S.A.

UNIFIED MICROSCOPIC DESCRIPTION OF THE LIGHTEST NUCLEI

R. G. Lovas, A. T. Kruppa

Kernforschungszentrum Karlsruhe, Institut für Kernphysik III
P.O. Box 3640, D-7500 Karlsruhe 1, FRG
and
Institute of Nuclear Research of the Hungarian Academy of Sciences, Debrecen, P. O. Box 51, H-4001, Hungary*

R. Beck** and F. Dickmann**

Kernforschungszentrum Karlsruhe, Institut für Kernphysik III
P.O. Box 3640, D-7500 Karlsruhe 1, FRG

We describe the nuclei with $A\leq 4$ by mixing harmonic-oscillator states of different sizes, and the nuclei with $5\leq A\leq 8$ as $\alpha+(A-4)$-type two-cluster systems with the full model space of the separate clusters included. The energies and radii obtained and a detailed test of the ^{6}Li wave function show that, apart from the high-momentum behaviour, the description is excellent.

A unified description of a range of nuclei is usually achieved by exploiting the symmetry underlying the similarities and correspondence in their properties. The case of the lightest nuclei is exceptional perhaps owing to their erratic behaviour relative to each other. Because of this, it is more straightforward to set up a unified model right at the microscopic level, where resort to non-trivial symmetry concepts does not seem to reduce the complexity of the problem.

We are concerned with nuclei of mass number $A\leq 8$. We concentrate on the g.s.'s, and present a detailed test of the wave function of ^{6}Li only. We assume that the states $\psi_m(b)$ of nuclei b with $2\leq A_b\leq 4$ can be described as superpositions of N_b antisymmetrized harmonic-oscillator states $\phi_n(b)$ of the $(0s)^{A_b}$ configuration,

---

*Present and permanent address.
**Deceased.

$$\psi_m(b) = \sum_n c_{mn}(b)\phi_n(b) \qquad (m=1,\dots,N_b), \tag{1}$$

the size of each state $\phi_n(b)$ being specific. The nuclei with $5 \leq A \leq 8$ are then assumed to consist of an α cluster and a cluster a composed of the rest of the nucleons:

$$\Psi_A = \Sigma_{k\ell} \mathcal{A}_{\alpha a}\{\psi_k(\alpha)\psi_\ell(a)\chi_{k\ell}(\vec{r}_{\alpha a})\}. \tag{2}$$

Here $\mathcal{A}_{\alpha a}$ is the intercluster antisymmetrizer. The amplitudes $c_{mn}(b)$ $(b=\alpha,a)$ and $\chi_{k\ell}$ are to be determined by diagonalizing the corresponding microscopic hamiltonians. Since ψ_m with $m>1$ may be regarded as the breathing modes built on ψ_1, the two-cluster model is called the breathing cluster model[1]).

To compensate for the truncation of the model spaces, we used such effective forces as reproduce the experimental g.s. energies and sizes of d, t, $\tau(\equiv {}^3He)$ and α. The g.s. energies converged with $N_d=5$, $N_t=N_\tau=4$, $N_\alpha=3$. We used central forces of the form

$$V(r) = (W+MP^r+BP^\sigma-HP^\tau)\sum_{i=1}^{2} V_i \exp(-r^2/a_i^2), \tag{3}$$

where P^r, P^σ and P^τ are the coordinate-, spin- and isospin-exchange operators. Two of the effective forces are specified in table 1, with $W+B+M+H=1$, and $\eta=4W-M+2B-2H$ adjusted to the energy of ^{6}Li $[\eta=\eta(^6\text{Li})]$ or ^{7}Li $[\eta=\eta(^7\text{Li})]$.

As are shown in table 2, the energies of d, t and α are virtually exact and the deviations of the radii from

Table 1. Parameters of the effective nucleon-nucleon potential.

No.	a_1 (fm)	a_2 (fm)	V_1 (MeV)	V_2 (MeV)	W+M	$\eta(^6\text{Li})$	$\eta(^7\text{Li})$
1	0.4	1.2	1748.53	-180.818	1.07188	1.16685	1.43328
2	0.4	1.4	1375.48	-121.899	0.981565	1.09891	1.25494

Table 2. Energies (in MeV) and rms charge radii (in fm) of $2 \le A \le 4$ nuclei

	E_d	R_d	E_t	R_t	E_τ	R_τ	E_α	R_α
Potential 1	-2.225	2.103	-8.482	1.765	-7.709	1.780	-28.296	1.643
Potential 2	-2.225	2.153	-8.482	1.823	-7.742	1.839	-28.296	1.687
Experiment[2])	-2.225	2.095	-8.482	1.70	-7.718	1.88	-28.296	1.674

Table 3. Energies of two-cluster nuclei in MeV

Potential	$E(^5He)$	$E(^6Li)$	$E^*(^6Li,L=2)$	$E(^7Li)$	$E(^7Be)$	$E(^8Be)$
1 for ^6Li	-26.46	-31.99	3.74	-37.64		-55.49
1 for ^7Li	-26.69	-33.41		-39.24	-37.61	-56.12
2 for ^6Li	-26.64	-31.99	3.76	-38.16		-55.79
2 for ^7Li	-26.81	-32.85		-39.24	-37.68	-56.49
Experiment	-26.08[a]	-31.99	3.59[a]	-39.24 -39.09[a]	-37.60 -37.46[a]	-56.50

[a]Weighted average of a spin-orbit multiplet.

experiment are about as large as the experimental errors. (Note, however, that R_t and R_τ cannot be reproduced simultaneously.) The calculated and measured energies of the two-cluster nuclei, including an excitation energy E*, are listed in table 3. It is seen that all energies can be reproduced simultaneously with a remarkable accuracy of ~1 MeV.

The rms charge radius of ^6Li is obtained to be 2.52 and 2.59 fm with potentials 1 and 2, respectively, while the measured value[2]) is 2.56 fm. The $B(E2,3^+\rightarrow 1^+)$ transition probability also agrees with the data within the experimental error[3]). The calculated monopole charge form factor

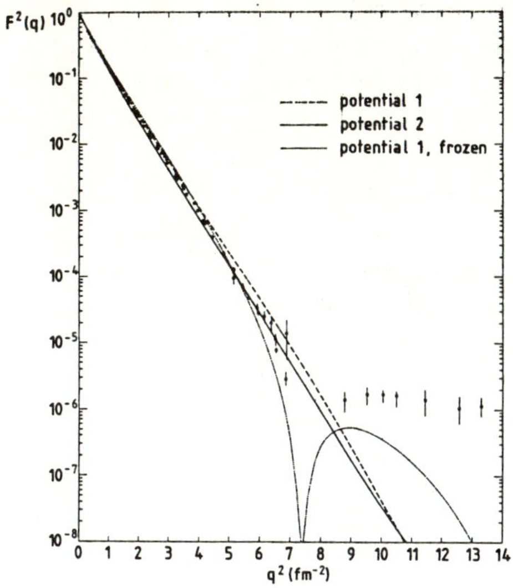

Fig. 1. Squared charge form factor of ^{6}Li. Experiment from ref.[4]).

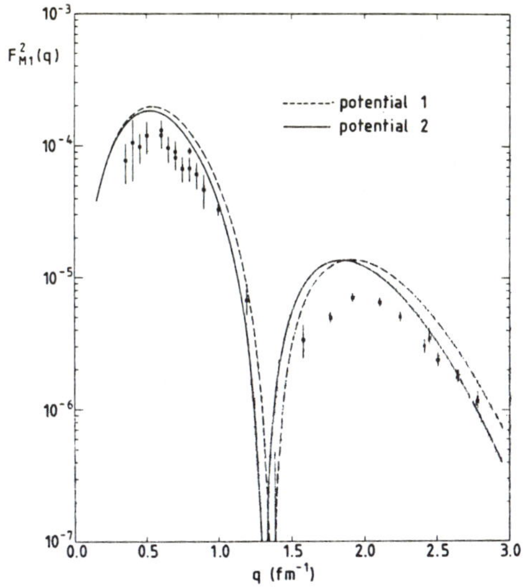

Fig. 2. Squared magnetic dipole form factor of ^{6}Li. Experiment from ref.[7]).

(fig. 1) shows an improvement on previous cluster-model calculations[5,6]) in the small-momentum region, but fails to reproduce the diffraction dip. [The charge form factor of the α particle calculated in our model behaves similarly[3]).] The calculated magnetic dipole form factor (fig. 2) also reproduces the data reasonably. The prediction for the $3^+ \rightarrow 1^+$ charge quadrupole transition form factor is even better[3]).

The two-fragment decomposition properties should provide the most stringent tests of the two-cluster model. The spectroscopic factors s(α+d) and s(t+τ) and the asymptotic

Table 4. Fragmentation properties of ^{6}Li

Potential	s(α+d)	$\bar{C}$(α+d)	s(t+τ)
1	0.926	3.14	0.574
2	0.929	3.31	0.580
Experiment	1.05 (ref.[8]))	2.15 (ref.[9]))	(0.73÷1.34)s(α+d) (ref.[10]))

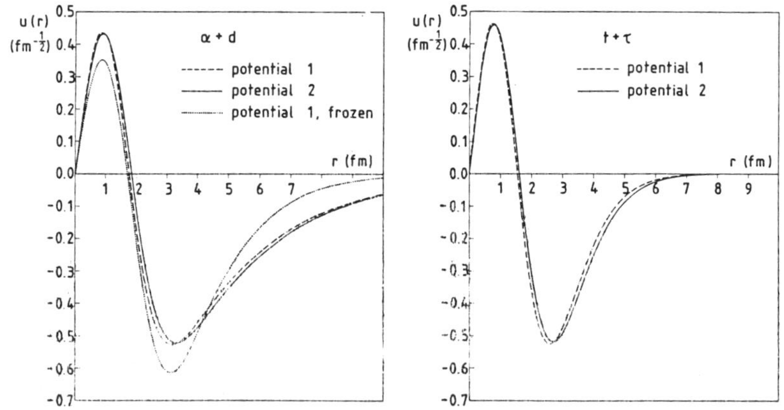

Fig. 3. The α+d and t+τ fragmentation amplitudes of ^{6}Li.

normalization constant $\bar{C}(\alpha+d)$ are shown in table 4. Unfortunately, because of uncertainties in the analyses, there is only one really reliable experimental value, that of $\bar{C}$, and its reproduction is of moderate quality. The amplitude functions u(r) belonging to $s(\alpha+d)$ [and $\bar{C}(\alpha+d)$] and to $s(t+\tau)$ are shown in fig. 3, while the $\alpha+d$ fragmentation strength $G^2(q)$, which is proportional to the squared Fourier transform of the former, is shown in fig. 4. Though this G^2 represents an improvement on the more conventional models[6]), it is still too large in the low-momentum region, which corresponds with the too large $\bar{C}(\alpha+d)$ obtained. An artificial way of shifting the weight of $G^2(q)$ to higher q values is through dropping the summation from eq. (2), i.e. freezing the clusters in their g.s.'s. We see from figs. 1 and 4 that this not only improves $G^2(q)$ but also brings about a diffraction dip in the charge form factor. Similar effects

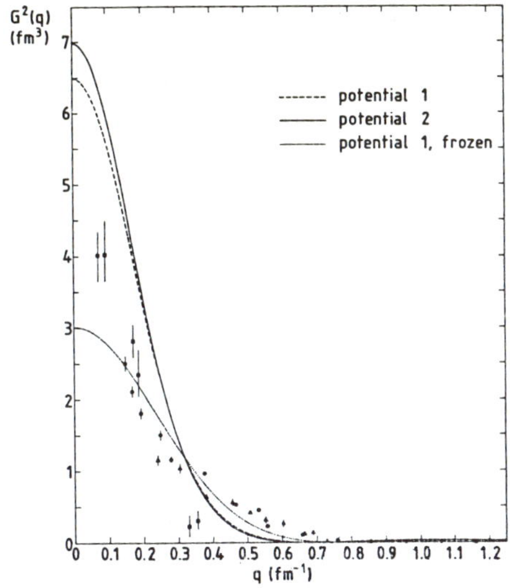

Fig. 4. The $\alpha+d$ fragmentation strength of ^{6}Li. Experiment from refs.[8,11]).

are likely to be achieved in a natural, though technically difficult, way by introducing short-range nucleon-nucleon correlations.

In summary, we attained a surprisingly coherent description of the A≤8 nuclei. The example of the most troublesome nucleus, ^{6}Li, shows that most of the details of the wave functions are also correct. Thus, despite the apparently erratic behaviour, at closer sight these nuclei appear to be remarkably similar and, indeed, perhaps even simple. In ^{6}Li, e.g., the amplitude square of the component that is orthogonal to the g.s. α+d configuration is only 0.085 (potential 2), and this value may be further reduced by short-range correlations. Where the wave function still needs improving is the high-momentum component, which is just the realm of short-range correlations and sub-nucleonic degrees of freedom.

REFERENCES

1. Beck, R., Dickmann, F. and Kruppa, A.T., Phys. Rev. C30, 1044 (1984).
2. Barrett, R.C. and Jackson, D.F., "Nuclear Sizes and Structure" (Clarendon, Oxford, 1977) p. 146.
3. Kruppa, A.T., Beck, R. and Dickmann, F., to be published.
4. Li, G.C. et al., Nucl. Phys. A162, 583 (1971); Suelzle, L.R. et al., Phys. Rev. 162, 992 (1967).
5. Kanada, H., Kaneko, T. and Tang, Y.C., Nucl. Phys. A389, 285 (1982).
6. Beck, R. and Dickmann, F., private communication.
7. Bergstrom, J.C., Kowalski, S.B. and Neuhausen, R., Phys. Rev. C25, 1156 (1982); Lapikás, L., Amsterdam (1978) unpublished.
8. Dollhopf, W. et al., Phys. Lett. 58B, 425 (1975).
9. Bornard, M.P. et al., Nucl. Phys. A294, 492 (1978).
10. Roos, P.G. et al., Nucl. Phys. A257, 317 (1976).
11. Bachelier, D., PhD. thesis, Orsay (1971) unpublished.

THE Sp(6, R) MODEL APPLIED TO ^{16}O AND ^{40}Ca

F. Arickx and J. Broeckhove

Rijksuniversitair Centrum
Dienst Teoretische en Wiskundige Natuurkunde
Groenenborgerlaan 171, B-2020 Antwerpen, Belgium

Abstract : A full symplectic model calculation using the Generator Coordinate method is presented for the spherical nuclei ^{16}O and ^{40}Ca. Skyrme-type as well as Gaussian-type interactions are used.

I. Introduction

The symplectic model has become increasingly popular during the last decade, because its algebraic structure yields a theory of nuclear collective phenomena within the framework of the microscopic shell model [1]. Various submodels have been constructed aimed at explaining specific collective excitations [2]. Parallel to the conceptual development, there has been significant progress in the computational techniques used to diagonalize semi-realistic effective interactions in the symplectic model spaces.

In this paper we apply the Sp(6, R) model to the light spherical nuclei ^{16}O and ^{40}Ca. We consider semi-realistic interactions that are familiar from the litterature, namely the Brink-Boeker force B1 [3] and the Skyrme forces MDI [4] and SkM [5]. We use the Generator Coordinate Method (GCM) as a technique to diagonalize the nuclear Hamiltonian associated with these forces in the symplectic modelspace.

The results that we obtain consist of (1) ground state properties which are of little interest since they do not improve on any existing calculations, and (2) giant monopole and quadrupole resonance properties which are to be compared with RPA results, and with experiment to the extent that data are available. In this paper our attention goes to the analysis of the giant resonance results in terms of symplectic submodels that correspond to restrictions on the deformed shapes that the nucleus may have (e.g. volume-conservation, axial symmetry,..etc.). The GCM is well-suited to such an analysis.

II. The GCM formulation of the Sp(6, R) model in spherical nuclei

Within the context of the symplectic model, the GCM was first used in the stretched symplectic approximation (originally called the Sp(2, R) model) [6] and later also in the SO(3)xD model [7]. Here we consider in detail the application of the full Sp(6, R) model to spherical nuclei. For such nuclei, the symplectic model space is built upon a bandhead with Sp(6, R) labels [w, w, w] (w equals 2.875 for ^{16}O and 9.875 for ^{40}Ca), and whose wave-function is the closed shell ground state configuration.

With respect to the sp(6, R) algebra (s is the particle index, i and j stand for x, y, z and b^+, b are harmonic oscillator creation and annihilation operators):

$$A^+_{ij} = \Sigma(s)\ b^+_{is}\, b^+_{js}$$

$$A_{ij} = \Sigma(s)\ b_{is}\, b_{js} \tag{1}$$

$$C_{ij} = \Sigma(s)\ (b_{is}\, b^+_{js} + b^+_{is}\, b_{js})$$

the bandhead $|0\rangle$ is a lowest weight state, i.e.

$$A_{ij}\, |0\rangle = 0$$

$$C_{ij}\, |0\rangle = w\, \delta_{ij}\, |0\rangle \tag{2}$$

The Sp(6, R) model space is generated by the repeated action of the raising operators upon $|0\rangle$ e.g.

$$|2\hbar\omega, L=0, M=0\rangle = (1/\sqrt{6w})\,(A^{+}_{xx} + A^{+}_{yy} + A^{+}_{zz})\,|0\rangle$$

$$|2\hbar\omega, L=2, M=0\rangle = (1/\sqrt{12w})\,(2A^{+}_{zz} - A^{+}_{xx} - A^{+}_{yy})\,|0\rangle \qquad (3)$$

for the states up to $2\hbar\omega$.

Rather than construct all basis states, the GCM approach is based on the observation [8] that the symplectic space can also be spanned by the states

$$\Psi_S(\mathbf{r}_1,\dots,\mathbf{r}_A) = (\det S)^{-2w}\;\Psi_0(S^{-1}\mathbf{r}_1,\dots,S^{-1}\mathbf{r}_A) \qquad (4)$$

where S is a 3-by-3 positive definite symmetric matrix. It deformes and rotates the initial Ψ_0 with scales and angles determined by decomposing $S = R\,D\,R^{-1}$ into rotation (R) and dilation (D) matrices. Thus any Ψ can be written as

$$\Psi(\mathbf{r}_1,\dots,\mathbf{r}_A) = \int \Pi dS_{ij}\, f(S)\,\Psi_S(\mathbf{r}_1,\dots,\mathbf{r}_A) \qquad (5)$$

and in this representation the Schrödinger equation is transformed into a Hill-Wheeler equation

$$\int \Pi dS_{ij}\,(\,\mathcal{H}(S,S') - E\,\mathcal{N}(S,S')\,)\,f(S) = 0. \qquad (6)$$

Formally this defines a 6-parametric GC problem but due to rotational invariance one can show that the angular dependence in $f(S) = f(R, D)$ must be that of the Wigner D-function, which transforms the problem to a 3 GC parameter (D) problem plus an angular momentum projection. The calculation of the kernels in equation (6) is exceedingly simple [8], making the whole scheme computationally feasible, despite the many-dimensional integrations involved in solving (6).

III. Results and discussion

In figures 1 and 2 we show the results of the full Sp(6, **R**) calculation for ^{16}O and ^{40}Ca and the forces MDI, SkM and B1. The spectra are given in terms of excitation energy with respect to the ground-state energy which in the case of ^{16}O is -139.0, -139.2 and -106.6 for the respective forces, and -410.4, -408.2 and -323.4 for ^{40}Ca.

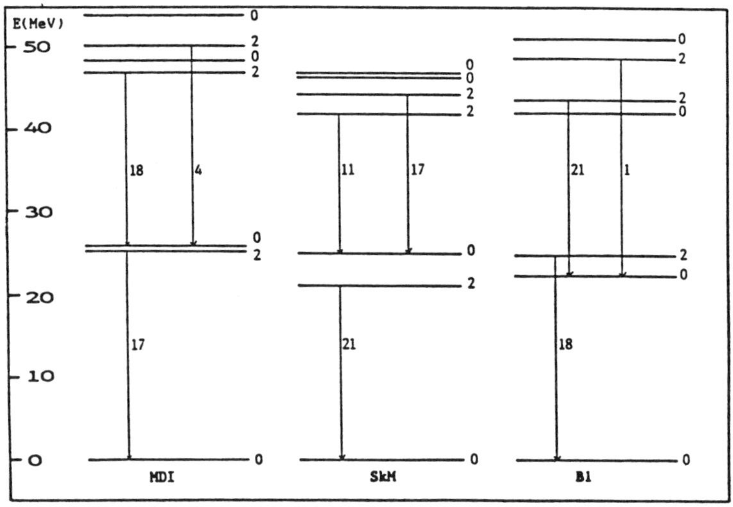

Figure 1 : Sp(6, **R**) spectra for ^{16}O with the interactions MDI, SkM and B1; BE(2) rates are given in units e^2fm^4.

For both nuclei one notices, apart from a reasonable description of the L=0 giant monopole resonance, that the various forces yield very different spectra for the higher lying L=0, and all L=2 states. In particular the splitting between the monopole and quadrupole resonance states is extremely force dependent; there is also apparent an inversion of the monopole and quadrupole resonance states with B1 in comparison to the Skyrme forces. Since the compressibility is taken into account in fitting the

Skyrme force parameters, the correct energy for the giant monopole resonance is to be expected, contrary to the quadrupole resonance energy. The latter observation is strengthed by inspection of the higher L=2 states and their BE(2) rates. It is clear that these forces are not well adapted to the description of such states.

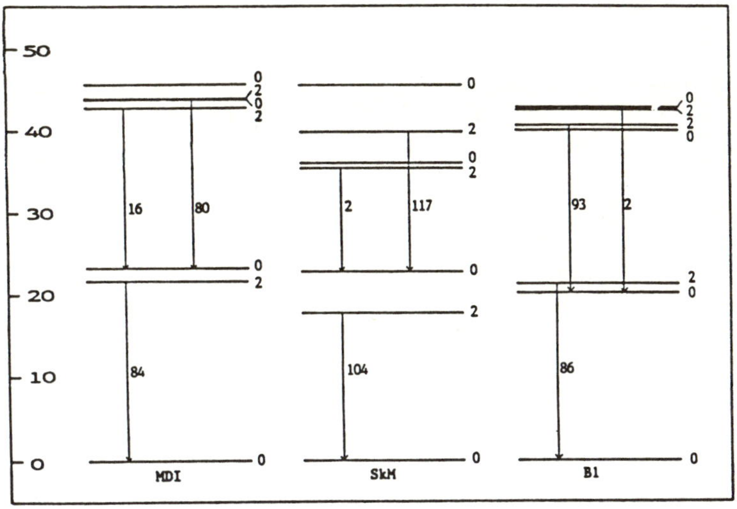

Figure 2 : Sp(6, R) spectra for ^{40}Ca with the interactions MDI, SkM and B1; BE(2) rates are given in units e^2fm^4.

In figure 3 we show the spectra obtained with the full Sp(6, R) model calculation on ^{40}Ca with the MDI force, compared to two sets of spectra derived from complementary modes of vibration. These have been considered earlier [9], [10] in an attempt to identify the vibrational eigenmodes of the nuclei. The first set comprises (1) the α-mode or breathing mode, obtained by restricting the generator coordinates to $D_{xx} = D_{yy} = D_{zz}$ in the dilational matrix D; (2) the β-mode which is the volume-conserving quadrupole mode, i.e. $D_{xx} = D_{yy}$, D_{zz} with $D_{xx}D_{yy}D_{zz}$ = constant; (3) the γ-mode, a non axial volume conserving mode given by D_{xx} = constant, $D_{xx}D_{yy}D_{zz}$ = constant. The second set consists of the Sp(2, R) stretching modes : (1) the z-mode, an axially symmetric stretching of the z-axis with $D_{xx} = D_{yy}$ = constant, and (2) the x-y-mode or axially symmetric stretching of the x-y plane described by $D_{xx} = D_{yy}$, D_{zz} = constant.

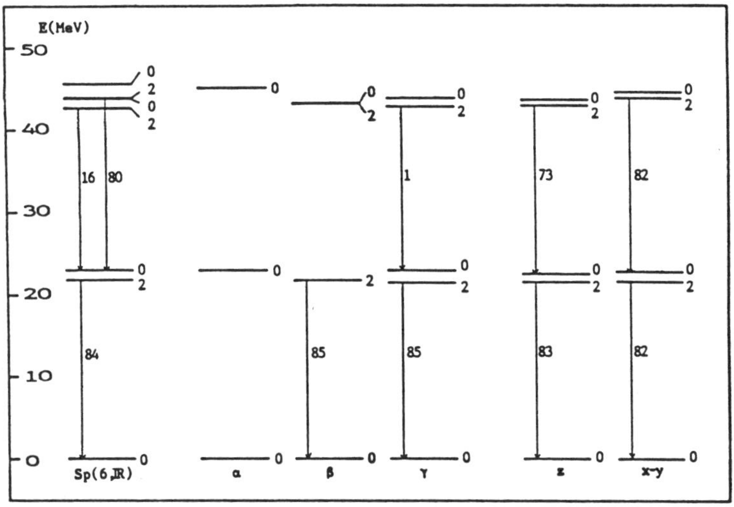

Figure 3 : Spectra of the submodels corresponding to the various geometric restrictions (see text); BE(2) rates are given in units e^2fm^4.

In figure 3 one immediately notices for the giant resonance energy region (20 - 25 Mev) the absence of the L=2 state in the α-mode, respectively the L=0 state in the β-mode. All other modes present a picture that is equivalent to that obtained with the full Sp(6, R) calculation, as well for energies as for BE(2) rates. Taking into account the pronounced $2\hbar\omega$ character of the giant resonance states the above observations can easily be explained. The $2\hbar\omega$ component of the various geometric modes lead to the following intrinsic states :

$$|\alpha, 2\hbar\omega\rangle - (A^+_{xx} + A^+_{yy} + A^+_{zz})\,|0\rangle$$

$$|\beta, 2\hbar\omega\rangle - (2A^+_{zz} - A^+_{xx} - A^+_{yy})\,|0\rangle$$

$$|\gamma, 2\hbar\omega\rangle - (A^+_{zz} - A^+_{yy})\,|0\rangle \qquad (7)$$

$$|z, 2\hbar\omega\rangle - (A^+_{zz})\,|0\rangle$$

$$|x\text{-}y, 2\hbar\omega\rangle - (A^+_{xx} + A^+_{yy})|0\rangle$$

Clearly all intrinsic states yield the same L=0 respectively L=2 state, and it is also evident that the α- and β-modes contain no L=2, respectively L=0 contribution. The absence of the L=2 and L=0 states in the α- and β-modes can also be concluded directly from the sphericity and volume conservation in the respective modes. As far as the higher lying states are concerned neither set of modes sufficiently fits any of the exact states, particularly in view of the BE(2) rates. Our conclusion from this analysis is that none of the geometric pictures provides an adequate description of the vibrational eigenstates.

References

[1] Rowe D.J., Rep. Prog. Phys 48 (1985) 1419
Carvalho J. et al., Nucl. Phys A452 (1986) 240
[2] Arickx F., Broeckhove J. and Deumens E., Nucl. Phys. A318 (1979) 269
Broeckhove J. and Van Leuven P., Phys. Rev. C29 (1984) 628
[3] Brink D. M. and Boeker E., Nucl. Phys. A91 (1967) 1
[4] Faessler A. et al., Nucl. Phys. A239 (1975) 477
[5] Krivine H. et al., Nucl. Phys. A336 (1980) 155
[6] Arickx F., Broeckhove J. and Deumens E., Nucl. Phys. A318 (1979) 269
Arickx F., Broeckhove J. and Deumens E., Nucl. Phys. A377 (1982) 121
[7] Vassanji M. G. and Rowe D.J., Phys. Lett. B125 (1983) 103
Vassanji M. G. and Rowe D.J., Phys. Lett. B127 (1983) 1
[8] Arickx F. and Broeckhove J., in preparation
[9] Arickx F., Broeckhove J. and Deumens E. Phys. Lett. B106 (1981) 275
Arickx F. et al., Nucl Phys. A398 (1983) 467
[10] Abgrall Y. et al., Nucl. Phys. A346 (1980) 431

COLLECTIVE M1 TRANSITIONS IN ^{46}Ti[a)]

O. Civitarese[b)], Amand Faessler, R. Nojarov[c)]
Institut für Theoretische Physik, Universität Tübingen, D-7400, Tübingen, FRG

Abstract A dominance of the spin contribution to the M1 transition matrixelements is found for the strong 1^+ state of ^{46}Ti at 4.32 MeV using quasiparticle RPA, while the giant 1^+ resonance at 26.8 MeV is essentially an orbital excitation.

A low-lying $K^{\pi}=1^+$ state, which is related to the ground state through a strong magnetic dipole transition was found recently[1] in ^{46}Ti (Table 1). We describe here the $K^{\pi}=1^+$ states in the quasiparticle RPA using an isospin-dependent quadrupole-quadrupole residual interaction. The single-particle basis is obtained from an axially-symmetric Woods-Saxon potential[2] with $V_s= -45$ MeV, $V_d=-55$ MeV, $a_n=a_p=0.67$ fm, $r_n=1.3$ fm, $r_p= 1.28$ fm, $\lambda_0^n=21.7$, $\lambda_0^p=17.2$, $\beta=0.27$. The density of the levels around the Fermi surface, which is very low in light nuclei, had to be increased in order to

a) Supported by the Alexander von Humboldt Foundation and the Bundesministerium für Forschung und Technologie.
b) Permanent address: Department of Physics, University of La Plata, Argentina.
c) Fellow of the A v Humboldt Foundation. Permanent address: Institute of Nuclear Research and Nuclear Energy, Bulgarian Academy of Sciences, Sofia 1784, Bulgaria.

TABLE 1. Orbital and spin contributions to the M1 transition matrix elements for ^{46}Ti.

	E [MeV]	B(M1) [μ_N^2]	orbital	spin
exp.	4.32	1.0(0.2)	1.12	0.93
th.	4.30	0.92	0.19	1.78
th.	26.79	0.18	0.87	0.004

get stronger M1 transition. The largest value was obtained with proton single-particle levels ε([321] 1/2)=-1.3 Mev, ε([321] 3/2)=-4.5 MeV and ε([330] 1/2)=-4.2 MeV. The BCS pairing gaps were: Δ_n=1.5 MeV, Δ_p=1.2 MeV and the gyromagnetic factors: $g_1(n)=0$, $g_1(p)=1$, $g_s(t_z)=0.7g_s^{free}(t_z)$.

We found a strong low-energy 1^+ state in ^{46}Ti (Fig.1 and Table 1), lying at E=4.3 MeV. It has predominantly isovector character and exhausts 21% of the energy-weighted sum rule for M1 transitions. The energy and the B(M1) value are both in a good agreement with the experimental data. The main components of the RPA wave function are 2qp proton pairs from the $1f_{7/2}$ shell: 78% [321] 1/2, [321] 3/2 and 14% [330] -1/2, [321] 1/2. The ratio "orbital/spin" between the orbital and spin contribution to the M1 transition probability B(M1)= (orbital+ spin)^2 $3/4\pi$ is 0.1 in our case (Table 1), while the experimental estimate[1], which is also model dependent, has a lower limit 1.2. A ratio 1.4 was found previously with a rotational model[3] but B(M1) was overestimated by a factor of four. In our case the main contribution to the matrix element comes from the spin-flip transition [321] 1/2$\rightarrow$[321] 3/2.

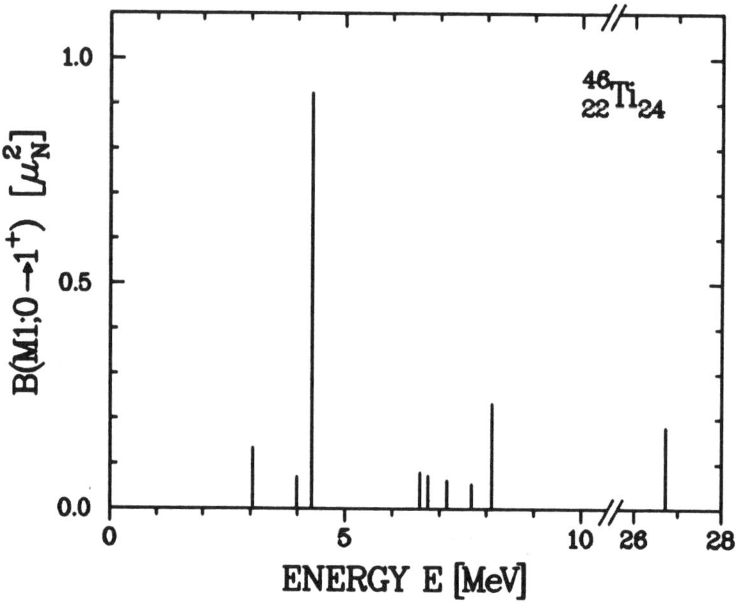

FIGURE 1. M1 strength distribution in ^{46}Ti (all the states with B(M1)↑ > 0.05 $\mu_N{}^2$ are displayed)

In contrast to the strong low-energy 1^+ state, the giant resonance at 26.8 MeV is an almost purely orbital excitation (Table 1). It is strongly collective (there is no qp pair with more than 10% contribution to the wave function) and consists of $\Delta N=2$ neutron and proton 2qp pairs.

REFERENCES

1. C. Djalali et al, Phys. Lett., 164B, 269 (1985)
2. R. Nojarov, J. Phys. G: Nucl. Phys., 10, 539 (1984).
3. L. Zamick, Phys. Lett., 167B, 1 (1986).

YRAST E2 STRENGTHS AND SIDE FEEDING TIMES IN ^{70}Se AND ^{72}Se*

J. HEESE, K.P. LIEB, L. LÜHMANN, F. RAETHER, B. WÖRMANN
II. Phys. Inst., Universität Göttingen, D-3400 Göttingen, Fed. Rep. Germany
D. ALBER, H. GRAWE
Hahn Meitner Institut, D-1000 Berlin, FRG
J. EBERTH, T. MYLÄUS
Inst. f. Kernphysik, Universität zu Köln, D-5000 Köln, Fed. Rep. Germany

Abstract Level energies and lifetimes of the ground band in ^{70}Se appear to be consistent with a transition from oblate to prolate shape at $\hbar\omega \cong 0.5$ MeV.- Experimental and statistical model estimates of side feeding times in ^{14}N and ^{36}Ar induced reactions are discussed.

INTRODUCTION

Shape coexistence in the light Se and Kr isotopes has been discussed for many years. Recent studies of aligned (1qp) and (2qp) bands and Hartree Fock Bogoliubov cranking calculations have given a clear picture of how the nuclear shape develops as function of the rotational frequency and nucleon number |BEN84, NAZ85, LUH85, LIE85|. These results have motivated us to resume the investigation of even-even nuclei near the N = Z line, in particular ^{70,72}Se and ^{74,76}Kr using ^{14}N, ^{16}O, ^{36}Ar and ^{40}Ca induced fusion reactions |HEE86, BER86|.

STRUCTURE OF ^{70}Se

Yrast states in ^{70}Se have been previously established via the reaction ^{58}Ni(^{14}N,pn) up to spin 8^+ resp. 7^- |WAD80, AHM81|; recently, the OSIRIS group has proposed yrast states up to spin 14^+ from their study of the reaction ^{40}Ca(^{36}Ar,α2p) |MYL86|. We have applied the recoil distance technique to both reactions using the 39 MeV ^{14}N beam of the Cologne tandem accelerator and the 115 MeV

^{36}Ar beam of VICKSI. Rolled 0.9 μm thin ^{nat}Ni foils and a 350 μg/cm^2 ^{nat}Ca layer sandwiched between two thin Au foils (for protection against oxidation) were used as plunger targets. In the ^{58}Ni + ^{14}N run the strong ^{70}Se lines were also measured in coincidence with a large volumen NE213 neutron filter; details of the plunger measurements and analysis are given in |HEE86|. The mean lives and electromagnetic transition strengths are summarized in Table 1.

TABLE I Lifetimes and B(Eλ) values in ^{70}Se

Transition			Lifetime τ (ps)			B(Eλ) ($e^2fm^{2\lambda}$)
E_γ	$I_i \rightarrow$	I_f	(a)	(b)	adopted	
945	2^+	0^+	1.3(3)	1.7(3)	1.5(3)	E2: $720(^{180}_{120})$
1094	4^+	2^+	1.5(5)	1.4(3)	1.4(3)	E2: $372(^{102}_{66})$
964	6^+	4^+	3.1(9)	4.7(8)	3.9(8)	E2: $251(^{66}_{43})$
1035	8^+	6^+	---	2.9(15)^c	<6	E2: >114
1601	2^+	0^+	4.7(13)	---	4.7(13)	E2: $8(^4_2)$
656	2^+	2^{+d}				E2: $700(^{370}_{210})$
1572	3^-	2^+	6.1(9)	---	6.1(9)	E1: $2.7(^5_3)10^{-5}$
1349	5^-	4^+	8.8(24)	13(5)^c	8.8(24)	E1: $2.9(^9_6)10^{-5}$

a Reaction ^{58}Ni(^{14}N,pn) at 39 MeV
b Reaction ^{40}Ca(^{36}Ar,α2p) at 115 MeV
c Effective lifetime not corrected for feeding
d B(M1) = $8.7(^{15}/_8)\ 10^{-3}\ \mu_N^2$ from E2/M1 mixing ratio $\delta = -1.0(^1/_2)$ |AHM81|.

The experimental B(E2) and $I_x(\omega)$ values of the stretched E2 ground band transitions are compared in Figure 1 with the even-even neighbors ^{72}Se and ^{68}Ge |HAM76, LIE77, HEL78, PAR83|. It is evident that the E2 strengths in the three nuclei behave quite differently, with B(E2) values decreasing in ^{70}Se, sharply increasing in ^{72}Se and being constant in ^{68}Ge. The reduction of B(E2) values in ^{70}Se by a factor of 3 emphasizes in a very drama-

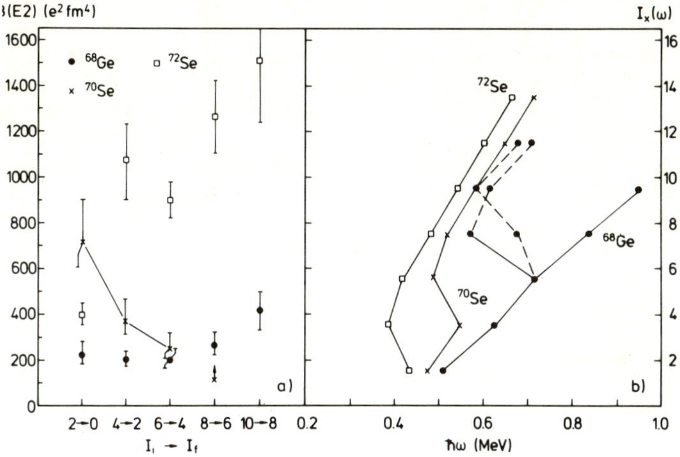

Fig.1: B(E2) and $I_x(\omega)$ values of the ground bands in ^{70,72}Se and ^{68}Ge |AHM81, LIE77, PAR83|.

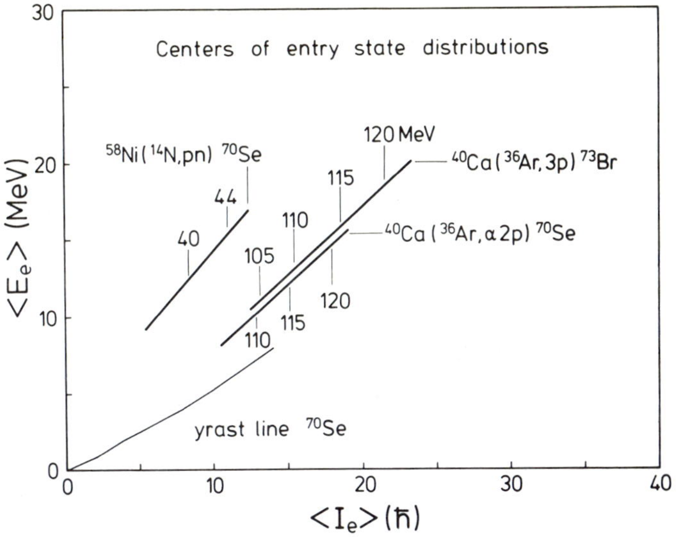

Fig.2: Centers of entry state distributions of the reactions ^{58}Ni(^{14}N,pn)^{70}Se, ^{40}Ca(^{36}Ar,α2p)^{70}Se and ^{40}Ca(^{36}Ar,3p)^{73}Br as function of the beam energy |HEE86|.

matic way the slight backbending of the energies visible between spin 4 and 6. Assuming axial symmetry for the ground and lowest 2^+ state, one finds $|\beta_2| = 0.30(3)$ from the measured $B(E2,2^+\rightarrow 0^+)$. Above spin 8^+ the ground band merges into a band with nearly constant moment of inertia, similar to the band in ^{72}Se which has prolate deformati. . $\beta_2 = 0.31(2)$ |LIE77, WOR85|. At $\beta_2 = +0.3$, proton and neutron $g_{9/2}$ (2qp) alignment can take place, leading to an aligned single-particle spin of $i_x \cong 8\hbar$, in contrast to the data. An alternative, although not unique interpretation of this band crossing effect is to assume a transition from an oblate ground state to a prolate s-band similar to ^{72}Se. Hartree Fock calculations including pairing correlations |NAZ85, NAZ86| indeed predict oblate deformation for the ^{70}Se ground band ($\beta_2 = -0.30$). Furthermore, many nuclei in this mass region are known to be γ-instable so that (1qp) and (2qp) alignment drives them to prolate shapes ($-30^\circ \leq \gamma \leq +10^\circ$ |LUH85, LUH86, WOR85|). Assuming strongly mixed wave functions for the 4^+ and 6^+ yrast states in ^{70}Se, one expects a cancellation in the E2 matrix elements as observed experimentally, but large $B(E2) \cong 1000\ e^2fm^4$ above spin 8^+. Lifetime measurements in this region are in progress to test our prediction.

ESTIMATES OF CONTINUUM FEEDING TIMES

In the analysis of the recoil distance data, we have assumed negligible side feeding times $\tau_f \ll \tau$. In the following discussion we present arguments which justify this assumption.

a) We first note in Table I that the lifetimes obtained in both reactions are in very good agreement with each others. Side feeding times in light heavy ion induced reactions have been found to be of the order of 0.05 - 0.15 ps |LIE77, LUH85, HLM78|. The present data give an upper limit of $\tau_f < 0.5$ ps for the ^{40}Ca + ^{36}Ar reaction, as long as the projectile energy is close to the Coulomb barrier.

b) Concurrently with the ^{70}Se data, we have also analyzed decay functions of the lowest three yrast transitions in ^{72}Se populated in the reaction ^{60}Ni(^{14}N,pn) and ^{40}Ca(^{36}Ar,4p). The deduced mean lives (see Table II) show again very good agreement for both reactions and with the results of most previous measurements |HAM76, LÖB75, HEL78, KIP79|.

c) A further experimental limit of τ_f for the ^{40}Ca + ^{36}Ar system comes from a DSA line shape analysis of yrast transitions in ^{73}Br excited in the reaction ^{40}Ca(^{36}Ar,3p) at 105 MeV beam energy. Side feeding times of τ_f(1861 keV, 17/2$^+$) = 0.20(15) ps and τ_f(2856 keV 21/2$^+$) = 0.5(3) ps have been deduced |HEE86|.

TABLE II LIFETIMES OF THE LOWEST YRAST STATES IN ^{72}Se OBTAINED IN DIFFERENT REACTIONS

Reaction; E_b (MeV)	862 2^+	1637 4^+	2467 6^+	Ref.
^{60}Ni(^{14}N,pn); 39	4.0(5)	2.7(3)	2.6(4)	|HEE86|
^{58}Ni(^{16}O,2p); 46-58	5.2(5)	2.7(3)	2.2(3)	|HEL78|
^{58}Ni(^{16}O,2p); 40-45	3.1(6)	1.2(3)	< 0.7^b	|LÖB75|
^{58}Ni(^{16}O,2p); 46	---	$4.5(^{15}_{10})$	2.6(7)	|HAM76|
^{58}Ni(^{16}O,2p); 42-50	4.8(6)	3.8(5)	1.6(5)	|KIP79|
^{40}Ca(^{36}Ar,4p); 115	3.4(5)	2.40(25)	2.5(3)^a	|HEE86|
Adopted value τ(ps)	4.3(5)	2.8(4)	2.3(2)	
B(E2) (e^2fm^4)	$399(^{52}_{41})$	$1044(^{174}_{130})$	902(86)	

a Effective value not corrected for feeding
b Not included in final average value

d) Besides the structure and density of levels above the yrast line, side feeding times mostly reflect the energy and angular momentum balance of the entry states and the discrete levels in consideration. The experimentally found small side feeding times in the mass A ≅ 80 region for projectiles A_1<20 rely on the fact that, on the average, few dipole transitions can make up for this balance |SIM77, HLM78|. The situation may, however, become less favorable for heavier projectiles which shift the entry state distribution closer to the yrast region. For the ^{40}Ca(^{40}Ca,3p)^{77}Rb reaction at 122 MeV beam energy, Lühmann et al. |LUH86| obtained τ_f = 0.4 - 1.0 ps.

We have performed Monte Carlo simulations of the particle and γ-ray flux in the two reactions leading to ^{70}Se, by means of the fusion-evaporation codes CASCADE |PUH77| and PACE, as function of the beam energy. The other evaporation residues discussed in this contribution, e.g. ^{73}Br + 3p and ^{72}Se + 4p, were also considered.

The entry state distribution of the ^{58}Ni(^{14}N,pn) reaction at 39 MeV shown in Figure 2 centers at $\langle E_e \rangle$ = 12 MeV and spin $\langle I_e \rangle$ = 8 $\hbar$. Few dipole and/or quadrupole transitions are sufficient to carry the system to the low lying yrast levels. The situation thus ressembles very much the one discussed previously |HLM78, SIM77|, where it has been shown that the spectral shape and multiplicity of the γ radiation as well as the fusion cross section and particle flux are well accounted for by the statistical model.

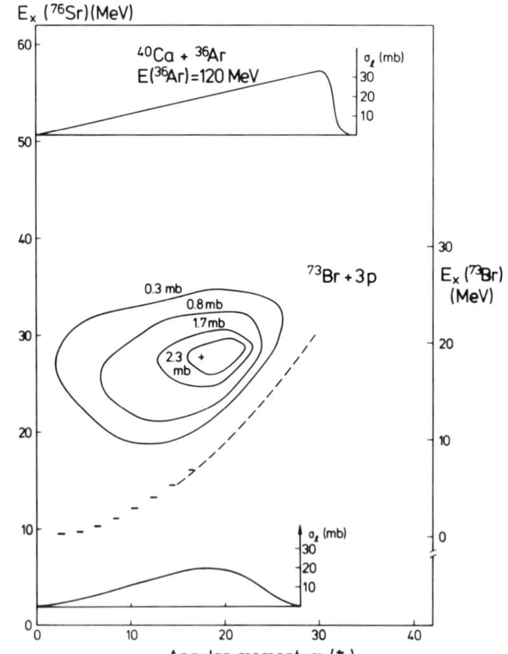

Fig.3: Fusion and entry state distribution of the reaction ^{40}Ca(^{36}Ar,3p)^{73}Br at 120 MeV |CHR86|.

Figure 2 also displays the centers of entry state distributions of the reactions $^{40}Ca(^{36}Ar,\alpha 2p)$ and $^{40}Ca(^{36}Ar,3p)$ as function of the beam energy. The complete entry state distribution of the latter reaction at 120 MeV beam energy is displayed in Fig. 3 together with spin distribution in the compound nucleus ^{76}Sr and the yrast line of ^{73}Br |WOR85,CHR86|.

References

|AHM81| A. Ahmed, et al., Phys. Rev. C24, 1486 (1981)

|BEN 84| R. Bengtsson and W. Nazarewicz, XIX Winter School on Physics, Zakopane/Poland (1984)

|BER 86| F. J. Bergmeister, et al., Intern. Symposium "Symmetries and Nuclear Structure" Dubrovnik, R.A.Meyer, ed. (1986) to be publ.

|CHR 86| F. Christancho and K. P. Lieb, to be publ.

|HAM 76| J. H. Hamilton, et al., Phys. Rev. Lett. 36, 340 (1976)

|HEE 86| J. Heese, et al., Z. Phys. in press

|HEL 78| H. P. Hellmeister, et al., Phys. Rev. C17, 2113 (1978)

|HLM 78| H. P. Hellmeister, K. P. Lieb and W. F. J. Müller, Nucl. Phys. A307, 515 (1978)

|LIE 77| K. P. Lieb and J. J. Kolata, Phys. Rev. C15, 939 (1977)

|KIP 79| V. G. Kiptilyi, et al., Bull. Akad. Sci. USSR, Phys. Ser. 43, 26 (1979)

|LIE 85| K. P. Lieb, L. Lühmann and B. Wörmann, Intern. Symposium "Recent Advances in the Study of Nuclei Off the Line of Stability" Chicago (1985) in press

|LÖB 75| K. E. G. Löbner, et al., Z. Phys. A274, 251 (1975)

|LUH 85| L. Lühmann, et al., Phys. Rev. C31, 828 (1985)

|LUH 86| L. Lühmann, et al., Europhys. Lett. 1, 623 (1986)

|MYL 86| T. Myläeus and J. Eberth, priv. communication

|NAZ 85| W. Nazarewicz, et al., Nucl. Phys. A435, 397 (1985)

|NAZ 86| W. Nazarewicz, priv. communication

|PAR 83| T. Paradellis and C. A. Kalfas, Phys. Rev. C25, 350 (1983); and references therein

|PUH 77| F. Pühlhofer, Nucl. Phys. A280, 267 (1977)

|SIM 77| J. J. Simpson, et al., Nucl. Phys. A287, 362 (1977)

|WAD 80| R. Wadsworth, et al., J. Phys. G6, 1403 (1980)

|WOR 85| B. Wörmann, et al., Z. Phys. A322, 171 (1985)

(n,n'γ) STUDY OF ^{94}Mo USING REACTOR FAST NEUTRONS

TAMAS BELGYA, GABOR MOLNAR, BÉLA FAZEKAS AND ÁRPAD VERES
Institute of Isotopes of the Hungarian Academy of Sciences, H-1525 Budapest, POB 77, Hungary

Abstract The ^{94}Mo vibrational-like nucleus was investigated via (n,n'γ) reactions induced by reactor fast neutrons. Level, decay scheme and multipole mixing ratios are presented and compared to previous model calculation.

INTRODUCTION

Our earlier works on the transitional ^{98}Mo and ^{100}Mo nuclei[1,2,3] pointed out that these nuclei have mixed structures of normal and intruder states. The heavier even-even Mo nuclei have mainly intruder low lying states while in the lighter even-even Mo nuclei the normal ones dominate and the intruder states increase as the number of neutrons decrease[1].

^{94}Mo has two neutrons outside the N=50 closed shell and two protons outside the Z=40 closed subshell. Since it has only a few valence nucleons it cannot be described well within the interacting boson model, but more successful calculations have been made within the framework of shell model[4,5]. Both the exact and weak-coupling shell model calculations made by Kumar et al. using ^{88}Sr core are in good agreement with the experimental low-lying energy spectrum. Adamides et. al. have made two calculations assuming ^{90}Zr and ^{88}Sr cores and concluded that more experimental information is required for more obvious comparison between experimental and theoretical levels.

Another interesting attempt to describe level energies, electromagnetic properties and spectroscopic factors for ^{94}Mo has been made by Paar[6] within the cluster vibrational model assuming ^{92}Mo as a vibrational core coupling to the two valence neutrons.

The (n,n'γ) reaction is highly suitable for studying these low-lying, low-spin states for it combines the nonselectivity of statistical excitation with the high precision of γ-spectroscopic methods. The recent experimental results on ^{94}Mo are discussed and compared with previous experiments and theoretical calculations.

The present experiment is a part of the studies having in view the examination of shape transitional even-even nuclei around A=100 mass number.

EXPERIMENTS AND RESULTS

The experimental set up is the same as described in ref.[2,7]. The target employed was 10g of molybdenum metal enriched to 88% in mass 94. Single γ-ray at angle 90° to the beam and angular distributions at six angles were measured using Ge(Li) detectors.

In the ^{94}Mo(n,n'γ) reaction over 60 γ-rays were identified. Almost all known γ-rays up to 3.5 MeV with spin values up to 6 were observed and placed, including some new ones, in the decay scheme (Figure 1). Precise level energies and intensities of the γ-rays are also presented. Three new levels have been introduced at 2993 keV, 3398 keV and 3456 keV level energies. Some of the new unplaced γ-transitions have higher than 1600 keV energies with low intensity which couldn't be placed in the decay scheme. It would be inter-

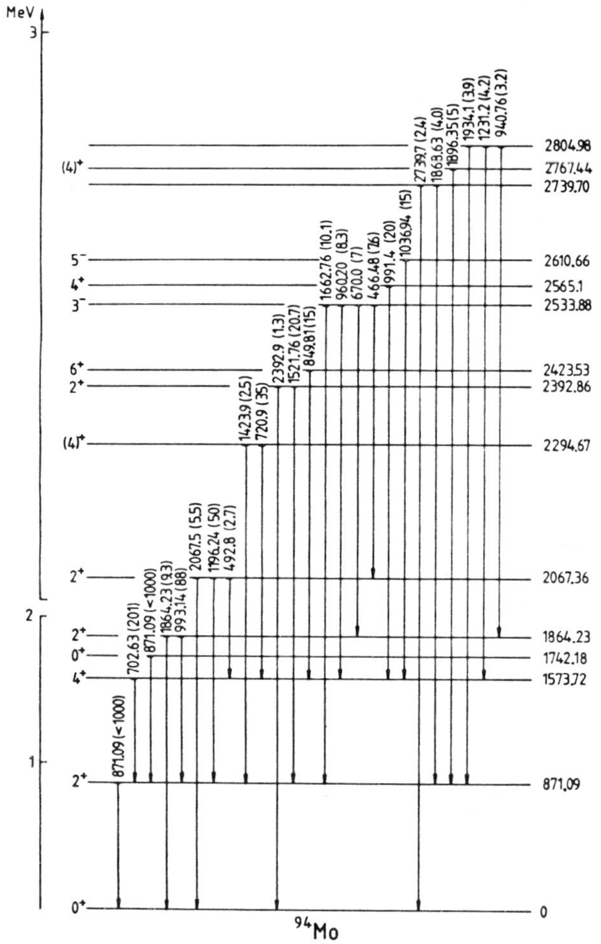

FIGURE 1/a. Level and decay scheme observed in the ^{94}Mo (n,n'γ) reaction.

esting to know whether one or more of them decay to the first 0^+ state. Due to the relatively large number of the high energy γ-transitions mentioned above and the level population requirements[2] it is evident that the level density in the range 3 MeV and 3.5 MeV is about twice of the

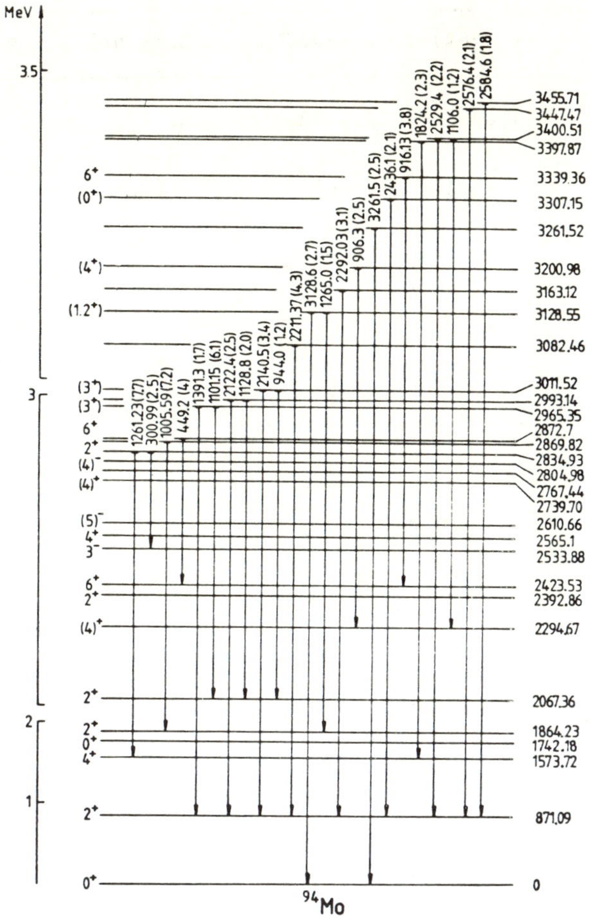

FIGURE 1/b. Level and decay scheme observed in the ^{94}Mo (n,n'γ) reaction. (continued)

known.

The results of the angular distributions measurements are presented in Table I. We expected more results from this measurement (for example the angular distributions of the peaks at 720 keV and 849 keV), but the complexity of the

TABLE I Angular distribution results for ^{94}Mo including cascade corrections

E_i(keV)	E_γ(keV)	J_i^π-->J_f^π	Multipol.	mix.ratio
1864.2	1864.3	2_2^+-->0_1^+	E2	--
1864.2	993.1	2_3^+-->2_1^+	E2/M1	$-1.52^{+0.33}_{-0.51}$
2067.5	1196.2	2_3^+-->2_1^+	E2/M1	$+0.00\pm0.05$ $+2.86^{+0.28}_{-0.26}$
2533.9	1662.8	3_1^--->2_1^+	M2/E1	$-0.03^{+0.02}_{-0.03}$ $-4.10^{+0.7}_{-1.5}$
2565.1	991.4	4^+-->4^+	E2/M1	$-0.07^{+0.12}_{-0.10}$ $+1.13\pm0.42$
2610.7	1036.9	$(5)^-$-->4^+	M2/E1	-0.03 ± 0.02 $-3^{+21}_{-\infty}$
2767.4	1869.4	$(4)^+$-->2^+	M3/E2	0.00 ± 0.7 -15^{+6}_{-66}
2834.9	1261.2	$(4)^-$-->4^+	M2/E1	-0.03 ± 0.05 $+1.08\pm0.30$

spectra did not allow us to determine more data. None of the spin assignments is inconsistent with the earlier assumption[8] although we could not do unique spin assumption on the 2611 keV, 2767 keV and 2834 keV levels. Multipole mixing ratios are presented for the most probable spin assignments. On the basis of the available earlier data on multipole mixing ratios[5,9] it seems to be more realistic if an average value is adopted for the multipole mixing ratios of the 2_2^+-->2_1^+ γ-transition and the 2_3^+-->2_1^+ γ-transition with the first value 0.00 ± 0.05. The second multipole mixing ratio of the $2_3^+ \to 2_1^+$ γ-transition could not be accepted because it would give a very large B(E2) value (200 Wu.) in comparison with the life time of this level $t = 0.046^{+0.014}_{-0.007}$ ps[5].

Finally the doublet at 993 keV assumed by Sugiyama et al.[10] has been separated into a lower energy transition starting from the level at 2565 keV with an energy 991 keV and a higher energy transition starting from the level at 1864 keV with an energy 993.1 keV.

COMPARISON WITH THE MODELS AND THE EARLIER EXPERIMENTS

Figure 2. shows the experimental and the theoretical[5,6,11] level schemes. It can be seen, that the shell model calculations with ^{88}Sr core (SM-2 and SM-3b) give much

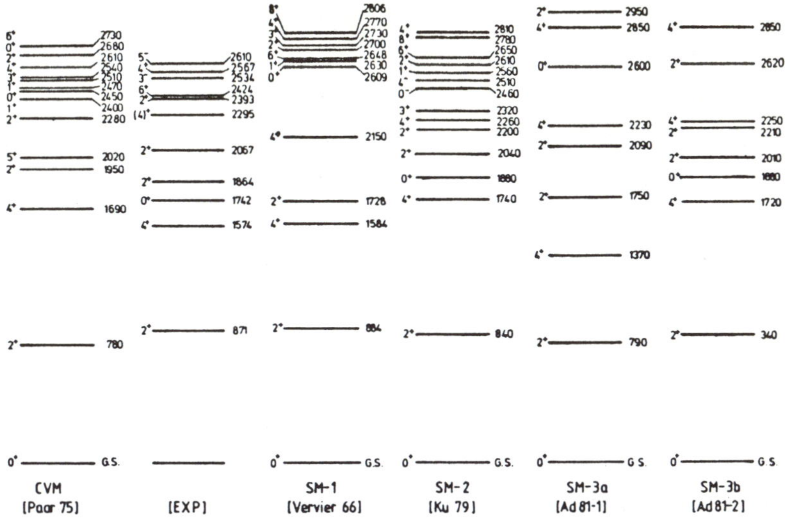

FIGURE 2. The experimental and theoretical level schemes of ^{94}Mo.

better level energy predictions than the other descriptions, particularly in case of the position of the first excited 0^+ state. Recently Wadswart et al.[12] found large two-proton hole component in the first excited 0^+ state of ^{94}Mo, which allows us to take into account the two-proton hole excitations across the Z=40 subshell.

The experimental and calculated[5,6] multipole mixing ratios and R_{i1} values are collected in Table II for $2_i^+ \rightarrow 2_1^+$ transitions. The multipole mixing ratio of the $2_2^+ \rightarrow 2_1^+$ transition observed in this work is about the average of the experimental results published by Aras and Adamides et al. The $2_3^+ \rightarrow 2_1^+$ multipole mixing ratio differs significantly from the value of Adamides et al. and gives a more realistic small B(E2). Despite the opposite sign of the experimental and theoretical values, in relation to the magnitudes of present multipole mixing ratios the first alternative description in calculation I. of Adamides et al. seems to be more probable and the second one may be ignored. This can also be seen in case of the R_{i1} values which contain the appropriate branching ratios and level energies as well. In this way these models, I. and II., cannot predict the sign of the multipole mixing ratios well, but they give a better approach for the B(E2) values. These facts and the extrapolation of the IBA calculation[1] for ^{94}Mo suggest the need of a new calculation with an extended shell model space based on ^{88}Sr core.

Another approach made by Paar[6] took into account directly the interplay of the vibrational and single particle degree of freedom of ^{94}Mo. This approach gives good predictions for the observed electromagnetic properties of the 2_1^+, 2_2^+ and 4_1^+ states of the ^{94}Mo nucleus (see Table II). Nevertheless the level scheme differs from the observed one and

TABLE II The experimental and calculated multipole mixing ratios

(E2/M1)	This work	Pre.exp.	CVM[6]	SM-I[5]	SM-II[3]
$2_2^+ \to 2_1^+$	$-1.52^{+0.33}_{-0.51}$	$-0.87^{+0.09}_{-0.17}$ 5	1.04[a]	+0.369	+0.148
		$-3.2^{+0.7}_{-0.9}$ 5		-0.106[b]	
		$-2.00^{+0.35}_{-0.18}$ 9			
$2_3^+ \to 2_1^+$	0.00 ± 0.05	$+0.62^{+0.21}_{-0.18}$ 5		-0.128	+0.176
				+0.444[b]	
$2_4^+ \to 2_1^+$		$-1.85^{+039}_{-0.057}$ d			
R_{21}	0.0053(2)	0.0087(5)[c]	0.03	0.02	0.05
		0.0041(2)[d]		0.52[b]	
		0.0052(21)[14]			
		0.0047(2)[e]			
R_{31}	2.26[f]	0.026(23)[5]		0.51	0.89
	0.42[g]			0.02[b]	
R_{41}		0.014(8)[d]			

$R_{i1}=B(E2,2_i^+ \to 0_1^+)/B(E2,2_i^+ \to 2_1^+)$, a/ absolute value, b/ alternative description of the 2_2^+ and 2_3^+ states, c/ calculated with $\delta=-0.87$, d/ calc. with $\delta=-3.2$, e/calc. with $\delta=-2.00$, f/ one σ lower limit, g/ two σ lower limit.

shows some similarity to shell model calculation I. of Adamides et al[5]. For instance the first excited 0^+ state is too high in energy. The calculation with the same unified model approach made by Paradellis et al.[13] in the zirconium region shows similar discrepancies in the theoretical and experimental level scheme. They pointed out that the lack of the proton cluster might be the cause of the low lying 0^+ state.

At last it would be interesting to see whether a shell model calculation with extended configuration space or a CVM calculation with the same extended configuration space can give better results for this nucleus. These calculations would give us more information about the role of the cluster field mechanism. Naturally some further experimental information is also necessary on the decay properties of the low lying states, particularly for the first excited 0^+ state.

REFERENCES

1. M. Sambataro and G. Molnár, Nucl. Phys., A376, 201 (1982).
2. G. Molnár, I. Diószegi, Á. Veres and M. Sambataro, Nucl. Phys., A403, 342 (1983).
3. R.A. Meyer, J. Lin, G. Molnár, B. Fazekas, Á. Veres and M. Sambataro, Phys. Rev., C29, 1839 (1984).
4. A. Kumar and R.K. Bansal, Nucl. Phys. G., 5, 265 (1979).
5. E. Adamides, L.D. Skouras and A.C Xenuolis, Phys. Rev., C23, 2016 (1981).
6. V. Paar, Z. Phis., 271, 11 (1974).
7. G. Molnár, B. Fazekas, L. Dabolczi, T. Belgya and Á. Veres, Reactor neutron beam facility for high-precision (n,n'γ) reaction spectroscopy, (Preprint IZIN-1, Budapest, 1986).
8. H.W. Müller, Nucl. Data Sheets, 44, 277 (1985).
9. N.K. Aras, E. Eichler and G.G. Chilosi, Nucl. Phys., A112, 609 (1968).
10. Y. Sugiyama and S. Kikuchi, Nucl. Phys., A264, 179 (1976).
11. J. Vervier, Nucl. Phys., 75, 17 (1966).
12. R. Wadsworth, M.D. Cohler, S.M. Lane, M.J. Smithson, D.L. Watson, Ronald E. Brown, D.M. Drake, J.C. Peng and J.W. Sunier, J. Phys. G., 10, 79 (1984).
13. Th. Paradellis, S. Hontzeas and H. Blok, Nucl. Phys., A168, 539 (1971).
14. J. Barette, A. Boutard and S. Monaro, Can. J. Phys., 47, 955 (1969).

COEXISTENCE OF SPHERICAL AND DEFORMED SHAPES AND THE DOUBLE SUBSHELL CLOSURE AT ^{96}Zr

G. MOLNÁR[*+], S. W. YATES[*], E. W. KLEPPINGER[*], T. BELGYA[+], B. FAZEKAS[+], A. VERES[+], and R. A. MEYER[‡]
[*]University of Kentucky, Lexington, KY 40506, USA; [+]Institute of Isotopes, Budapest, H-1525, Hungary; [‡]Lawrence Livermore National Laboratory, Livermore, CA 94550, USA

Abstract The coexistence of spherical shell model and intruder deformed states is established on the basis of inelastic neutron scattering (INS) data for doubly closed subshell ^{96}Zr. It is shown that the four-particle, four-hole intruder band fosters ground state deformation in the heavy Zr nuclei.

The occurrence of a double subshell closure at ^{96}Zr (Z=40, N=56) has long been expected to have an important influence on the spherical to deformed nuclear shape transition in the A~100 mass region.[1] While the recent discovery[2] of a positive-parity four-particle, four-hole (4p-4h) intruder band built on the first excited 0^+ state of ^{96}Zr established an important analogy with the coexistence of spherical shell model and deformed intruder states at double major shell closures, e.g. at ^{16}O and ^{40}Ca, many questions have remained unanswered. First, the coexisting spherical shell model states should also be characterized in order to see the extent to which other orbitals are involved. Second, the nature of the deformation associated with the 4p-4h band is not clear; strong excitation in the alpha-pickup reaction might suggest dipole deformation,[3] provided the requisite negative-parity states can be found. Finally, the influence of the double subshell closure on the level structures of the neighboring nuclei and the relationship to the shape coexistence observed in the transition from spherical to deformed ground states should be examined. The present paper, based partly on new spectroscopic information on ^{96}Zr levels from inelastic neutron scattering (INS) studies,[4] is an attempt to elucidate the aforementioned problems.

The double subshell closure effect at ^{96}Zr, i.e. at Z=40 and N=56, becomes evident from nucleon binding energy as well as first 2^+ state energy systematics. Single proton and neutron transfer studies have shown that the ground state of ^{96}Zr is characterized by the nearly complete filling of the $2p_{1/2}$ proton and the $2d_{5/2}$ neutron orbitals. These observations are well accounted for by semiempirical shell model calculations[5] using active $2p_{1/2}$, $1g_{9/2}$ proton and $2d_{5/2}$, $3s_{1/2}$ neutron orbitals outside an ^{88}Sr core. There is a basic difference between ^{96}Zr and the other magic Zr isotope, ^{90}Zr, for which the ground state and the first excited 0^+ state are mixtures of $2p_{1/2}$ and $1g_{9/2}$ proton configurations.[6] In ^{90}Zr the ground state occupation probability of the $\pi 2p_{1/2}$ orbital is much smaller[5] than in ^{96}Zr.

As is the case with ^{90}Zr, the first excited state of ^{96}Zr is a 0^+ state. This state behaves quite differently, however, in that it is strongly excited in multiparticle transfer reactions, most notably the (d,^{6}Li) alpha-pickup reaction.[7] As our inelastic neutron scattering (INS) studies have shown,[2] a positive-parity band can be

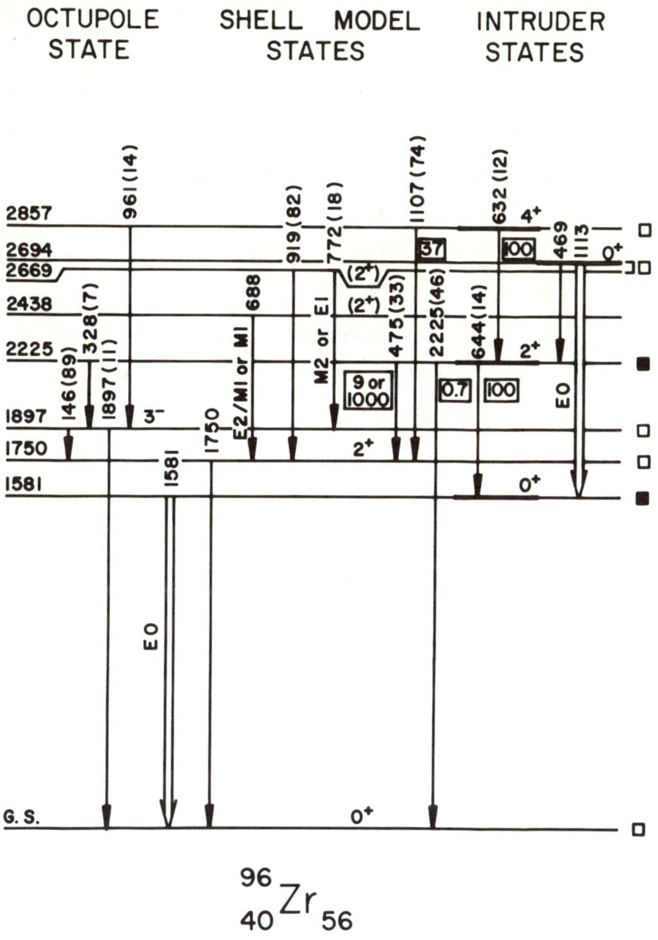

FIGURE 1. Partial level scheme of ^{96}Zr. Numbers in boxes are B(E2) ratios; filled (open) squares denote strong (weak) excitation in the $(d,{}^{6}Li)$ reaction.

associated with the 0_2^+ state. Figure 1 illustrates how this intruder band is distinguished from the remainder of the observed states on the basis of the γ-decay pattern. These other states, except for the 3^- octupole level, should be the shell model states associated with the spherical ground state configuration. Even though some of the spin assignments remain tentative, it is rather straightforward to associate them with the lowest 2^+ states predicted by the shell model calculations of Gloeckner[5].

To examine alpha clustering as a possible interpretation for the observed 4p-4h intruder band built on the 1581-keV 0_2^+ state, we have searched for a 1^- state decaying into the 2225-keV 2^+ band member. The 2694-keV state seemed a likely candidate but, as Figure 2 clearly illustrates, the inelastic neutron scattering cross section inferred from the (n,n$'\gamma$) reaction yield is consistent only with a 0^+ spin-parity assignment. The recent observation in an in-beam (t,pe$^-$) reaction experiment of an E0 transition from this state to the 0_2^+ state[8] confirms our assignment. Moreover, decay of the newly observed 0_3^+ state to the 0_2^+ level instead of the ground state is in accordance with the ^{98}Zr data.[9] This suggests that the deformation associated with the 0_2^+ state of ^{96}Zr is of the conventional quadrupole type, while the ground state is spherical with nearly pure doubly closed subshell structure.

Data on ^{97}Zr suggest that, as with the $\nu 2d_{5/2}$ orbital, the $3s_{1/2}$ neutron orbital is well separated and an approximate double subshell closure occurs also at ^{98}Zr. Indeed, our level scheme for ^{96}Zr exhibits a great similarity to the low-lying levels[9] of ^{98}Zr, at least where

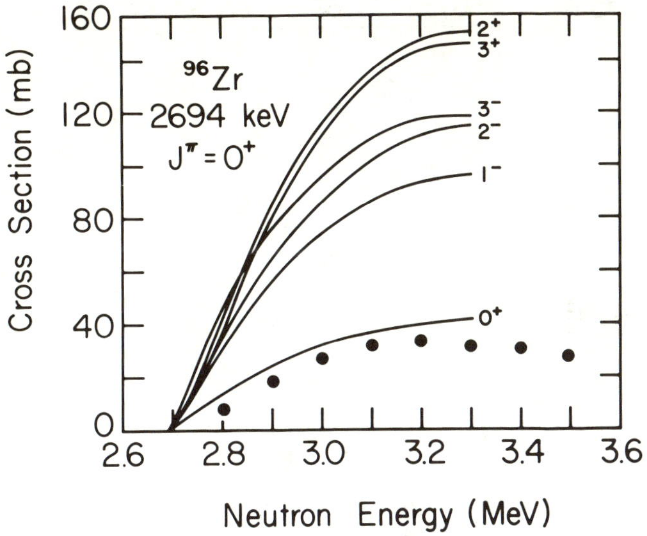

FIGURE 2. Inelastic neutron scattering excitation function of the 2694-keV level of ^{96}Zr as compared with Hauser-Feshbach-Moldauer calculations for various spin-parity assumptions.

the 0^+ and 2^+ states are concerned. Hence, the transition from spherical ^{96}Zr to deformed ^{100}Zr appears to proceed through an intruder mechanism involving proton and neutron particle-hole pair excitations across subshells. The driving force, as shown by Federman and Pittel[10], is the strong interaction between protons and neutrons occupying the highly overlapping $\pi g_{9/2}$ and $\nu g_{7/2}$ orbitals. When the shell gaps are small, the intruder energy may become negative[11] leading to the interchange of spherical ground and deformed intruder states, as implied in Figure 3 for the N=60 Zr and Sr isotones. The systematics of the 0_2^+

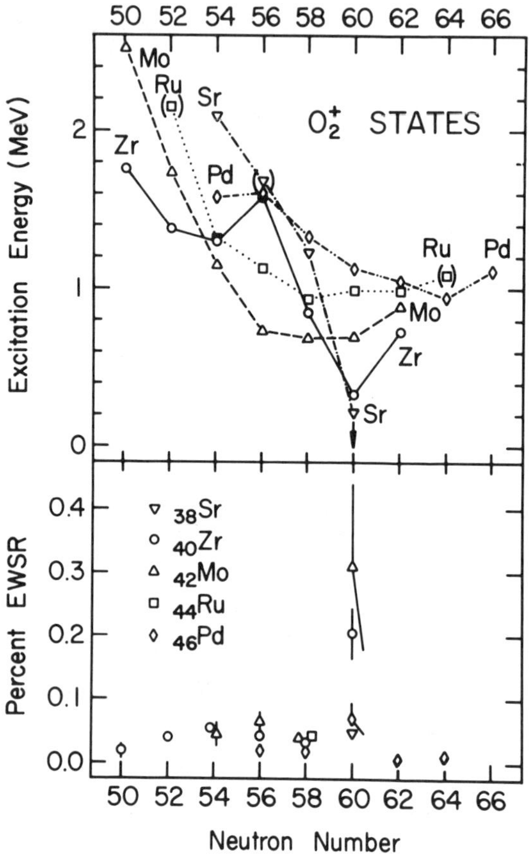

FIGURE 3. Energy systematics of the 0_2^+ states of even-even nuclei in the A~100 region (upper part). Percent depletion of the energy-weighted sum rule, EWSR = $E(0_2^+)\rho^2(E0;0_2^+ \rightarrow 0_1^+)2.88A^{5/3}/Z^2$ for the same nuclei (lower part).

states energies also demonstrate the shift of the minimum towards the middle of the 50-82 major neutron shell as the proton number increases beyond the Z=40 closed subshell value. The energy-weighted electric monopole transition strength of the $0_2^+ \rightarrow 0_1^+$ transitions clearly demonstrates strong configuration mixing only at N=60 where the coexisting structures have the smallest energy separations.

In conclusion, we have shown that even if the double subshell closure at Z=40 and N=56 tends to hinder the onset of deformation for ^{96}Zr, it still becomes possible by the neutron-proton intruder mechanism. The excited deformed configuration coexists with a spherical ground state configuration and becomes the deformed ground state as more neutrons are available to occupy the critical $\nu g_{7/2}$ spin-orbit partner orbital. More detailed studies of the coexisting states of ^{96}Zr and possibly of ^{98}Zr would lead to a better understanding of the deformation mechanism in the A~100 mass region as well as of the neutron-proton interaction in heavy nuclei where the unlike nucleons occupy different shell model orbitals.

We wish to thank E. A. Henry for providing his data prior to publication. This work has been supported in part by National Science Foundation grant No. INT-8512479.

REFERENCES

1. R. K. Sheline, I. Ragnarsson, and S. G. Nilsson, Phys. Lett. 41B, 115 (1972).
2. G. Molnár, S. W. Yates, and R. A. Meyer, Phys. Rev. C 33, 1843 (1986).
3. F. Iachello, in Proceedings of the Fourth Int. Conf. on Clustering Aspects of Nuclear Structure and Nuclear Reactions, edited by J. S. Lilley and M. A. Nagarajan (Reidel, Dordrecht, 1985) p. 101.
4. S. W. Yates and G. Molnár, in Proc. Int. Research Conf. on Nuclear Structure, Reactions, and Symmetries, Dubrovnik, Yugoslavia (1986), to be published.

5. D. Gloeckner, Phys. Lett. 42B, 381 (1972) and Nucl. Phys. A253, 301 (1975).
6. J. Vervier, Nucl. Phys. 75, 17 (1966).
7. A. Saha, G. D. Jones, L. W. Put, and R. H. Siemssen, Phys. Lett. 82B, 208 (1979); A. M. van den Berg, A. Saha, G. D. Jones, L. W. Put, and R. H. Siemssen, Nucl. Phys. A429, 1 (1984).
8. E. A. Henry, in Proc. Int. Research Conf. on Nuclear Structure, Reactions, and Symmetries, Dubrovnik, Yugoslavia (1986), to be published.
9. K. Kawade, G. Battistuzzi, H. Lawin, H. A. Selič, K. Sistemich, F. Schussler, E. Monnand, J. A. Pinston, B. Pfeiffer, and G. Jung, Z. Phys. A304, 293 (1982).
10. P. Federman and S. Pittel, Phys. Rev. C 20, 820 (1979).
11. K. Heyde, P. Van Isacker, R. F. Casten, and J. L. Wood, Phys. Lett. 155B, 303 (1985).

SUBSHELL CLOSURE INFLUENCE ON INTRUDER STATES IN EVEN-EVEN NUCLEI +

N. Kusnezov[*,1] and R.A. Meyer[2]

1) Lockheed Missiles and Space Co., Palo Alto, CA 94304
2) Lawrence Livermore National Laboratory, Livermore, CA 94550

Abstract The occurrence of a 14.3-ns 0^+ isomer in ^{102}Pd and associated states in ^{104}Pd and ^{106}Pd are discussed within the framework of intruder states that occur lowest in energy at the middle of regions of subshell closure. Its properties are shown to be consistent with a $2d_{3/2}$-proton, $2d_{5/2}$-neutron configuration. The possibility of an $1h_{11/2}$-proton, $1h_{9/2}$-neutron intruder band in the closed proton subshell nucleus $^{146}Gd_{82}$ is discussed.

The recent study of the low-energy excited states in ^{102}Pd and ^{104}Pd revealed several uncommon features. First, an excited 0^+ state was traced from its lowest energy in ^{102}Pd to its highest known energy in ^{106}Pd [LUO86]. Such a dependence of increasing energy with increasing mass is the opposite of the expected trend of states in this mass region of Pd isotopes. For ^{102}Pd, the 0^+ level possessed a 14.3-ns half-life and was almost degenerate with a second 0^+ level. The deexcitation properties of this state were quite different. For example, the dominant deexcitation was by a 65.1-keV E2 transition which possessed a B(E2) value of 96 SPU. This is in contrast to the limit of $<4x10^{-4}$ for the E2 deexcitation path to the first 2^+ vibrational state. This latter highly-hindered E2 transition, in combination with an observed E0, yielded one of the largest known X_{211} values known even though the monopole matrix element was small for this mass region. Also, a 2^+ level 350 keV higher in energy was observed to deexcite to the 0^+ band head with a relative B(E2) value 240 times greater than that observed for the

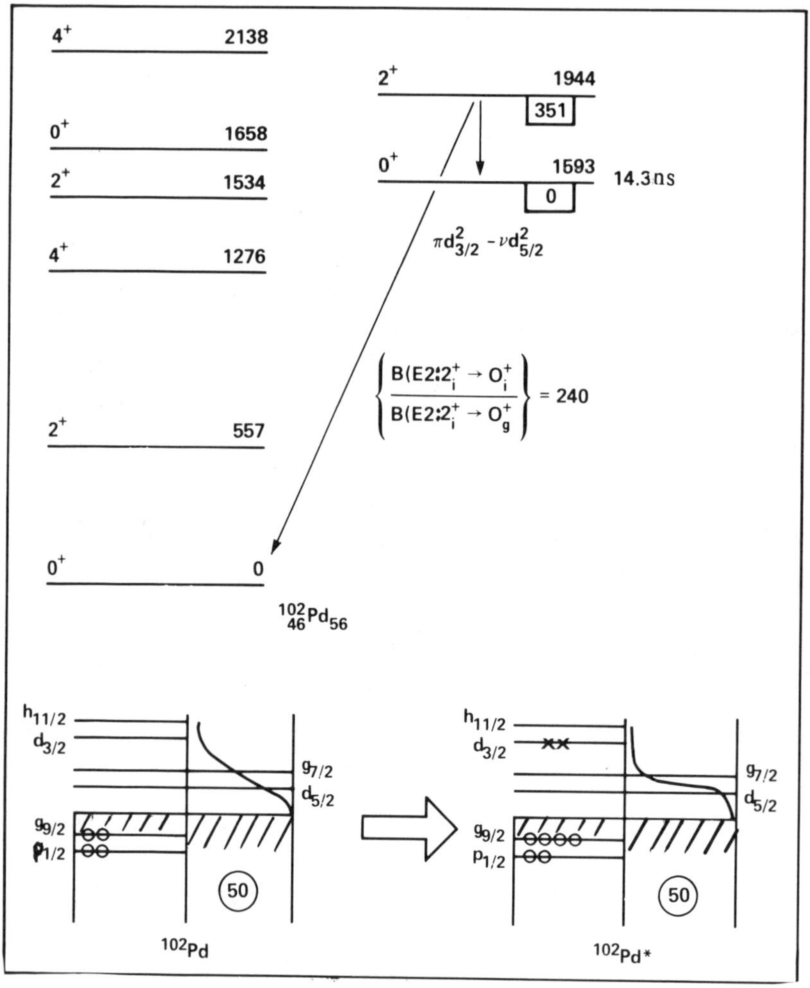

Fig. 1) Level diagram of ^{102}Pd showing the categorization into two sets of levels: states that are mostly vibrational in nature (left-hand set) and states that are mostly of $d_{3/2}$-proton intruder character (right-hand set) {data taken from [LUO86]}. The insert at the bottom of the figure schematically shows the excitation mechanism.

transition to the ground state. These features are within the framework of known properties of intruder states and the strong interaction between cross-shell $2d_{3/2}$ protons (**p**) and $2d_{5/2}$ neutrons (**n**) which are filling along with the $1g_{7/2}$ orbital in this mass region of the Pd isotopes.

Federman and Pittel [FED79] have discussed the role of the $^3S_1$ force and spin-orbit partners $\{n_{l,j}=l+1/2(\mathbf{p})$ & $n_{l,j}=l-1/2(\mathbf{n})\}$ in being capable of breaking pairing and producing deformation in nuclei. They have shown the dominant role played by the interaction of $1d_{3/2}$ protons with $1d_{5/2}$ neutrons in producing relatively low-energy or even ground-state rotational bands in light nuclei. Thus, ^{16}O has an intruder rotational excited-state band [BRO66,MOR56]. Aprahamian and coworkers [APR85] have shown that intruder states should occur lowest in energy near midshell and recent evidence has shown that they occur lowest at mid-subshell [MEY82,HEY83]. For the light Pd nuclei we might expect that a $d_{3/2}$-proton intruder state which is interacting with the available $d_{5/2}$ neutrons would occur mid-subshell at N~56. Further, we know that $2d_{5/2}$- and $1g_{7/2}$-proton intruder states occur at an energy as low as ~1 MeV in the Cd and Pd nuclei [KUS86,MEY85]. Thus we might expect that if the proton-$d_{3/2}$ intruder state were to occur, its energy could be estimated to be at ~1.5 MeV in ^{102}Pd.

The properties of ^{102}Pd may be accounted for by the occurrence of a $2d_{3/2}$-**p** intruder configuration similar in nature to the $1g_{9/2}$-**p**, $1g_{7/2}$-**n** intruder structure that coexists in double closed subshell ^{98}Zr [MEY86]. In Fig. 1 we show the levels of ^{102}Pd categorized into two positive-parity sets, where the major deexcitation strength occurs within the set. The lower part part of the figure shows the conceptual mechanism for the occurrence of the ^{102}Pd intruder band (labeled $^{102}Pd^*$), where a pair of particles is promoted into the $d_{3/2}$ orbital and induces polarization of the neutrons into a higher $d_{5/2}$ occupancy. The two 0^+ states could well be nearly degenerate in the unperturbed system. The strong interaction would then be expected to lower the 0^+ intruder band head energy. Also, the energy of the 2^+ member could increase through its interaction with a lower energy but near-lying 2^+ vibrational level. Both of these effects would then account for the 350-keV value of $E_{int.}(2^+) - E_{int.}(0^+)$, which is slightly greater than the value extrapolated from the known $pd_{3/2}$-$nd_{5/2}$ bands in light nuclei or the moment of inertia of other intruder bands known in the heavier Pd and Cd nuclei [MEY85].

The observed transition strengths for the ^{102}Pd intruder band deexcitations are indicative of strong mixing. Kusnezov and coworkers [KUS86] have shown that when the intruder band drops to an energy near the two-phonon vibrational triplet, that it strongly interacts with the vibrational multiplet. This alters the observed deexcitation pattern in a significant way. Thus, the mixed band head can posses an allowed deexcitation to a 2^+ vibrational state while the unperturbed band head would be expected to have a highly hindered, if not forbidden, deexcitation. Cancellation between E2 matrix elements in the mixed configurations can give rise to retarded transitions that would be allowed in an unperturbed configuration. In this context, we can account for the 17 SPU observed for the intraband transition from the 2^+ member (see [LUO86]) and the 96 SPU for the deexcitation of the band head.

Recently, Mann and coworkers [MAN86] have identified several 0^+ states in ^{146}Gd by performing (p,t) transfer reaction studies on the exotic target ^{148}Gd (see [LAN86]). They show that the first 0^+ level (at 2165 keV) is the 2-proton pairing vibration state and that the second 0^+ level (3020 keV) is the 2-neutron pairing vibration state. Using a simplified approach, they cannot account for a third 0^+ level (4534) and its related 2^+ level 192 keV higher in energy. However, the lowest 4p4h intruder state can occur at an energy roughly equal to the sum of the 2-neutron state plus the proton subshell gap less the neutron-proton interaction energy. This can be estimated to be ~4620 keV, close to the energy of the third

0^+ level. A test for the candidacy of the 4726-keV 2^+ level being the 2^+ member of a 4p4h intruder band in ^{146}Gd is the strength with which it is populated in the (p,t) reaction. We can expect that the (p,t) cross section ratio for the population of the 0^+ and 2^+ members of the 4p4h band should be the same as the ratio for the population of the 2-neutron pairing vibration levels. The ratio for the 4p4h band candidates is 1.3±0.5 similar to the ratio of 1.1±0.4 for the population of the 2-neutron pairing vibration levels. Further, the mass adjusted moment of inertia taken from the 4p4h intruder band in ^{16}O gives an expected 2^+-0^+ energy difference of 199 keV (c.f. 192 keV for ^{146}Gd). If this band is the equivalent of the 4p4h $1g_{9/2}$-proton, $1g_{7/2}$-neutron intruder configuration that was found at the Z=40, N=56/58 subshell closure (for details see [MEY86,MOL87]), then we can expect that its makeup is mainly $1h_{11/2}$-proton, $1h_{9/2}$-neutron. Casten et al. [CAS81], have shown that such a configuration plays a significant role in the occurrence of deformation in this region.

REFERENCES

+ Work supported in part, by U.S.D.O.E. at LLNL (contract W-7405-ENG-48) and by U.S.N.S.F. (Grant Nr. YOR 83/074).
* Participating guest at LLNL.

APR85 A. Aprahamian, et al., Phys. Rev. Lett. 140B 22 (1985).
BRO66 G.E. Brown and A.M. Green, Nucl. Phys. 75 401 (1966).
CAS81 R.F. Casten, et al., Phys. Rev. Lett. 47 1433 (1981).
FED79 P. Federman and S. Pittel, Phys. Rev. C 20 820.
HEY83 K. Heyde, et al., Physics Reports 102 291 (1983).
KUS86 D.F. Kusnezov, et al., in "Nuclear Structure, Reactions, and Symmetries", R.A. Meyer and V. Paar, Eds., World Scientific Pub. Pte. Ltd., Singapore, p. 783 (1986).
LAN86 R.G. Lanier, et al., in "Recent Advances in the Study of Nuclei off the Line of Stability", R.A. Meyer and D.S. Brenner, Eds., Am. Chem. Soc., Washington, D.C. p. 476 (1986)
LUO86 M. Luontama, et al., Zeit. Phys. 324 317 (1986).
MAN86 L.G. Mann, et al., 1986 Nuclear Chemistry Div. Ann. Rpt., LLNL, UCAR 10062/86, p.4-43 (1986).
MEY82 R.A. Meyer, et al., Phys. Rev. C 25 682 (1982).
MEY85 R.A. Meyer, Hyperfine Interactions 22 385 (1985).
MEY86 R.A. Meyer, et al., Phys. Lett. 177 271 (1986).
MOL87 G. Molnar, et al., This Book, p. 237 (1987).
MOR56 H. Morinaga, Phys. Rev. 101 254 (1956).

EVIDENCE FOR QUASI-PARTICLE PHONON MIXING IN ODD-ODD DEFORMED NUCLEI +

R.G. LANIER, C.M. LEDERER*, AND R.A. MEYER

Nuclear Chemistry Division
Lawrence Livermore National Laboratory
Livermore, California 94550

The nuclear levels in odd-odd ^{172}Tm were studied by gamma-ray spectroscopy using ^{172}Er sources that were repeatedly eluted to remove the ^{172}Tm daughter. Analysis of these data are used to suggest that quasi-particle phonon coupling alter the band properties.

The details of the ^{170}Tm level structure have been reinvestigated by Dewberry et al.[1,2]. However, the ^{172}Tm partner has not been studied in such detail.[3] We measured gamma-ray singles, Compton suppression, and gamma-gamma coincidence spectra of ^{172}Er sources that were purified of daughter Tm activity using of ion-exchange techniques[4]. The decay scheme which we can establish for the decay of ^{172}Er to levels of ^{172}Tm is given in Fig. 1. New levels at 145, 239, and 714 keV are established by coincidence relationships while levels at 479, 496, 526, and 797 keV are established by energy sums and differences.

When we compare the known structure of ^{170}Tm to the level properties of the lowest energy 0^- and 1^- bands of ^{172}Tm, we find anomalies in the variations of simple nuclear properties. The K = 2 & 3 band heads taht form the {[411 1/2][512 5/2]} configurational pair exhibit a splitting of approximately 230 keV in both nuclei. For this pair the Coriolis interaction can affect the observed splittings. The energy of the $I^{\pi}K = 3^-2$ rotational state can be perturbed through the Coriolis interaction with the 3^-3 band head, but we

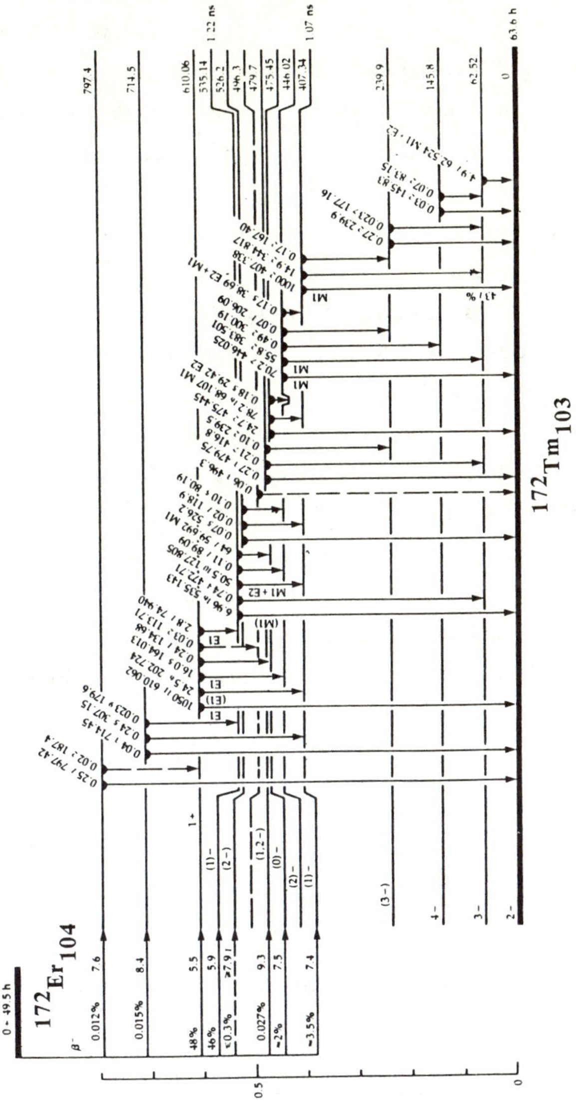

Fig. 1 Decay scheme for ^{172}Er to levels of ^{172}Tm.

calculate this effect to be very small (~1 keV). This, and the fact that the $\triangle K=1$ off-diagonal matrix element associated with the Coriolis coupling to either the 3^-0 of 3^-1 states of the {[411 1/2][521 1/2]} configuration is expected to be zero, suggests that the relative energy of the 3^-2 rotational state should be similar in both nuclei. This is what is observed. Thus, the observed properties of the {[411 1/2][512 5/2]} configuration in both ^{170}Tm and ^{172}Tm are virtually identical.

For similar reasons, we might expect to observe little or no difference when we compare the properties of the K=0 bands in both nuclei. What we observe, however, is that there is a significant difference in the residual-interaction matrix elements derived from the experimental splittings and in the moment-of-inertia parameters for the these bands. We suggest that this difference involves different degrees of quasiparticle-phonon mixing in the $K=0^-$ bands in both nuclei. While the {[411 1/2][521 1/2]}1^-, 0^- configurational pair make up the ground- and first-excited intrinsic state of ^{170}Tm, the configurational pair occurs at ~500 keV in ^{172}Tm. At the latter energy this configuration can easily interact with core phonon excitations; thus, one would expect larger perturbations on the ^{172}Tm K = 0,1 structure when compared with that of ^{170}Tm. In fact, large phonon admixtures in one-quasiparticle states which involve either the proton [411 1/2] orbital or the neutron [521 1/2] orbital are distinctive characteristics[5] of nuclei in this mass region.

REFERENCES

+ Supported by U.S.D.O.E. contract Nr. W-7405-Eng-48.
* LLNL Participating Guest. Permanent address: Universitywide Energy Research Group, Univ.of California, Berkeley, CA 94720.
1. R.A. Dewberry, et al., Phys. Rev. C 24 1643 (1981).
2. L.R. Greenwood, Nucl. Data Sheets 15 497 (1975).
3. M.R. Schmorak and R.L. Auble, Nucl. Data Sheets 15 371 (1975).
4. R.G. Lanier, UCAR 10062-83/1, LLNL Report, p. 191 (1983).
5. M.E. Bunker and C.W. Reich, Rev. Mod. Phys. 43 348 (1971).

THE APPLICATION OF THE DDM TO THE EVEN ISOTOPES OF CADMIUM AND PALLADIUM

J COPNELL

MAPS, University of Sussex, Brighton, BN1 9QH, England.

K KUMAR

Physics, Dept., Tennessee Technological University, Cookeville, TN 38505, USA

Abstract The Dynamic Deformation Theory is applied to the even isotopes of cadmium and palladium, producing potential energy surfaces of these nuclei as functions of beta, gamma deformation. Values of predicted energy levels, transition probabilities, and mixing ratios, are compared with a compilation of experimental results, and various trends across each range of isotopes are compared with the O(6), U(5), and SU(3) descriptions of nuclei.

INTRODUCTION

The DDM is discussed at length elsewhere[1,2], and so we will not dwell on it here except to note that the collective trial wavefunction, A_{IK}, is expanded as

$$A_{IK} = \sum_{p} ap \; B_{PIK} \qquad (1)$$

where B_{PIK} is a function containing the product of the following:

a polynomial function of β^2 and $\beta^3 \cos 3\gamma$, a tensor function of β , γ , of rank I, obtained from the coupling of quadropole tensors, and a gaussian of the form; $\exp(-\frac{1}{2} D^2)$, where D^2 can be chosen (in the computer code), to be either

A) β^2/b^2 where b is the gaussion range, chosen by a minimization of the final first two 0^+ states.

B) $(\beta^2-\beta_0^2)^2/b^2$ where β_0 is the beta deformation at which the minimum in the potential occurs.

or C) The sum of 2 gaussians, one of each of the above types. The purpose of the gaussian is to ensure that the wave-function vanishes for large deformation.

All three varients were applied to the Cd isotopes, in order to find the best solution, if indeed, they were significantly different. The best solution was decided upon, on the basis of 3 criteria, firstly, producing levels in an acceptable order, secondly, producing energy levels and B(E2)'s that were not ridiculously awry, and thirdly, giving a small theoretical error in the level energies. In all cases, the three criteria were in agreement. It was found that a viable solution could not be found for the latter case, but for the former two cases, solutions were found which appeared to be 'equally' correct but which were at varience with one another. This led to the development of a new method of solving the Collective Schrödinger Equation which involves replacing the gaussian in the trial wavefunction (for the excited states), with the ground state wave-function found by the completely numerical method discussed in depth in ref. 2). The potential and inertial functions however, are not affected by these variations, since they depend on the single particle motions only.

RESULTS

Henceforth, the DDM with the gaussian as in case A and B above, and the new version will be referred to as DDM(sph), DDM(def), and DDM(NS86) respectively.

One of the main approximations of the DDM is the adiabatic approximation, that the collective frequencies are much less than the single particle frequencies. We can estimate the validity of this, by employing the rotational model, which gives the relations

$(41\ A^{-1/3})/E0^+_\beta$ = N Beta for β-vibrations which should be >>1, and $3 \times (41\ A^{-1/3})/E2^+_g$ = I max for rotations.

The former is plotted in figure 1. for even isotopes of Cd, and it can be seen that DDM(Sph) and DDM(NS86) are good, whilst DDM(Def) is somewhat more dubious especially towards the middle of the neutron shell (N = 66, A = 114), although it is still quite reasonable. Imax from DDM(Def) was found to be approximately equal to that obtained when the experimental value of the first 2^+ energy was used, and all variations gave Imax far greater than could be profitably studied with the model.

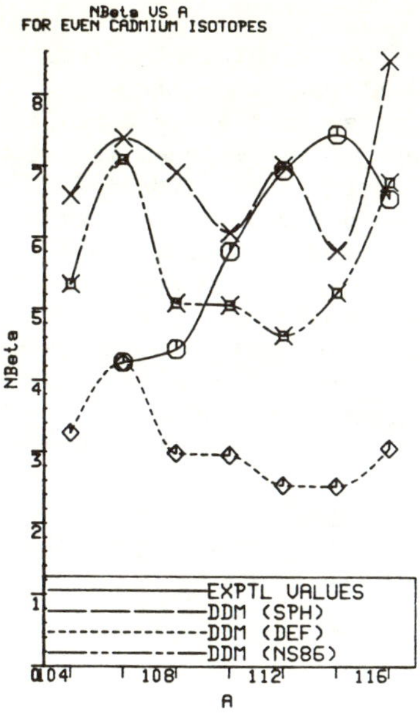

Figure 1

NBeta vs. A for Cd.

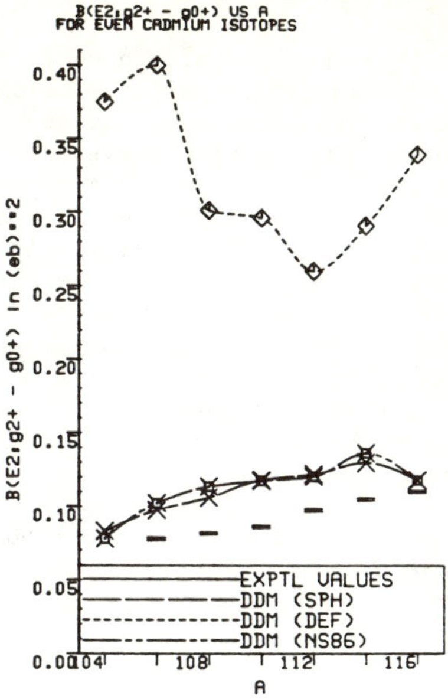

Figure 2

B(E^{2^+} $-g^{0^+}$) vs. A for Cd.

These two results can be simply understood in terms of the predictions of the energy levels. It was found that DDM(Sph) produced a ground band with a moment of inertia that was too high by a factor ≈2-3, but reproduced other bands fairly well, and most notably gave good values of transition probabilities. DDM(Def) on the other hand, gave the ground band very well, pushed other bands too high, and could not reproduce transition probabilities at all. DDM(NS86) gave values in remarkable agreement with DDM(Sph) considering the different method used, but with the theoretical errors in the level energies somewhat larger, probably due to the numerical method for the calculation of the ground state wave-function. (See figure 2 and also figure 1 of reference 2)).

These considerations lead to the conclusion that DDM(Sph) is the 'correct' solution and the fit to the experimental ground band energies produced in DDM(Def) is a spurious result. However, it must be noted that DDM(NS86) obliviates the necessity for any such ambiguity in the future.

Potential Energy Plots

The potential energy used in the DDM is calculated from the deformed single particle motion, at each node on a β,γ mesh. These values have been plotted in figures 3 and 4 for for ^{110}Cd, and for the whole range of Cd isotopes in figures 5 and 6. The zero point motion energy for ^{110}Cd is 4.77 MeV and is only just larger than the prolate-oblate energy difference. Hence, from figures 3 and 4, it can be seen that ^{110}Cd is only just γ-unstable. From figures 5 and 6, it can be seen that the cadmium isotopes are spherical at N = 56, 58 (A = 104, 106), become more prolate deformed as N increases to the centre of the shell (N = 66, A = 114), and then less again. ^{112}Cd is given to be rotational. This is confirmed from figures such as 3 and 4 for the whole set of Cd isotopes.

The potential energy plots for palladium give a similar behaviour-spherical at N = 50, A = 96, through γ-unstable, and becoming rotational for ^{108}Pd and ^{110}Pd.

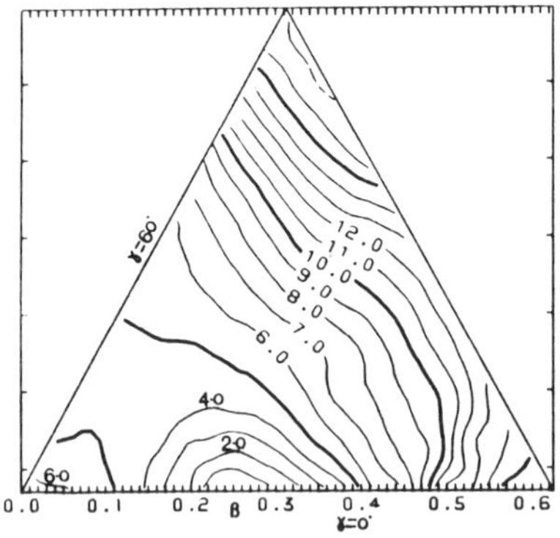

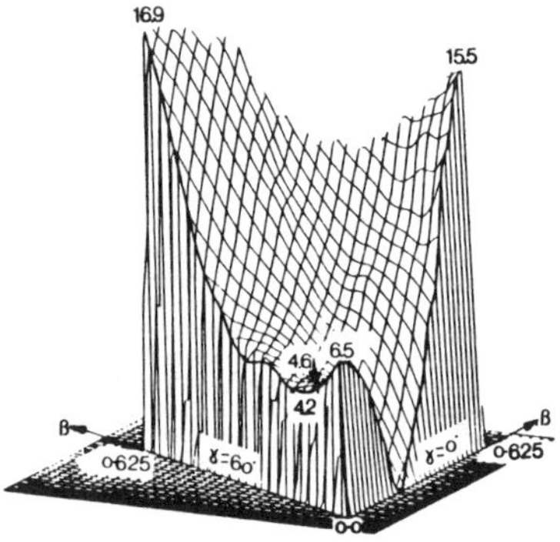

Figure 3

Figure 4

Potential energy surfaces of ^{110}Cd (EZPM = 4.77 MeV).

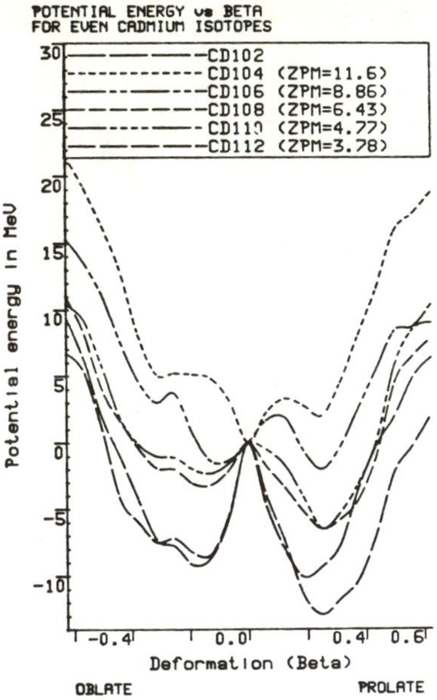

Figure 5

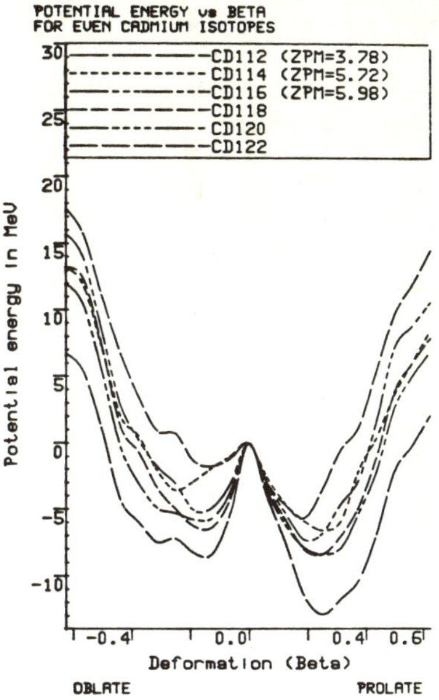

Figure 6

Potential energy of the Cd isotopes for $\gamma = 0$, $-0.625 \leq \beta \leq 0.625$. Normalized to the spherical shape.

Quadropole moments of 2^+_g

From figure 7, it can be seen that the DDM reproduces the experimental values[3] fairly well for Cd. The magnitude is too large, probably due to the moment of inertia of the ground band being calculated too large, but the sign is correct (-ve), and agrees with the potential energy surfaces since this implies a prolate deformation. The same is true of Pd isotopes.

Mixing Ratios

Delta values, where there is experimental data[3)] to compare them with, are all predicted to be of the correct sign and the right order of magnitude by DDM(Sph), although all are too large $\delta(2^+\gamma\to 2^+_g)/E(e+\gamma-2^+_g)$ for Cd isotopes is plotted in figure 2 of reference 2). DDM(NS86) appears to predict the experimental trend quite well too, as well as the order of magnitude and the sign. Insufficient data was available to compare the DDM with other mixing ratios. Transitions from K = 0 band to K = 0 bands have δ positive both experimentally and theoretically. This type of behaviour was noted but not explained in reference 4).

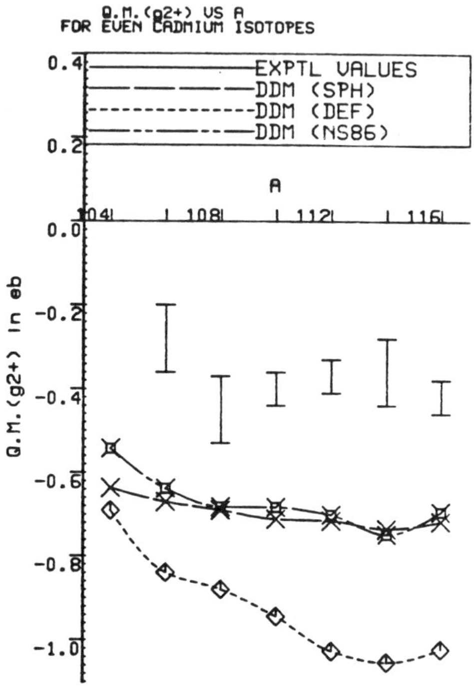

Figure 7

Q.M.$(2^+{}_g)$ for Cd.

Transition Probabilities

These were found to be consisten' with experiment using DDM(Sph) and DDM(NS86). As in the case of the mixing ratios above, DDM(Def) gave unacceptable results. See figure 2 and also figure 2 in reference 2).

DISCUSSION

The DDM(Sph) version of the DDM gives results in good agreement with the experimental data of Cd and Pd isotopes. For Cd, it predicts nucleii ranging from vibrational/γ-unstable at ^{104,106}Cd through γ-unstable to

^{112}Cd which is given as rotational, and then back to γ-unstable again as N increases. ^{110}Cd is probably rotational too since the zero point motion is only just in excess of the prolate-oblate potential barrier. From the potential energy surfaces, there is predicted a large change in behaviour between ^{102}Cd (un-stable) and ^{104}Cd (vibrational), but this is not reflected in other data. The deformation in the model appears to be too large for these nucleii (from the energy of the first 2^+ and other data), and the magnitude of the excess increases towards ^{114}Cd, the middle of the neutron shell. The δ-values are all in sign agreement with experiment, and they are all negative except for $2^+\beta \rightarrow 2^+g$ and $2^+\beta\beta \rightarrow 2^+g$ transition (2nd, 1st, 3rd and 1st K = 0 bands respectively), which also agrees with experiment. From 35 tests and graphs of DDM results vs. experiment, only 6 showed a contrary trend to the above.

For Pd isotopes, the results were equally impressive, and supported the conclusions from the potential energy plot data.

It should be noted, that DDM(NS86) agrees with DDM(Sph) in all salient points but appears to do moderately better in some cases (e.g. Q.M.(2^+g)), whilst also removing the ambiguity inherent in the former method.

We acknowledge the interest shown in this work by Professor J P Elliott of Sussex University.

1. Kumar K, "Nuclear Models and the Search for Unity in Nuclear Physics" 1984.

2. Kumar K, this conference.

3. References to experimental data are omitted due to their number-mostly NDS.

4. Hamilton, W D and Kumar, K, J. Phys. G5, 1567 (1979).

THE STRUCTURE OF ODD-MASS Sb NUCLEI NEAR ^{132}Sn

C.A. Stone and W.B. Walters
Dept. of Chemistry, University of Maryland, College Park, MD, U.S.A.

Abstract We have studied the decay of Sn-127m, Sn-129m,g, and Sn-131m,g into states of odd-mass Sb. Level schemes are presented for the decay of the low-spin ($3/2^+$) Sn isomers. The complete $\pi g_{7/2}\otimes 2^+$ multiplet has now been identified in Sb-127,129,131 and this has led to the identification of this multiplet in Sb-123,125. These results now offer an attractive test for shell model calculations in the region of Sn-132.

INTRODUCTION

The level schemes of odd-A Sb can provide valuable information on nuclear structure near the Z=50 shell closure. The strength of this shell closure suggests that these nuclei can be described as a single proton, the 51st proton, coupled to a neutron core. Two types of states can be seen in these nuclei: single particle states and states belonging to multiplets formed by the proton interacting with an excited neutron core. Examples of single particle states are the ground state and first excited state in odd-mass $^{123-131}$Sb. The ground state spin is $7/2^+$, corresponding to the proton in a $g_{7/2}$ state. Close to the ground state, at 798 keV in ^{131}Sb, is a $5/2^+$ state formed by the promotion of the $g_{7/2}$ proton to the $d_{5/2}$ state.

Neutron core excitations in the odd-mass Sb nuclei will be analogous to those seen in the adjacent even-even Sn nuclei. The first excited state is a 2^+ state and in ^{130}Sn this lies at 1217 keV. Interaction of the odd proton, in ^{131}Sb, with this 2^+ state will produce states ranging in spin from $3/2^+$ to $11/2^+$ for the $\pi g_{7/2}\otimes 2^+$ multiplet and $1/2^+$ to $9/2^+$ for the $\pi d_{5/2}\otimes 2^+$ multiplet. In ^{131}Sb, a two neutron-hole nucleus, the multiplets should be separated by ca. 800 keV. Removing neutron pairs from the nucleus will cause the $5/2^+$ single particle state to drop in energy. The center of gravity of the multiplets will move closer together and the levels will become more strongly mixed. Observing how these levels change as neutron pairs are removed will provide more details of the residual interactions.

Huck et al.[1] have reported limited results for ^{129}Sb. Two level schemes were given, corresponding to the decay of the low-spin ($3/2^+$) and high-spin ($11/2^-$) ^{129}Sn isomers. Data on this decay were not extensive enough to assign spins to the low-lying states. Auble et al.[2] studied ^{129}Sb using a ^{130}Te(d,^{3}He) reaction. Their work has established the spins of three low-spin states. The 2.2 hour half life of the $11/2^-$ isomer in ^{127}Sn has made this nucleus easily studied by chemically separating Sn from fission products. Apt and Walters[3] have constructed an extensive high-spin level scheme for ^{127}Sb. Additional information on ^{127}Sb has been given by the work of Auble et al.[2] They identified three excitation states that are similar to those found in ^{129}Sb.

We have performed extensive studies on the decay of ^{131m,g}Sn, ^{131m,g}Sn, and ^{127m}Sn into excited states of odd-A Sb. These studies were undertaken to identify positive parity states belonging to the $\pi g_{7/2} \otimes 2^+$ multiplet and the $\pi d_{5/2} \otimes 2^+$ multiplet. A second goal of our studies was to determine the spin of the isomeric states near 2 MeV in ^{129,131}Sb and to define the high-spin excitation states above these isomers. In this paper, we present results on the low-spin systems for ^{127,129,131}Sb. More extensive results on these nuclei will be given in future articles on both the low-spin and high-spin systems.

EXPERIMENTAL MEASUREMENTS

Experiments were performed using the TRISTAN mass separater at Brookhaven National Laboratory's High Flux Beam Reactor. Samples of mass-separated activity were produced by depositing the ion beam on an aluminized mylar tape. After a fixed deposition time, a tape sequencer moved the samples to a remote counting position where the data were collected.

Two types of data were collected for these studies: gamma-gamma coincidences and time-dependent or gamma-multiscaling (GMS) data. Four-detector gamma-gamma coincidences were measured and recorded on magnetic tape for later processing. Details of the four-detector coincidence counting system can be found reference 4. GMS data were used to follow the decay of gamma activity in a mass chain and identify gamma rays which belong to the decay of a particular nucleus. For the A=131 chain, GMS data were also used to determine the half lives of the Sn isomers. A second set of experiments was performed to measure conversion electrons in the A=129 and A=131 mass chains.

RESULTS

$^{131}Sb_{80}$, A TWO NEUTRON-HOLE SYSTEM

Nearly 200 gamma rays were identified as belonging to the decay of ^{131}Sn. Of these, 45 have been assigned to the decay of the $3/2^+$ Sn isomer and 80 have been assigned to the decay of the $11/2^-$ isomer. One difficultly with the ^{131}Sn decay is that the two isomers have similar half lives. GMS data indicate they are 55.8±0.5 sec. for the $11/2^-$ isomer and 58.4±0.6 sec for the $3/2^+$ isomer. The remaining 75 gamma rays have not been seen in any coincidence gates and can thus not be assigned to the decay of a particular ^{131}Sn isomer.

A preliminary level scheme for the decay of the $3/2^+$, ^{131}Sn, isomer is shown in figure 1. The $11/2^+$ member of the $\pi g_{7/2} \otimes 2^+$ multiplet is also shown for completeness. The spins of the $11/2^+$ level and the $9/2^+$ level were determined from the decay of the $11/2^-$,^{131}Sn, isomer. They are fed by a pair of gamma rays that depopulate the $15/2^-$ isomeric state at 1676 keV. The strongest of these gamma rays is the 450.6 keV transition, 750 units compared to 20 units for the decay to the $9/2^+$ state. Conversion electron data indicate the 450.6 keV gamma ray is an M2 transition. The 1481.0 keV level has been assigned a spin of $7/2^+$. Depopulation of this level goes primarily to the $7/2^+$ ground state, the $5/2^+$ state at 798.5 keV and the $9/2^+$ excited state. No gamma

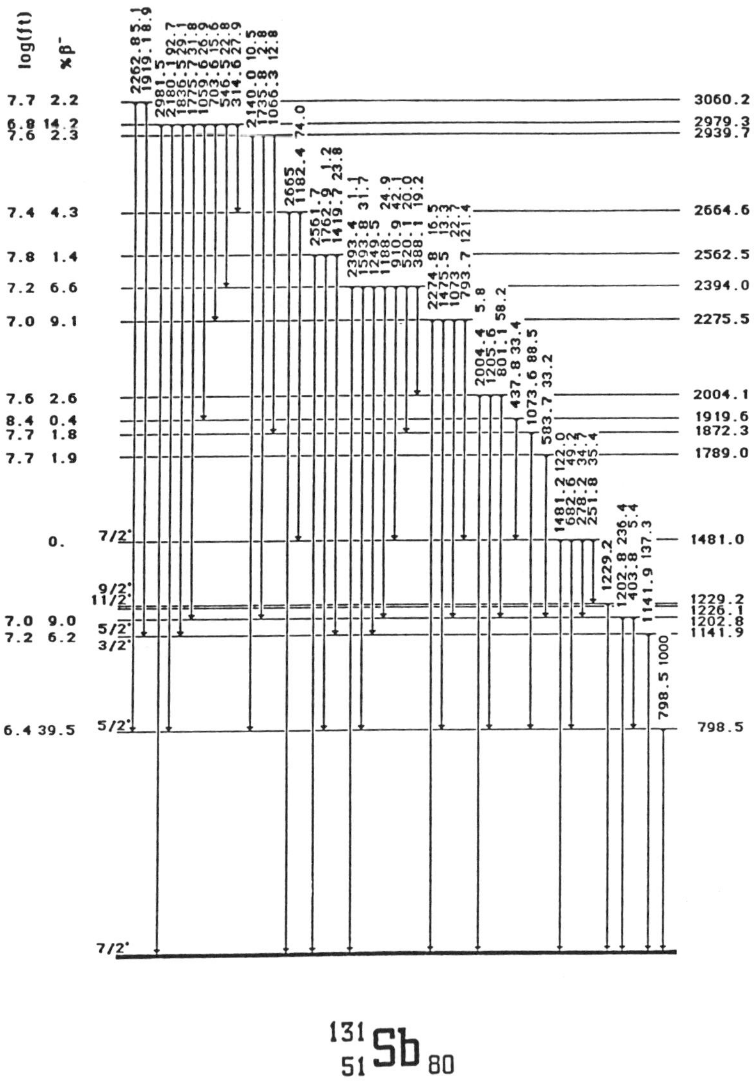

Figure 1. Proposed level scheme for ^{131}Sb.

feeding was observed to the $11/2^+$ or the $3/2^+$ states.

$^{129}Sb_{78}$, A FOUR NEUTRON-HOLE SYSTEM

Results for the decay of ^{129}Sn are shown in figure 2. We have identified 31 gamma rays from the decay of the $3/2^+$ isomer and ca. 60 gamma rays from the decay of the $11/2^-$ isomer. The difference between the half lives of the isomers is now large enough that gamma rays can be easily assigned to a particular isomer. Six gamma rays have been found to have a two-component decay, those from the 1849.9 keV excitation state and those from the 1053.8 keV state. These excitation states probably have a spin of $7/2^+$.

Levels at 1128.6 and 1161.6 keV have spins of $11/2^+$ and $9/2^+$, resp.. Spin assignments for these levels were based on the decay of the $15/2^-$ and $19/2^-$ isomers in the high-spin system. The $15/2^-$ isomer feeds both states with similar intensity, 88 units for the 1128.6 keV level and 21 units for the 1161.6 keV level. The $19/2^-$ isomer is a 17 min. β^--decaying isomer with a single gamma branch to the 1128.6 keV level. Conversion electron data indicate this transition has an M4 multipolarity and support the $11/2^+$ assignment.

The 913.6 keV state has been assigned a spin of $3/2^+$. There is a 3.2% β^- branch to this level which limits the spin to less than $7/2^+$. Reaction studies by Auble indicate that its spin is $3/2^+$. The level at 1252.4 keV has a single gamma branch to ground. It is fed by four gamma rays but has little β^- feeding. Since the 1252.4 keV gamma ray does not have a long-lived component in its decay, the spin must be $<7/2^+$. The level at 1503.8 keV is analogous to the $7/2^+$ level at 1481.0 keV in ^{131}Sb. Based on the gamma feeding by the 1503.8 keV level, we have assigned a spin of $5/2^+$ to the 1252.4 keV level. The 1493.6 keV state decays to both the $5/2^+$ state at 645.3 keV and the $3/2^+$ keV state at 913.6 keV. No gamma ray branch to ground has been seen, an upper limit of 1.0 has been assigned. Results by Auble et al.[2] assign a spin of $1/2^+$ to this state.

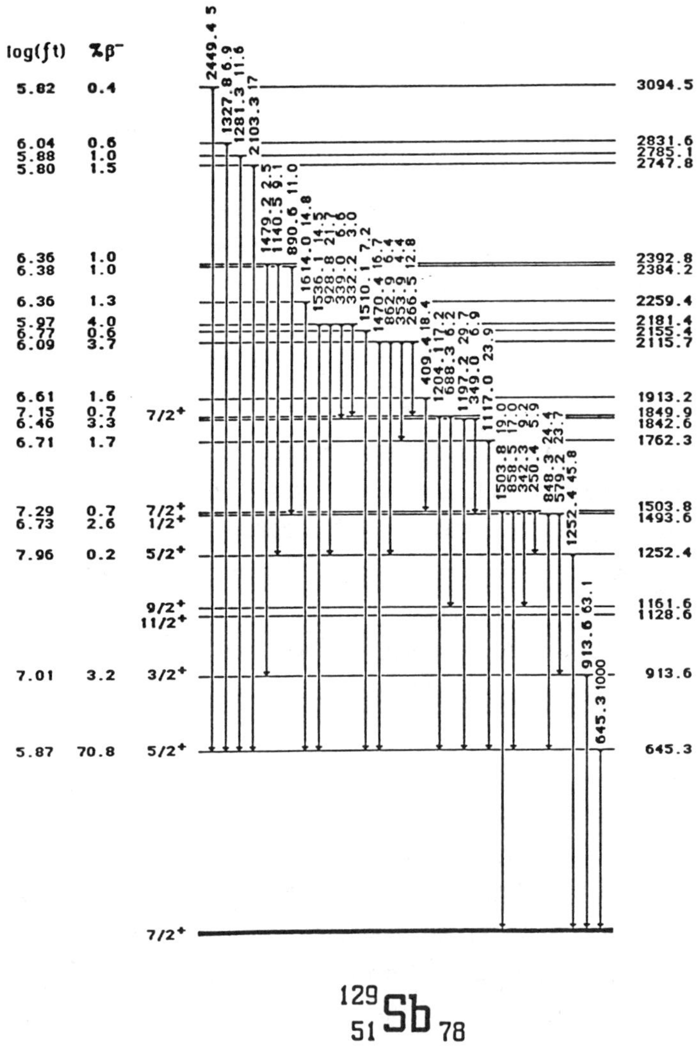

Figure 2. Proposed level scheme for ^{129}Sb.

$^{127}Sb_{76}$, A SIX NEUTRON-HOLE SYSTEM

Data on ^{127}Sb were limited. Q_β has now dropped to 2.9 MeV and fewer states are accessible in β^- decay. 16 new gamma rays were identified from the decay of the $3/2^+$, ^{127m}Sn, isomer and this has established 11 new excitation states in this system. Our level scheme for this system is shown in figure 3.

Reaction data[2] indicate that the level at 1185.6 keV has a spin of $1/2^+$. This level decays to both the $5/2^+$ level at 490.6 keV and to the level at 765.2 keV. The spin of the 765.2 keV state is probably $3/2^+$. This assignment is based primarily on the systematics of the $3/2^+$ state in ^{129,131}Sb and on the feeding from the $1/2^+$ level. Spins of the $9/2^+$ and $11/2^+$ states were assigned by Apt and Walters. They are consistent with the systematics that are seen in the decay of ^{129m,131m}Sn. This is also true for their assignment of $9/2^+$ to the 1584.3 keV level and $7/2^+$ for the 1471.2 keV level.

DISCUSSION

Systematics for the odd-mass Sb isotopes show a very regular trend. This is summarized in figure 4. The removal of two neutrons from the N=82 shell (^{131}Sb) causes a large number of states to be formed. Several of these states will belong to the $\pi g_{7/2}\otimes 2^+$ or $\pi d_{5/2}\otimes 2^+$ multiplets. In ^{131}Sb, the $\pi g_{7/2}\otimes 2^+$ multiplet is nearly degenerate, lying near 1200 keV. Somewhat higher in energy are members of the $\pi d_{5/2}\otimes 2^+$ multiplet and of these the $1/2^+$, $5/2^+$, and $9/2^+$ members have probably been seen.

The $5/2^+$ single particle state drops in energy as neutron pairs are removed from the nucleus. This causes the $\pi d_{5/2}\otimes 2^+$ multiplet to move closer to the $\pi g_{7/2}\otimes 2^+$ multiplet. The effect can be seen on the $9/2^+$ and $11/2^+$ members of the $\pi g_{7/2}\otimes 2^+$ multiplet. The $9/2^+$ level from the $\pi d_{5/2}\otimes 2^+$ multiplet lies at 1890 keV in ^{131}Sb. This moves down in energy at a somewhat lower rate than the $5/2^+$ single particle state. The $11/2^+$ and $9/2^+$ state in the $\pi g_{7/2}\otimes 2^+$ multiplet lie at a nearly constant energy with the $9/2^+$ level at a slightly higher energy. By

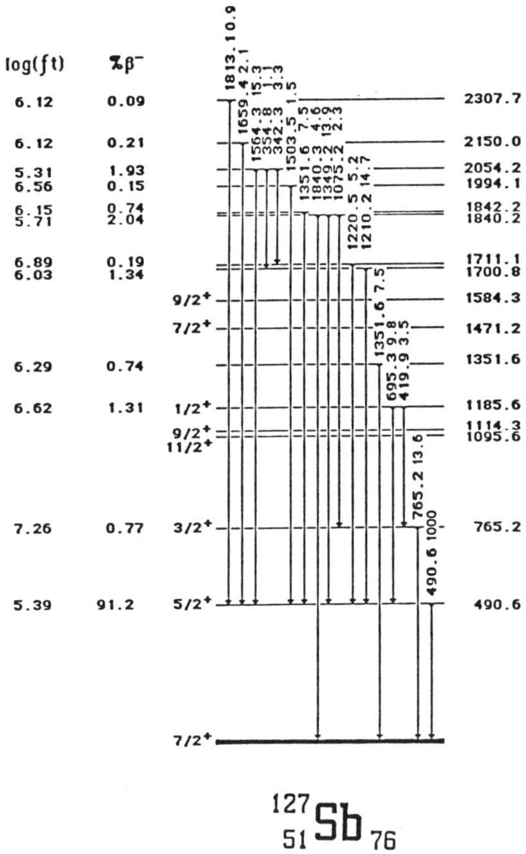

Figure 3. Proposed level scheme for ^{127}Sb.

Figure 4. Systematics of the odd-mass Sb isotopes.

^{123}Sb, the two 9/2$^+$ states have moved close enough together that their repulsion pushes the first 9/2$^+$ state below the 11/2$^+$ state. Levels in the $\pi g_{7/2}\otimes 2^+$ multiplet expand from a nearly degenerate multiplet (^{131}Sb) to one covering over 1 MeV (^{123}Sb). The 3/2$^+$ state is the only member of the multiplet which moves to a lower energy. The 5/2$^+$, 7/2$^+$ members move up in energy, becoming closer together as states from the $\pi d_{5/2}\otimes 2^+$ multiplet begin to repel them.

CONCLUSIONS

Our studies on the odd-A Sb nuclei have revealed valuable systematics on how the neutron core changes as neutron pairs are removed from the nucleus. We now have identified all members of the $\pi g_{7/2}\otimes 2^+$ multiplet in $^{127\text{-}131}$Sb. The systematics on these nuclei have given us a better understanding of results reported for ^{123}Sb and ^{125}Sb. In addition, we have identified several members of the $\pi d_{5/2}\otimes 2^+$ multiplet. Further analysis of our data will probably locate the remaining members of this multipet.

The known systematics on the odd-A Sb nuclei now offer an exciting test for shell model calculations near ^{132}Sn. The nuclear systems are very simple since there is only a single proton coupled to the neutron core. Yet, in their simplicity, finer details of the effective interactions are being studied than could be seen in the even-even Sn isotopes. Fitting a multiplet of states from the coupling of the proton to a neutron-hole configurations will give a greater challenge to shell model calculations. Also important will be the wealth of new information on the high-spin systems that we identified and will report in future articles.

REFERENCES

1. H. Huck, M.L. Perez, and J.J. Rossi, Phys. Rev. **C26**, 621 (1982).
2. K.E. Apt and W.B. Walters, Phys. Rev. **C9**, 310 (1974).
3. R.L. Auble, J.B. Bull, and C.B. Fulmer, Nuc. Phys. **A215**, 383 (1973).
4. A. Wolf, C. Chung, W.B. Walters, R.L. Gill, M. Shmid, H.-I. Liou, M. Stelts, R. Chrien, G. Peaslee, and D.S. Brenner, Nucl. Instr. Meth., **206**, 397 (1983).

THE ON-LINE TEMPERATURE NUCLEAR ORIENTATION OF Xe AND Cs ISOTOPES

T.L. SHAW, V.R. GREEN, C.J. ASHWORTH, J. RIKOVSKA, N.J. STONE
Clarendon Laboratory, Oxford;

P.M. WALKER,
SERC Daresbury Laboratory;

I.S. GRANT,
University of Manchester.

Abstract Using the technique of low temperature nuclear orientation (LTNO), the ground state magnetic moments of ^{121}Xe and ^{119}Xe have been measured to be $|\mu|$ = 0.65(3)nm and 0.59(6)nm. These values are consistent with a Nilsson configuration n[413 5/2] with moderate deformation. In addition, from nuclear orientation and relaxation studies of Cs isotopes, we have been able to deduce the moment of ^{118}Cs^{m} to be 5.4(1.1)nm. This large value clearly shows the presence of the proton [404 9/2] orbital.

INTRODUCTION

We have performed a series of experiments in the mass region A=110=122 using the Daresbury On-line Isotope Separator (DOLIS), which is on-line to the Nuclear Structure Facility (NSF) 20 MV tandem accelerator. This paper deals with measurements made on neutron deficient Xe and Cs isotopes in this region. The isotopes were produced by incidence of $^{32-34}$S ions (with energies between 150 and 175 Mev) on a ^{93}Nb target. The reaction products were mass separated and then accelerated to 60 keV before implantation into a polarized iron foil, maintained at temperatures $\geq$ 10 mk during implantation by a $^3He/^4He$ dilution refrigerator. The interaction of the nuclear magnetic moment with the magnetic hyperfine field of the implanted nuclei results in large degrees of nuclear polarization, measured by the anisotropic γ radiation emitted following the β decay of the implanted nuclei. The γ ray anisotropies were measured using Ge(Li) detectors placed along and perpendicular to the orientation axis (θ = 0, $\pi/2$), which is defined by a 0.7T field applied to the Fe foil. The γ ray angular distribution may be written as:

$$W(\theta) = 1 + f \sum_{k=2,4} B_k(I,\mu,T)A_kU_kQ_kP_k(\cos\theta)$$

The parameters which depend only on the properties of the implanted nuclei are the fraction in good sites f, which is the fraction of implanted nuclei which experience the full hyperfine field and the orientation parameters B_k, which depend on the spin I and magnetic moment μ of the implanted nuclei, as well as on the temperature T.

The other parameters U_k,A_k and Q_k are constants for each γ transition, determined by the nuclear level scheme of the daughter nucleus, the multipolarity of radiation involved, and the detector solid angle at the source. The magnetic moments in this paper have been deduced from the temperature dependance of the γ ray anisotropy, which is defined to be:

$$E(T) = \frac{W(0)}{W(90)} - 1$$

normalized to yields at 1K (where there is no nuclear orientation). This takes into account any variations in source strength during implantation.

$^{121}Xe \rightarrow {}^{121}I$ DECAY

The level scheme for ^{121}I fed by the β decay of ^{121}Xe with ground state spin $5/2^+$ is well established(1). Fig. 1 shows the temperature dependance of the anisotropy E(T) for the 445 keV transition in $^{121}I(7/5^+ \rightarrow 5/2^+)$. The mixing ratio δ(E2/M1) for this transition has been measured as -0.8(4)(2), and more recently, -0.78(20)(3). Taking this latter value we have fitted the data in fig. 1, taking B_{hf}=160T(4), to obtain f = 43(3)% and $|\mu|$ = 0.65(3)nm.

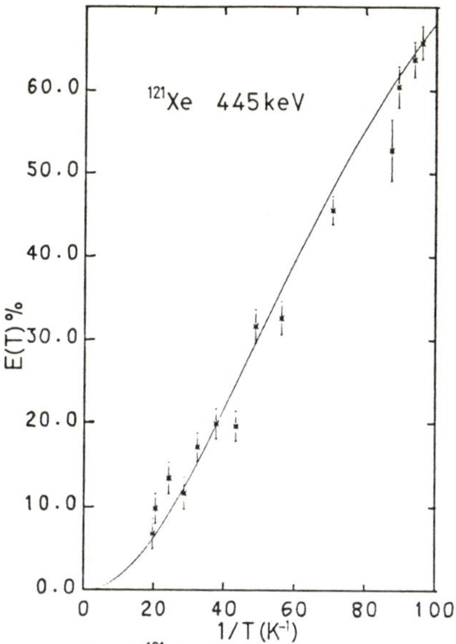

Fig.1. ^{121}Xe temperature dependence

$^{119}Xe \rightarrow {}^{119}I$ DECAY

Less is known about the decay of $^{119}Xe \rightarrow {}^{119}I$, but Marguier(5) et al. postulate a $5/2^+$ assignment for the ground state spin of ^{119}Xe. Fig. 2 shows the temperature dependance of the anisotropy E(T) for the 462 keV transition in ^{119}I. A fit to this data yields $|\mu|$ = 0.59(6)nm.

In this case not enough is known about the transition to extract f. However, taking it to be the analogue of the 445 keV transition in ^{121}I(5) and assuming the same fraction in good sites to be the same as above we can estimate a mixing ratio δ(E2/M1) of -0.4(2) for this transition.

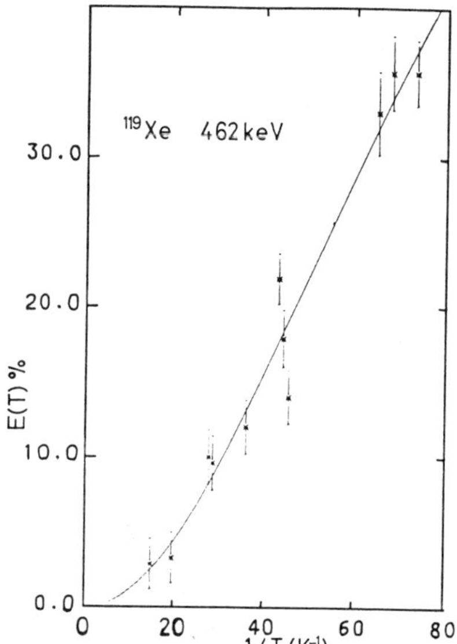

Fig.2. ^{119}Xe temperature dependence

DISCUSSION OF Xe MOMENTS

Chowdhury[6] et al. have observed collective features in both ^{119}Xe and ^{121}Xe and we use a Nilsson model description of the spin 5/2 ground states with parameters $\mu = 0.45$ and $k = 0.0637$[7,9]. There are three possible Nilsson orbitals which can give rise to a spin 5/2 state. They are the [413 5/2], the [532 5/2] close to the Fermi surface and the [402 5/2] at slightly higher energy. We have calculated the magnetic moments for these orbitals as a function of deformation (using the effective neutron g-factors, $g_s = 0.6g_s^{free}$, $g_1 = 0$, and $g_r = Z/A$) and the results are shown in fig. 3 along with the experimental value for ^{121}Xe. The negative parity [532 5/2] orbital clearly gives too small a moment to fit and can be ruled out. However, it is not easy to distinguish between the other two possibilities.

Evidence for the [402 5/2] orbital comes from Mueller[7] et al. who have measured the magnetic moment of $^{123}Ba^g$ to be -0.687(18), which they explain in terms of a n[402 5/2] configuration. In order to get this orbital close to the Fermi surface they need to invoke γ deformation of about 20° (which has little influence on their theoretical moment of -0.5).

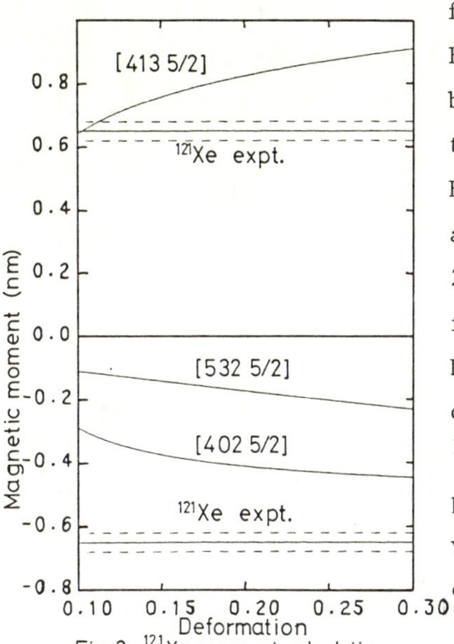

Fig.3. ^{121}Xe moment calculation

There is also evidence for the assignment [413 5/2]. Based on the comparison between experimental and theoretical nuclear moments Ekstrom(7) et al. have assigned the configuration $2^+\{p[404\ 9/2]n[413\ 5/2]\}$ for the odd-odd $^{120}Cs^g$. However, it should be pointed out that the large moment of $^{120}Cs^g$ is mainly due to the proton orbital, and is not very sensitive on the choice of neutron orbital. Sofia (9) et al. have studied the decay of $^{121}Cs \rightarrow {}^{121}Xe$ and they assign the [413 5/2] orbital to 5/2 ground state with $\beta = 0.16 \rightarrow 0.19$ (with the same Nilsson parameters as above).

Our measured value for ^{121}Xe is consistent with this latter assignment of [413 5/2], although, we cannot rule out the [402 5/2] alternative as we only know the magnitude of the moment and not its sign.

Cs RELAXATION AND MOMENT MEASUREMENTS

From nuclear orientation measurements of $^{119}Cs^m$, $^{120}Cs^g$ and $^{121}Cs^m$ (whose magnetic moments have all been measured) we have been able to determine the hyperfine field of Cs<u>Fe</u> to be $B_{hf} = 40.8(0.7)$ Tesla(10). When we come to look at

$^{118}Cs^m$, with a shorter half-life (17s) than the above Cs isotopes, it becomes necessary to take into account the spin-lattice relaxation time. Because the half-life is short not all the nuclei have time to reach thermal equilibrium with the lattice before they decay. This results in observed effects being attenuated compared to the full thermal equilibrium effects, and for a detailed analysis a knowledge of the relaxation mechanism is essential.

We recently performed an experiment where we measured the spin-lattice relaxation of $^{121}Cs^m$($t_{\frac{1}{2}}$=121s). We implanted a short burst (4s) of $^{121}Cs^m$ into a cooled Fe foil (at ∿ 8mK) and observed the nuclei relax to the lattice temperature. To build up sufficient statistics this was repeated many times. Fig. 4 shows data for the strong 179 keV transition, produced in the $^{121}Cs^m$ decay. The dot-dashed line represents the relative source strength as a function of time, with implantation starting at time zero. The data points represent the mean count rate (for 2s counting times) in the axial directions (θ = 0) and their deviation from the dot-dashed line is a measure of the nuclear orientation. The fitted curve corresponds to a Korringa constant, C_k of 0.059(15) sK, using $|\mu|$ = 5.4nm[11] and B_{hf}=41T. For different isotopes in

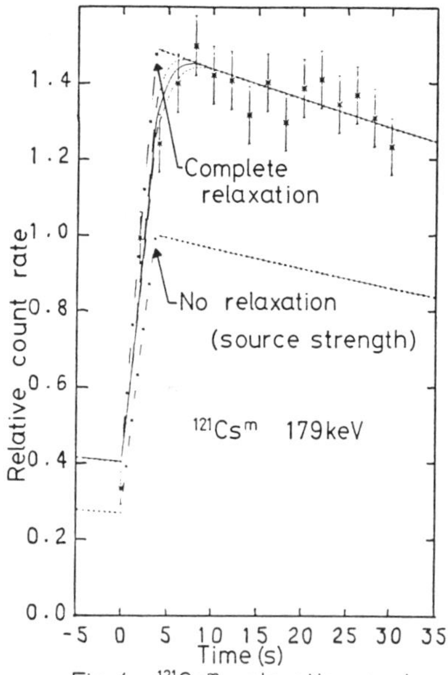

Fig.4. $^{121}Cs^m$ relaxation expt.

the same host the product $r = \gamma_n^2 C_k$ is constant (assuming magnetic interactions to be dominant), where $\gamma_n = \mu\mu_n/hI$, and for $^{121}Cs^m\underline{Fe}$ we find r to be $5.1(1.4).10^{-15}$ T^2sK^{-1}.

Fig. 5 shows the temperature dependence of the anisotropy E(T) for the 586 kev ($6^+ \rightarrow 4^+$) transition populated in the decay of $^{118}Cs^m$. In fitting this data we have used $r = 5.1(1.4).10^{-15}$, with B_{hf}=41T and I=6, and we extract a moment of 5.4(1.1) nm (the value of the moment changes very little if fits are performed for spins 7 or 8). The dotted line shows the predicted effect for full thermal equilibrium (i.e. full relaxation before decay), with all other parameters the same. The large moment identifies the presence of the proton [404 9/2] orbital.

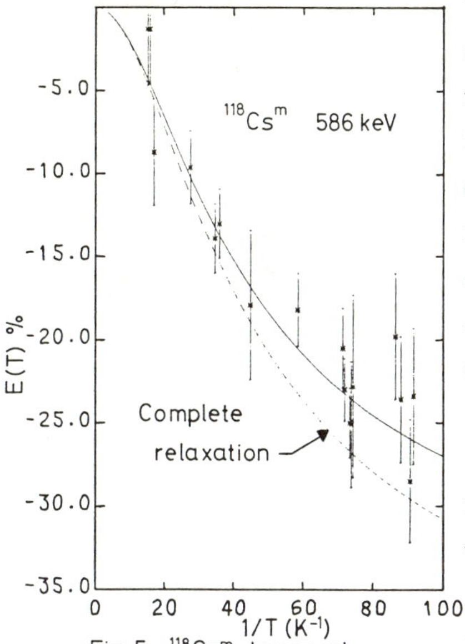

Fig.5. $^{118}Cs^m$ temperature dependence

References

1. T. Tamura et al. Nucl. Data Sheets 26 (1979) 385
2. U. Hagemann et al. Nucl. Phys. A389 (1982) 341
3. M. Gai et al. Phys. Rev. C26 (1982) 1101
4. E. Schoeters et al. Hyp. Int. 2 (1976) 365
5. G. Marguier et al. Report LYCEN/8518 (1985)
6. P. Chowdhury et al. Phys. Rev. C23 (1981) 733
7. C. Ekstrom et al. Phys. Lett. 76B (1978) 565
8. A.C. Mueller et al. Nucl. Phys. A403 (1983) 234
9. K. Sofia et al. Phys. Rev. C24 (1981) 1615
10. T.L. Shaw et al, to be published
11. C. Thibaut et al. Nuclear Physics A367 (1981) 1.

QUASIPARTICLE-CLUSTER-VIBRATION MODEL (QCVM) INTERPRETATION OF ODD-MASS N = 82 NUCLEI

EMILIAN DRĂGULESCU, VLADIMIR PAAR*, MARIN IVASCU, COSTEL PETRACHE, DOREL POPESCU, GHEORGHE SEMENESCU AND KLAAS ALLAART**
Central Institute of Physics, R-76900 Bucharest, ROMANIA
*Faculty of Science, University of Zagreb, YU-4100 Zagreb, YUGOSLAVIA
**Natuurkunding Laboratorium, Vrije Universiteït, 1081 HV Amsterdam, THE NETHERLANDS

Abstract The properties of the odd mass N=82 isotones are calculated in the QCVM. The whole sequence of N=82 isotones from ^{135}I to ^{149}Ho with 3,5,7,9,11,13,15,17 particles (protons outside doubly magic core $^{132}_{50}Sn_{82}$) was described using practically the same model parameters in the calculations only the number of valence-shell protons is changed. The resulting spectra and electromagnetic properties for positive and negative parity states of odd mass N=82 isotones are also compared with experimental data equally well as large shell-model calculations.

1. INTRODUCTION

The N=82 isotones have been extensively studied both experimentally and theoretically. The odd-even N=82 isotones have been described by shell-model calculations assuming a doubly magic Z=50, N=82 core [1-6].

In this description, the protons which are added in excess of Z=50, occupy the following single-particle configurations $1g_{7/2}$, $2d_{5/2}$, $2d_{3/2}$, $3s_{1/2}$ and $1h_{11/2}$. The low spin even-parity states in ^{141}Pr have been calculated by Wildenthal [2] taking into account the shell-model excitations $(1g_{7/2}, 2d_{5/2})^{Z-50}$ and $(1g_{7/2}, 2d_{5/2})^{Z-51}$ $(3s_{1/2}, 2d_{3/2})$. [2] The two body part of the hamiltonian was parametrized in terms of the modified surface delta interaction. This calculation was later extended to the high-spin states by Prade et al. [3,4] and Jäger [5] in ^{143}Pm and ^{141}Pr nuclei. In the framework of the space restricted to the configurations $(1g_{7/2}, 2d_{5/2})^{Z-51}$ $(1h_{11/2})^1$ the negative-parity high-spin states of odd-even N=82 isotones have been also calculated.[6]

The odd-even N=82 nuclei have been also treated in the framework of the quasiparticle Tamm-Dancoff approximation[7-9] by diagonalizing residual interaction on the quasi-particle space consisting of one and three proton quasiparticles.

In recent years there has been considerable interest in the study of odd-mass N=82 nuclei. This interest has been stimulated by extensive experimental studies of the doubly magic nuclei $^{132}_{50}Sn_{82}$ and $^{146}_{64}Gd_{82}$. It is interesting therefore to observe in surrounding nuclei ^{132}Sn and ^{146}Gd assumed as cores, the coupling of various particle and hole states to these core.

2. CALCULATION AND QCVM DISCUSSION

In the present work we are describing the whole sequence of odd mass N=82 isotones; $^{135}_{53}I$, $^{137}_{55}Cs$, $^{139}_{57}La$, $^{141}_{59}Pr$, $^{143}_{61}Pm$, $^{145}_{63}Eu$, $^{147}_{65}Tb$ and $^{149}_{67}Ho$ in the framework of quasi-cluster-vibration model (QCVM).[14] In this model the number-projected one- and three-quasiparticle states are coupled to the quadrupole phonons. Thus, we assume that three of the valence shell particles may occur in any possible configuration within the valence shell particles of the same kind occur as superfluid Cooper pairs. On the other hand, it is assumed that all the other relevant degrees of freedom are described by phonons, similarly as in the cluster-vibration-model CVM[17].

In QCVM one may attempt to describe the sequence of nuclei with 3,5,7,9,... particles beyond a closed shell using in principle the same model parameters in the hamiltonian. Such a QCVM calculation was performed for the odd Zn isotopes.[14,15].

In its simplest version the QCVM hamiltonian reads[14,15]

$$H_{QCVM} = \sum_{\alpha} \varepsilon_{\alpha} a^+_{\alpha} a_{\alpha} - \frac{1}{4} G \sum_{a,b} (2j_a+1)^{1/2} (2j_b+1)^{1/2} A^+_{00}(a\,a) \bullet$$

$$\bullet A_{00}(bb) + \hbar\omega_2 \sum_{\mu} (b^+_{\mu} b_{\mu} + \tfrac{1}{2}) - a\, 4\pi \sum_{\mu} (b^+_{\mu} + \tilde{b}_{\mu}) \bullet$$

$$\bullet \sum_{\alpha\,\beta} < \alpha\, | Y^*_{2\mu} |\, \beta > a^+_{\alpha}\, a_{\beta} \qquad (1)$$

with the standard notation

$$A^+_{00}(aa) = \sum_{m_a} < j_a m_a j_a - m_a \,|00 > a^+_{\alpha}\, a^+_{\bar{\alpha}} \quad ;$$

a^+_{α} and b^+_{μ} denote the quasiparticle and phonon creation operators, respectively.

The QCVM hamiltonian (1) can be extended by adding the anharmonic terms based on the SU(6) symmetry and by expressing the quasiparticle-vibration interaction in terms of SU(6) boson generators, analogously as for the case of coupling one particle to quadrupole vibrations[19].

If we adopt the IBA representation [20,21] as a familiar keyword, the QCVM with the hamiltonian (1) would appear as a harmonic SU(6) limit of an extended IBFM, with three fermions (BCS quasiparticles) coupled to s, d - bosons.

Energy systematics for the first 2_1^+ and 3_1^- states in N=82 isotones (^{132}Sn to ^{148}Dy) is summarized in Figure 1.

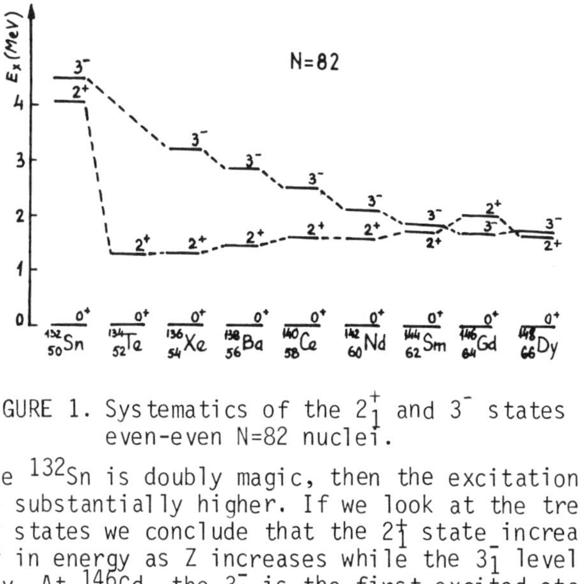

FIGURE 1. Systematics of the 2_1^+ and 3^- states in even-even N=82 nuclei.

Because ^{132}Sn is doubly magic, then the excitation energies became substantially higher. If we look at the trend of 2^+ and 3_1^- states we conclude that the 2_1^+ state increase gradually in energy as Z increases while the 3_1^- level drops rapidly. At ^{146}Gd, the 3_1^- is the first-excited state and the 2_1^+ state is located at unusually high energy; this fact was interpreted as a closure at Z=64.

In the present calculations the proton single-particle states are: 0, 0.96, 2.8, 3, and 3.2 MeV for the $2d_{5/2}$ $1g_{7/2}$, $1h_{11/2}$, $3s_{1/2}$ and $2d_{3/2}$ configurations, respectively. Here, the energies of $2d_{5/2}$, $1g_{7/2}$ and $1h_{11/2}$ configurations have been taken from the experimental spectrum of ^{133}Sb from ref.[22,23] and the relative positions of $3s_{1/2}$ and and $2d_{3/2}$ with respect to $1h_{11/2}$ were taken from ref. [5] The pairing force strength is G = 0.25 MeV the phonon energy $\hbar\omega$ = 1.6 MeV and the coupling strength a = 0.3 MeV.

These magnitudes are consistent with the magnitudes of the corresponding parameters used in the previous CVM calculations.[17)]

The calculated low energy spectra of some sequence of odd-mass N=82 nuclei is presented in Figure 2. It should be

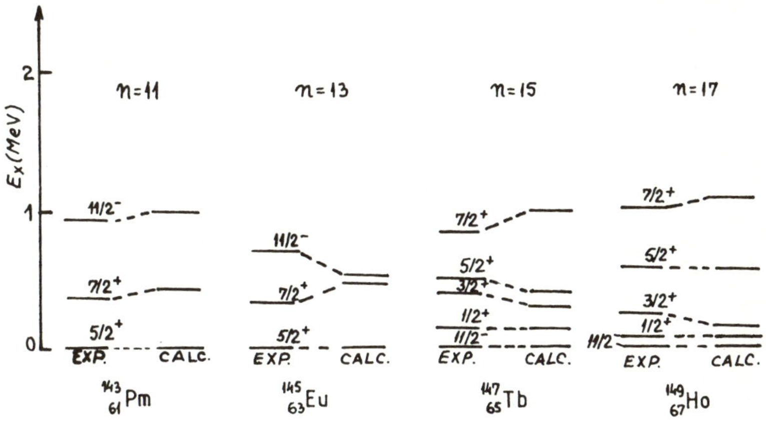

FIGURE 2. Comparison of experimental and QCVM calculated energies for the first states in odd-even N=82 nuclei (with n = 11, 13, 15, 17 valence-shell particles)

noted that the only difference between the calculations for the above nuclei is in the number of valence-shell protons which is variable.

In order to take into account the increased collectivity and the Z=64 closure (^{146}Gd), for ^{145}Eu (n=13), ^{145}Tb (n=15) and ^{149}Ho (n=17), it was necessary to increase the particle-vibration coupling strength from 0.4 MeV to 0.5 MeV respectively. In addition, the pairing strength G is decreased at 0.14 MeV value (calculated from the relation G=23/A and suggested by the calculations done by Chasman[26)]) for ^{147}Tb and ^{149}Ho nuclei in order to reproduce the first positive-parity experimental levels. Our calculations suggest for Z>64, N=82 nuclei the ground state with $11/2^-$ spin.

Those detailed QCVM calculations for the ^{141}Pr nucleus (n=9) are presented in Figure 3 and Table 1 in comparison with experimental data [25)]. The available shell-model results for ^{141}Pr nucleus are also given [4,5)].

As seen from Figures 2,3 and Table 1 the agreement

TABLE 1 Theoretical (QCVM+SHELL-MODEL) electromagnetic properties of ^{141}P and comparison with the experimental data

Transition	$B(E2)\downarrow$ $(e^2\ b^2)$			$B(M1)\downarrow$ (μ_N^2)		
$J_i^\pi \rightarrow 5/2^+$	SM	QCVM(9p)	EXP.	SM	QCVM(9p)	EXP.
$1/2_1^+$	0.025	0.053	0.057 ± 0.015			
$3/2_1^+$	0.026	0.090	0.062 ± 0.006			
$3/2_2^+$	0.0011	0.00012	≤ 0.006			
$5/2_2^+$	0.0102	0.041	0.035 ± 0.006			
$5/2_3^+$	0.0009	0.026	0.025 ± 0.006			
$7/2_1^+$	0.00044	0.00010	0.00108 ± 0.00008	0.0032	0.0106	0.0047
$7/2_2^+$	0.000019	0.010	0.0082 ± 0.0015			
$7/2_3^+$	0.0072	0.052	-			
$9/2_1^+$	0.033	0.038	0.037 ± 0.005			
$9/2_2^+$	0.0122	0.013	0.015 ± 0.005			

State J_i^π	$Q(eb)$			$\mu(\mu_N)$		
	SM	QCVM(9p)	EXP.	SM	QCVM(9p)	EXP.
$5/2_1^+$	-0.02	-0.171	-0.078	1.5	4.0	4.3
$7/2_1^+$	0.42	0.294		0.72	2.23	2.78
$11/2_1^-$	6.65	7.0	7.20			

between QCVM calculations and experiment is rather good; the energy spectrum and the electromagnetic properties are well reproduced. From the comparison of the QCVM results and the shell-model results, the QCVM energy spectra and electromagnetic properties are in a slightly better agreement with experimental data. This feet is explained due to the well choosing of the degrees of the QCV model (quasiparticles and quadrupole vibrations). An extensive study for the properties of odd-even N=82 isotones was accomplished.[27)]

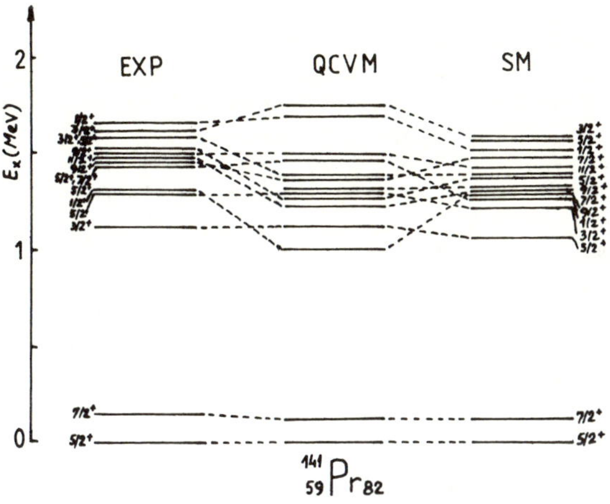

FIGURE 3. Comparison of the experimental level energies in ^{141}Pr nucleus with QCVM (n=9 particles outside the $^{132}_{50}Sn_{82}$ core) and shell-model calculations.

3. CONCLUDING REMARKS

The application of the QCVM to the whole sequence of odd-even N=82 isotones (spectra and electromagnetic properties by changing only the number of valence-shell particles) demonstrates the real physical basis and usefulness of this model.

The QCVM results are at least as good as the shell-model (efforts are much less for the QCVM than for the very large basis shell model calculations): we suggest that the QCV model is to be preferred as a predictive tool in nuclear spectroscopy of nuclei away from closed shells.

REFERENCES

1. B. H. Wildenthal, Phys.Rev.Lett., 22, 1118 (1969).
2. B. H. Wildenthal, Phys.Lett. 29B, 274 (1969).
3. H. Prade, L. Käubler, U. Hagemann, H. U. Jäger, M. Kirchbach, L. Schneider, and F. Stary, Nucl.Phys., A333, 33 (1980).
4. H. Prade, W. Enghardt, H.U. Jäger, L. Kaubler, H. J. Keller, and F. Stary, Nucl.Phys.,A370, 47 (1980).
5. H. U. Jäger, private communication.
6. W. Enghardt, and H. U. Jäger, Nucl.Phys., A438, 148 (1985).
7. M. Waroquier and K. Heyde, Nucl.Phys., A144, 481 (1970).
8. N. Freed and W. Miles, Phys.Lett., 32B, 313 (1970)..
9. N. Freed and W. Miles, Nucl.Phys., A158, 230 (1970).
10. K. Kawade, K. Sistemich, G. Battistuzi, K. Shizuma, and J. Blomqvist, Z.Phys., A308, 33 (1982).
11. T. Björnstad et al., Report CERN-EP/85-201 (1985).
12. P. Kleinheinz, R. Broda, P. J. Daly, S. Lunardi, M. Ogawa, and J. Blomqvist, Z.Phys., A290, 279 (1979).
13. P. J. Daly, P. Klenheinz, R. Broda, S. Lunardi, H. Backe, and J. Blomqvist, Z.Phys., A298, 173 (1980).
14. K. Allaart, P. Hofstra, and V. Paar, Nucl.Phys., A366, 384 (1981).
15. A. Van Engmond, P. Hofstra, E. Boeker, K. Allaart and V. Paar, Z. Phys., A300, 323 (1981).
16. K. Allaart, P. Hofstra, and V. Paar, Int. Phys. Conf., 62, 54 (1982).
17. V. Paar, Nucl.Phys., A211, 29 (1973).
18. D. Janssen, R. V. Jolos, and F. Dönau, Nucl.Phys.,A224 93 (1974).
19. V. Paar, S. Brant, L. F. Canto, G. Leander, and M. Vouk, Nucl.Phys., A378, 41 (1982).
20. A. Arima and F. Iachello, Phys.Rev.Lett., 35, 1069 (1975).
21. F. Iachello and O. Scholten, Phys.Rev.Lett., 43, 679 (1979).
22. K. Sistemich, W. D. Lauppe, T. A. Khan, H. Lawin, H. A. Selic, J. P. Bocquet, A. Monnaud, and F. Schussler, Z.Phys., A285, 305 (1978).
23. J. Blomqvist, A. Kerek, and B. Fogelberg, Z.Phys., A314, 199 (1983).
24. V. Paar and S. Brant, Nucl.Phys., A303, 96 (1978).
25. E. Drăgulescu, M. Ivaşcu, I. Gurgu, C. Petrache, D. Popescu, G. Semenescu, G. Wesselborg, and V. Paar, submitted to Nucl.Phys. A.
26. R. R. Chasman, Phys. Rev. C21, 456 (1980).
27. E. Drăgulescu, V. Paar, M.Ivaşcu, C. Petrache,D.Popescu, G. Semenescu, and K. Allaart, to be published.

THE FIRST EXCITED 0^+ STATE IN ^{144}Nd

M.K. HARDER†, W.D. HAMILTON†, G. COLVIN+ AND A.R.H. SUBBER†

† Physics Division, University of Sussex, Brighton BN1 9QH, U.K.

+ Institut von Laue-Langevin, 38042 Grenoble Cedex, France.

Abstract An electron spectrum for ^{144}Nd has been obtained following thermal neutron capture by ^{143}Nd at the ILL, Grenoble. A transition at 2084.38 keV has been seen, confirming the presence of an excited 0^+ level at this energy. Other observed transitions which feed and depopulate this level are discussed. The X(E0/E2) value is given.

INTRODUCTION

The nucleus ^{144}Nd (N=84) lies at the boundary of a transition region where the nucleus shape changes from spherical (vibrational) to axially deformed. This is a sensitive area to test different nuclear models, as many properties of the nucleus may be expected to change rapidly at the onset of deformation. For this reason ^{144}Nd has been the subject of several experiments which have attempted to determine its decay scheme.

Even after several such experiments, which successfully enriched the scheme, a problem has remained in identifying the first excited 0^+ state. Krane et al[1] reviewed the situation, and by re-analysing previous work they proposed that the level at 2084 keV was the first excited 0^+ state.

This paper reports on new internal-conversion measurements carried at the Institute Laue-Langevin, Grenoble. An electron transition was seen at 2084.38 keV which confirms the level predicted by Krane et al. The excellent resolution of the spectrometer[2] also gave two transitions in the region of 1388 keV, which may account for some of the problems experienced in specifying the 2084 level in previous investigations.

DATA ANALYSIS AND DISCUSSION

A spectrum was obtained of the electrons emitted after thermal neutron capture by enriched ^{143}Nd (76.2%) in Nd_2O_3 form. The data show a transition of 2084.38 keV, as shown

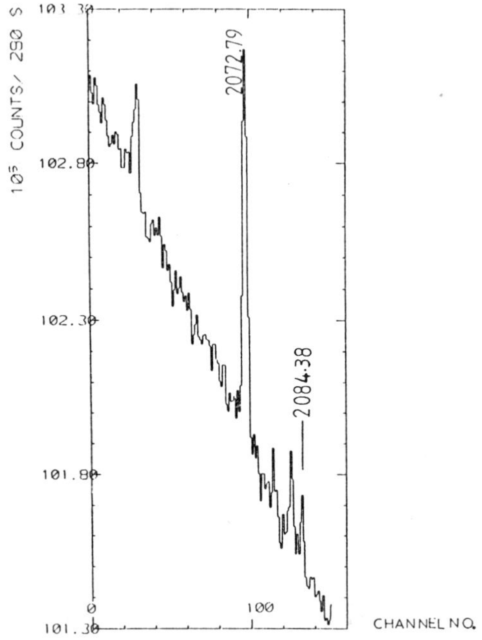

FIGURE 1. Electron spectrum around 2 MeV, showing line at 2084 keV.

in fig.1. This transition is pure E0 as no corresponding γ-ray transition was seen in (n,γ) measurements by Snelling and Hamilton[3] which used a similarly enriched target. The line was calibrated with respect to stronger lines of the ^{144}Nd spectrum, which were themselves calibrated against the $2_1^+ \rightarrow 0$, $4_1^+ \rightarrow 2_1^+$, $6_1^+ \rightarrow 4_1^+$, $2_2^+ \rightarrow 0$ lines given by Snelling and Hamilton.

A level had been previously proposed at 2084 keV [4,5,6], but the spin-parity was tentatively assigned as 2^+. Krane et al then proposed the spin assignment at 0^+. The transition does not fit any of the levels given by Snelling and Hamilton, nor those listed in a summary by Tuli[7], except as a transition to the ground state. The 0^+ state at 2084.38 keV is thus confirmed, as shown in fig. 2a.

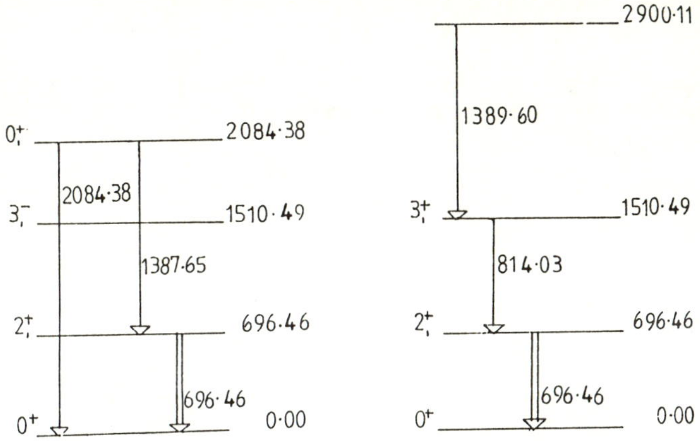

FIGURE 2a. The position of the 2084.38 and 1387.65 electron transitions

FIGURE 2b. The position of the 1389.60 line, referred to by previous authors as the 1388 line.

By comparing fig.2b with fig.2a, it is possible to understand some of the previous difficulties. The 1388 keV line, not previously known to be a doublet, was assigned by earlier authors as shown in fig.2b. In the work of Snelling and Hamilton, this was supported by coincidences obtained with gates at 696 and 814 keV.

Due to the resolution of the (n,e^-) experiment it was possible to see that two electron lines exist in the region of 1388 keV, at 1387.65 keV and at 1389.62 keV. These transitions were found not to fit between any other levels in the decay scheme other than as shown in figs. 2a,2b. Placed as such they provide no conflict with the coincidence measurements.

The feeding of the 2084 level is more difficult to trace. The ground state of the target nucleus, ^{143}Nd, is $7/2^-$, and the capture state has been shown to be predominantly 3^- [8]. This level is not seen to be fed by a primary γ-ray[9,10]. However, the (n,e^-) measurements have shown a new transition at 968.75 keV which implies a feeding of the 2084 level from the 3052.80 level, as shown in fig.3. This transition does not fit between any of the other levels, and its position as given is supported by the coincidence spectra of Snelling and Hamilton.

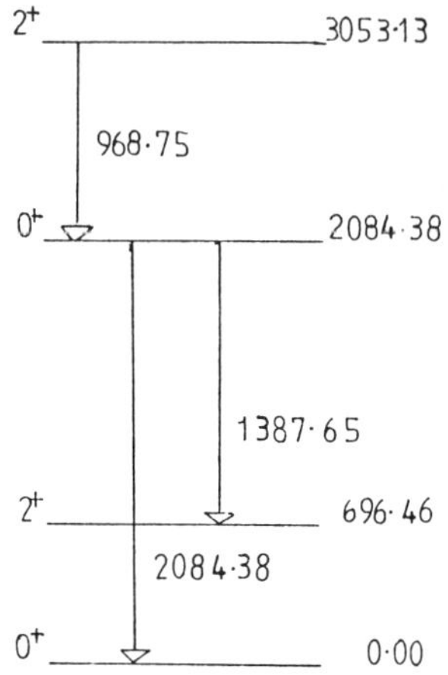

FIGURE 3. Population and de-population of the 2084.38 keV level.

If this 3053 keV level is itself fed directly from 3^- capture state, its spin assignment would most likely be 2^+, giving a 3^- - 2^+ - 0^+ cascade.

Table I. X(E0/E2) values for E0 transitions.

Transition (keV)	Initial. - Final levels	Level spins	Relative e-intensity x 10^{-3} (a)	Relative γ-intensity (b)	α_K (expt) x 10^{-3}	α_K (theoretical) α_K(E1) x 10^{-3}	α_K(M1) x 10^{-3}	α_K(E2) x 10^{-3}	δ (c)	X(E0/E2)
980.48	2294 1314	4^- 4^+	709 (2)	35.5 (2)	3.06 (17)	0.81	3.1	1.95	-0.54<6	0.058 (14)
1136.74	2451 1314	4^+ 4^+	44 (2)	19.7 (2)	2.24 (12)	0.64	2.3	1.45	-1.3 (4)	0.101 (4)
1376.12	2072 696	2^- 2^-	36 (2)	22.1 (2)	1.61 (9)	0.45	1.45	0.94	$0.31 \begin{Bmatrix} +0.4 \\ -0.3 \end{Bmatrix}$	$0.61 \begin{Bmatrix} +43 \\ -20 \end{Bmatrix}$
1671.79	2367 696	2^+ 2^+	29.0 (15)	28.4 (2)	1.02 (6)	0.335	0.92	0.66	0.17 (8)	$2.2 \begin{Bmatrix} +50 \\ -19 \end{Bmatrix}$
2084.38	2084 0	0^+ 0^+	0.80 (13)	——	——				——	0.095 (18)

(a) Intensities normalized to pure E2 transitions.
(b) Intensities normalized to the 696 KeV transition.
(c) Quoted from Snelling and Hamilton, Ref. 3.

The X(E0/E2) value for the $0^+ \rightarrow 0^+$ transition is given in table 1 together with the values of E0 components in other $\Delta I = 0$, $I_i = I_f \neq 0$, transitions. The results of a fuller analysis of the data from the (n,e^-) experiment will be published shortly.

REFERENCES

1. K.S. Krane, S. Raman, and F.K. McGowan, Phys. Rev., C27, 2863 (1983).
2. W. Mampe, K. Schreckenbach, P. Jeuch, B.P. Maier, F. Braumandl, J. Larysz and T. von Egidy, Nucl. Instr. Meth., 154, 127 (1978).
3. D.M. Snelling and W.D. Hamilton, J. Phys. G : Nucl. Phys., 9, 763 (1983).
4. S. Raman, Nucl. Phys., A107, 402 (1968).
5. M. Behar, Z.W. Grabowski and S. Raman, Nucl. Phys., A219, 516 (1974).
6. E.A. Andreev, V.K. Basenko and S.P. Sitko, Sov. J. Nucl. Phys., 23, 6 (1976).
7. J.K. Tuli, Nuclear Data Sheets 27, 97 (1979).
8. S. Raman, G.G. Slaughter, J.A. Harvey, E.T. Jurney, D.A. McClure, J.C. Wells Jr and J. Lin, in "Proceedings of the Second International Symposium on Nuetron Capture Gamma Ray Spectroscopy and Related Tlpics", edited by K. Abrahams, F. Strecher-Rasmussen, and P. van Assche (Reactor Centrum Nederland, Petten, 1975), p.562.
9. S. Raman, O. Shahal, D.A. McClure and M.J. Kenny, in "Neutron Capture Gamma Ray Spectroscopy and Related Topics, 1981", edited by T. von Egidy, F. Gonnenwein and B. Maier (Institute of Physics, Bristol, 1981), p.435.
10. H. Postma and E.R. Reddingus, Physica, 34, 541 (1976).

A SEMIEMPIRICAL SHELL MODEL CALCULATION IN THE N=82, Z=64 REGION

J. KOWNACKI
Inst. of Exp. Physics, Univ. of Warsaw, Poland

C. BAKTASH
Oak Ridge National Lab., Oak Ridge, TN 37831, USA

Abstract The aim of the present work was to determine a consistent set of single-particle energies and two-body interaction matrix elements that are relevant in the region around ^{146}Gd nucleus. The resulting values were used to calculate the medium- and high-spin states that are fed by the decay of the 550 ns isomer in ^{147}Gd.

The ^{147}Gd nucleus[1] has 1 valence neutron relative to ^{146}Gd, which is assumed to be a doubly-magic core nucleus.[2] In an attempt to understand and describe the states observed between the 550 ns $I=49/2^+$ and 27 ns, $I=27/2^-$ isomers, a semi-empirical shell-model calculation along the lines developed by Talmi[3] and Blomqvist[4] has been performed.

The method requires knowledge of empirical single-particle energies and multi-particle interaction in nuclei around ^{146}Gd. One has to stress here that in this kind of shell-model calculations, the main problem is clearly the evaluation of the matrix elements of the residual interaction between many particle states, ψ_J. The method rests on the assumption that the many particle matrix elements $\langle \psi_J | V | \psi_{J'} \rangle$ can be reduced to sums over two-particle interaction matrix elements (plus core polarization). The single-particle energies and the empirical 2-nucleon interaction energies of the nuclei necessary for the present calculation, are summarized in tables (1-6). For the proper mass scaling the mass excesses are taken from the experimental data when available, and from the systematics or shell model analysis[4] otherwise. The energies of the excited states in ^{147}Gd corresponding to the following configuration have been calculated:

$$\pi(h^2_{11/2}\, d^{-2}_{5/2})\, \nu(f_{7/2}) \qquad 27/2^- \leqslant I \leqslant 35/2^-$$

$$\pi(h^2_{11/2}\, d^{-1}_{5/2}\, g^{-1}_{7/2})\, \nu(f_{7/2}) \qquad 27/2^- \leqslant I \leqslant 39/2^-$$

$$\pi(h^2_{11/2}\, d^{-2}_{5/2})\, \nu(i_{13/2}) \qquad 29/2^+ \leqslant I \leqslant 41/2^+$$

$$\pi(h^2_{11/2}\, d^{-1}_{5/2}\, g^{-1}_{7/2})\, \nu(i_{13/2}) \qquad 35/2^+ \leqslant I \leqslant 45/2^+$$

Some of the calculated energies relative to the excitation energy of the $I^{\pi} = 27/2^-, \nu(f_{7/2})\pi(h^2_{11/2}\; d^{-2}_{5/2})$ level (3,582 MeV) are compared with the experimental values (Bakander[1]) in Fig. 1.

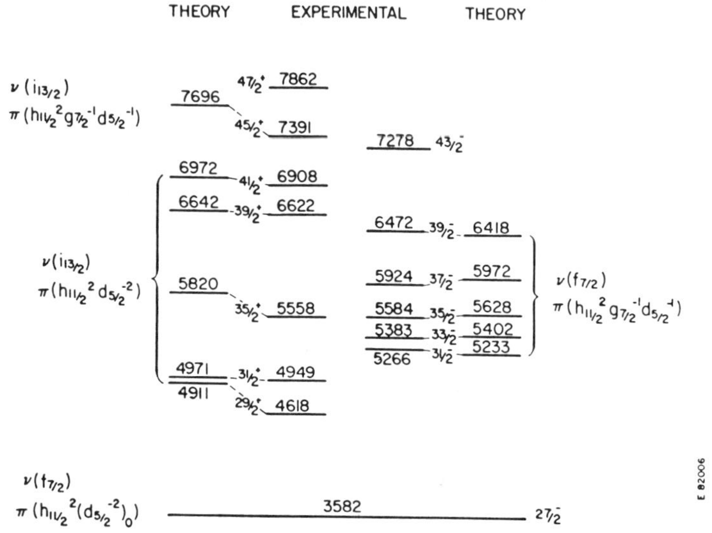

Fig. 1. Partial level diagram of ^{147}Gd. Experimental and calculated levels are compared.

Based on the close correspondence between the calculated and measured excitation energies of the yrast states we ascribe the following configurations to the levels shown in Fig. 1. The negative parity states, $31/2 \leqslant I \leqslant 39/2$ (shown to the right in the figure) are assigned to the $\nu(f_{7/2})\pi(h^2_{11/2}\ g^{-1}_{7/2}\ d^{-1}_{5/2}$ multiplet. The positive-parity states $29/2 \leqslant I \leqslant 41/2$ (left) are reproduced by coupling the $\pi(d^{-2}_{5/2}\ h^2_{11/2})$ configuration to the "effective" $i_{13/2}$ neutron orbital. The states at 7,391 MeV ($45/2^+$) and 7,278 MeV ($43/2^-$) are not uniquely established. The $45/2^+$ state may be of the type $\nu(i_{13/2})\pi(h^2_{11/2}\ d^{-1}_{5/2} g^{-1}_{7/2})$.

The largest spin value attainable in coupling one valence neutron to 2-proton particle-hole excitations is 45/2. Higher spin positive - parity states (such as $47/2^+$ level at 7,826 MeV) may be formed by seven-quasiparticle states of the type $\nu(f_{7/2})\pi(d^{-2}_{5/2}$ $g^{-1}_{7/2}\ h^{3}_{11/2})$. One would like to stress here that only one set of effective 2-body interaction energies was used in the present calculations.

The authors wish to acknowledge the many stimulating discussions with S. Bjornholm, J. Borggreen, J. Pedersen, and G. Sletten. One of us (J.K.) thanks the Polish Ministry of Science, High Education and Technics for partial financial support under contract CPBP 01.06.13.01. Oak Ridge National Laboratory is operated by Martin Marietta Energy Systems for the U.S. Department of Energy under Contract No. DE-AC05-84OR21400.

REFERENCES

1. J. Pedersen, B.B. Back, F.M. Bernthal, S. Bjornholm, J. Borgreen, O. Christensen, F. Folkmann, B. Herskind, T.L. Khoo, M. Neiman, F. Puhlhofer and G. Sletten, Phys. Rev. Lett. **39** (1977) 990.
   O. Bakander, C. Baktash, J. Borggreen, J.B. Jensen, J. Kownacki, J. Pedersen, G. Sletten, D. Ward, H.R. Andrews, O. Hausser, P. Skenved and P. Taras Nucl. Phys. **A389** (1982) 93.
2. P. Kleinheinz, M. Ogawa, R. Broda, P.J. Daly, D. Haenni, H. Beuscher and A. Kleinrahm, Z. Phys. **A286** (1978) 27.
3. A. de-Shalit and I. Talmi. Nuclear Shell Theory (Academic Press, N.Y., 1963).
4. J. Blomqvist, Symposium on High-Spin Phenomena in Nuclei, Argonne National Lab., March 15-17, 1979.
   J. Blomqvist, I. Bergstrom, C.J. Herrlander, C.G. Linden and K. Wikstrom, Phys. Rev. Lett. **38** (1977) 534.
5. R.C. Pardo et al., Rep. MSUCL no **304**, (1979).

TABLE 1

Single particle energies around ^{146}Gd nucleus

Nucleus	Orbit	Excit.energy E_j	g.s.separation energy E_j^g
$^{147}_{64}Gd_{83}$	$2f_{7/2}$	0	-7455±50
	$1i_{13/2}$	997	-6458
	$3p_{3/2}$	1153	-6302
	$1h_{9/2}$	1398	-6057
	$2f_{5/2}$	1847	-5608
$^{145}_{64}Gd_{81}$	$3s_{1/2}^{-1}$	0	+11297±70
	$2d_{3/2}^{-1}$	27	+11324
	$1h_{11/2}^{-1}$	748	+12045
	$2d_{5/2}^{-1}$	1015	+12312
$^{145}_{63}Eu_{82}$	$2d_{5/2}^{-1}$	0	+5449±40
	$1g_{7/2}^{-1}$	329	+5878
	$1h_{11/2}^{-1}$	717	+6266
	$3s_{1/2}$	809	+6358
	$2d_{3/2}^{-1}$	1042	+6591
$^{147}_{65}Tb_{82}$	$1h_{11/2}$	0	-1781
	$3s_{1/2}$	(30)	-1751
	$2d_{3/2}$	253	-1498
	$2d_{5/2}$	354	-1399
	$1g_{7/2}$	719	-1032

TABLE 2. Effective 2-proton interaction energies in ^{148}Dy.

The values of ΔJ^{exp} for $\pi(h^2_{11/2})$ configuration have been extracted using two values of mass excesses.

Main config.	J^π	ϵ_J^{exp}	E_J 2 part sep.en.	E_J^0 0-order sep.en.	ΔJ^{exp} $E_J-E_J^0$	$\Delta J(\delta)$
$\pi(h_{11/2}^2)$	0^+	0	-6086*	2x(-1781) = -3562	-2524	-2801
	2^+	1678	-4408		-846	-1124
	4^+	2428	-3658		-96	-374
	6^+	2733	-3353		+209	-70
	8^+	2834	-3252		+309	+31
	10^+	2920	-3166		+396	+117
$\pi(h_{11/2}^2)$	0^+	0	-5941**	2x(-1703) = -3406	-2535	
	2^+	1678	-4263		-857	
	4^+	2428	-3513		-107	
	6^+	2733	-3208		+198	
	8^+	2834	-3107		+299	
	10^+	2920	-3021		+385	
$\pi(h_{11/2}s_{1/2})$	5^-	2349	-3737	-3532	-205	
	6^-	2382	-3704			-172
$\pi(h_{11/2}d_{3/2})$	4^-	2678	-3408	-3279		-129
	5^-	2777	-3309			-30
	6^-	2765	-3321			-42
	7^-	2739	-3347		-68	

* The calculated mass excess of ^{146}Gd: -7595 (30) from J. Blomqvist et al.[4].

** The experimental mass excess value taken from Pardo et al.[5].

In the further calculations the E_J^g = -6086 keV was used to reproduce the energy of the states in ^{145}Eu, ^{146}Gd, ^{147}Tb, ^{148}Dy, ^{149}Ho and ^{150}Er;

TABLE 3. Effective proton-protonhole interaction energies ΔJ in ^{146}Gd.

$E_J^g = 0$

Main config.	J^π	ϵ_J^{exp}	E_J 2 part sep.en.	E_J^0 0-order sep.en.	ΔJ^{exp} $E_J - E_J^0$	ΔJ(δ)	Comments ΔJ*
$1h_{11/2}2d_{5/2}^{-1}$	3^-	1579		+3668	-2089	-1605	-1700*
	4^-	2918*				-676	-750*
	5^-	2658			-1010	-910	-900*
	6^-	3268*				-573	-400*
	7^-	2982			-686	-690	-590*
	8^-	3183			-485	-489	-390*
$1h_{11/2}1g_{7/2}^{-1}$	2^-	3497*		+3997		-500	-500*
	3^-	3787*				-650	-210*
	4^-	3867*				-81	-130*
	5^-	3507*				-93	-490*
	6^-	3497*				-46	-500*
	7^-	3397*				-165	-600*
	8^-	3294			-703	-23	-650*
	9^-	3428			-569	-431	-500*

* Interaction energies obtained from the best fit to ^{144}Sm, ^{145}Eu, ^{146}Gd and ^{147}Tb nuclei. Also energy of levels are calculated by means of these ΔJ* values.

TABLE 4. Effective 2-proton holes interaction energies ΔJ in $^{144}_{62}$Sm.

$E_J^g = +8710$ keV

Main config.	J^π	ϵ_J^{exp}	E_J 2 part sep.en.	E_J^0 0-order sep.en.	ΔJ^{exp} $E_J - E_J^0$	ΔJ(δ)	Comments ΔJ*
$2d_{5/2}^{-2}$	0^+	0	+8710	+10899	-2180	-1820	-1100*
	2^+	1660	+10370		-528	-514	-100*
	4^+	2190	+10900		+2	+16	0*
$2d_{5/2}^{-1}1g_{7/2}^{-1}$	1^+	2884*		+11227		+349	+367*
	2^+	2884*				+359	+377*
	3^+	2797*				+260	+280*
	4^+	2682*				+147	+165*
	5^+	2767*				+236	+250*
	6^+	2323 (2417*)			-294	-118	-100*
$1g_{7/2}^{-2}$	0^+	2617⁺		+11556		-229	
	2^+	2562⁺				-284	
	4^+	3023⁺				+177	
	6^+	3449⁺				+603	

* Interaction energies tested for high spin states in ^{146}Gd and ^{145}Eu. Also energies of levels are calculated using these ΔJ* values.

⁺ The excitation energy of level ϵ_J^{exp} marked by * or ⁺ is the calculated one.

TABLE 5. Effective proton-neutron interaction energies in ^{148}Tb.

E_J^g = -9761 keV

Main config.	J^π	ε_J^{exp}	E_J 2 part sep.en.	E_J^0 0-order sep.en.	ΔJ^{exp} $E_J - E_J^0$	$\Delta J(\delta)$	Comments ΔJ^*
$\pi h_{11/2} \nu f_{7/2}$	2^+					-1614	-1614*
	3^+					-89	-89*
	4^+					-621	-611*
	5^+					-173	-173*
	6^+					-347	-347*
	7^+	368	-9393		-157	-306	-306*
	8^+	(490)**	-9271		-35	-184	-113*
	9^+	0	-9761	-9236	-525	-746	-600*

* Interaction energies adopted on the basis of the best fit to the high spin states in ^{149}Dy and ^{148}Tb as well as to the states in ^{147}Gd below 27 ns isomer state.

** One has to stress here that the excitation energy of the 8^+ state is not known experimentally

TABLE 6. Effective proton hole-neutron interaction energies in ^{146}Eu.

E_J^g = -1797 keV†

Main config.	J^π	ε_J^{exp}	E_J 2 part sep.en.	E_J^0 0-order sep.en.	ΔJ^{exp} $E_J - E_J^0$	$\Delta J(\delta)$	Comments ΔJ^*
$\pi d_{5/2}^{-1} \nu f_{7/2}$	1^-	385	-1412	-2006	+594	+449	544*
	2^-	230	-1567		+439	+294	389*
	3^-	116	-1681		+325	+180	275*
	4^-	0	-1797		+209	+64	159*
	5^-	(80)	-1717		+289	+144	100*
	6^-	275	-1522		+484	+339	434*
$\pi d_{5/2}^{-1} \nu i_{13/2}$	4^+					+256	
	5^+					+31	
	6^+					+98	
	7^+					+64	
	8^+					+48	
	9^+					+163	

† The value E_J^g = -1942 keV was deduced when using Blomqvist et al.[4)] shell model analysis.

* Interaction energies adapted from the fit to ^{148}Tb, ^{145}Sm and ^{144}Pm high-spin states.

DESCRIPTION OF LOW-LYING STATES IN DEFORMED NUCLEI WITHIN THE QUASIPARTICLE-PHONON NUCLEAR MODEL

V.O. NESTERENKO
Joint Institute for Nuclear Research, Dubna, USSR

ABSTRACT

The energies and wave functions of the low-lying non-rotational hexadecapole states with $K^{\pi} = 4^{+}$ and 3^{+} in deformed doubly even nuclei in the region $158 \leqslant A \leqslant 188$ are calculated within the quasiparticle-phonon nuclear model (QPNM). Fixed values of the constants of isoscalar hexadecapole forces $\mathcal{X}_0^{(43)}$ and $\mathcal{X}_0^{(44)}$ were used for the calculations. In general the calculated properties of $K^{\pi} = 4^{+}$ and 3^{+} states are in qualitative agreement with the available experimental data.

1. INTRODUCTION

The low-lying quadrupole and octupole vibrational states are correctly described in the RPA and well studied experimentally in doubly even nuclei[1-3]. There are numerous experimental data on the low-lying $K^{\pi} = 3^{+}$ and 4^{+} states (K is momentum projection onto the nuclear symmetry axis) in the rare-earth region[2,4-22]. The states with $K^{\pi} = 4^{+}$ in Gd and Dy isotopes have large two-quasiparticle components[2,4-8]. The mixing of neutron and proton two-quasiparticle components in the 3^{+} states has been observed in Yb isotopes in transfer reactions[12,13]. According to the experimental data[20] there are strong E4 transitions in the excitation of $I^{\pi} K = 4^{+}4$ states in Os isotopes in (α, α') reaction. They indicate the hexadecapole structure of these states.

The investigation of 3^+ and 4^+ states is necessary for determination of the role of hexadecapole forces and for solving the problem of existence of low-lying two-phonon collective states in deformed nuclei. For this aim the results of calculations within QPNM[1,23,24] of the energies, structure and B(E4)-values for the low-lying nonrotational $K^\pi = 3^+$ and 4^+ states in the rare-earth region are considered in this talk. Structure and collectivity of these states are discussed. The talk is based on the results obtained with V.G.Soloviev, A.V.Sushkov and N.Yu.Shirikova in work[25].

2. DETAILS OF CALCULATIONS

The Hamiltonian of the QPNM consists of an average field of the neutron and proton systems as the Saxon-Woods potential, pairing and the multipole isoscalar and isovector forces[1,24]. The wave function includes one- and two-phonon components and the Pauli principle is taken into account in the two-phonon components.

A phonon basis is constructed in the RPA. The multipolarities $\lambda\mu$ =20,22,30,31,32,43,44 were taken into account. Hexadecapole states with $K^\pi = 3^+$ and 4^+ are generated by forces with $\lambda\mu$ = 43 and 44. It should be noted that the description of all the hexadecapole states has been obtained with fixed values of the constants of the isoscalar hexadecapole interaction, $\varkappa_0^{(43)} = 0.015$ fm^2/MeV and $\varkappa_0^{(44)} = 0.022$ fm^2/MeV. The calculations showed that the role of isovector forces is negligible. The constants of other residual multipole forces were adjusted so as to reproduce experimental energies of the lowest nonrotational states. The single-particle spectrum was taken from the bottom of the potential well up to +5 MeV. The pairing interaction constants were chosen according to pairing ener-

gies. The B(E4) -values were calculated in the adiabatic approximation with $e_{eff.} = 0.1$. Other details of calculations are given in ref.[25].

3. PROPERTIES OF $K^{\pi} = 3^+$ AND 4^+ STATES

The results of calculations for the low-lying states with $K^{\pi} = 3^+$ and 4^+ and available experimental data[2,4-22] are shown in the table. It includes the experimental energies, state structure and contribution of one-phonon components as well as the values calculated within the QPNM, namely reduced probabilities of E4 transitions to 4^+K states from the ground one, contributions of one-phonon $\lambda\mu i$ (i is the number of phonon in the RPA)and two-phonon configurations and the largest two-quasiparticle components.

The calculations showed the following general picture for hexadecapole 3^+ and 4^+ states in the region $158 \leqslant A \leqslant 188$. The Gd and Dy isotopes contain almost two-quasiparticle 4^+ states with an energy from 1.2 to 1.9 MeV and $B(E4) \leqslant 0.5$ spu. The nuclei $^{168,170}Er$ $^{168,170,172,174}Yb$ and $^{174,176,178}Hf$ have sufficiently collective 3^+ states with an energy 1.2 - 1.7 MeV and B(E4) = 0.8 - 1.6 spu, and with the largest collectivity for Yb isotopes. The collectivity of 4^+ states begins to increase from heavy Hf isotopes to Os isotopes and it appears to be large (B(E4) ~ 4 spu) in Os isotopes. The calculations reproduce the dominating two-quasiparticle components of 3^+, 4^+ states obtained experimentally as well as the mixing of neutron and proton two-quasiparticle components of 3^+ states in Yb isotopes[12,13]. In those nuclei where the low-lying $3^+$and 4^+ states have not been observed experimentally the calculated energies of the lowest two-quasiparticle states with $K^{\pi} = 3^+$ and 4^+ are usually higher than 2 MeV.

Table. Results of calculations for low-lying $K^{\pi} = 3^{+}$ and 4^{+} states in rare-earth nuclei

Nucleus	K^{π}_{n}	Experiment		Calculations in the QPNM	
		E, MeV	Structure	E, MeV. B(E4), spu.	Structure
1	2	3	4	5	6
^{158}Gd	4^{+}_{1}	1.380	(t,α),pp411↑+413↓is large	1.2	441 95%,{221,221} 3.3%
				0.4	441: pp413↓+411↑85%
					523↓+521↑13%
	4^{+}_{2}	1.920	(d,p),nn521↑+523↓is large	1.7	442 98%
				0.01	442: nn523↓+521↑87%
					pp413↓+411↑12%
^{160}Gd	4^{+}_{1}	1.184		1.2	441 93%, 442 3%,{221,221} 1%
				0.3	441: nn523↓+521↑53%
					pp413↓+411↑46%
^{158}Dy	4^{+}_{1}	1.895	logft=4.9:nn521↑+523↓is large	2.2	441 92%, 442 6%
				0.4	441: pp413↓+411↑36%
					nn523↓+521↑28%
^{160}Dy	4^{+}_{1}	1.694	logft=4.7:nn521↑+523↓is large	1.7	441 98%,{221,221} 0.6%
				0.1	441: nn521↑+523↓85%
	4^{+}_{2}	2.096	logft=5.8	2.0	442 85%,{201,442} 6%
				0.1	442: nn642↑+651↑74%
					nn523↓+521↑8%
^{162}Dy	4^{+}_{1}	1.535	(d,t),nn521↑+523↓is large	1.7	441 98%
				0.2	441: nn523↓+521↑92%
^{164}Dy	4^{+}_{1}	2.294	logft = 5.6	2.2	441 99%
				0.3	441: pp413↓+411↑66%
					nn523↓+521↑30%
^{166}Er	4^{+}_{1}	1.976		2.0	441 95%,{221,221} 0.8%
				2.4	441: nn523↓+521↑52%
^{168}Er	3^{+}	1.653		1.6	431 99%

1	2	3	4	5	6
	4_1^+	2.055	no large two-phonon comp.	0.8 2.0 3.3	431: nn512↑+521↓90% 441 94%,{221,221}1% 441: nn514↓+521↓18% pp413↓+411↑ 9%
^{170}Er	3_1^+	1.217		1.2 0.6	431 99% 431: nn512↑+521↓94%
^{168}Yb	3_1^+	1.452	(α,2n) } nn 20% g_K- factor } pp 80%	1.6 1.6	431 98% 431: pp404↓-411↓61% nn512↑+521↓13%
^{170}Yb	3_1^+	1.470		1.3 1.4	431 99% 431: nn512↑+521↓ 66% pp404↓-411↓ 24%
^{172}Yb	3_1^+	1.172	(d,t),(d,p),nn512↑+521↓75% (p,α) } nn512↑+521↓73% magn.mom. } pp404↓-411↓27%	1.3 1.3	431 99% 431: nn512↑+521↓68% pp404↓-411↓20%
	3_2^+	1.663	(d,t),(d,p),nn512↑+521↓is notic.	1.6 0.4	432 100% 432: pp404↓-411↓52% nn512↑+521↓30%
	4_1^+	2.073	(p,α),pp404↓+411↓is noticeable	1.94 $3\cdot10^{-4}$	441 99% 441: pp404↓+411↓99%
^{174}Yb	3_1^+	1.606		1.4 1.3	431 99% 431: nn514↓-521↓45% pp404↓-411↓39%
	3_2^+	2.016	(p,t)	1.8 0.1	432 100% 432: nn514↓-521↓52% pp404↓-411↓40%
^{174}Hf	3_1^+	1.303		1.3 1.0	431 99% 431: nn512↑+521↓75%
^{176}Hf	3_1^+	1.578	(d,t),nn514↓-521↓is large	1.3 1.2	431 99% 431: nn514↓-521↓72% pp404↓-411↓12%
	4_1^+	1.888	(d,t),nn514↓+521↓is large	1.7	441 100%

1	2	3	4	5	6
^{178}Hf	4_1^+	1.514	(d,p)	0.01 1.8	441: nn514↓+521↓99% 441 95%, {221,221} 3%
	3_1^+	1.828		0.4 1.7	441: nn514↓+510↑ 89% 431 99%
^{184}W	3_1^+	1.425		1.0 1.2	431: pp404↓-411↓79% 431 96%
^{186}Os	4_1^+	1.352	B(E4) is large in (α, α')	0.8 1.2 4.0	431: nn503↑+510↑91% 441 90%,{201,441}3% 441: pp402↑+402↓56% nn514↓+510↑13%
^{188}Os	4_1^+	1.280	B(E4) is large in (α, α')	1.0 4.0	441 89%,{201,441}3% 441: pp402↑+402↓62%

In ^{166,168}Er the calculations with $\varkappa_0^{(44)}=0.022$ fm^2/MeV provide energies of 4_1^+ states by 0.5 MeV higher than the experimental values and B(E4)< 0.1 spu. In order the energies of the 4_1^+ states in ^{166,168}Er be equal to 2 MeV, one has to take $\varkappa_0^{(44)}= 0.035$ fm^2/MeV. But in this case the collectivity of the 4_1^+ state of these nuclei is explicitly overestimated.

So, except for the 4^+ states in ^{166,168}Er the calculated properties of hexadecapole states with $K^\pi = 3^+$ and 4^+ in the region $158 \leqslant A \leqslant 188$ are in qualitative agreement with the available experimental data and this description has been obtained with the fixed values of the constants $\varkappa_0^{(43)}$ and $\varkappa_0^{(44)}$.

All the calculated states 4_1^+ and 4_2^+ have small two-phonon components, especially $\{221, 221\}$. The present calculations confirm the conclusion made in refs.[23,26] about the absence of collective low-lying two-phonon states in deformed nuclei.

States with $K^\pi = 3^+$ and 4^+ have been investigated in some nuclei within the interacting boson model (IBM), particularly, when the g-boson is introduced (see, e.g. refs.[27-28]). However within the IBM there is no general description of 3^+ and 4^+ states in the rare-earth region. The comparison of the results of calculations for hexadecapole states in some nuclei within QPNM and IBM is given in refs.[24,29].

REFERENCES

1. Soloviev V.G., "Theory of Complex Nuclei", Moscow, Nauka (1971), transl. Pergamon Press (1976).
2. Grigoriev E.P. and Soloviev V.G., "Structure of Even Deformed Nuclei", Moscow, Nauka (1974).
3. Ivanova S.P. et al., Part. Nucl. 7, 450 (1976).

4. Greenwood R.C. et al., Nucl.Phys. A304, 327 (1978).
5. Burke D.G. et al., Nucl.Phys. A366, 202 (1981).
6. Anderson D.L. et al., Phys.Rev. C18, 383 (1978).
7. Ramsoy T. et al., Nucl.Phys. A438, 301 (1985).
8. Warner D.D. et al., Phys.Rev. C27, 2292 (1983).
9. Panar J.D. and Burke D.G., Can. J.Phys. 57, 1999(1979).
10. Davidson W.F. et al., Can. J.Phys. 62, 1538 (1984).
11. Bondarenko V.A. et al., Izv.Akad.Nauk SSSR, Ser.Fiz. 46, 2080 (1982).
12. Walker P.M. et al., Phys.Lett. 116B, 393 (1982).
13. Burke D.G. and Blezius J.W., Can.J. Phys. 60, 1751 (1982).
14. Kostov L.K. et al., Nucl.Phys. A406, 541 (1983).
15. Greenwood R.C. and Reich C.W., Phys.Rev. C23, 153 (1981).
16. Browne E., Nucl.Data Sheets, 41, 511 (1978).
17. Horen D.J. and Harmatz B., Nucl.Data Sheets 19, 383 (1976).
18. Greenwood L.R., Nucl.Data.Sheets. 13, 549 (1974).
19. Martin M.J. and Stelson P.H., Nucl.Data Sheets, 21, 1 (1977).
20. Burke D.G. et al., Phys.Lett. 78B, 48 (1978).
21. Bagnell R.D. et al., Phys.Lett. 66B, 129 (1977).
22. Borghols W.T.A. et al., Phys. Lett. 152B, 330 (1985).
23. Soloviev V.G. and Shirikova N.Yu., Yad.Fiz. 36, 1376 (1982).
24. Soloviev V.G., In: Proc.Intern.School on Nucl.Struct. (Alushta, 1985) D4-85-851, Dubna, 8 (1985).
25. Nesterenko V.O. et al., Preprint JINR, E4-85-856, Dubna (1985).
26. Soloviev V.G. and Shirikova N.Yu., Z.Phys. A301, 263 (1981).
27. Van Isaker P. et al., Nucl.Phys. A380, 383 (1982).
28. Yoshinaga N. et al., Phys.Rev.Lett., 56,1116 (1986).
29. Soloviev V.G.,Preprint JINR, E4-85-878,Dubna (1985).

LOW-LYING $K^{\pi}=1^+$ STATES IN DEFORMED RARE-EARTH NUCLEI[a)]

O. Civitarese[b)], Amand Faessler, R. Nojarov[c)]
Institut für Theoretische Physik, Universität Tübingen, D-7400, Tübingen, FRG

Abstract Low-lying magnetic dipole $K^{\pi}=1^+$ states were studied in ^{154}Sm, ^{156}Gd, ^{158}Gd, ^{164}Dy, ^{168}Er and ^{174}Yb using the quasiparticle Random Phase Approximation. The model hamiltonian includes an isospin-dependent quadrupole-quadrupole interaction and a mean field as given by an axially symmetric Woods-Saxon potential. The energies and B(M1) values agree well with the experimental data.

The $K^{\pi}=1^+$ states, lying at about 3 MeV excitation energy, which were recently observed experimentally [1] in the above six rare-earth nuclei and which are connected to the ground state by strong magnetic dipole transitions, are described here in the framework of the quasiparticle RPA. The model hamiltonian includes an isospin-depen-

a) Supported by the Alexander von Humboldt Foundation and the Bundesministerium für Forschung und Technologie.
b) Permanent address: Department of Physics, University of La Plata, Argentina.
c) Fellow of the A v Humboldt Foundation. Permanent address: Institute of Nuclear Research and Nuclear Energy, Bulgarian Academy of Sciences, Sofia 1784, Bulgaria.

dent separable quadrupole-quadrupole interaction:

$$H=H_o - \frac{1}{2}\sum_{m,t_z,t_{z'}} \chi_{t_z t_{z'}} \hat{Q}^+_{2m}(t_z)\, \hat{Q}_{2m}(t_{z'}),$$

$$t_z=n,p$$

$$\chi_{nn}=\chi_{pp}\ , \quad r=(\chi_{nn}-\chi_{np})/(\chi_{nn}+\chi_{np}) \qquad (1)$$

H_o being the single quasiparticle term, i.e. the mean field with pairing correlations taken into account through the BCS approximation. The reduced probability for an M1 transition from the ground state to the ν-th excited 1^+ state is:

$$B(M1,\nu)\uparrow = \left(\frac{3}{4\pi}\right)\langle 1^+_\nu \| \hat{M}_1 \| 0 \rangle^2,$$

$$\hat{M}_{1m}=\sum_{t_z}\left(g_1(t_z)\hat{1}_{1m}(t_z) + g_s(t_z)\hat{s}_{1m}(t_z)\right) \qquad (2)$$

where $\hat{M}_{1m}$ are the tensor components of the magnetic dipole operator. The single-particle basis is obtained from an axially-symmetric Woods-Saxon potential with parameters from Ref. 2 and a proton spin-orbit coupling constant λ_o=15-19. The deformation and gap parameters are taken from Ref. 3. The gyromagnetic factors are: $g_1(n)=0$, $g_1(p)=1$, $g_s(t_z)=0.7g_s^{free}(t_z)$. The experimental energies of the 1^+ states (Figure 1) are reproduced with a mean overestimation of about 0.2 MeV. The experimental B(M1)↑ values, ranging from 0.8 to 1.5 μ_N^2, are reproduced very well - almost within the limits of the experimental uncertainties, which are about 0.3 μ_N^2 (Figure 2). We found that the low-lying 1^+ states, which concentrate an appre-

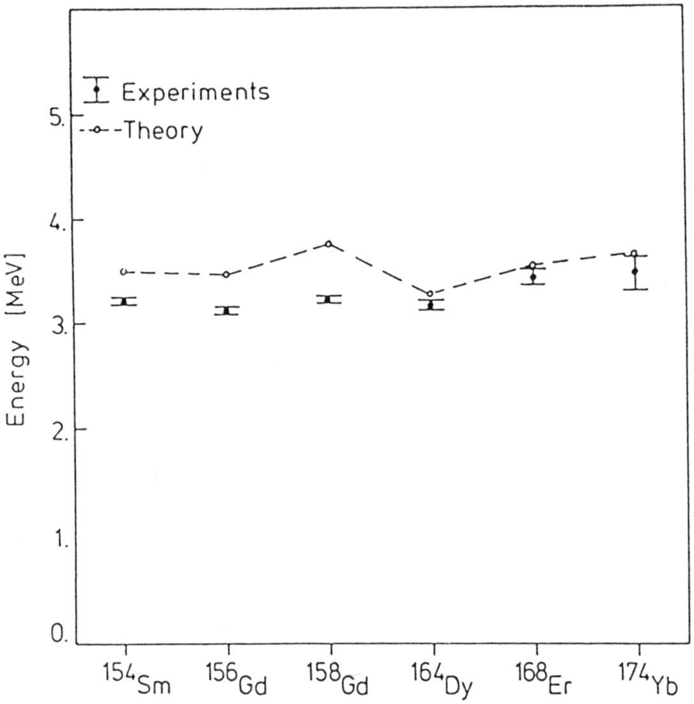

FIGURE 1. Theoretical and experimental[1] energies of the strong 1^+ states.

ciable M1 strength comparable with the experimental findings, have predominantly isovector character. All of them are built up on a two-quasiproton pair coming mainly from the $1h_{11/2}$ shell. In all these states the spin contribution to the magnetic dipole transition matrix element is in phase and comparable in magnitude to the orbital contribution. A reduction of the ratio between the

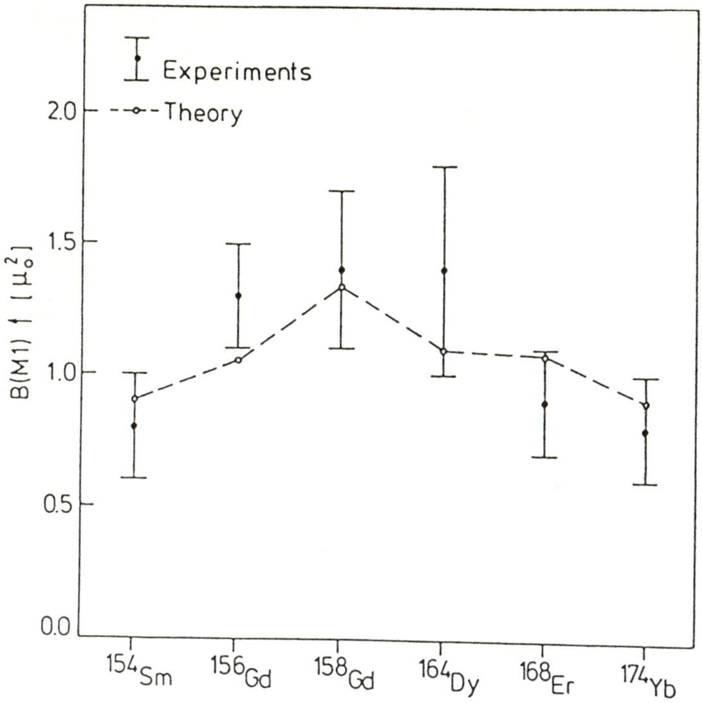

FIGURE 2. Theoretical and experimental[1] B(M1) values of the strong 1^+ states.

isovector and the isoscalar strength of the residual interaction from -3.5 to -0.5 does not change appreciably the above results.

REFERENCES

1. D. Bohle et al., Phys. Lett., 148B, 260 (1984).
2. R. Nojarov, J. Phys. G: Nucl. Phys. 10, 539 (1984)
3. R. Nojarov, Z. Bochnacki and A. Faessler, Z. Phys. A324 (1986) in press.

COLLECTIVE STATES IN ^{230}Ra

W. KURCEWICZ and E. RUCHOWSKA
Institute of Experimental Physics, University of Warsaw, Pl-00-681 Warsaw, Poland

P. HILL and N. KAFFRELL
Institut für Kernchemie, Universität Mainz, D-6500 Mainz, Federal Republic of Germany

G. NYMAN
Department of Physics, Chalmers University of Technology, S-41296 Göteborg, Sweden

The ISOLDE Collaboration
CERN, Geneva, Switzerland

Abstract The γ-rays following the β^--decay of ^{230}Fr have been investigated by means of γ-ray singles including multispectrum analysis and γγ coincidence measurements using Ge(Li) detectors. Most of the observed transitions could be placed in a level scheme comprising 23 new excited states of ^{230}Ra, 10 of them grouped into the $K^\pi=0^+$ ground-state band, the $K^\pi=0^-$ band with its 1^- state at 710.9 keV and a $K^\pi=2^+$ γ-vibrational band with its band head at 734.8 keV.

INTRODUCTION

In the even isotopes of radium and thorium with neutron numbers from 134 to 138 the $K,J^{\pi}=0,1^{-}$ states occur at an unusually low excitation energy. These states can be interpreted as related to a ground-state octupole deformation[1,2].

Recently, a number of experimental and theoretical investigations have produced evidence for stable octupole deformation in nuclei with neutron number $N \simeq 136$ (see e.g. refs.[3,4] and references quoted therein).

In order to extend the knowledge on collective states in even Ra isotopes[1,2,5] we have undertaken a study of the β^{-}-decay of ^{230}Fr into levels of ^{230}Ra. No information on the excited states in ^{230}Ra was available prior to this investigation.

EXPERIMENTAL PROCEDURE

Francium activities were produced in spallation reactions of ^{238}U induced by 600 MeV protons from the CERN Synchro-cyclotron. The nuclide ^{230}Fr was obtained as a mass-separated beam from the online separator[6,7] ISOLDE II. The beam was collimated in order to reduce contaminations from adjacent masses, and then deflected onto a movable tape that was operated in a start-stop mode to

carry the activity into the counting position, which was equipped with a small plastic scintillator and two Ge(Li) detectors (active volumes: 32 and 180 cm^3; energy resolution at 1332 keV: 1.75 and 1.8 keV, respectively). For background reduction the γ-rays were gated by the signals from the plastic scintillator used as ß-ray detector.

A rather strong contamination from 13-s ^{211}Ra and its daughter products were observed. The Ra activity was able to pass the mass separator as an ionized $^{211}Ra^{19}F$-molecule.

The half-life of 19.1(5) s for ^{230}Fr has been obtained by means of multispectrum analysis using cycles of 20 s collection time followed by 6x7 s measuring time.

For the energy calibration ^{230}Fr was measured simultaneously with a set of γ-ray standards (^{57}Co, ^{60}Co, ^{85}Sr, ^{88}Y, ^{109}Cd, ^{113}Sn, ^{137}Cs, ^{139}Ce and ^{203}Hg). Precisely known γ-lines from natural background and from ^{211}Ra descendants[8] were used as internal energy standards. The energies as well as the intensities of the γ-ray transitions assigned to the decay of ^{230}Fr are listed in table I. The γγ coincidence experiments have been performed simultaneously with the singles γ-ray measurements.

Table I Energies, intensities and placements in the level scheme of γ-rays following the decay of ^{230}Fr

E_γ /keV	I_γ	Initial level	E_γ /keV	I_γ	Initial level
57.4(1)	6.1(6)	57.4	832.4(3)	3.7(4)	1211.7
129.1(1)	100(7)	186.5	835.5(3)	11.8(12)	893.0
192.5(1)	17.0(17)	379.0	840.4(5)	2.8(4)	
264.5(2)	3.6(4)	1144.3	847.2(3)	5.9(7)	1033.8
265.5(2)	2.3(3)	1033.8	863.5(3)	2.7(3)	1897.1
323.1(2)	3.5(3)	1033.8	886.0(2)	4.8(5)	
333.1(2)	2.7(3)	1522.5	898.7(2)	8.1(8)	2043.4
336.2(2)	2.1(2)		958.0(2)	6.5(7)	1144.3
338.9(3)	1.0(2)	1188.7	965.8(2)	4.9(5)	
375.8(2)	5.8(6)	1144.3	971.8(2)	15.0(15)	1158.4
385.5(3)	1.6(2)		976.3(3)	8.2(8)	1033.8
500.8	$<$1.2	879.8	986.5(3)	1.6(2)	
548.0(3)	3.0(3)	734.8	1002.2(3)	13.5(14)	1188.7
553.2(2)	4.4(4)[a)]	932.1	1004.3(3)	11.2(11)	1897.1
581.9(1)	16.6(12)	768.4	1018.8(3)	4.5(5)	
587.0(2)	7.8(8)	1466.8	1025.4(2)	7.3(7)	1211.7
599.3(1)	17.3(12)	785.8	1033.8(5)	3.0(5)	1033.8
619.8(3)	3.7(4)		1049.2(3)	10.0(7)	
633.3(2)	5.2(5)		1086.9(5)	3.0(5)	1144.3
653.4(1)	23.0(16)	710.9	1094.8(3)	18.0(18)	1281.0
663.2(1)	29.7(21)	849.8	1101.2(2)	14.1(11)	1158.4
677.4(1)	61(4)	734.8	1106.5(3)	4.3(5)	
693.3(1)	23.2(16)	879.8	1111.8(3)	6.6(7)	1897.1
699.1(2)	6.5(5)		1128.8(2)	21.5(15)	1897.1
706.5(1)	16.3(11)	893.0	1154.6(2)	15.4(11)	1341.1
711.0(1)	24.4(60)[a)]	710.9	1162.6(2)	24.6(17)	1897.1
711.0(1)	124(12)[a)]	768.4	1191.6(5)	4.1(5)	
728.4(1)	66(4)	785.8	1219.0(5)	4.7(5)	2005.0
734.9(1)	54(4)	734.8	1223.4(3)	14.1(14)	1281.0
745.4(3)	14.8(22)	932.1	1230.8(5)	2.8(3)	
754.0(5)	4.4(11)	1522.5	1236.6(3)	10.7(11)	2005.0
765.5(3)	4.1(5)	1144.3	1259.1(5)	1.6(3)	
792.4(2)	18.4(13)	849.8	1270.2(2)	10.8(11)	2005.0
801.9(3)	1.7(3)		1274.6(3)	4.8(7)	2043.4
811.9(3)	6.8(7)	1522.5	1294.0(2)	12.5(13)	2005.0

Table I, continued

E_γ /keV	I_γ	Initial level	E_γ /keV	I_γ	Initial level
1326.1(3)	2.3(4)		1744.4(5)	4.2(4)	
1330.2(5)	1.8(5)		1750.5(5)	2.4(4)	
1349.6(5)	2.0(4)		1793.7(5)	2.6(4)	
1365.1(5)	2.4(5)		1839.3(5)	3.2(5)	1897.1
1465.0(5)	1.2(3)	1522.5	1857.8(3)	8.1(8)	2043.4
1473.7(5)	1.8(6)		1869.4(3)	23.7(24)	
1495.3(5)	4.6(7)		1880.1(3)	9.6(10)	
1519.9(5)	2.1(4)		1883.9(5)	2.3(5)	
1545.4(5)	6.9(7)		1887.9(5)	3.8(6)	
1562.4(5)	4.8(7)		1908.3(5)	1.9(3)	
1601.3(5)	2.3(4)		1927.4(5)	4.6(5)	
1605.1(5)	3.4(5)		1947.7(3)	14.5(15)	2005.0
1610.0(5)	3.5(5)		1987.4(5)	7.1(7)	
1615.8(5)	4.0(6)		2013.7(5)	6.8(7)	
1662.6(5)	4.2(6)		2041.5(5)	3.1(3)	
1702.0(3)	8.2(8)		2093.1(5)	7.7(8)	
1710.2(2)	15.6(16)	1897.1	2232.6(5)	3.0(3)	
1737.5(5)	4.7(5)		2376.5(5)	4.1(4)	

[a] Data taken from our coincidence measurements.

RESULTS AND DISCUSSION

About half of the observed γ transitions following the β^--decay of ^{230}Fr were placed in a decay scheme based on coincidence results and energy balances (see column 3 in table I). The spin-parity assignments for the excited states in ^{230}Ra can be suggested from a comparison with the lighter isotopes ^{224,226,228}Ra (refs.[1,2,5]). At an energy below 1 MeV ten excited states in ^{230}Ra have been grouped into three rotational bands.

These include the ground-state $K^{\pi}=0^{+}$ band with a moment-of-inertia parameter A equal to 9.57, the $K^{\pi}=0^{-}$ band with its 1^{-},3^{-} and 5^{-} members at 710.9, 768.4 and 879.8 keV, respectively, and an A-parameter equal to 5.75, and a $K^{\pi}=2^{+}$ band with its band head at 734.8 keV and an A-parameter equal to 8.50. From the E_{4^+}/E_{2^+} energy ratio of 3.25 calculated for the members of the ground-state band it may be concluded that ^{230}Ra is a better rotator than the lighter Ra isotopes. The energy splitting of 699.4 keV between the $K^{\pi}=0^{-}$ and $K^{\pi}=0^{+}$ bands in ^{230}Ra is much higher than in ^{224}Ra and ^{226}Ra (ref.[1]) and suggests zero octupole deformation for ^{230}Ra.

This work was supported in part by the Polish Ministry of Science, Higher Education and Technics under the contract CPBP 01.06 and the German Bundesministerium für Forschung und Technologie.

REFERENCES

1. W. Kurcewicz et al., Nucl. Phys. A356, 15 (1981) and references therein.
2. W. Kurcewicz et al., Proc. 4th Int. Conf. on Nuclei Far from Stability, Helsingør 1981, CERN 81-09, 1981, p. 649.
3. G.A. Leander, in Nuclear Structure 1985, edited by R. Broglia, G.B. Hagemann and B. Herskind (Elsevier, 1985) p. 249.
4. W. Nazarewicz, ibid. p. 263.
5. E. Ruchowska et al., Nucl. Phys. A383, 1 (1982).
6. H.L. Ravn et al., Nucl. Instr. Meth. 139, 267 (1976).
7. L.C. Carraz et al., Nucl. Instr. Meth. 158, 69 (1979).
8. U. Reus and W. Westmeier, At. Data Nucl. Data Tables 29, 1 (1983).

Chapter 4

OCTUPOLE AND DIPOLE MODES IN NUCLEI

Moderator:

F.J.W. HAHNE
Institute for Theoretical Nuclear Physics
Univeristy of Stellenbosch
Stellenbosch 7600, South Africa

A STUDY OF OCTUPOLE BANDS WITHIN THE INTERACTING BOSON MODEL

A. F. BARFIELD*
University of Arizona, Tucson, AZ 85721, USA

Abstract An IBM-1 plus f-boson octupole model is applied to nine deformed rare-earth nuclei. Limited results are given for five of these nuclei. Detailed spectra and B(E3)'s are presented for ^{168}Er. The model works well for eight of the nine nuclei studied.

INTRODUCTION

Octupole bands in nuclei are identified experimentally by enhanced E3 transition rates between the ground state and the $I^{\pi} = 3^{-}$ member of the band. It is often found that a known octupole state is also strongly populated in single-nucleon transfer reactions.[1,2] Thus, such collective states have a complex nature, with one or more large 2qp components.

In this paper, we report on a study of octupole bands in deformed rare-earth nuclei, carried out within the Interacting Boson Model (IBM).[3] While this model has been widely applied to positive parity collective states in nuclei, previous applications to octupole states have been limited in scope.[4,5] In the present study, the method is applied systematically to a wide range of nuclei within the deformed region, in an attempt to obtain a global fit with smoothly varying parameters.

---

*Present address: Universität Siegen, D-5900 Siegen, FR Germany.

THEORY

Octupole states are described in the IBM by adding an f boson (L=3) to the usual s-d model space.[3] This is done in the IBM-1 framework, in which neutron and proton degrees of freedom are not separately distinguished. The total number of bosons, N, is conserved;

$$N = n_s + n_d + n_f \quad , \tag{1}$$

where n_s, n_d, and n_f are the numbers of s, d, and f bosons, respectively, and $n_f = 0$ or 1. The Hamiltonian for the combined system is

$$H = a_1 L_d \cdot L_d + a_2 Q_d \cdot Q_d + \varepsilon_f \hat{n}_f$$

$$+ A_1 L_d \cdot L_f + A_2 Q_d \cdot Q_f + A_3 :E^{\dagger}_{df} \cdot E_{df}: \quad , \tag{2}$$

where the operators are defined by

$$L_d = \sqrt{10}(d^{\dagger} \times \tilde{d})^{(1)} \, , \quad Q_d = s^{\dagger} \times \tilde{d} + d^{\dagger} \times s + \chi(d^{\dagger} \times \tilde{d})^{(2)} \quad ,$$

$$L_f = 2\sqrt{7}(f^{\dagger} \times \tilde{f})^{(1)} \, , \quad Q_f = -2\sqrt{7}(f^{\dagger} \times \tilde{f})^{(2)} \quad ,$$

and

$$E^{\dagger}_{df} = \sqrt{5}(d^{\dagger} \times \tilde{f})^{(3)} \quad . \tag{3}$$

The first two terms in Eq. (2) describe the positive parity core and the third term is the f-boson Hamiltonian. The remaining terms describe the interaction between the f boson and the s-d core. The last term in Eq. (2) is called the exchange term; it can be shown to arise naturally from an octupole-octupole interaction.[6,7] The E3 transition operator can be written

$$T^{(E3)} = e_3 \, [(s^\dagger \times \tilde{f} + f^\dagger \times s) + \chi_3 \, (d^\dagger \times \tilde{f} + f^\dagger \times \tilde{d})]^{(3)} \quad . \qquad (4)$$

The selection of octupole bands from among the low-lying experimental negative parity bands is made on the basis of known B(E3) transition rates. No attempt is made to fit low-lying 2qp $K^\pi=3^-$ bands; these are manifestly outside of the IBM model space.

RESULTS AND DISCUSSION

The nuclei considered in the study are ^{154}Sm, ^{156}Gd, ^{158}Gd, ^{158}Dy, ^{162}Dy, ^{168}Er, ^{172}Yb, ^{178}Hf, and ^{182}W. The negative parity parameters and low-lying octupole bandhead energies are presented in Table I for four of these nuclei. The octupole effective charge, e_3 is constant for all of the nuclei studied. Thus, the calculated B(E3)'s are effectively determined by one parameter, χ_3. The variation in this parameter reflects different ways in which the E3 strength is distributed among the low-lying experimental octupole bands. The f-boson energy, ϵ_f, and the f-sd quadrupole and exchange strengths, A_2 and A_3, are found not to vary smoothly with mass number and, thus, cannot be predicted for neighboring nuclei by simple extrapolation or interpolation. In view of the known dominance of many octupole states by 2qp components, this is not surprising.

More detailed results are given for the nucleus ^{168}Er, as an example. This nucleus has been particularly well-characterized experimentally.[2,9,12-14] It is known that the lowest $K^\pi = 0^-, 1^-$, and 2^- bands have octupole collectivity, along with several higher-lying bands, and that the $K^\pi = 3^-, 4^-$, and 6^- bands between ~1 and 2 MeV are 2qp in nature. The bandhead energies for these 2qp bands are shown in a box in Fig. 1, along with a collective 3^- state

TABLE I Negative parity parameters and octupole bandhead energies for four nuclei. All energies are in keV.

	^{158}Gd	^{168}Er	^{172}Yb	^{178}Hf
Parameters				
ε_f	1260	1340	1250	750
A_1	10	0	0	0
A_2	- 60	- 42	- 57	- 37
A_3	- 120	- 135	- 145	- 150
χ_3	- 0.3	4.1	3.5	1.7
$e_3(e^2b^3)$	0.07	0.07	0.07	0.07
Bandhead energies				
$K^\pi = 0^-$				
calc	1207	1798	1582	d
exp	1263[a]	1786[b]	1600[c]	
$K^\pi = 1^-$				
calc	1026	1344	1142	1308
exp	977[a]	1358[b]	1155[c]	1310[e]
$K^\pi = 2^-$				
calc	1726	1577	1753	1259
exp	1793[a]	1569[b]	1757[c]	1260[e]

[a] Ref. 8.
[b] Ref. 9.
[c] Ref. 10.
[d] Not known experimentally.
[e] Ref. 12.

at 2269 keV. The theory predicts a second relatively low-lying $K^\pi = 1^-$ band. Three other higher-lying calculated bandheads are also shown, in a box, including that for an even-spin $K^\pi = 0^-$ band ($IK^\pi = 00^-$).

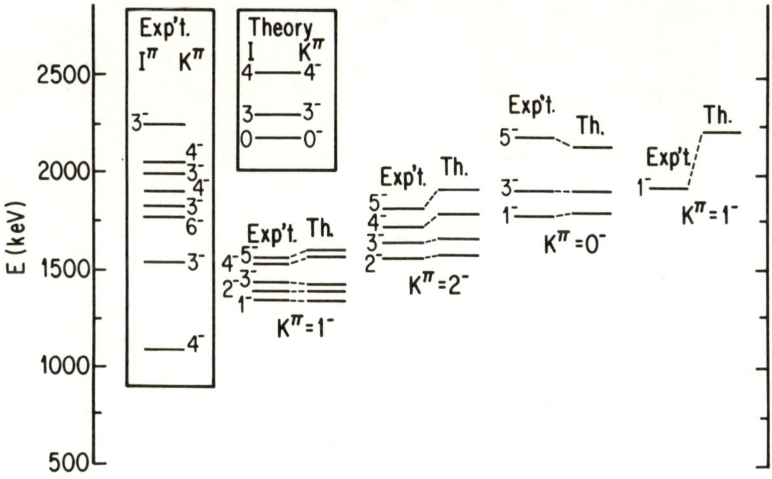

FIGURE 1. Negative parity spectra for ^{168}Er. The data are from Refs. 9, 12, and 13.

Table II presents B(E3)'s for low-lying negative parity bands in ^{168}Er.

Table II B(E3; $0^+_1 \to 3^-$) for ^{168}Er. Excitation energies are given in MeV and transition rates in e^2b^3. The data are from Ref. 14.

K^π	IBM E(3⁻)	IBM B(E3)	Exp E(3⁻)	Exp B(E3)
1^-	1.43	0.046[a]	1.43	0.046(5)
3^-	b		1.54	0.003(1)
2^-	1.67	0.066	1.63	0.058(6)
3^-	b		1.83	0.007(2)
0^-	1.91	0.038	1.91	0.023(3)

[a]Normalized to experiment.
[b]2qp band, not in IBM space.

In addition to ^{168}Er, seven of the other eight nuclei studied are also reasonably well-described by the theory, although the data base is not always extensive enough to tie down the parameters.[6,15] In contrast, the octupole states for the nucleus ^{162}Dy cannot be reproduced by the present model.[6,15] Figure 2 shows the negative parity spectra resulting from two attempts to fit this nucleus. Fitting to the $K^\pi = 2^-$ and 0^- bands results in a low-lying 1^- band, while fitting to the 2^- and 1^- bands results in a high-lying 0^- band. It is clear that degrees of freedom outside of the IBM-1 model space are needed in order to obtain the band ordering $K^\pi = 2^-, 0^-, 1^-$.

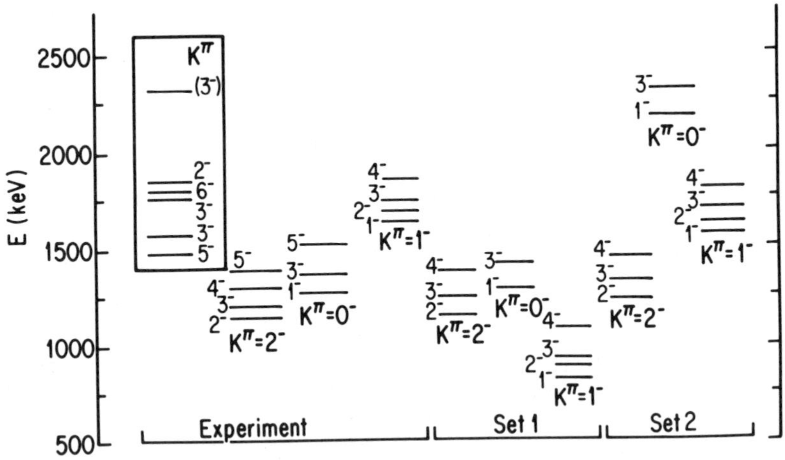

FIGURE 2. Negative parity spectra for ^{162}Dy. The experimental negative parity spectra and two attempts to fit the octupole bands are shown. All three experimental bands are known to have octupole collectivity. The data are from Refs. 16 and 17.

CONCLUSION

Nucleus by nucleus, with the exception of ^{162}Dy, the model gives a reasonable description of the states fitted. In addition, it predicts higher-lying band structure. On a global scale, the parameter sets are similar but cannot be predicted for neighboring nuclei. More experimental data are needed, both to test the predictions of the model and to extend the scope of the calculations.

This work was done in collaboration with J. L. Wood and B. R. Barrett and was supported in part by the US National Science Foundation under Grant No. PHY-84-05172 and by the US Department of Energy under contracts DE-AS05-80ER10599 and DE-AC05-76OR00033.

REFERENCES

1. R. A. O'Neil and D. G. Burke, Nucl. Phys. A182, 342 (1972).
2. D. G. Burke et al., Nucl. Phys. A445, 70 (1985); and references therein.
3. A. Arima and F. Iachello, Phys. Lett. 57B, 39 (1975); Ann. Phys. 111, 201 (1978).
4. O. Scholten, F. Iachello and A. Arima, Ann. Phys. 115, 325 (1978).
5. R. De Leo et al., Phys. Lett. 165B, 30 (1985); and references therein.
6. A. F. Barfield, Ph.D. Thesis, University of Arizona (1986).
7. A. F. Barfield and O. Scholten, to be published.
8. R. C. Greenwood et al., Nucl. Phys. A304, 327 (1978).
9. W. F. Davidson et al., J. Phys. G7, 455 (1981).
10. J. R. Cresswell et al., J. Phys. G7, 235 (1981)
11. B. Fogelberg and A. Backlin, Nucl. Phys. A171, 353 (1971).
12. P. O. Tjom and B. Elbek, Nucl. Phys. A107, 385 (1968).
13. D. G. Burke, B. L. W. Maddock, and W. F. Davidson, Nucl. Phys. A442, 424 (1985).
14. I. M. Govil et al., Phys. Rev. C33, 793 (1986).
15. A. F. Barfield, J. L. Wood, B. R. Barrett, and O. Scholten, to be published.
16. C. A. Fields et al., Nucl. Phys. A389, 218 (1982).
17. D. D. Warner et al., Phys. Rev. C27, 2292 (1983).

INTRINSIC REFLECTION ASYMMETRY IN THE LIGHT RARE EARTHS

J.D. ROBERTSON and W.B. WALTERS
Dept. of Chemistry, Univ. of Maryland, College Park, MD 20742

Abstract An investigation of the β^- decay of ^{145}Cs into levels of ^{145}Ba was made at TRISTAN. A level scheme for ^{145}Ba up to 785 keV is proposed. The study was undertaken to search for evidence of intrinsic reflection asymmetry in the structure of ^{145}Ba. The octupole deformed rotational structure which is seen in the light odd-A actinides was not found in ^{145}Ba.

INTRODUCTION

In a 1985 Physical Review Letter, Leander et al. proposed that nuclei in the vicinity of ^{145}Ba may be a new region of intrinsic reflection asymmetry similar to the light-actinide region.[1] Three reasons are given in the letter as to why the light rare earths may be a new region of octupole deformation. First, the energy of the 1^- state in ^{146}Ba at 738 keV, in units of the first 2^+ state, is as low as the 1^- states in the light Ra and Th isotopes. It was the discovery of these low-lying negative parity states in the light actinides that first led to the suggestion that they were octupole deformed. Secondly, the Strutinsky calculations of Nażarewicz et al. indicate that Z=56 as well as Z or N=88-90 is an optimal particle number for octupole deformation.[2] And finally, the difference observed between the experimental and theoretical masses of Moller and Nix in the Ra and Th region is also found in the vicinity of ^{145}Ba.[3] Like the light-actinide region, this difference can be accounted for by the extra 1 MeV of binding energy obtained by including octupole correlations in the mass calculations.

The ground state (g.s.) spin, spectroscopic quadrupole moment (Q_s), and magnetic dipole moment of ^{145}Ba have been measured at Isolde.[4] From the g.s. spin of 5/2 and Q_s of 1.15 b, Leander et al. calculate that ^{145}Ba may be octupole deformed with $\beta_3 \approx 0.08$.[1] These results, however, are not conclusive. The same calculation for the dipole moment can not distinguish between zero and non-zero octupole deformation. In addition, one can account for the g.s. spin of 5/2 in the Nilsson model with no octupole deformation by allowing the odd neutron to occupy the [523]5/2 orbital.

Another indication of octupole deformation in this region would be the presence of a pair or more of opposite parity rotational bands in the structure of ^{145}Ba. More specifically, these bands would have a small energy splitting and would be connected by enhanced E1 transitions. In addition, the decoupling parameters for any opposite parity K=$1/2^{\pm}$ bands should be equal in magnitude but opposite in sign. This type of octupole deformed rotational structure is observed in both ^{225}Ra and ^{227}Ac.[5,6] In a previous study of ^{145}Ba, Rapaport et al. found one E1 transition but did not identify any rotational bands.[7] In their study, however, only 31 γ rays were assigned to the β^- decay of ^{145}Cs and the structure of ^{145}Ba was only determined up to 780 keV by energy sum relationships. A later investigation of the decay of ^{145}Cs defined the structure of ^{145}Ba up to 2.8 MeV.[8] Unfortunately, no new spins or parities could be assigned from the latter study. Thus, to determine whether or not ^{145}Ba does contain parity doublets similar to those observed in the light actinides, we undertook a study of the β^- decay of ^{145}Cs into levels of ^{145}Ba.

RESUTLS

A total of 117 γ-rays have been assigned to the decay of ^{145}Cs and a level scheme for ^{145}Ba has been constructed up to 1.3 MeV. The low-lying portion of the level scheme is presented in Figure 1. The g.s. negative parity assignment is based upon the systematics of the N=89 isotones and the Z=56 isotopes; both ^{149}Nd[9] and ^{151}Sm[10] have a g.s. $J^{\pi}=5/2^-$ while the g.s. J^{π} of ^{139}Ba=$7/2^-$[11], ^{141}Ba=$3/2^-$[12], and ^{143}Ba=$5/2^-$.[1] Only three E1 transitions were identified in the decay of ^{145}Cs which allowed us to assign opposite parity (to that of the g.s.) to only four levels; the 319-, 435-, 547- and 785-keV levels. A complete description of the experiment and the results will soon be published.

CONCLUSION

One interpretation of the low-lying structure observed in ^{145}Ba is to group the levels at 112-, 175-, 277-, and 416-keV into a K=$1/2^-$ band with a rotational parameter of 20 keV and a decoupling parameter (a) of -2. The rotational parameter for the proposed band is the right magnitude as the rotational parameters for the well defined rotational bands in the N=89 isotones ^{153}Gd and ^{155}Dy vary from 7 to 20 keV.[13] Likewise, a value of a=-2 is consistent with a decoupled band built upon the [530]1/2 neutron orbital.[14] There are, however, two major problems with this interpretation of the structure found in ^{145}Ba. First, and perhaps most seriously, we find that it is impossible to construct rotational bands on any of the four positive parity states observed. The octupole deformed rotational structure which is seen in the light odd-A actinides is not found in ^{145}Ba. Secondly, the decoupling parameter of -2 for the proposed band indicates that ^{145}Ba is strongly quadrupole deformed with $\beta_2 \approx 0.3$.[14] The systematics of the N=89

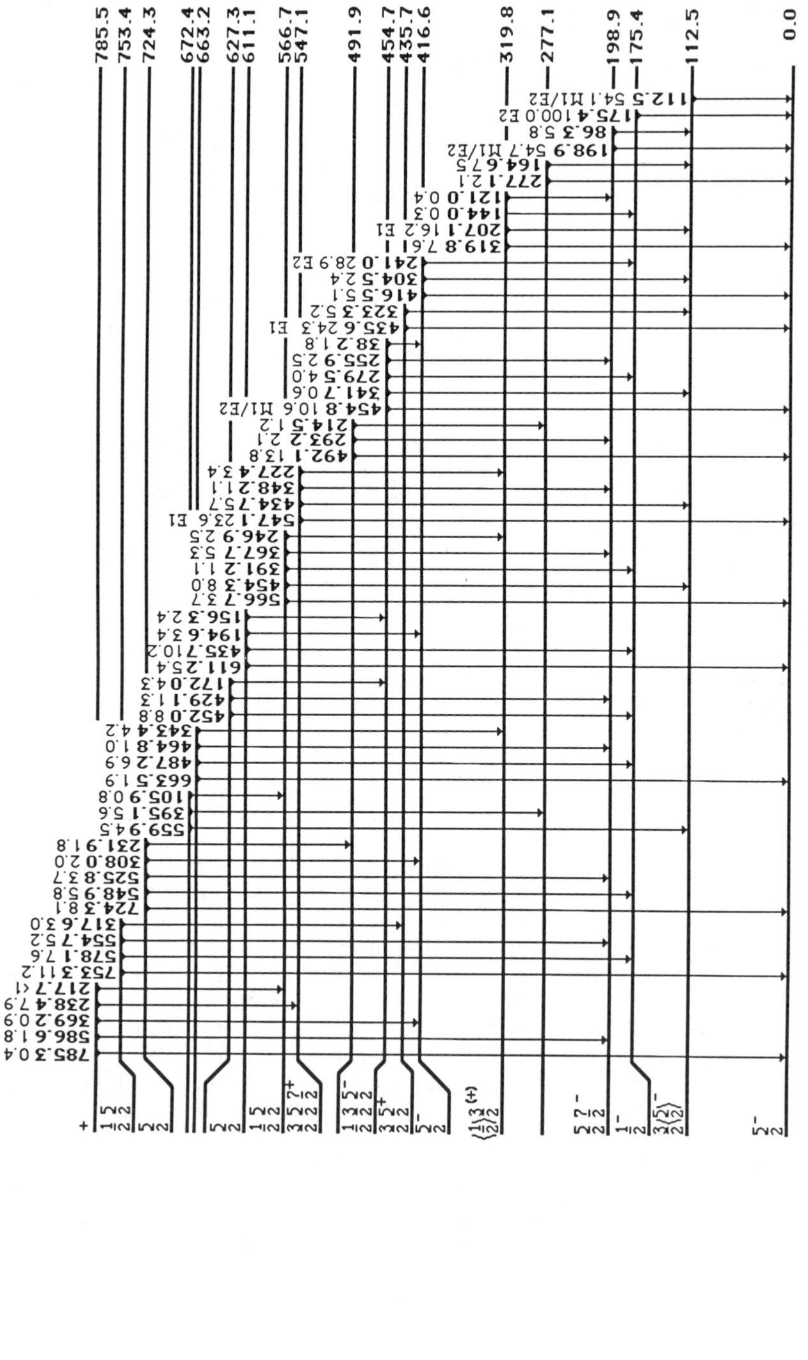

$^{145}_{56}$Ba$_{89}$

Figure 1. A portion of the level scheme proposed for ^{145}Ba.

isotones indicate, however, that ^{145}Ba is only weakly quadrupole deformed as the rotational structure found in ^{153}Gd and ^{155}Dy clearly disappears in moving across the N=89 isotones towards ^{145}Ba. In addition, the deformation parameters calculated from the lifetimes of the first 2^+ states in ^{144}Ba and ^{146}Ba indicate that ^{145}Ba is only weakly quadrupole deformed; the β_2/β_{2sp} ratio for ^{144}Ba and ^{146}Ba is 6.8 and 7.5 respectively compared to the ratio of 12.2 and 10.7 found for the quadrupole deformed nuclei ^{154}Gd and ^{156}Dy.[13] This discrepancy in the decoupling parameter is even worse when octupole deformations are included in the calculations. Leander et al. calculate that a should be approximately +1.4 for the [530]1/2 neutron orbital.[1]

Because ^{145}Ba is not strongly deformed, an alternative interpretation of the structure is to treat the low-lying levels as collective vibrations. The energies of the first group of negative parity levels are clustered around the 181-keV energy of the first 2^+ state in the ^{146}Ba core. Moreover, the energies of the second group of negative parity states are clustered around the 513-keV 4^+ level in the core. The positive parity states, in contrast, clearly lie below the 738-keV 1^- level in ^{146}Ba. However, these levels could be pushed down by some type of intermediate particle-core interaction involving the h11/2 proton and h9/2 neutron spin-orbit partner orbitals. A second, and more serious problem, with a vibrational interpretation of the low-lying structure is the $5/2^-$ ground state. If ^{145}Ba is a spherical vibrator, then one would expect the g.s. to be $7/2^-$ or $9/2^-$. The g.s. of $5/2^-$ could arise from a three-hole f7/2 cluster similar to the $(f7/2)^3$ cluster states found in the N=85 isotones. The $5/2^-$ state of the cluster is then pushed below the $7/2^-$ and $9/2^-$ states in the same way that the $3/2^-$ state is pushed below the $7/2^-$ state in the N=85 isotones.[15] A relatively strong interaction, however, would be required to account for the $1/2^-$ 175-keV level with an $(f7/2)^3$ cluster ground state.

Although the structure of ^{145}Ba remains unexplained, a few conclusions can be drawn by examining the systematics of this transitional region. The drop in energy of the first 2^+ state in the N=88 isotones from 344 keV in ^{152}Gd to 199 keV in ^{144}Ba indicates that the N=88 isotones become more deformed as one moves towards Ba. In contrast, however, the disappearance of clear rotational structure from ^{153}Gd to ^{145}Ba and the decrease in the density of levels across the N=89 isotones suggests that the lighter N=89 isotones are less quadrupole deformed. This discrepancy can be resolved by treating the N=89 isotones as a hole in N=90 rather than a particle attached to N=88. Figure 2 is a plot of the ratio of the energy of the first 4^+ state to the energy of the first 2^+ state of the even-even nuclei from N=84-92. The large gap that occurs between N=88 and N=90 indicates that the Z=64 subshell closure disappears as the h9/2 neutron and the h11/2 proton spin-orbit partner orbitals both begin to fill and strongly interact. Thus, the minima in the energy of the first 2^+ state for the N=88 isotones at Z=56 simply reflects the Z=64 subshell closure. On the other hand, the systematics of the N=89 isotones indicate that the odd particle in the N=89 nuclei aligns with the more deformed N=90 system where midshell is Z=66.

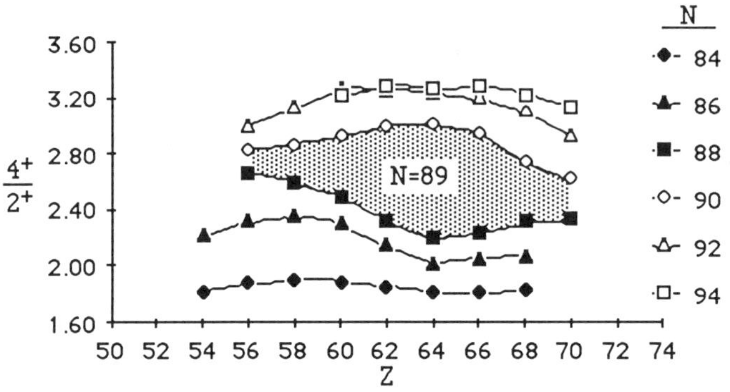

Figure 2. The ratio of the energies of the first 4^+ to the first 2^+ states as a function of Z for constant N.

The low-lying structure of the N=89 isotones ^{145}Ba, ^{147}Ce, and ^{149}Nd is very similar and is difficult to describe in the framework of the standard quadrupole deformed Nilsson model. This suggests that nuclei in the vicinity of ^{145}Ba form a region of weak deformation separate from the main region of quadrupole deformation found in the heavier lanthanides. As already mentioned, these nuclei lie in the large gap created by the disappearance of the Z=64 subshell closure (Figure 2). In addition, there is a large structure change that occurs in neutron-rich nuclei between N=85 and N=87. The neutron-rich N=85 nuclei from Z=54-64 have a low-lying triplet of levels with spins of $3/2^-$,$5/2^-$, and $7/2^-$ which are separated from the higher lying states by a large energy gap. (This structure is well described by the cluster vibrational model proposed by Paar et al.)[15] When only two more neutrons are added, however, this energy gap completely disappears and an unexplainable $1/2^-$ state appears at low-energy. This $1/2^-$ state drops from 576 keV in ^{151}Gd to 34 keV in ^{143}Ba. Moreover, the systematics of the 1^- states in the N=88, 90, and 92 nuclei run counter to what is expected for this region. In Figures 3a and 3b, the energies of the first 3^- and 1^- states of the even-even nuclei in this region are plotted against Z for constant values of N. The drop in energy of the 3^- state towards higher Z for N=84 and N=86 simply reflects the increase in deformation in moving away from the closed shell at Z=50. Likewise, the minima in the 3^- energy should be near Z=66 for N≥90 and at Z=57 for N=88; midshell for these isotones. The shift in the minima below Z=66 for N=90-94 is due to the filling of both neutron and proton orbitals which have large Y_{30} matrix elements; near Z=56, the $\pi d5/2$ and $\pi h11/2$ orbitals and near N=88, the $\nu f7/2$ and $\nu i13/2$ orbitals. The systematics of the first 1^- states in N=86 and N=88 can also be explained in terms of shell closures as the energy of the 1^-

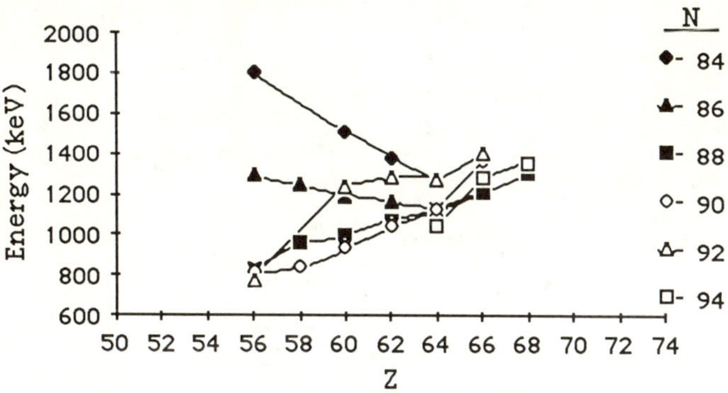

Figure 3a. The energies of the first 3^- state as a function of Z for constant N.

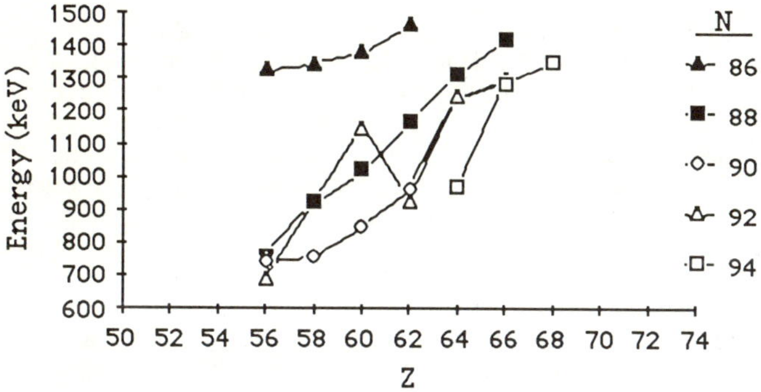

Figure 3b. The energies of the first 1^- state as a function of Z for constant N.

state for both isotones is lowest at midshell. The systematics of the first 1^- states for N≥90, on the other hand, run counter to what one expects.If the 1^- state is produced by the interaction of the octupole vibration with the quadrupole deformed nucleus, then one would expect this state to be lowest when this interaction is the strongest (when the nucleus is most deformed). This, however, does not appear to be the case as the minima in the energy of the first 1^- states are well below midshell (Z=66) for N≥90.

In conclusion, therefore, we would suggest that the key to understanding this region is to first gain an understanding of the negative parity states found in these even-even transitional nuclei. It will be important to know whether or not these negative parity states can be described as octupole vibrations or whether one must include octupole deformation and/or dipole or cluster modes.

1. G.A. Leander, W. Nazarewicz, P. Olanders, I. Ragnarsson, J. Dudek, Phys. Lett. 152B, 284 (1985).
2. W. Nazarewicz, P. Olanders, I. Ragnarsson, J. Dudek, G.A. Leander, P. Moller, E. Ruchowska, Nucl. Phys. A429, 269 (1984).
3. G.A. Leander, R.K. Sheline, P. Moller, P. Olanders, I. Ragnarsson, A.J. Sierk, Nucl. Phys. A388, 452 (1983).
4. A.C. Mueller, F. Buchinger, W. Klempt, E.W. Otten, R. Neugart, C. Ekstrom, J. Heinemeier, Nucl. Phys. A403, 234 (1983).
5. R.K. Sheline et al., Phys. Lett. 133B, 13 (1983).
6. R.K. Sheline and G.A. Leander, Phys. Rev. Lett. 51, 359 (1983).
7. M.S. Rapaport, G. Engler, A. Gayer, I. Yoresh, Z. Physik A305, 359 (1982).
8. H. Dejbakhsh, Ph.D. Thesis, Univ. of Oklahoma 1985.
9. J.A. Pinston, R. Roussille, G. Sadler, W. Tenten, J.P. Bocquet, B. Pfeiffer, D.D. Warner, Z. Physik A282, 303 (1977).
10. W.B. Cook, M.W. Johns, G. Lovhoiden, J.C. Waddington, Nucl. Phys. A259, 461 (1976).
11. L.K. Peker, Nucl. Data Sheets 32, 42 (1981).
12. L.K. Peker, Nucl. Data Sheets 45, 69 (1985).
13. Table of Isotopes, 7'th ed., edited by C.M. Lederer and V.S. Shirley (Wiley, New York, 1978).
14. J.M. Irvine, Nucler Structure Theory, (Pergamon, New York, 1972)
15. V. Paar, G. Vanden Berghe, C. Garrett, J.R. Leigh, G.D. Dracoulis, Nucl. Phys. A350, 139 (1984).

INFLUENCE OF BROKEN PAIR CONFIGURATIONS ON OCTUPOLE STATE PROPERTIES*

R.P. YAFFE+

Chemistry Department, San Jose State University
San Jose California 95114

D.R. NETHAWAY, A.L. PRINDLE AND R.A. MEYER

Lawrence Livermore National Laboratory
Livermore, California 94550

Gamma-ray spectroscopy measurements of ^{168}Er levels populated in the beta decay of ^{168}Tm, combined with a reanalysis of Coulomb excitation and inelastic scattering data, provide level lifetime information. These results support octupole state Coriolis coupling calculations. Comparison of the deexcitation rates of the two neutron [{521 1/2][633 7/2]} broken pair configuration is made and some evidence for K mixing among the octupole bands is presented.

Recently, Burke and coworkers[1] showed that the 3^- octupole state in ^{168}Er has a nearly pure two neutron {[521 1/2][633 7/2]} configuration. This supports the calculations of Neergaard and Vogel[2] (NV) who calculate that this broken pair contributes to 99% of the makeup of the 3^- octupole state. Also, recent neutron-capture gamma-ray completness studies[3,4] have established all of the octupole states and give all band members below approximately 2 MeV. The relative ordering and energy of the K^π=0^-, 1^-, 2^-, 3^- octupole states again confirm the prior calculations of NV. Unfortunately, analysis of Coulomb excitation data misassigned the K values of 3^- states and early (d,d') studies[4] could not assign 3^- values to weakly excited states. This has led to

the suggestion that the NV calculations were not applicable to ^{168}Er[5]. Using our previously reported[6] detailed spectroscopy measurements in combination with a reanalysis of both the Coulomb excitation [7] and the (d,d') results, we find that all the data do support the Coriolis redistribution of B(E3)↑ strength in the Coulomb excitation of ^{168}Er. The new B(E3)↑ values in combination with our measured E3 branching ratios then allow us to calculate the level half lives and determine absolute deexcitation rates for transitions from the octupole 3^- band members. These transition strengths along with the beta transition strengths allow us to investigate the properties of the two neutron {[521 1/2][633 7/2]} configurational partners and find some evidence for K mixing among the octupole bands.

From a purely phenomological view, the octupole states in deformed nuclei should be collective. However, Neergaard and Vogel[2] (NV) have shown that low-energy two-quasiparticle configurations can dominate the configurational make up of octupole states. In addition, NV have shown that the mixing brought about by the Coriolis interaction can alter the properties of the octupole states such as the B(E3)↑ values to members with 3^-K (herein we use I^{π}K). Davidson and coworkers[3] have used neutron-capture gamma-ray techniques to establish all the known levels below approximately 2 MeV in ^{168}Er. We find good agreement when we compare the known octupole band members with the calculations of NV. This suggests that, for this nucleus, the NV calculations provide a good picture of the octupole states. Unfortunately, published Coulomb excitation data[5] have been used to suggest that the B(E3)↑ strength is in disagreement with these calculations. However, McGowan[7] has reanalysed the Coulomb excitation data with the present knowledge of the location of octupole members (the presently known 3^-1 member at 1431 keV was originally thought to be the 3^-0 band member). We have compared the NV calculations with and without Coriolis coupling to the reanalyzed B(E3)↑ values taken

from McGowan's reanalysis[6] and a reanalysis of the (d,d') experiments of Tjom and Elbek[4]. As can be seen in Fig. 1, the Coriolis coupled calculations of NV agree with the measured B(E3)↑ values, particularly for the two lowest energy octupole bands. Further support comes from recent transfer reaction studies[1,8,9] which establish a {[521 1/2][633 7/2]} two neutron configuration as making up ~99% of the 3^-3 octupole band head; again consistent with the calculations of NV. Thus the octupole two-quasiparticle components calculated by NV that are given in Table 1 can be taken as a good indication of the major components in these bands.

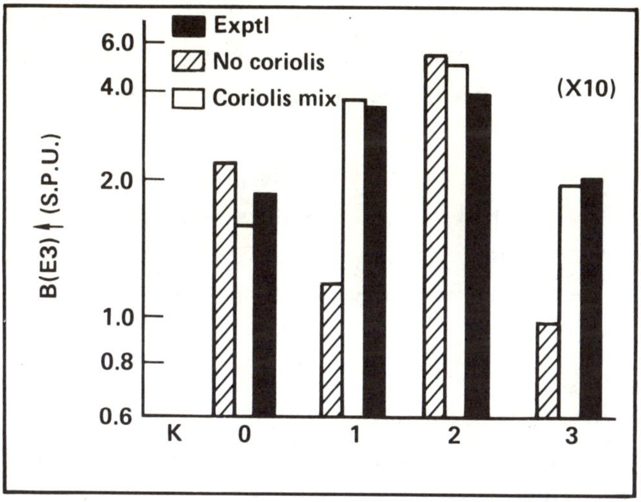

Fig.1 Comparison of the B(E3)↑ strength calculated using the model of Neergaard and Vogel[2] with experimental values.

Table 1. Octupole State Components

K-Value[a]	Configuration	%
K=0 (1786/1913)[b]		
proton	[404 7/2][523 7/2]	21
neutron	[514 7/2][633 7/2]	38
	[512 5/2][642 5/2]	11
	[521 1/2][651 1/2]	6
K=1 (1358/1431)		
proton	(see note c)	<1
neutron	[512 5/2][633 7/2[	89
	[523 5/2][633 7/2]	6
K=2 (1569/1633)		
proton	[411 3/2][523 7/2]	44
neutron	[521 3/2][633 7/2]	27
	[512 5/2][624 9/2]	7
K=3 (1431/1542)		
proton	(see note c)	<1
neutron	[521 1/2][633 7/2]	99

a) The components listed in this table were calculated by P. Vogel using the model discribed in Ref. 2.
b) The values given in parenthesis are the calculated value of the band head over the experimentally observed value.
c) There are no proton configurations in this band that are predicted to have over a 1% component.

The 3^- spin partner of the two neutron {[521 1/2][633 7/2]} configuration has been determined to dominate the K=3 octupole state with its band head at 1542 keV.[1,8,9] Using the coupling rules of Gallagher[10] we should expect its K=4 partner to occur at a lower energy. Both g-factor[11] measurements and transfer reaction studies[1,8,9] have shown that the K=4 band at 1094 keV with a half life of 107 ns has a ~70% {[521 1/2][633 7/2]} two-neutron component with the majority of the residual component (25%) coming from the {[523 7/2][411 1/2]} two-proton configuration. Using our data, and the reanalysed B(E3)l data, we have determined the deexcitation properties of these band heads which are given in Table 2.

Table 2. Hindrance Factors[a] for Transitions from the Neutron {[521 1/2][633 7/2]} Configurational Band Heads.

To	From					
	4-4			3-3		
	E1	M2	E3	E1	M2	E3
0^+0						4.7
2^+0	1.4e5	700	2.0e7			
4^+0	2.6e9	e4-e5		2.0e6	200-1400	
2^+2		112		4.9e4		
3^+2	4.5e6			3.2e5		
4^+2	7.2e6			1.1e5	~20	
				M1	E2	
4^-4				67	4	
5^-4					0.5	

a The values are given as, for example, 1.4e5 signifies 1.4×10^4 hindrance over the Weisskopf estimate.

The mixing of higher-K two-quasiparticle states into rotational members of octupole states has been shown to effect the properties of octupole states in light Er nuclei by Fields et al.[12]. Thus the mixing of higher K octupole states into lower K octupole band members might be expected to occur in ^{168}Er as well. Such mixing may account for the unexpected 10 fold jump in beta strength within the K=1 octupole band. Because there is a two unit change in K value for beta transitions from 3^+3 ^{168}Tm to band members of the K=1 octupole band, the beta transtions must proceed via unique first forbidden (UFF) beta decay.[13] The UFF transitions are governed by a single matrix element; hence they are not subject to cancellations between different beta matrix elements such as found in ordinary first forbidden beta transitions.[13] The UFF $\log f_1 t$ value is 10.20 for population of the K=1 band head. When a statistical factor for the difference in substates populated is included, this value is consistent with the known $\log f_1 t$ value of 9.40 for the ^{167}Tm to ^{167}Er UFF beta transition between ^{167}Tm

with a [411 1/2] ground state and the [521 5/2] excited state in ^{167}Tm (c.f. this is the basic transition from ^{168}Tm with a configuration of {**p**[411 1/2]**n**[633 7/2]}[14] to the ^{168}Er 1^-1 octupole state with a 89% two-neutron {[512 5/2][633 7/2]} configuration). Strictly, only the UFF process should contribute to the population of other band members even though the spin difference would allow a first forbidden transitions to occur. The high $\log f_1 t$ value of 10.18 for the population of the 2^-1 member is consistent with this picture. However, if there is K-mixing into a band member then first forbidden beta transitions become possible. This may be the cause of the factor of approximately 10 increase in the beta strength to the 3^-1 member over that to the 2^-1 member ($\log ft$ = 9.73 or $\log f_1 t$ = 9.38). The K=3 octupole band is near in energy and, as discussed above, has a nearly pure two neutron {[521 1/2][633 7/2]} configuration which is populated with a logft of 6.26. Thus only a small admixture of the K=3 band into the 3^-1 band member could account for the increase in beta population of the 3^-1 state. A more complete version of this will be presented in the literature.[15]

+ Participating guest at LLNL, Livermore, CA 94550
* This work supported, in part by Associated Western Universities and in part by U.S. D.O.E. under contract Nr. W-7405-Eng-48

1) D.G. Burke, et al., 1985 McMaster Univ. Annual Rpt. (1985) 78.
2) K. Neergaard and P. Vogel, Nucl. Phys. A145 33 (1970).
3) W.F. Davidson, et al., J. Phys. G 7 455 and 843 (1981).
4) P.O. Tjom and B. Elbek, Nucl. Phys. A107 385 (1968).
5) F.K. McGowan, et al., Nucl. Phys. A297 51 (1978).
6) R.A. Meyer, et al.,LLNL FY82 Nucl.Chem.Div. Ann.Rpt.,32(1982).
7) F.K. McGowan, ORNL, (Private communication, 1985).
8) D.G. Burke et al., Can. J. Phys. (in press 1986).
9) D.G. Burke et al., Nucl. Phys. A442 424 (1985).
10) C.J. Gallagher, Jr., Phys. Rev. 126 1525 (1962).
11) A. Furusawa, et al., Phys. Rev. C 21 2575 (1980).
12) C.A. Fields et al., Phys. Rev. C 26, 290 (1982).
13) E.J. Konopinski, "The Theory of Beta Radioactivity", Interscience Publ., J. Wiley, N.Y. (1966).
14) Z. Preibsz, et al., Nucl. Phys. A201 486 (1973).
15) R.A. Meyer, et al., Phys.Rev. C (sub.) UCRL-94316(1986)

SEARCH FOR TWO-OCTUPOLE-PHONON 0^+ STATE IN ^{208}Pb

R. Julin, J. Kantele, J. Kumpulainen, M. Luontama, A. Passoja, W. Trzaska, and E. Verho

Department of Physics, University of Jyväskylä, SF-40100 Jyväskylä, Finland

J. Blomqvist

Research Institute of Physics, S-10405 Stockholm 50, Sweden

Abstract In-beam conversion-electron spectra up to 6.5 MeV of energy, from bombardments of ^{207}Pb with 10 MeV deuterons and ^{208}Pb with 17.3 MeV protons, were measured. Many new transitions from known states at 4 to 6.5 MeV excitation to the ground state in ^{208}Pb were identified. The E0 lines from the pairing vibrational 4863 keV and 5233 keV 0^+ states are strong but no evidence for an E0 transition from the sought-after two-octupole-phonon 0^+ state was found.

The 3^- state in ^{208}Pb at 2615 keV is the best-known example of one-phonon octupole vibration. However, the validity of the simple vibrator picture is limited in view of the fact that no multi-phonon states of octupole character in ^{208}Pb or in any other even-even nucleus have been observed. Recently, states of two-octupole-phonon excitation coupled to a neutron or to a neutron pair have been found in ^{147}Gd [KLE82] and in ^{148}Gd [LUN84].

In ^{208}Pb, the two-octupole-phonon quartet of states with $I^{\pi} = 0^{+}$, 2^{+}, 4^{+} and 6^{+} should lie at about twice the energy of the 3^{-} state, i.e., in the 5.2 MeV region. Attempts have been made to identify these states but only tentative results have been reported [MAR83,LAN77]. Such states are expected to be populated very weakly in nuclear reactions which makes their identification difficult.

From the experimental point of view, the decay properties of the 0^{+} member of the multiplet play an interesting role. A rough estimate reveals the somewhat surprising fact that the K conversion of the E0 transition to the ground state could represent of the order of 10% branch of the total deexcitation of the state. This expectation might offer a method for the identification of the octupole 0^{+} state. However, obvious problems arise in the detection of high-energy conversion electrons.

Three 0^{+} states are expected to lie in the 5 MeV region in ^{208}Pb. Two of them have been firmly identified in two-neutron transfer studies as being situated at 4859 keV and 5236 keV [IGO71]. The 4859 keV state has been interpreted unambiguously as the neutron pairing vibrational state. The 5236 keV state was first suggested [IGO71] to be the 0^{+} member of the two-octupole-phonon quartet, because of the closeness to the unperturbed two-phonon energy. However, in [BLO70] it is shown that the proton pairing vibrational state lies at 5.3 MeV. Thus the two-phonon state has still remained unobserved.

Our combination electron spectrometer including a Siegbahn-Slätis type magnetic lens and a Ge(HP) detector (80 mm^2 x 5 mm), renders it possible to detect electrons up to 7 MeV with an energy resolution of about 2.5 keV at 1 MeV and about 8 keV at 5 MeV. With this spectrometer, we obtained a conversion-electron spectrum from a bombardment of a 4 mg/cm^2 thick enriched ^{207}Pb (89%) target with 10 MeV deuterons; the

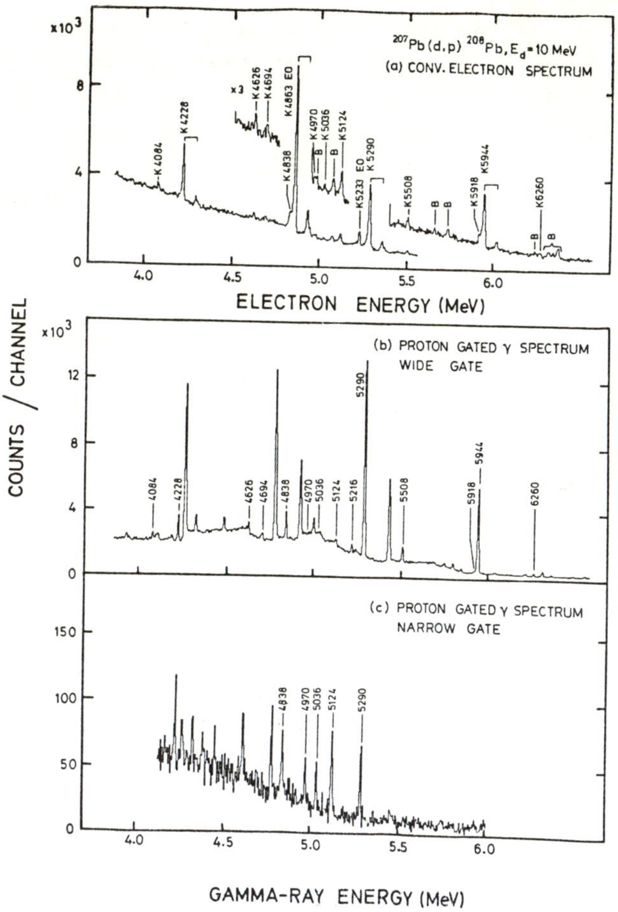

Fig. 1. High-energy parts of conversion-electron and proton-gated γ-ray spectra from a bombardment of ^{207}Pb with 10 MeV deuterons. In (a), B indicates background lines. The gate settings for (b) and (c) correspond excitation energies of 2.5-8 MeV and about 5 MeV in ^{208}Pb, respectively.

spectrum is shown partially in Fig 1a. Conversion-electron peaks corresponding to known ground-state E1 transitions from 1^- states at 4841, 5291, 5507, 5943, and 6257 keV, and a known E2 transition from the 4086 keV 2^+ state populated in the $^{207}Pb(d,p)^{208}Pb$ reaction [LED78], are identified in this spectrum. These lines, together with high-energy lines from a ^{66}Ga source, were used for the energy calibration, which gives an accuracy of about 5 keV.

For the identification of other new conversion-electron lines in the spectrum of Fig. 1a, we performed a gamma-proton coincidence experiment with the same reaction. Fig. 1b shows the high-energy part of the γ-ray spectrum gated with the part of the Si(Li) spectra which corresponds to excitation energies of 2.5 to 8 MeV in ^{208}Pb. By setting narrower gates on the particle spectrum, it was possible to identify weak lines, as well (Fig. 1c).

With the kind of analysis described, two new E0 lines corresponding to transition energies of 4863±5 keV and 5233±5 keV in Pb are identified in the electron spectrum of Fig. 1a. These energies agree well with those for the known neutron and proton pairing vibrational 0^+ states. All the other unknown lines of this electron spectrum are also visible in the γ-ray experiment. We identify them with previously unobserved ground-state M2, E3, or E2 transitions from known levels in ^{208}Pb [LED78]. No candidate for an E0 transition from the sought-after two-octupole-phonon 0^+ state could be identified.

Although (d,p) is not a good reaction for populating a two-octupole-phonon 0^+ state, this state could perhaps be observed via mixing with the pairing vibrational 0^+ states. The sensitivity of conversion-electron spectrometry in the search for a 0^+ state in ^{208}Pb is demonstrated in this measurement: the total cross sections for the

K-conversion electron production of the 4863 keV and 5233 keV E0 transitions, derived from the spectrum of Fig 1a, are about 5 μb and 0.5 μb, respectively. In the case of the pairing-vibrational states, the E0 transition to the ground state should dominate the decay of these states [MAR83]. Consequently, total cross sections for the population of these states in the (d,p) reaction at 10 MeV are thus roughly 15 μb and 2 μb, respectively.

In another experiment, we measured an electron spectrum from the bombardment of a 5 mg/cm^2 thick enriched (98%) ^{208}Pb target with 17.3 MeV protons. This energy corresponds to an IAR energy in the inelastic proton scattering, which could favor the population of the two-octupole-phonon 0^+ state in ^{208}Pb [BLO70]. The high-energy part of the spectrum shown in Fig. 2 exhibits again the 4863 keV and 5233 keV E0 transitions from the pairing-vibrational 0^+ states. In this case, the K conversion electron production cross sections are about 10 μb and 6 μb, and the total cross sections for the level populations are about 30 μb and 20 μb, respectively.

The four known E1 transitions from the 4841 keV, 5291 keV, 5507 keV and 5943 1^- levels and the E2 transition from the 4086 keV 2^+ level to the ground state can also be identified in this spectrum. The K lines of the 4230 keV and 5922 keV transitions indicated in the spectrum obviously correspond to the 4228 keV and the 5918 keV transitions observed in the electron and γ-ray spectra from the (d,p) reaction.

Also in this case, we fail to observe a candidate for an E0 transition from the expected two-octupole-phonon 0^+ state. If the K conversion of the E0 transition from this state to the ground state represents an about 10% branch of the total deexcitation, we obtain a limit of about 2 μb for the population of a possible two-octupole-phonon 0^+ state in ^{208}Pb in inelastic proton scattering at 17.3 MeV.

This limit is roughly constant for the whole region from 4 to 6 MeV, since the decrease of the E0 branch with increasing energy is associated with a decreasing background. Of course, this limit is not valid if the line happens to coincide with a known line in the spectrum.

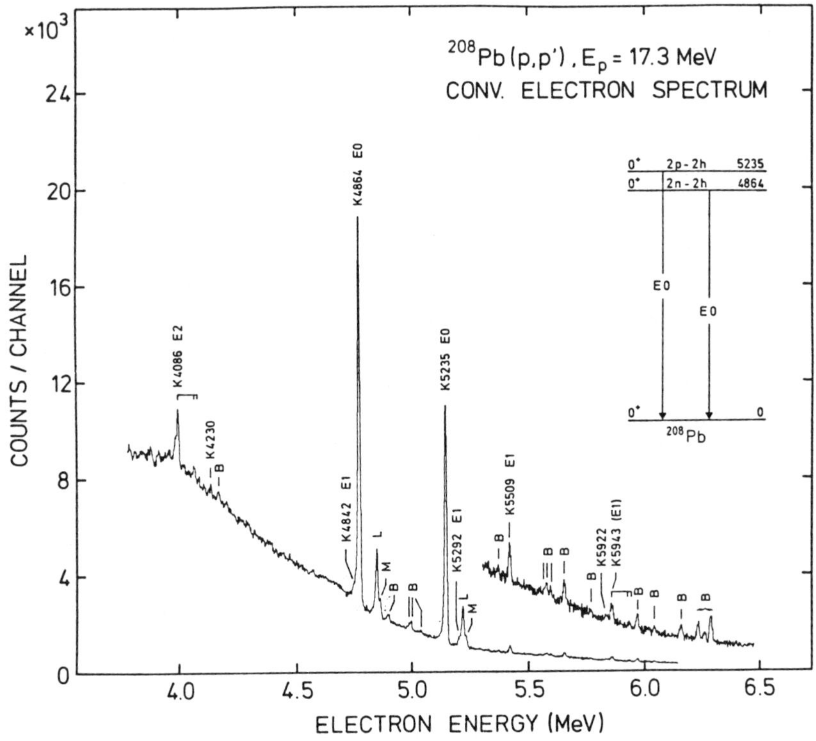

Fig 2. High-energy part of conversion-electron spectrum from a bombardment of ^{208}Pb with 17.3 MeV protons.

References

[BLO70] J. Blomqvist, Phys. Lett. 33B, (1970) 541

[IGO71] G.J. Igo, P.D. Barnes and E.R. Flynn, Ann. of Phys. 66, (1971) 60

[KLE82] P. Kleinheinz, J. Styczen, M. Piiparinen, J. Blomqvist and M. Kortelahti, Phys. Rev. Lett. 48, (1982) 1457.

[LAN77] S. Landowne, C.H. Dasso, R.A. Broglia and A. Winther, Phys. Lett. 70B, (1977) 292.

[LED78] Table of Isotopes, edited by C.M. Lederer and V.S. Shirley (Wiley, New York, 1978), 7th ed.

[LUN84] S. Lunardi, P. Kleinheinz, M. Piiparinen, M. Ogawa, M. Lach and J. Blomqvist, Phys. Rev. Lett. 53, (1984)1531.

[MAR83] M.A.J. Mariscotti, D.R. Bes, S.L. Reich, H.M. Sofia, P. Hungerford, S.A. Kerr, K. Schreckenbach, D.D. Warner, W.F. Davidson and W. Gelletly, Nucl. Phys. A407, (1983) 98.

SEARCH FOR STABLE OCTUPOLE DEFORMATION IN THE ACTINIDE REGION: 223 ≤ A ≤ 229

M.J.G.BORGE[1],D.G.BURKE[1,2],F.CALAPRICE[3],H.GIETZ[4],P.HILL[4], N.KAFFRELL[4],W.KURCEWICZ[5],G.LØVHØIDEN[6],S.MATTSSON[7], R.A.NAUMANN[1,3],K.NYBØ[6],G.NYMAN[7] and T.F.THORSTEINSEN[6]

1) ISOLDE, CERN,1211-Geneva 23,Switzerland
2) McMaster University,Hamilton,Ontario,Canada
3) Princeton University,Princeton,N.J. 08540,U S A
4) Institut Kernchemie,Univ.Mainz,Mainz, Germany
5) Wydziat Fizyki, Uniw.Warszawski,Warsaw,Poland
6) Fysisk Institutt, Univ.Bergen,Bergen,Norway
7) Dept. of Physics,Chalmers Univ. of Tech., Göteborg,Sweden

Abstract A miniorange-type electron spectrometer has been constructed and used to study conversion electrons and γ-e^- coincidences following the β-decay of neutron rich francium and radon isotopes. We report the results obtained so far for ^{223}Fr, ^{227}Ra and ^{229}Ra. Multiscaling analysis of ^{223}Rn gave a new value of the half-life of 23.2±0.4 min. These measurements allow us to assign 32 new γ lines following the decay of ^{223}Ra and 37 γ lines following the decay of ^{229}Fr. For ^{227}Ra, the multipolarities of ~30 transitions have been established. The results support the placement of levels and transitions in the earlier level scheme, but require changes in the previously assigned parities for four of the levels. Also, one E0 transition was identified.

INTRODUCTION

There has recently been a great deal of interest in the possibility that stable reflection-asymmetric shapes may exist for nuclei in the actinide region near A ~ 225. The available experimental evidence[1-6] indicates a strong probability that such shapes exist in ^{223,225,227}Ra, but the level schemes for most odd-A nuclides in this region are not

well established. In many cases the assignments of spins, parities, and other quantum numbers to the levels have been made on the basis of systematics in neighbouring nuclides, or by arguments which are strongly model-dependent. Therefore it is important to test the experimental foundation on which the claims for observed octupole deformations are based. One expected feature of an octupole-deformed nucleus is that it should have parity-doublets[4,7], i.e. close-lying bands with the same K-value but opposite parity, and there should be enhanced E1 and E3 transitions connecting these bands. It is therefore important to establish the parities of levels and the multipolarities of transitions in nuclides of the A ~ 225 region.

For this purpose, a miniorange-type electron spectrometer has been constructed. The present report describes the spectrometer system and presents briefly the results obtained so far for ^{223}Fr, ^{227}Ra and ^{229}Ra.

EXPERIMENTAL SET-UP

Francium (radon) isotopes were produced in UC (ThC) target by spallation reactions induced by 600 MeV protons from the CERN Synchro-cyclotron. After ionization in a plasma discharge ion source (for rare gases) or in a surface exchange ion source (for alkalies) the nuclei were mass-separated in the ISOLDE on-line separator. The beam was collimated to minimize contamination from adjacent masses and collected on the aluminium-coated Mylar tape of a tape transport system. This system was operated in a start-stop mode so that the sample could be collected for a preselected time and then transported to the source position of the miniorange electron spectrometer for the counting period. The physical arrangement of the apparatus is shown in Figure 1. The detector in the electron spectrometer was a 300 mm^2 Si(Li) device with a thickness of 2 mm, cooled to liquid nitrogen temperature by means of a copper cold finger. An absorber block

placed between the source and detector shields the latter from direct X-rays and γ-rays. A set of permanent magnets positioned around the spectrometer axis deflects electrons from the source to the detector with efficiencies depending on the electron energy and the magnet configuration.

Several different magnet types were tested. The set finally chosen consisted of three equally spaced flat rectangular magnets 5 cm × 2 cm × 0.5 cm with the long dimension oriented radially, giving the optimum performance for a distance of ~ 15 cm between the detector and the source and with the magnets situated midway. The overall energy resolution was ~ 2.0 keV full width at half-maximum (FWHM). Energy calibration was established using known lines from ^{152}Eu and ^{207}Bi sources.

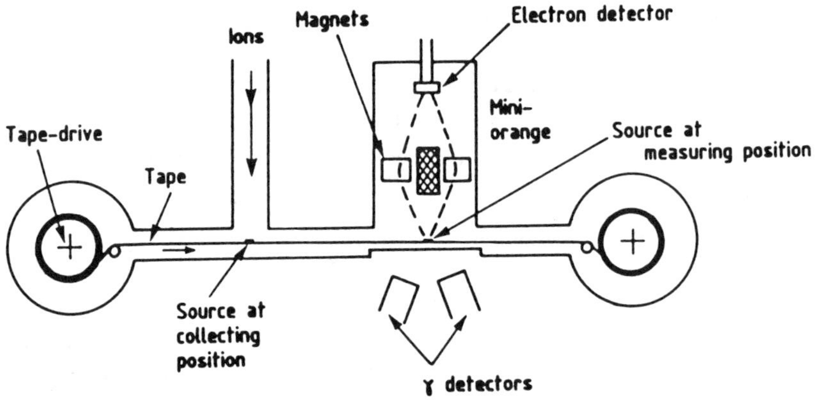

FIGURE 1. Arrangement of the experimental set-up

Two high-purity Ge photon detectors were also placed near the source position of the spectrometer, as shown in fig. 1, permitting γ-e$^-$ and γ-γ coincidences to be recorded simultaneously. These detectors had active volumes of 68 cm^3 and 88 cm^3, and typical resolutions (FWHM) of 0.8 keV at 122 keV and 1.8 keV at 1.3 MeV. Energy and efficiency calibrations for these detectors were determined with ^{133}Ba, ^{152}Eu,

^{207}Bi and ^{241}Am sources and also with a commercially obtained absolutely-calibrated reference source of different radionuclides. Coincident events between any pair of the three detectors were selected by standard hardware circuitry, with the information from the analog-to-digital and time-to-amplitude converters being stored on magnetic tape in event-recorded mode.

EXPERIMENTAL RESULTS

^{223}Fr

Singles gamma-ray and conversion-electron spectra were measured from ^{223}Rn samples which were collected and counted for periods of 200 sec each. A multiscaling experiment was also performed by collecting a sample for 15 min and recording the singles γ-ray and electron spectra in 5 min intervals for a period of an hour. It was found that most of the prominent lines decayed with a half-life that was only about half the value of 43±5 min reported earlier[8] for the ^{223}Rn half-life obtained by a milking technique. The new value of the ^{223}Rn half-life deduced from these data is 23.2±0.4 min, which is also consistent with the growth and decay curves observed for some of the known ^{223}Ra lines associated with the ^{223}Fr daughter activity. Prominent gamma transitions observed in ^{223}Fr following the β-decay of ^{223}Rn are listed in Table I. The multipolarities indicated in this table were obtained from measured K and L internal conversion coefficients, with the designation M1+E2 used when the experimental value was between the theoretical values[10] for E2 and M1 transitions.

^{227}Ra

^{227}Fr activity was collected for periods of 50 sec and measured for periods of 200 sec. During a data collection time of ≈ 17 hours, approximately 10^6 γ-e^- coincidences were

TABLE I Prominent γ-rays in the $^{221}Rn \rightarrow {}^{221}Fr$ Decay

ENERGY (keV) Present[a)]	Previous'	Relative $I\gamma$	Multi-polarity
69.0	69.21 ±0.04		
106.9	106.79 ± 0.06	~ 5	
121.6	121.53 ± 0.04	6	
140.9		26	E1
147.6	147.48 ± 0.04	4	
158.9		4	
160.4	160.26 ± 0.05	5	
162.5		4	
171.8	171.90 ± 0.08	12	E1
173.9		4	
206.3		13	E1
216.0		4	
243.3		4	
360.4		4	M1 + E2
416.0		55	M1 + E2
458.1		6	M1 + E2
461.1		8	M1 + E2
491.4		19	M1 + E2
522.5		9	M1 + E2
~ 547		20(multiplet)	
~ 565		14(multiplet)	
591.8		≡100	M1 + E2
621.5		15	M1 + E2
635.2		76	M1 + E2
648.7		29	M1 + E2
654.0		44	M1 + E2
723.0		23	
818		6	
882		4	
948		4	
988		4	
1240		2	
1297		4	
1318		~ 4	
1403		4	
1408		9	
1461		4	
1765		1	

a) Estimated uncertainties on the energies are ± 0.3 keV for values up to 450 keV, ± 0.5 keV for those between 450 and 800 keV, and ± 1.5 keV for higher energy lines.

obtained. An examination of all the γ-γ and γ-e^- data confirms the placement of transitions and levels shown in the previously reported level scheme[11]. Therefore the important contribution made by the present measurements is new information in the form of conversion coefficients. These were obtained from the singles electron spectra, but for some lines which were not resolved in the singles spectra the coincidence results were used.

The results obtained for the 42 transitions studied are shown elsewhere[12]. One interesting observation is that the 379.2 keV transition has a value of α_K of 2.4±0.8 that is much larger than expected[10] for an E1(α_K=0.017), E2(α_K=0.044) or M1(α_K=0.32) transition. Since it competes with several E1 and M1 transitions in the decay of the 675.9 keV level, a higher order multipolarity is unlikely. Therefore the 379.2 keV transition is assumed to have an E0 component. The high value of α_K for this line is also confirmed by coincidence data with the 175.9 keV line.

The new multipolarity information has shown that the parities previously proposed[11] for the levels at 384.3 keV, 438.8 keV, 475.0 keV and 731.6 keV are incorrect. However, all the new data are consistent with the earlier suggested assignments for low-lying bands that have been used to argue for the presence of an octupole deformation. To this extent the new results support the existence of such a deformation.

^{229}Ra

A multiscaling experiment was performed by using cycles of 50 sec collection time followed by 10×10 sec measuring time. In such a way 75 samples were collected and recorded with the electron detector and 60 samples were recorded with one of the Ge detectors. 14 of the more prominent γ lines were used to determine a half-life of 50.2±0.4 sec in agreement with the previous value[13] of 48±2 sec. The combined information of the multiscaling measurements together with γ-γ coincidences and comparison of present relative intensities

of the lines with our previous data allow us to identify 37 new γ-lines as following the decay of ^{229}Fr.

A preliminary level scheme has been outlined based on coincidence relations. Tentative spin and parity assignments have been made from analogy with the isotone ^{231}Th. The analysis of the coincidence data is still in progress, so it is too early to draw any conclusion from them.

REFERENCES

1. I.Ragnarsson, S.G.Nilsson and R.K.Sheline, Phys. Rep. 45, 1 (1978).
2. I.Ragnarsson, Phys. Lett. 130B, 353 (1983).
3. R.K.Sheline, D.Decman, K.Nybø, T.F.Thorsteinsen, G.Løvhøiden, E.R.Flynn, J.A.Cizewski, D.G.Burke, G.Sletten, P.Hill, N.Kaffrell, W.Kurcewicz, G.Nyman and G.Leander, Phys. Lett. 133B, 13 (1983).
4. G.A.Leander and R.K.Sheline, Nucl. Phys. A413, 375 (1984).
5. K.D.A.Wendt, Doctoral Dissertation, J. Gutenberg Univ., Mainz (1985).
6. R.K.Sheline, Phys. Letts. 166B, 269 (1986).
7. I. Ahmad, R.R.Chasman, J.E.Grindler and A.H.Friedman, Phys. Rev. Lett. 52, 503 (1984).
8. F.D.S.Butement and V.J.Robinson, J. Inorg. Nucl. Chem. 26, 1 (1964).
9. S.K.Vasiliev, B.S.Dzhelepov, R.B.Ivanov, M.A.Mikhailova, A.V.Mozhzchin and B.I.Shestakov, Izv. Ak. Nauk. SSSR, Ser. Fys. 45, 1895 (1981).
10. R.S.Hager and E.C.Seltzer, Nucl. Data Tables A4, 1 (1968).
11. T. von Egidy, G.Barreau, H.G.Börner, W.F.Davidson, J.Larysz, D.D.Warner, P.H.M. van Assche, K.Nibø, T.F.Thorsteinsen, G.Løvhøiden, E.R.Flynn, J.A.Cizewski, R.K.Sheline, D.Decman, D.G.Burke, G.Sletten, N.Kaffrell, W.Kurcewicz, T.Bjørnstad and G.Nyman, Nucl. Phys. A365, 26 (1981).
12. "Measurements of multipolarities in ^{227}Ra as tests of evidence for stable octupole deformation in ^{227}Ra" M.J.G.Borge et al., manuscript in preparation.
13. L.C.Carraz, S.Sundell and H.L.Ravn, Nucl. Instrum. and Methods 158, 69 (1979).

STUDY OF INTERNAL CONVERSION ELECTRONS IN ^{228}Ra

M. J. G. BORGE[1], D. G. BURKE[1,2], H. GIETZ[3], P. HILL[3], N. KAFFRELL[3], W. KURCEWICZ[4], G. LØVHØIDEN[5], S. MATTSSON[6] R. A. NAUMANN[1,7], K. NYBØ[5], G. NYMAN[6] and F. THORSTEINSEN[5] and the ISOLDE collaboration

1. ISOLDE, CERN, 1211-Geneva 23, Switzerland
2. McMaster University, Hamilton, Ontario, Canada
3. Institut für Kernchemie, Universität Mainz, D-6500 Mainz, Germany
4. Institute of Experimental Physics, University of Warsaw, 00-681 Warsaw, Poland
5. Fysisk Institutt, Universitetet i Bergen, N-5000 Bergen, Norway
6. Dept. of Physics, Chalmers Univ. of Technology, S-41296 Göteborg, Sweden
7. Princeton University, Princeton, N. J. 08540, USA

Abstract Spin values and parities for the collective excited states of ^{228}Ra have been determined by performing conversion electron measurements using a mini-orange β spectrometer. The level structure of ^{228}Ra is interpreted in terms of nuclear models.

INTRODUCTION

For nuclei in the region around neutron number N ≈ 134 both experimental results[1,2] and theoretical predictions[3-5] suggest that reflection asymmetric ground-state equilibria may occur. This work describes a new study of the decay of ^{228}Fr to neutron-rich ^{228}Ra, which is predicted to be on the border-line of the region of octupole deformed nuclei.

Studies of the low-lying levels of ^{228}Ra have recently been performed by Ruchowska et al.[6], and van den Berg et al.[7]. These investigations produced a detailed level scheme of ^{228}Ra.

However, the tentative spin-parity assignments given by Ruchowska et al.[6] were made by comparison with the lighter isotopes ^{224,226}Ra and by analysis of the reduced branching ratios. Evidence for a strongly excited 0^+ state at E_X = 0.8 MeV in this nucléus was found by van der Berg et al.[7], using the ^{232}Th(d,^{6}Li)^{228}Ra reaction.
The aim of this work was to determine spins and parities for the lower excited states of ^{228}Ra by performing conversion electron measurements. The measured properties for collective excitations of ^{228}Ra are briefly discussed in terms of nuclear models.

EXPERIMENTAL METHODS AND RESULTS

Sources of ^{228}Fr were produced in a spallation reaction of ^{238}U induced by 600 MeV protons from the CERN Synchrocyclotron and mass-separated in the ISOLDE II on-line isotope separator facility. A selected ion beam was directed via an external beam-line into the measuring equipment. In addition to methods used in a previous work[6], a mini-orange β spectrometer[8] was used to measure conversion electrons. This spectrometer combined a liquid-nitrogen cooled Si(Li) detector (300 mm^2 area, 2 mm depletion depth and 2.0 keV resolution) with a filter of permanent magnets around a central absorber. The low and high-energy parts of the conversion electron spectrum of ^{228}Fr are presented in fig. 1.
In order to check whether the stronger conversion lines belonged to the ^{228}Fr decay, multispectra measurements were also performed. The observed half-lives of the L,M-64, L,M-140, L-171, K-676, K-707 and K-721 keV conversion electron lines were in agreement with the expected value[6] of 38 ± 1 sec.
Precisely known transition energies from ^{228}Fr [ref. 6] were used as internal energy standards. The transmission curve of mini-orange spectrometer was obtained using conversion lines from sources of ^{133}Ba, ^{152}Eu and ^{207}Bi.
The experimental conversion electron subshell ratios establish E2 multipolarity for the 63.83 keV ($2^+ \longrightarrow 0^+$) and 140.86 keV ($4^+ \longrightarrow 2^+$) transitions. For the 63.83 keV transition the

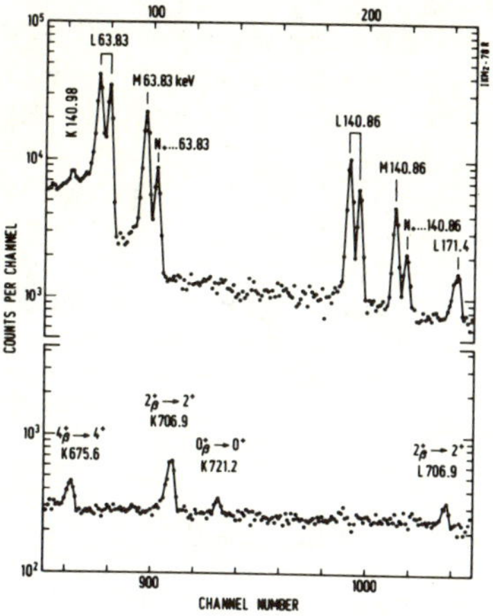

FIGURE 1. Low and higher energy portions of the conversion electron spectrum from the decay of ^{228}Fr to ^{228}Ra.

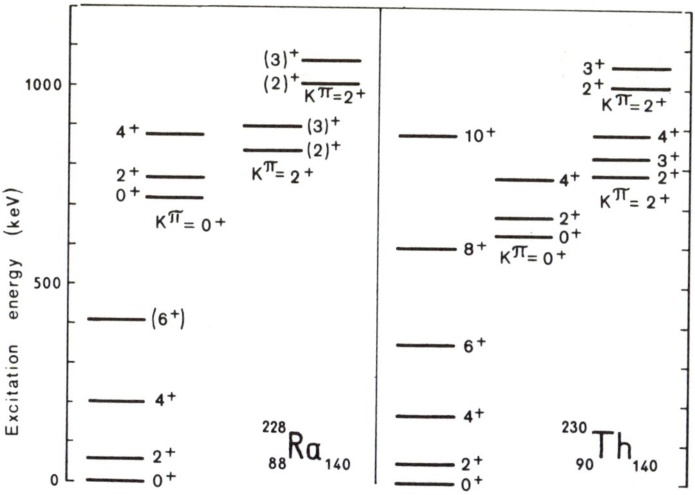

FIGURE 2. Partial level schemes of ^{228}Ra and isotone ^{230}Th.

experimental electron intensity ratios (L_1 + L_2) : L_3 : M : (N + ...) = 100 : (83.5 ± 2.3) : (60 ± 4) : (18.7 ± 2.0) could be compared with theoretical[9] values of 100 : 82.5 : 50 : 15.6 for E2 multipolarity. For the 140.86 keV transition the experimental and theoretical E2 intensity ratios are 100 : (57.4 ± 3.4) : (39.6 ± 2.7) : (13.6 ± 1.6) and 100 : 56.9 : 37.7 : 15.1, respectively, indicating E2 multipolarity. By adopting the theoretical E2 conversion coefficient, $\alpha_{M+N+...}$ = 0.507, for the 140.86 keV transition and taking relative electron and gamma [6] intensities, one can evaluate the conversion coefficients for observed transitions. The resultant value of 9 ± 2 for the L conversion coefficient of the 171.4 keV transition, compared to the theoretical values of 0.02, 0.63, 0.54 for E1, E2 and M1 multipolarities, respectively, establishes a major E0 component in this transition. A summary of the α_K values for measured transitions is presented in table 1.

DISCUSSION

The interpretation of the even parity states of ^{228}Ra is presented in fig.2 besides the partial level scheme of ^{230}Th. All the new data support the spin values tentatively assigned in the previously-reported decay scheme [6].

The K^π = 0^+ ground-state band.
The assignments of I^π = 2^+ and 4^+ to the 63.83 and 204.69 keV levels, respectively, are based on the E2 multipolarities of the 63.8 and 140.9 keV transitions. However, the tentative assignment of I^π = 6^+ remains for the 411.67 keV state.

The K^π = 0^+ excitation.
The K = 0 assignment for the 721.17, 770.70 and 880.30 keV levels is clearly established by their E0 decays (see table 1) to the 0^+, 2^+ and 4^+ levels of the ground-state band. A dimensionless ratio of the E0- and E2-transition probabilities X(E0/E2) has been evaluated for the transitions investigated in this work along the same line as [ref. 10]. The values of X for the 721.2, 706.9 and 675.6 keV E0 transitions are 0.07 ± 0.02, 0.11 ±

0.01 and 0.20 ± 0.08, respectively. The experimental results should be compared with the theoretical X_β value of 0.19 in the case of beta vibration, where $X_\beta = 4 \cdot \beta_0^2$ and the quadrupole deformation parameter $\beta_0 = 0.22$ [ref. 11]. Recently, α-like clustering effects were introduced to describe the structure and dynamics of heavy actinide nuclei[12]. It is interesting to compare the results here to the cluster model prediction of X ≈ 0.25, which is approximately constant for different numbers of bosons. Notice that the experimental X value cannot discriminate between these two very different theoretical descriptions.

The $K^\pi = 2^+$ excitations

Two excited $K^\pi = 2^+$ bands, with bandheads at 846.2 and 1013.3 keV, connected by rather strong E0 transitions, were proposed by Ruchowska et al.[6]. From the present work positive parity has been established for the 846.2 keV level, interpreted as a γ-vibrational state. The present data established an E0 component in the 171.4 keV transition connecting the $I^\pi = 3^+$ members of the $K^\pi = 2^+$ bands. The level at 1013.3 keV is assigned as an $I^\pi K = 2^+2$ state by analogy with levels in the isotone ^{230}Th [ref. 13], and extends the systematics of the second $K^\pi = 2^+$ states observed in heavier actinide nuclei[14]. The nature of the second low-lying collective $K^\pi = 2^+$ band is not very clear. Its excitation energy is much lower than predicted for the two-phonon β+γ state in the simple model of harmonic nuclear vibrations. These second low lying $K^\pi = 2^+$ states need further study.

The $K^\pi = 0^-$ band.

The E1 multipolarities for the 332.9 and 410.4 keV transitions indicate negative parities for the 474.14 and 537.57 keV levels, tentatively assigned previously in [ref. 6]. The measured properties of ^{228}Ra are similar to those observed in the isotone ^{230}Th, and suggest zero octupole deformation for ^{228}Ra.

Recently, the properties of collective states in the N = 134 region have also been described in terms of the multiphonon method[15] as well as in an α-cluster picture [12].

TABLE I The relative intensities of the K conversion lines, internal conversion coefficients α_K and multipolarity assignments for measured transition in ^{228}Ra.

Transition energy [a] (keV)	I_K	α_K [c]	Multi-polarity
140.86±0.02	540±80	0.27±0.05(0)	E2
332.91±0.05	31.0±4.8	3.2±0.7(-2)	E1
410.40±0.06	40.7±5.1	1.7±0.3(-2)	E1
675.6±0.5	55.5±5.0	1.3±0.3(0)	E0+E2
706.9±0.1	100±6	5.5±0.8(-1)	E0+E2
721.2±0.5 [b]	16.2±4.7	≥1.4(0)	E0
782.3±0.1	9.3±3.9	1.2±0.5(-2)	E2

a) Data from [ref.[6]].
b) Data from present study.
c) Numbers in parentheses are powers of ten.

REFERENCES

1. W. Kurcewicz et al., Nucl. Phys. A356 (1981) 15.
2. R. K. Sheline et al., Phys. Lett. 133B (1983) 13.
3. A. Gyurkovich et al., Phys. Lett. 105B (1981) 95.
4. P. Möller and J. R. Nix, Nucl. Phys. A361 (1981) 117.
5. W. Nazarewicz et al., Nucl. Phys. A429 (1984) 269 and references quoted therein.
6. E. Ruchowska et al., Nucl. Phys. A383 (1982) 1.
7. A. M. van den Berg et al., Nucl. Phys. A422 (1984) 45.
8. J. van Klinken et al., Nucl. Instr. Meth. 151 (1978) 433.
9. F. Rösel et al., Atomic Data and Nucl. Data Tables 21 (1978) 291
10. A. V. Aldushchenkov and N. A. Voinova, Nucl. Data Tables 11 (1972) 299.
11. A. S. Goldhaber and G. Scharff-Goldhaber, Phys. Rev. C17 (1978) 1171.
12. H. J. Daley and B. R. Barrett, Nucl. Phys. A449 (1986) 256.
13. W. Kurcewicz et al., Acta Phys. Pol. B2 (1971) 451.
14. A. Bohr and B. R. Mottelson, "Nuclear Structure", vol. 2 (Benjamin, Massachusetts, 1975), Ch. 6.
15. R. Piepenbring, Phys. Rev. C27 (1983) 2968.

ON THE BASIC ASSUMPTIONS, EXPERIMENTAL VERIFICATION, AND INTERPRETATION OF A NEW ALPHA-CLUSTER MODEL

H.J. DALEY
SERC, Daresbury Laboratory, Warrington WA4 4AD, U.K.

Abstract The basic assumptions made in applying the α-cluster nuclear vibron model to actinide and rare earth nuclei are discussed in this article. It is argued that there is experimental support for those assumptions and additional experimental tests of the model are suggested.

INTRODUCTION

There has recently been much experimental and theoretical interest concerning the basic structure of many nuclei in the actinide region.[1,2,3] There are two schools of thought that have formed regarding the structure of those nuclei: i) α-clustering effects are important,[4] and ii) static octupole deformation occurs.[5]

The interacting boson model (IBM) was recently extended[6] such that nuclear clustering effects could also be included within the same IBM framework. This extended version of the IBM is referred to as the nuclear vibron model (NVM). The NVM was then specialized to the case of α-clustering and applied in a phenomenological manner to the actinide mass region[7,8] in an attempt to find experimental support for the general hypothesis that α-clustering effects are important in heavy nuclei. The patterns which emerged from the analysis of the actinides were found to

manifest themselves in some rare earth nuclei also,[9] thus providing additional important information. The most basic assumptions of this α-cluster model[7,8] distinguish it from other models.[1,5,10] These fundamental assumptions will be outlined, and experimental support for them will be offered in the next section.

BASIC ASSUMPTIONS AND EXPERIMENTAL SUPPORT

Two basic assumptions have been made in applications of the α-cluster NVM to the actinide and rare earth regions.[7,8,9,11] These will be referred to as i) the coexistence hypothesis, and ii) the dynamic dipole hypothesis. In the former case, it is hypothesized that the usual IBM-1 configuration is lowest in energy, coexists and mixes with an excited configuration which differs somewhat in structure due to the presence of an α-cluster. The former configuration is the 0α configuration and the latter the 1α configuration. In the second case it is hypothesized that the dipole mode of the dinulcear system is of importance in lowest order and it is of the dynamic, or vibrational, type; i.e., it is assumed that there is no permanent dipole deformation in the 1α configuration.

Let us consider some consequences of these two hypotheses by first considering the ^{156}Gd case. In Figure 1a experimental energy levels are plotted. Notice that the $K^\pi = 0_1^+$, 0_3^+, and 2_1^+ bands all have almost identical moments of inertia, but the moment of inertia of the $K^\pi = 0_2^+$ band is 10% larger. The 0_2^+ band is assumed to be primarily the lowest energy band of the 1α configuration and will be called the α-band.

Such α-bands have also been identified in many actinide nuclei.[8,11] The $K^{\pi} = 0_3^+$ and 2_1^+ bands are assumed to be primarily IBA-1 β- and γ-bands, respectively. If one now couples the one phonon dipole mode of the dinuclear system to the α-band one obtains a 1^- singlet, followed by a $(1^-,2^-,3^-)$ triplet, then a $(3^-,4^-,5^-)$ triplet, etc. as one goes up the α-band. The groups of negative parity states should have the same splitting as the α-band; however, all negative parity states may be shifted in energy together, according to the value of the dipole phonon energy (see Figure 1b). To the right in Figure 1b the same states are shown with an additional $L_a \cdot L_b$ splitting within each triplet. The agreement between the experimental and the theoretical symmetries supports the

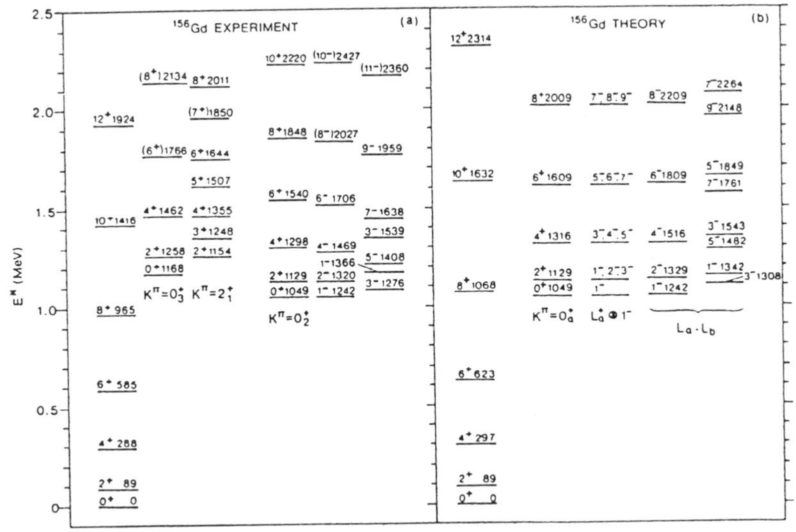

FIGURE 1. Experimental a) and theoretical b) energy levels[9] for ^{156}Gd. The experimental $K^{\pi} = 2_1^+$ band was shifted up by 104 keV and the negative parity states were shifted down by 193 keV. The true unshifted energies are given in keV above the corresponding state.

dynamic dipole hypothesis. The deviations of the experimental bands from pure L(L+1) splittings are not relevant to the features which are being emphasized here and therefore the SU(3) symmetry breaking is being ignored. The negative parity states, which result from α-clustering, should γ-decay more strongly to the α-band than to the ground band since the α-band is mostly, but not purely, 1α. Experimentally one finds $B(E1;2_2^- \rightarrow 2_2^+)/B(E1;2_2^- \rightarrow 2_1^+) \approx 5$ which offers strong support for the coexistence hypothesis. Additional support has also been presented[8] for the strong coupling limit in which $K^\pi = 0^-, 1^-$ bands form.

An estimate of the B(E1) for the $n_p = 1$ to $n_p = 0$ transition assuming two-protons and two-neutrons all moving in phase with an RMS amplitude approximately equal to the radius of ^{156}Gd yields $B(E1;1^- \rightarrow 0^+) \approx 4 \times 10^{-2}$ W.u. The corresponding value determined empirically with the α-cluster NVM is $3 \pm 1 \times 10^{-2}$ W.u. which strongly supports the α-cluster-like picture.[9]

FUTURE DIRECTIONS

Additional Experimental Tests

The best test of the coexistence hypothesis would be to determine the experimental ratio $B(E1;0_2^+ \rightarrow 1_1^-)/B(E1;0_1^+ \rightarrow 1_1^-)$ for the ^{224}Ra nucleus. This is where many of the simplifying assumptions are expected to be most valid. If such ratios prove to be too difficult to obtain experimentally, the ratio $B(E1;1_2^- \rightarrow 0_2^+)/B(E1;1_2^- \rightarrow 0_1^+)$ in ^{224}Ra and/or ^{226}Ra should also provide an important test.

A Survey of Mass Regions

A survey of various nuclear mass regions is in progress in an attempt to identify 1α configurations from the published data. It is of interest to assess whether this is a general phenomenon which occurs in many mass regions or not. The α-bands have been located in many actinide nuclei[8] and in the ^{156}Gd nucleus.[9] Additional candidates for α-bands are the $K^{\pi} = 0^+$ bands at 1275 keV in ^{160}Dy and 1246 keV in ^{164}Er. A more detailed analysis of these nuclei, based on the α-cluster NVM, is in progress. It is

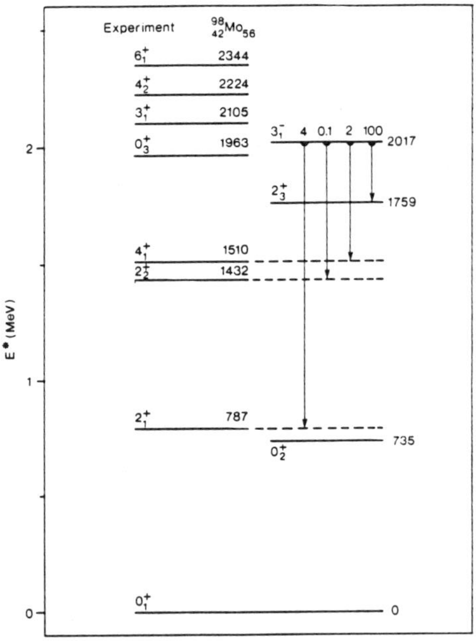

FIGURE 2. The experimental energy levels[12] for ^{98}Mo are given in keV above each state. The 0_2^+, 2_3^+ and 3_1^- states displaced to the right in the figure are suggested to be members of the 1α configuration while the rest are 0α states. The numbers above the 3_1^- states are B(E1) values relative to $B(E1;3_1^- \rightarrow 2_3^+) \equiv 100$.

hoped that experimental studies of these nuclei will also be undertaken. In addition to the actinide and rare earth regions there are some very interesting features of the mass A ≈ 100 region (see Figure 2).

Alpha-Particle Knockout Reactions

The (p,α) reaction[13] appears to proceed via a knockout mechanism in many nuclei.[14] Preliminary studies suggest that the α-cluster NVM may prove to be very useful in extracting nuclear structure information from such experiments.

An Optical Model Potential

A more macroscopic approach, on the nuclear scale, to the NVM type of clustering is being developed in terms of the collective a-oscillations of Satchler.[15] Low energy monopole[16] and dipole[17] a-oscillations have been observed experimentally; therefore, it is plausible that such concepts may be useful in approximating clustering effects in heavy nuclei.

Dipole Versus Octupole Modes

The dipole and octupole interpretations are not mutually exclusive, but there could be differences if one multipolarity predominates over the other. Differences between the dipole and octupole phonon models were discussed in references 8 and 9. It has been suggested, based on some theoretical considerations, that nuclei in the ^{224}Ra mass region may have static octupole deformed ground states.[5] If one recalls that the onset of quadrupole deformation results in B(E2)'s which are

enhanced by at least an order of magnitude relative to the corresponding collective phonon values, then one should expect to see a similar strong enhancement of the B(E3)'s over the collective octupole phonon value as a consequence of octupole deformation.[18] This is therefore a crucial experimental test of whether or not some actinide nuclei do indeed have static octupole deformed ground states.

The α-cluster NVM does not include an octupole mode directly. The model could be extended to include an f*-boson corresponding to an $L = 3^-$ a-oscillation and/or an f-boson[10] (R-oscillation) if there is experimental evidence that demands it. However, at present the coupling of the dipole and quadrupole modes is in best agreement with the data in the regions that have been studied to date.[8,9,11] The question of how important f*- and/or f-bosons are is still an open one which is being investigated.

CONCLUSIONS

The fundamental assumptions of the α-cluster NVM were presented and it was argued that there is experimental support for those assumptions. An attempt was made to highlight the differences between this model and other models, and additional experimental tests were suggested. It was argued that there is experimental support for an α-particle-like interpretation of the dipole mode. Efforts are still being made to determine the limitations of this new model which seems to have a fairly large range of applicability. It was suggested that this model may prove to be very useful in the analysis of the data from various

types of reactions. Experimentalists and theorists are now both left with several challenges. Due to the successful phenomenological stage of this research, a more microscopic understanding of the model is now very desirable.

ACKNOWLEDGEMENTS

I have had the pleasure of collaborating with B.R. Barrett, R. Bonetti, M. Gai, F. Iachello and M.A. Nagarajan on various aspects of this project. I would also like to thank D. Goutte and G. Molnar for interesting discussions concerning new applications, and T. Otsuka for informing me of his very interesting microscopic p-boson calculations prior to publication.

REFERENCES

1. K. Neergard and P. Vogel, Nucl. Phys., A149, 209 (1970); K. Neergard and P. Vogel, ibid., 217.
2. W. Kurcewicz et al., Nucl. Phys., A304, 77 (1978).
3. R.F. Casten et al., Phys. Lett., 40B, 333 (1972).
4. F. Iachello and A.D. Jackson, Phys. Lett., 108B, 151 (1982).
5. A. Gyurkovich et al., Phys. Lett., 105B, 95 (1981); G.A. Leander et al., Nucl. Phys., A388, 452 (1982); G.A. Leander and R.K. Sheline, Nucl. Phys., A413, 375 (1984).
6. H.J. Daley and F. Iachello, Ann. Phys., 167, 181 (1986).
7. H. Daley and F. Iachello, Phys. Lett., 131B, 281 (1983).
8. H.J. Daley and B.R. Barrett, Nucl. Phys., A449, 256 (1986); H.J. Daley, Proc. Conference on Nuclear Structure with Heavy Ions, Legnaro, 27-31 May 1985, ed. R.A. Ricci and C. Villi (Italian Physical Society, Bologna, Italy 1986) p.129.

9. H.J. Daley and M.A. Nagarajan, Phys. Lett., 166B, 379 (1986).
10. O. Scholten, F. Iachello and A. Arima, Ann. Phys., 115, 325 (1978).
11. H.J. Daley, Ph.D. Thesis, University of Arizona (1984), unpublished.
12. R.A. Meyer et al., Phys. Rev., C29, 1839 (1984).
13. P.E. Hodgson, Proc. Second La Rabida Summer School on Nuclear Physics, La Rabida, Spain, 24 June - 5 July 1985, ed. M. Lozano and G. Madurga (World Scientific, Singapore 1985), p.439.
14. R. Bonetti and L. Colli-Milazzo, Phys. Rev., C32, 1869 (1985).
15. G.R. Satchler, Nucl. Phys., A100, 481 (1967).
16. H.P. Morsch and P. Decowski, Phys. Lett., 82B, 1 (1979).
17. P. Decowski, H.P. Morsch and W. Benenson, Phys. Lett., 101B, 147 (1981).
18. K. Böning et al., Phys. Lett., 161B, 231 (1985).

DESCRIPTION OF COLLECTIVE MAGNETIC DIPOLE MODE IN TERMS OF THE SDG IBM

YOSHIMI AKIYAMA
Physics Department, College of Humanities and Sciences, Nihon University, Tokyo 156, Japan

Abstract A general account of the magnetic dipole mode in terms of the sdg interacting boson model-2 is given. Three states of $I^{\pi}=1^{+}$ are predicted for the magnetic dipole excitations from the ground state. Measurements of the electromagnetic decay of these states to various $I^{\pi}=0^{+}$ states are pointed out to be valuable to establish this theoretical description.

INTRODUCTION

The collective magnetic dipole (M1) mode in strongly deformed nuclei has been successfully interpreted[1-2] by means of the interacting boson model-2 (IBM-2), which distinguishes the proton boson from the neutron boson. The original form of the IBM-2, which takes only the s- and d-bosons, can be generalized to include higher bosons. With this generalization, a straightforward extension[3-4] of the sd description of the M1 mode can be made, but this is not all that can be done. There appears a new possibility,[5] under which the M1 excitation is symmetric in the proton and neutron bosons. Therefore, this new type of the M1 mode persists within the framework of the IBM-1, where the proton and neutron bosons are not distinguished each other. This M1 mode is explained in the framework of the IBM-2 with the s-, d- and g-bosons (sdg IBM-2), where the traditional M1 description itself is seen to split into two cases. Three collective M1 states are predicted here. According to recent experimental information,[6] there exist more than one M1 states in some nuclei. The sdg IBM-2 theory may open a possibility to theoretically understand this experimental finding.

M1 OPERATOR

The M1 transition operators are assumed in the first approximation to be one-body operators in bosons. They are tensors of rank one. There are four tensor operators of rank one in the sdg IBM-2, which are classified according to their SU(2) ×U(15) tensor characters. The group SU(2) describes the proton-neutron degree of freedom, and the group chain U(15) ⊃SU(3) is used to describe the s-, d- and g-bosons.

The creation and annihilation operators of the s-, d- and g-bosons (l=0,2,4) are denoted as

$$b^{\dagger}_{slm} \quad \text{and} \quad \tilde{b}_{-sl-m}=(-1)^{\frac{1}{2}-s+l-m}b_{slm}, \tag{1}$$

where $s=+\frac{1}{2}$ and $-\frac{1}{2}$ stand for the proton and neutron bosons. The one-body SU(2)×U(15) tensor operator is defined by

$$T(F_0s_0(\lambda_0\mu_0)K_0L_0M_0)=(b^{\dagger}\tilde{b})^{F_0s_0(\lambda_0\mu_0)K_0L_0M_0}, \tag{2}$$

where the tensor characters in SU(2) are expressed by the F-spin, F_0=0 and 1, and its projection s_0. The SU(3) tensor character is denoted by $(\lambda_0\mu_0)$. For the rank-one tensor, L_0=1, $(\lambda_0\mu_0)$=(1,1) and (3,3), and K_0=1. For our later use, explicit reference to s_0 and M_0 is not necessary. The SU(2) ×U(15)tensor operator relevant to the M1 transition is simply denoted as $T(F_0(\lambda_0\mu_0))$ hereafter. The operator T(0(1,1)) is just the angular momentum operator, and does not induce the M1 transitions. There are one F-scalar and two F-vector operators for the M1 transition in the sdg IBM-2 description. The F-scalar one can give rise to the M1 transition also within the framework of the sdg IBM-1. Its SU(3) tensor character is (3,3). The two F-vectors are classified as the (1,1) and (3,3) tensors. The (1,1) operator exactly corresponds to the M1 operator in the sd IBM-2 description, but the two (3,3) operators do not have their counterparts in the sd IBM-2. Classification of these M1 operators is shown in Table I. A general form of the M1 transition operator is expressed by a linear combination of these three:

$$\underline{\underline{T}} = a\, T(0(3,3)) + b\, T(1(1,1)) + c\, T(1(3,3)). \tag{3}$$

$I^{\pi}=1^{+}$ STATE

The states of N bosons are classified by the product group SU(2)×U(15). Because of the Bose-Einstein statistics, each state has the same Young diagrams for the SU(2) and SU(15) symmetries. Therefore, one Young pattern label [G] is enough to specify the symmetry of the bosons. The IBM-1 description

TABLE I Classification of the M1 operators and $I^{\pi}=1^{+}$ states excited from the ground state ([G]=[N] and ($\lambda\mu$)=($4N$,0)).

M1 operator F_0	($\lambda_0\mu_0$)	State excited [G]	($\lambda\mu$)
0	(3,3)	[N]	($4N$-6,3)
1	(1,1)	[N-1,1]	($4N$-2,1)
1	(3,3)	[N-1,1]	($4N$-6,3)*

*This representation appears twice in the [N-1,1] symmetry. The first one is defined as the one whose I=1 state is given by operating the M1 operator to the ground state. Then the other is free of the M1 excitation.

can be obtained by restricting to [G]=[N]. Further classifications are made by using the group reduction chain SU(15)$\supset$ SU(3)$\supset$O(3). The ground state of heavily deformed nuclei belongs to the [N] and ($4N$,0) representations of the SU(15) and SU(3) groups in the SU(3) limit. With the M1 operators listed in Table I, the selection rules given by the SU(2) and SU(3) groups predict the $I^{\pi}=1^{+}$ states, excited from the ground state, as also listed in Table I. If $N \geq 4$, two states labelled as ($4N$-6,3) appear in the [N-1,1] symmetry due to the SU(3)-degeneracy. One of them is defined as an M1 state, which is produced by operating the F_0=1, ($\lambda_0\mu_0$)=(3,3) tensor operator to the ground state. Then the other is free of the M1 excitation. Thus, the M1 states correspond one to one to the M1 operators defined in the preceding section.

Theoretically, one of the important points is the prediction of energies of the M1 states. To get an idea, let us assume an SU(3)-preserving Hamiltonian

$$H = x\,C + y\,L^2 + z\,M, \qquad (4)$$

where the C is the Casimir operator of SU(3) and the L is the angular momentum operator. The M is the Majorana force, whose eigenvalue depends only upon the symmetry [G]. Without the Majorana force, the SU(3) Casimir operator determines the sequence of the SU(3) bands. Then the lowest I=1 state belongs to the [N-1,1], ($4N$-2,1) representation in the sdg description, while it belongs to the SU(6) symmetry [N-1,1] and the SU(3) symmetry ($2N$-2,1) in the sd one. In either case, it is located in between the band-heads of the ground and β (or γ) bands. A rather high excitation energy of the M1 state can be interpreted as due to the strong Majorana com-

ponent of the boson interaction in both of the IBM-2 descriptions. With such a strong Majorana force in the sdg IBM-2, however, the lowest I=1 state would be the one with $[N]$ and $(4N-6,3)$, whose energy is about 1.5 times that of the band-head of the β or γ band according to the Hamiltonian (4). Experimentally the collective M1 state is not observed at such a low excitation energy, and this might be taken as an objection to the sdg IBM. By introducing SU(3)-breaking interactions into the Hamiltonian, however, the sdg IBM allows much more flexible descriptions of the energy bands compared to the sd IBM. Indeed, recent calculations[5] of ^{168}Er in terms of the sdg IBM-1 predict an I=1 state at E_x=3.8 MeV. This state contains 61 % of the (58,3) K=1 component. Therefore, here appears a new possibility of identifying this state to an experimentally observed level of strong M1 strength.

An important problem is to estimate the strength of the M1 excitations. Unfortunately precise information on the M1 transition operator is not available yet. Here we simply assume that the coefficients a, b and c in eq.(3) are all equal to unity. Then the B(M1) values for the excitation of the [16] (58,3), [15,1] (62,1) and [15,1] (58,3) states in ^{168}Er are 3.5, 0.9 and 1.7, respectively. Thus these three types of excitation are all important from the phenomenological point of view. A microscopic study of the M1 interaction would be very valuable in this respect.

M1 DE-EXCITATION

If the electromagnetic decay of the M1 states can be observed, one may obtain much deeper understanding of the M1 mode. The transitions from the M1 states to various $I^\pi=0^+$ states are of particular interest, because they are prohibited in the E2 transition. Therefore, one can expect that rather precise experimental information concerning the M1 interaction might be obtained. A theoretical consideration follows.

The operator for the M1 excitation, given by eq.(3), also determines the M1 decay which we now consider. We will discuss the effects of each term in it separately. The initial state with I_i=1 belongs to the SU(3) representation $(\lambda_i\mu_i)$, which is equal to $(4N-2,1)$ or $(4N-6,3)$. The final state with I_f=0 is assumed to belong to the symmetry $[N]$, and further to the SU(3) representation $(\lambda_f\mu_f)$, which is restricted to $\lambda_f+2\mu_f=4N$ for the following consideration. Selection rules are predicted as shown in Table II. The transition from the $[G_i]=[N]$ state to the $[G_f]=[N]$ states with the F_0=0, (3,3) operator can also be realized within the framework of the sdg IBM-1.

TABLE II Selection rules for the de-excitation of the M1 states.

M1 tensor F_0 $(\lambda_0\mu_0)$	Initial state $[G_i]$ $(\lambda_i\mu_i)$	Final state, $[G_f]=[N]$, $(\lambda_f\mu_f)$ $(4N,0)$	$(4N-4,2)$	$(4N-8,4)^2$	$(4N-12,6)^3$
0 (3,3)	$[N]$ $(4N-6,3)$	×	×	×	×
	$[N-1,1]$ $(4N-2,1)$		Not allowed		
	$[N-1,1]$ $(4N-6,3)$		Not allowed		
1 (1,1)	$[N]$ $(4N-6,3)$		×	×	
	$[N-1,1]$ $(4N-2,1)$	×	×		
	$[N-1,1]$ $(4N-6,3)$		×	×	
1 (3,3)	$[N]$ $(4N-6,3)$	×	×	×	×
	$[N-1,1]$ $(4N-2,1)$	×	×	×	
	$[N-1,1]$ $(4N-6,3)$	×	×	×	×

It is worthwhile to also consider the sd IBM-2 description of the M1 decay. Here only one M1 state exists, belonging to the $[N-1,1]$, $(2N-2,1)$ representation. Its M1 decay to the $I_f=0$ states can be possible only when $(\lambda_f\mu_f)=(2N,0)$ or $(2N-4,2)$ because the M1 operator is restricted to the $(\lambda_0\mu_0)=(1,1)$ one. Therefore, one obtains an important conclusion that with only s- and d-bosons the $I^\pi=0^+$ states fed from the collective M1 state are restricted to two ones in the SU(3) limit, one belonging to the ground state band and the other to the β band. The same situation can occur in the sdg IBM-2 only when the M1 operator solely consists of the $F_0=1$, (1,1) tensor operator, i.e., $a=c=0$ in eq.(3). In general, many $I^\pi=0^+$ states are expected to be fed, as can be seen from Table II.

One of the most interesting points in the sdg theory is that the $[N]$, $(4N-6,3)$ state is an M1 state under the IBM-1 description. Under this description the B(M1) ratios of the M1 decay are calculated for ^{168}Er and shown in Table III. Apart from the decay to the ground state, a very large B(M1) value is predicted to the second member of the (52,6) states. Unfortunately its excitation energy is expected to be much higher than that of the M1 state, and this transition would not occur. Thus this information can not be used as a test of the sdg IBM-1 description. The decay to the β band is predicted to be weak, its B(M1) value being only 3.3 % of the ground state value. In the sd IBM-2 description the corresponding ratio is 3.6 %, which is rather very close to the sdg IBM-1 prediction given here. The B(M1)'s to the (56,4)

TABLE III *B*(M1) ratios for the decay of the M1 state of ^{168}Er in the sdg IBM-1 description. The *B*(M1) to the ground state is normalized to 100.

$(\lambda_f\mu_f)$	(64,0)	(60,2)	(56,4)		(52,6)		
			w=0	w=1	w=0	w=0	w=1
B(M1)	100	3.3	0.9	0.3	0.2	85.3	0.0

states in the sdg IBM-1 are vanishingly small. Thus the predictions of these two approaches for the M1 decay are surprisingly similar to each other. However, it should be noticed that in the third approach, i.e., the sdg IBM-2, different predictions can be made, because then all the three terms in eq.(3) contribute to the M1 strengths, as can be seen from Table II. If experimental *B*(M1)'s for the β or (4*N*-8,4) bands are much different from those estimated here, it might be a strong support for the sdg IBM-2 description.

CONCLUDING REMARK

We have seen that the prediction of the M1 mode in terms of the sdg IBM-2 is very much different from that of the sd IBM-2. Three M1 states are predicted by the sdg IBM-2, while only one state appears in the sd IBM-2. It is interesting to note that Bohle et al.[6] report that two strong M1 peaks are observed in (e,e') experiments of ^{164}Dy and ^{174}Yb. This information seems to suggest that the collective M1 state is not single, in contrast to the sd IBM-2 prediction. More detailed studies, both experimental and theoretical, are needed to justify the sdg IBM-2 interpretation.

The M1 transition operator consists of the three tensor operators. In other words, there are three parameters *a*, *b* and *c* in eq.(3) to specify the M1 interactions. Determination of these three is the central problem in the study of the collective M1 mode. Microscopic studies of the M1 interactions are awaited. The possibility of the [*N*], (4*N*-6,3) state as an M1 state depends on its M1 excitation strength. If g-factors of the d- and g bosons are exactly equal, this strength vanishes. Measurement of the decay properties of the M1 states proposed here could provide us with the best experimental test of the theory developed here. If this kind of experiments is carried out, we can expect to determine the three parameters directly from the experimental data.

REFERENCES

1. F. Iachello, Nucl. Phys., A358, 89c (1981).
2. A.E.L. Dieperink, Prog. Part. Nucl. Phys., 9, 121 (1983).
3. S. Pittel, J. Dukelsky, R.P.J. Perazzo and H.M. Sofia, Phys. Lett., 144B, 145 (1984).
4. B.R. Barrett and P. Halse, Phys. Lett., 155B, 133 (1985).
5. N. Yoshinaga, Y. Akiyama and A. Arima, Phys. Rev. Lett., 56, 1116 (1986).
6. D. Bohle, G. Küchler, A. Richter and W. Steffen, Phys. Lett., 148B, 260 (1984).

PHOTON SCATTERING IN THE REGION OF GIANT DIPOLE RESONANCE

S. TURRINI
Dipartimento di Fisica dell'Università,
Via Irnerio 46, I-40126 Bologna, ITALY

G. MAINO and A. VENTURA
ENEA, Divisione Fisica e Calcolo Scientifico,
Via Mazzini 2, I-40138 Bologna, ITALY

Abstract Photon scattering by non-magic even-even nuclei is investigated within the framework of the interacting boson model for the excitation of the giant dipole resonance. Elastic scattering includes the Delbrück contribution at the lowest perturbative order, in addition to the classical Thomson term.

INTRODUCTION

The interacting boson model (IBM)[1] has been extended in recent years to the treatment of collective particle-hole excitations across a shell closure, particularly the giant dipole resonance (GDR), where calculations have been carried out at the IBM limit symmetries[2-4] and in transitional regions[5-8]. The IBM parameters are so adjusted as to reproduce the total photoabsorption cross section, where the GDR is clearly observed. Inelastic photon scattering[3,7,8] can be evaluated without further adjustments.

As for the elastic scattering, two non-nuclear contributions play an important rôle in this energy region: the classical Thomson scattering by a point-like charge equal to the nuclear charge, Ze, and the radiative correction due to vacuum polarization effects, known as Delbrück scattering[9].

The purpose of the present work is to emphasize the rôle of non-nuclear effects in elastic scattering and to comment on inelastic scattering as a test of nuclear structure models.

THE NUCLEAR MODEL

The excitation of a GDR and its coupling to low-energy collective modes is described in IBM language by means of a P boson, with spin-parity $J^{\pi} = 1^-$, coupled to the usual s and d bosons responsible for low-lying positive-parity states. The s-d-P Hamiltonian has been discussed by a number of authors[2-8] and is merely reproduced here without further comment:

$$\hat{H} = \hat{H}(s,d) + E_P \hat{n}_p + b^{(0)}[(P^+ x\tilde{P})^{(0)}(d^+ x\tilde{d})^{(0)}]^{(0)} + b^{(1)}[(P^+ x\tilde{P})^{(1)}(d^+ x\tilde{d})^{(1)}]^{(0)} + b^{(2)}[(P^+ x\tilde{P})^{(2)} [(d^+ x\tilde{s} + s^+ x\tilde{d})^{(2)} + \chi(d^+ x\tilde{d})^{(2)}]]^{(0)}. \quad (1)$$

Its diagonalization by means of the basis states $|\Psi\rangle = (s^+)^{n_s}(d^+)^{n_d}(P^+)^{n_P}|0\rangle$ yields the energies, E_k, of the GDR components. The E1 transition from the m-th low-lying state of spin-parity I_m^+ to the k-th 1^- GDR state is given by the reduced matrix element of the dipole operator, $\hat{T}(E1)$:

$$\langle 1_k^- \| \hat{T}(E1) \| I_m^+ \rangle = \alpha_1 \langle 1_k^- \| (P^+ + P) \| I_m^+ \rangle. \quad (2)$$

The GDR widths, $\Gamma_k = \Gamma(E_k)$, which cannot be evaluated in the frame of the model, are assumed to depend on energy according to the power law $\Gamma(E_k) = \kappa E_k^{\gamma}$ [5,7,8]. The P-boson parameters in (1), the effective boson charge, α_1, in (2) and the coefficients of Γ are so adjusted as to reproduce the experimental photoabsorption cross-section in the GDR region.

Inelastic scattering and nuclear contribution to the elastic scattering can thus be evaluated without adjustable parameters, thus providing a stringent test of the nuclear structure model, since the Thomson and Delbrück amplitudes have been calculated and coherently added.

ELASTIC SCATTERING

The scattering cross sections are computed in the long wavelength limit, $R/\lambda\!\!\!^{-} \ll 1$, where R is the nuclear radius and $\lambda\!\!\!^{-}$ the wavelength of incident radiation.

In the elastic case, the well-known formulae[10] containing Thomson and nuclear contributions are modified so as to include the Delbrück effect. By letting λ (λ') =

± 1 be the circular polarization of the incident (scattered) photons and $K_{00}^{\lambda\lambda'}$ the total amplitude for elastic photon scattering by an even-even nucleus, the differential cross section for unpolarized photons can be written as:

$$\frac{d\sigma_{el}}{d\Omega}(\omega,\theta) = \frac{1}{2}\sum_{\lambda\lambda'} |K_{00}^{\lambda\lambda'}|^2 =$$

$$\frac{1}{2}\sum_{\lambda\lambda'} \{ |A^D_{\lambda\lambda'}(\omega,\theta)|^2 +$$

$$|A^T + A^N(\omega)|^2 \cdot [\tfrac{1}{4}(1+\cos^2\theta) + \tfrac{1}{2}\lambda\lambda'\cos\theta] + A^{DNT}(\lambda,\lambda',\omega,\theta)\}. \qquad (3)$$

Here, $A^D_{\lambda\lambda'}(\omega,\theta)$ is the Delbrück amplitude for photons of frequency ω scattered at angle θ, evaluated as in ref.[11]. A^T is the usual Thomson amplitude, $A^T = -Z^2e^2/(AMc^2)$, where AM is the nuclear mass. The nuclear, or Raman amplitude, A^N, reads:

$$A^N(\omega) = \frac{1}{3}\frac{(\hbar\omega)^2}{\hbar^2c^2}\sum_n |\langle 1_n^- \| \hat{T}(E1) \| 0_1^+\rangle|^2 \cdot [\frac{1}{E_n-\hbar\omega-i\Gamma_n/2} +$$

$$\frac{1}{E_n+\hbar\omega+i\Gamma_n/2}] . \qquad (4)$$

Finally, A^{DNT} is the interference of Delbrück, Raman and Thomson terms:

$$A^{DNT}(\lambda, \lambda', \omega,\theta) = 2\,[\mathrm{Re}\,A^D_{\lambda\lambda'}(\omega,\theta)\,[A^T + \mathrm{Re}\,A^N(\omega)] -$$

$$\mathrm{Im}\,A^D_{\lambda\lambda'}(\omega,\theta) \cdot \mathrm{Im}\,A^N(\omega)]\, d^1_{-\lambda,-\lambda'}(0,\theta,0). \qquad (5)$$

Here, $d^1_{-\lambda,-\lambda'}$ is a Wigner rotation matrix, with elements:

$$d^1_{1,1} = d^1_{-1,-1} = \cos^2(\theta/2)\ ;\ d^1_{1,-1} = d^1_{-1,1} = \sin^2(\theta/2).$$

INELASTIC SCATTERING

Let ω' be the frequency of the emitted photon leaving the nucleus in an excited state of spin-parity I_f^π. The

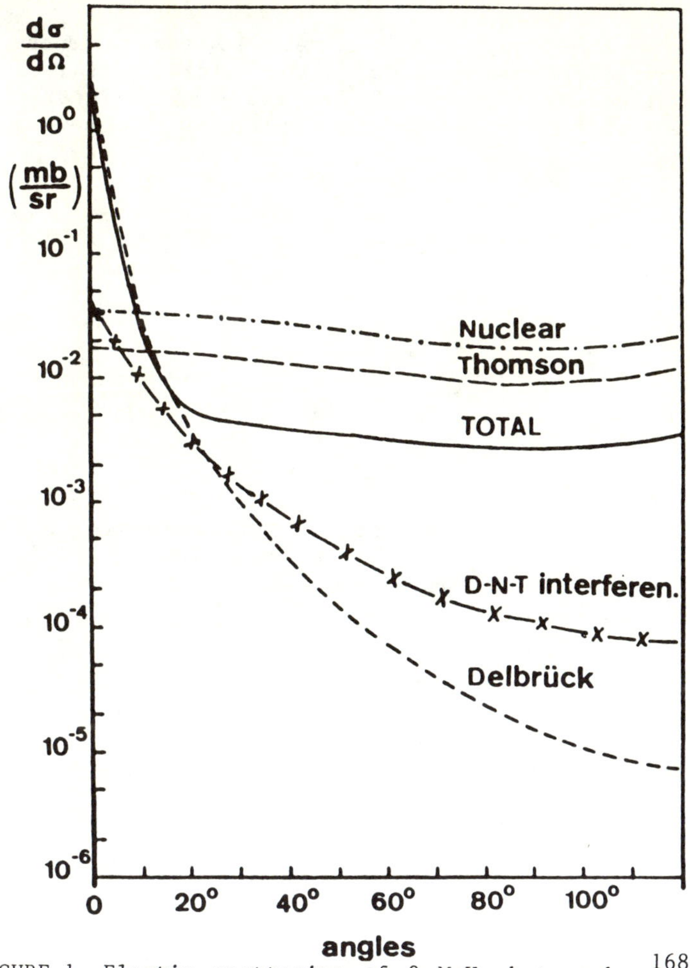

FIGURE 1. Elastic scattering of 9 MeV photons by ^{168}Er versus scattering angle.

differential cross section for unpolarized photons is written in the form[10]:

$$\frac{d\sigma_{in}}{d\Omega}(\omega, \omega', \theta) = \frac{\omega'}{\omega} \mid P_J \mid^2 \delta_{JI_f} \, g_J(\theta), \qquad (6)$$

$$P_J = \frac{e^2}{[3(2I_f+1)]^{\frac{1}{2}}} \frac{\omega\omega'}{c^2} \sum_n <I_f^+ \| \hat{T}(E1) \| 1_n^- ><1_n^- \| \hat{T}(E1) \| 0_1^+ > x$$

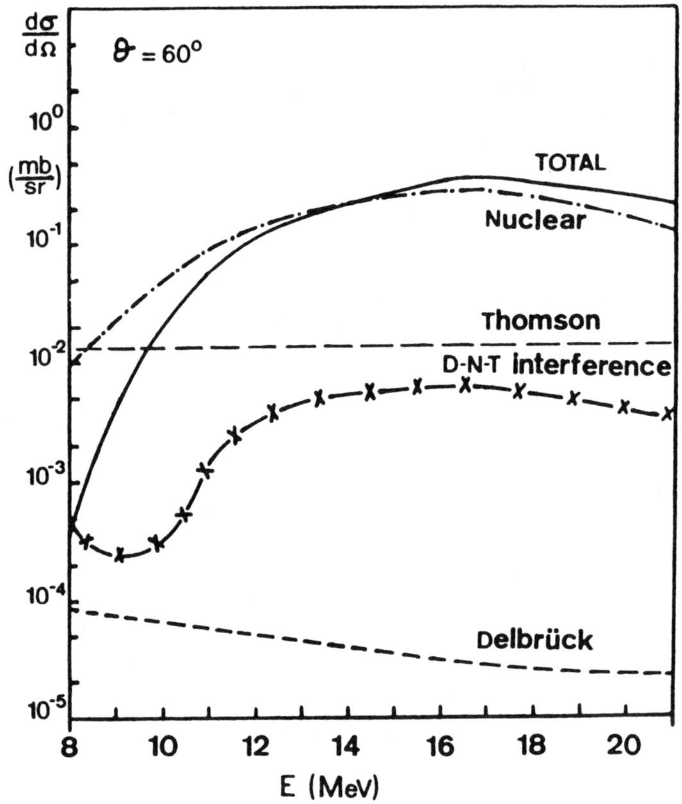

Figure 2. Elastic scattering at $\theta=60°$ versus incident photon energy (MeV).

$$\left[\frac{1}{E_n+\hbar\omega'+i\Gamma_n/2}+\frac{(-1)^J}{E_n-\hbar\omega-i\Gamma_n/2}\right] , \qquad (7)$$

is the nuclear polarizability; the angular distribution, $g_J(\theta)$, is given here in the cases J = 0 and J = 2:

$$g_0(\theta) = \frac{1}{6}(1 + \cos^2\theta) \ ; \ \ g_2(\theta) = \frac{1}{12}(13 + \cos^2\theta). \qquad (8)$$

RESULTS AND COMMENTS

We present here a few results on photon scattering by ^{168}Er, a strongly deformed nucleus already considered in ref. 3 in the SU(3) limit of the model. The IBM parameters adopted in the present work and listed in Table I allow us to fit the experimental photoabsorption[12] cross section with good accuracy and predict elastic and inelastic scattering cross sections.

Since the elastic cross section has already been evaluated in ref.[3] at large scattering angle, where the Delbrück effect is negligible, we present here calculations at small and intermediate angles. Fig. 1 shows the elastic cross section, its three components and the absolute value of the interference term (whose sign is negative) as functions of scattering angle at fixed photon energy, $\hbar\omega$ = 9 MeV, while Fig. 2 gives the same quantities versus photon energy at fixed scattering angle, $\theta = 60°$. In this case the main effect of the Delbrück amplitude is to reduce the total cross section, owing to the interference term.

TABLE I IBM parameters

$a_0^{(*)}$=30.0 keV ;	$a_1^{(*)}$=10.5 keV ;	$a_2^{(*)}$=-8.0 keV ;
$\chi = -\sqrt{7}/2$;	E_p = 14.05 MeV ;	$b^{(0)}$ = 0. ;
$b^{(1)}$ = 0.50 MeV ;	$b^{(2)}$ = 0.28 MeV ;	α_1 = 8.82 e.fm ;
κ = 0.045 MeV$^{1-\gamma}$;	γ = 1.7 .	

(*) Parameters of the Hamiltonian $\hat{H}(s,d)$[1].

As for inelastic scattering, our results are in reasonable agreement with ref.[3], so that it is not necessary to reproduce them here. Comparison with the experimental results of ref.[13] seems to encourage further applications of the IBM to analysis of photon scattering in the GDR region. Accurate calculation of the Delbrück contribution to the elastic channel and particularly of its interference with nuclear and Thomson amplitudes, as described in Section 3., is needed for suitable comparison with experimental information at small and intermediate angles.

Useful conversations with prof. B. De Tollis are gratefully acknowledged.

REFERENCES

1. For a recent review see: A. Arima and F. Iachello, Adv. Nucl Phys., 13, 139 (1984).
2. I. Morrison and J. Weise, J. Phys., G8, 687 (1982).
3. F.G. Scholtz and F.J.W. Hahne, Phys. Lett., 123B, 147 (1983).
4. D.J. Rowe and F. Iachello, Phys. Lett., 130B, 231 (1983).
5. G. Maino, A. Ventura, L. Zuffi and F. Iachello, Phys. Rev., C30, 2101 (1984).
6. F.G. Scholtz, Phys Lett., 151B, 87 (1985).
7. G. Maino, A. Ventura, L. Zuffi and F. Iachello, Phys. Lett., 152B, 17 (1985).
8. G. Maino, A. Ventura, P. Van Isacker and L. Zuffi, Phys. Rev., C33, 1089 (1986).
9. V. Costantini, B. De Tollis and G. Pistoni, Nuovo Cimento, A2, 733 (1971).
10. J.M. Eisenberg and W. Greiner, Nuclear Theory II: Excitation Mechanisms of the Nucleus, (North-Holland Publ. Co., 1970), p. 111 ff.
11. B. De Tollis and L. Luminari, Nuovo Cim., A81, 633 (1984).
12. G.M. Gurevich, L.E. Lazareva, V.M. Mazur, S. Yu. Merkulov, G.V. Solodukhov and V.A. Tyutin, Nucl. Phys., A351, 257 (1981).
13. A.M. Nathan and R. Moreh, Phys. Lett., 91B, 38 (1980).

DIRECT vs STATISTICAL DECAY OF NUCLEAR GIANT MULTIPOLE RESONANCES*

MAHIR S. HUSSEIN
Instituto de Física da Universidade de São Paulo,
Caixa Postal 20516 - 01498 - São Paulo, SP - BRASIL

Abstract A theoretical framework for the description of the decay of giant multipole resonances is developed. Besides the direct decay, both the pre-equilibrium and statistical (compound) decays are taken into account in a consistent way. It is shown that the statistical decay of the GR is not necessarily correctly described by the Hauser-Feshbach theory owing to the presence of a mixing parameter, which measures the degree of fragmentation. Applications are made to several cases.

The study of the decay properties of giant multipole resonances (GR) is of paramount importance for the unraveling of their dynamical, microscopic structure. Since giant resonances are located at high excitation energies, they mainly decay by particle emission. Treated as isolated resonances, the GR are characterized by a total average width composed of two pieces: the "escape width", $\Gamma^{\uparrow}$, which represents the coupling of the GR to the continuum, and the spreading width, $\Gamma^{\downarrow}$, that measures the degree of fragmentation of the strengh due to coupling to complex intrinsic nuclear configurations (e.g. 2p-2h)[1,2]. Of course, whereas the first stage of the reaction, namely the giant resonance population, is a very coherent process, in which 1-particle 1-hole configurations act in phase, the other, more complicated stages, are complex enough to call for a statistical treatment.

* Supported in part by the CNPq and the ICTP, Trieste.

It has so far been a common practice to analyze the particle spectra originating from the decay of GR with one of two extreme models, which ignore completely the intermediate, pre-equilibrium stages[3,4]. These models either assume the dominance of $\Gamma^{\uparrow}$, namely the GR decays predominantly "directly", or the predominance of $\Gamma_{H.F.}$, which imlies necessarily that the fragmentation of the resonance into the complex background is complete. In this last case the Hauser-Feshbach theory is utilized in the analysis[4].

Recently a combination of both decay mechanisms has been suggested by Beene et al.[5] in their analysis of the gamma decay of the giant quadrupole resonance of ^{208}Pb. These authors write for the gamma branching ratio P, the following

$$P = P_d + P_c = \frac{\Gamma^{\gamma}_{GR}}{\Gamma^{\downarrow}} + \left\langle\frac{\Gamma^{\gamma}_{c}}{\Gamma_c}\right\rangle \tag{1}$$

The notation in Eq. (1) is obvious. After making the reasonable approximation $\langle\Gamma^{\gamma}_{c}/\Gamma_{c}\rangle \cong \langle\Gamma^{\gamma}_{c}\rangle/\langle\Gamma_{c}\rangle$, Beene et al. then proceed with the calculation of this term utilizing experimentally deduced neutron strength functions from resonance studies in the system $n+^{207}Pb$. Their aim was the estimation of the compound nucleus average width $\langle\Gamma_c\rangle$ since $\langle\Gamma^{\gamma}_{c}\rangle$ is basically known.

A more convenient approach was adopted by Dias et al.[6] throught the identification

$$2\pi \langle\Gamma_c\rangle \rho_c = \sum_{\alpha} T_{\alpha} \tag{2}$$

where T_{α} is the neutron transmission coefficients, and ρ_c is the compound nucleus density of states. The neutron channel is certainly the dominant one at the excitation

energy considered, namely $E^* = 11$ MeV. The result of the calculation of Ref. 6 is that $\Gamma_c(E^*{=}11\ \mathrm{MeV}, 2^+) \cong 8.5$ keV. This is about an order of magnitude larger than the value obtained by Beene et al.[5].

Irrespective to the different starting points of Beene et al.[5] and Dias et al.[6], their final conclusion comes out basically the same, namely that the compound nucleus contribution to P, namely $P_c \equiv \langle\Gamma_c^\gamma\rangle/\langle\Gamma_c\rangle$ is quite appreciable and corresponds roughly to about 40%. The above example is but one of several showing the necessity of considering both direct and compound decay mechanisms of the GR. In fact another case studied recently by our group[7], namely the fission decay probabilities of the giant quadrupole and monopole resonances in ^{236}U through the (γ,f) reaction, also exhibit this feature at E_γ above the fission threshold.

The above findings motivated us to develop a detailed theory of the decay of nuclear giant resonances. Such a theory[8] goes beyond the one underlying Eq. (1) in the sense that we take P_G and P_c as not completely independent due to the presence of a mixing parameter which measures the degrees of fragmentation of the GR into the complex compound nucleus background. Unitarity as well as some basic ideas borrowed from pre-equilibrium studies[9] have been used to obtain the major result of our theory namely the following expression for the energy averaged partial cross section.

$$\bar{\sigma}_{ab} = \bar{\sigma}^G_{ab} + \mu\tau^G_a \frac{\tau^C_b + \mu\tau^G_b}{\Sigma_c(\tau^C_c + \mu\tau^G_c)} \tag{3}$$

where the first term corresponds to the short-time-delay component of the process, usually called the "direct" contribution, and the second being the long-time-delay compo-

nent, invariably called the "compound" contribution. This compound contribution is clearly different from the usual Hauser-Feshbach form, owing to the presence of the mixing parameter μ and two kinds of transmission coefficients; the GR one, τ^G, and the genuine compound one, τ^C. Dividing by τ_a^G, we obtain the equivalent to Eq. (1)

$$P = P_d + \mu \frac{\tau_b^C + \mu\tau_b^G}{\sum_c \tau_c^C + \mu\tau_c^G} \qquad (4)$$

Clearly the expression for P_c is more involved than the simple one used earlier for the analysis of the γ-decay of the in ^{208}Pb.

A careful analysis employing Eq.(4), could furnish important information about the mixing parameter μ. As an example we show in Fig. 1 an analysis performed on the neutron decay spectrum from the E0 giant resonance in ^{208}Pb[4].

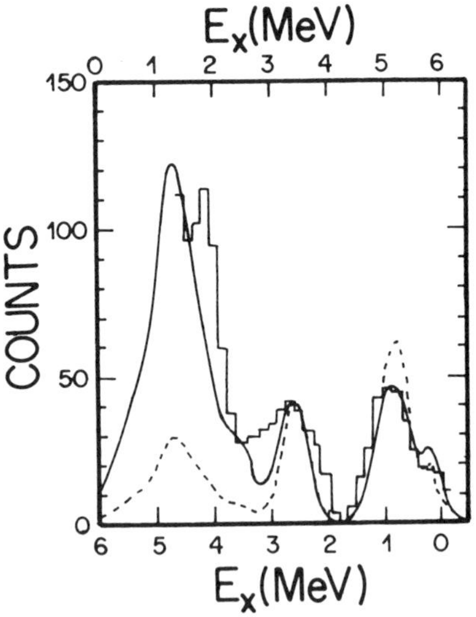

FIGURE 1. The histogram is the measured neutron decay spectrum from the ^{208}Pb (Ref. 4). The two curves shown by the full line (μ=1) and dashed line (μ=0.5) are the predicted spectrum using equation 6 taking into account the resolution of the experiment (500 keV). Each of the 141 neutron groups is represented by a gaussian with FWHM=500 keV (see Ref. 4 for more details). Both spectra (μ=1 and μ=0.5) are normalized to the number of neutrons in the interval between 3-4 MeV.

Two values of μ were considered, $\mu=1$ and $\mu=0.5$. The direct piece of the decay was estimated using the result of Kuchnir et al.[10] and de Haro et al.[11], whereas the statistical piece was calculated in accordance with Eq. (4) using for τ_b^C the Hauser-Feshbach model. Clearly for $\mu=1$ a renormalization of the calculation of Ref. 4 has to be made in order to account for the "indirect" CN decay exemplified by $(\mu=1)\tau_c^G$, whose value was taken to be $2\pi\,\Gamma_b^G\,\rho_G$ with $\rho_G \cong \mathrm{MeV}^{-1}$. It is obvious from the figure that the GMR in ^{208}Pb does not accomodate appreciable direct decay piece since the more likely case is $\mu=1$, in complete agreement with the conclusions of Ref. 4. Analysis of cases which clearly contain both compound and direct contributions are being carried out by our group using Eq. (3).

Before ending, we dwell a little on a possible generalization of Eq. (3) to incorporate the contribution arising from pre-equilibrium emission (e.g. from the 2p-2h stage). This is easily accomplished using the nested doorway approach of Ref. 9. The important new features are that the cross section is now composed of three distinct pieces, and the mixing parameter μ is divided into three terms. Namely[9,12].

$$\bar{\sigma}_{ab}=\sigma_{ab}^G+(1-\mu_2)\mu_1\tau_a^G\,\frac{\tau_b^{Pr}+\mu_1\tau_b^G}{\sum_{c''}(\tau_{c''}^P+\mu_1\tau_{c''}^D)}$$

$$+\,(\mu_1\mu_2+\mu')\tau_a^G\,\frac{\tau_c^C+\mu_2\tau_b^{Pr}+(\mu_1\mu_2+\mu')\tau_b^G}{\sum_c(\tau_c^C+\mu_2\tau_c^{Pr}+(\mu_1\mu_2+\mu')\tau_c^G)} \qquad (5)$$

with

$$\sigma^G_{ab} = (1-\mu_1-\mu')\tau^G_a \frac{\tau^G_b}{\sum_c \tau^G_c}$$

In the above μ_1 measures the mixing of GR with the 2p-2h states, which can be evaluated using the extended RPA approach, μ_2 refers to the mixing of the 2p-2h with the compound nuclear states and μ' refers to the mixing of the GR directly with the compound states, which may be set equal to zero for all practical purposes. The transmission coefficient related to the GR (1p-1h), the pre-equilibrium stage (2p-2h) and the compound stage are called τ^G, τ^{Pr} and τ^C, respectively. It is important to note here that unitarity is preserved both in Eqs. (3) and (5) in the sense that by summing over the final channels b, we obtain

$$\sum_b \bar{\sigma}_{ab} = \tau^G_a \tag{6}$$

irrespective of the detailed nature of the decay.

Using our time-delay arguments again, we identify the third term in Eq. (5) with the "compound" contribution. The calculation of this term is certainly not possible wiht the Hauser-Feshbach model, as one has to differentiate among the three types of transmission coefficients. We are presently studying the feasibility of applying Eq. (5) for data analysis.

REFERENCES

1. A. Van der Woude - Progress in particle and Nuclear Physics, 1986, in press.
2. G.F. Bertsch, P.F. Bortignon and R.A. Broglia - Rev.Mod. Phys. 55 (1983) 287.
3. L.S. Cardman - Nucl.Phys. A354 (1981) 173C.
4. H. Dias and E. Wolynec - Phys.Rev. C30 (1984) 1164.

5. J.R. Beene, G.F. Bertsch, P.F. Bortignon and R.A. Broglia - Phys.Lett. 164B (1985) 19.
6. H. Dias, M.S. Hussein, B.V. Carlson and A.C. Merchant - Phys.Lett. B (1986), in press.
7. H. Dias, J.D.T. Arruda-Neto, B.V. Carlson and M.S. Hussein - IFUSP/P-581 (1986), to be published.
8. H. Dias, M.S. Hussein and S. Adhikari - IFUSP/P-553 (1986), to be published.
9. M.S. Hussein and K.W. McVoy - Phys.Rev.Lett. 43 (1979) 1645.
10. F.T. Kuchnir et al. - Phys.Rev. 161 (1967) 1236.
11. R. de Haro, S. Krewald and J. Speth - Nucl.Phys. A388 (1982) 265.
12. W.A. Friedman, M.S. Hussein, K.W. McVoy and P.A. Mello Phys.Rep. 77 (1981) 47.

Chapter 5

HEAVY-ION INVESTIGATIONS

Moderator:

G. LØHØIDAN
Fysisk Institutt
Allegatan 55
5014 Bergen, Norway

DEPENDENCE OF HEAVY-ION NEUTRON TRANSFER ON DISTANCE AND SPIN*

J.O. RASMUSSEN and M.A. STOYER
Nuclear Science Division, Lawrence Berkeley Laboratory
University of California, Berkeley, California 94720
M.W. GUIDRY
Physics Department, University of Tennessee,
Knoxville, TN 37916 and Oak Ridge National Laboratory,
Oak Ridge, TN 37830
L.F. CANTO and R. DONANGELO
Instituto de Fisica, Universidade Federal do Rio de
Janeiro, Rio de Janeiro, Brazil

Abstract We review the "slope anomaly" data for 2-neutron transfer between heavy ions. Where one or both partners are deformed, the 2-neutron transfer always shows about the same dependence on distance-of-closest approach D as does 1-neutron transfer. Where both partners are spherical, the 2-neutron transfer falls off as the square of the 1-n transfer with one known exception $^{208}Pb(^{64}Ni,^{62}Ni)^{210}Pb$. We propose that the enhancement of 2-n transfer comes from the pairing attraction between the neutrons and that the enhancement requires availability of high-ℓ orbitals to form a tight wave packet in angle. A simple model is derived for numerical estimation. Prospects for testing Coriolis antipairing by heavy ion 2-n transfer are re-examined.

There is an unresolved puzzle regarding the dependence of 2-neutron transfer on distance of closest approach of heavy nuclei. We gave a brief review of the situation for deformed nuclei last year.[1]

*This work was supported in part by the Director, Office of Energy Research, Division of Nuclear Physics of the Office of High Energy and Nuclear Physics of the U.S. Department of Energy under Contract DE-AC03-76SF00098, and in part by a cooperative research grant INT-8302853 of the National Science Foundation and the Brazilian CNPq.

From early Breit-Ebel theory[2] the general logarithmic dependence of 1-neutron transfer cross sections on distance of closest approach D is understood. Indeed the slopes generally agree with the square-barrier WKB tunneling exponent depending on neutron separation energy and neutron mass.

Work of von Oertzen et al.[3] for Sn-on-Sn isotopes and by Sapotta et al.[4] for Ni on ^{208}Pb show the expected slope behavior for nearly all these systems with spherical partners. Work of various authors working with deformed systems (cf. Ref. 1) also show the expected slope behavior for 1-neutron transfer.

The puzzle arises when examining the slopes of 2-neutron transfer. In all cases studied for one or both partners being deformed the 2-neutron slopes are comparable to the 1-neutron slopes. The simple theoretical prediction, whether sequential or simultaneous 2-neutron transfer dominates, is for the 2-neutron slope to be about twice that of the 1-neutron. That is, the 2-neutron transfer probability should fall off with distance exponentially and as the square of the 1-neutron transfer probability. In almost all cases with spherical partners the expected double slope appears, except that the addition of 2-neutrons to ^{208}Pb shows the anomalously small slope of the deformed cases, whereas the other transfers removing neutron pairs from ^{208}Pb all show the expected behavior of double slope.

TABLE I The parameter c_x, which describes the slope of the transfer probability with the distance D (Ref. 4).

x=	-2n	-1n	+1n	+2n	+3n
$^{208}Pb + ^{58}Ni$			1.04	1.96	3.2
$^{208}Pb + ^{64}Ni$	1.04	1.01	1.21	2.33	3.4

We show in our Table I the slopes determined by Sapotta et al.[4] (their Table V). Only the addition of two neutrons to ^{208}Pb has the anomalously low slope.

In order to characterize the anomalously small slopes quantitatively, Himmele et al.[5] used the language of an effective excitation energy of the transferred pair. That excitation terminology has been used in later papers of others, though whether the excitation is thought of as real or virtual is not generally defined. Anyway, the work of Ref. 4 and the recent results of Guidry and collaborators with Ni or Sn-on-Dy isotopes seem to show[6] mostly "cold" transfer at and near the yrast line. Thus, there cannot be much real excitation if the excitation picture has any validity.

We shall below argue that the enhanced 2-neutron tunneling, where observed, is mainly a pairing effect. Why should enhanced tunneling (small slopes) be seen in some cases and not in others? Based on the present limited data set we postulate that to realize pairing-force energy lowering during simultaneous 2-neutron transfer at least one partner must have available near the Fermi surface two or more high-j orbitals. With such high-j orbitals available the neutron pair can form a tight correlation in angle, thus exploiting the short-range pairing force.

Examination of a Nilsson level diagram will show that for deformed rare-earth or actinide nuclei the various j orbitals are split widely by the quadrupole field; there are always at least two high-j orbitals available to form a tight angular wave packet. For the transfer on ^{208}Pb the addition of neutrons goes into $g_{9/2}$, $i_{11/2}$, and $j_{15/2}$ region, whereas the removal of neutrons must use the $p_{1/2}$, $f_{5/2}$, and $p_{3/2}$, primarily.

To be somewhat more quantitative in making this argument we note that zero-angle lobe of a Legendre function squared has an angular width of about 1/L radians. At moderately large L the Legendre function can be approximated as a cosine wave within a $1/(\sin\Theta)^{1/2}$ envelope. Thus, a linear combination of at least two high-L orbitals is needed to cancel the side lobes and realize a wave packet of width 1/L. If one high-J orbital were sufficient to form a tight wave packet, then some of the Sn-on-Sn two-neutron transfers of Ref. 3 would show slope anomalies by virtue of the $h_{11/2}$ orbitals.

Consider the two-neutron transfer as a two-dimensional barrier penetration problem in the center-to-center direction coordinates. The value of the angular integral will fix the depth of the pairing potential in the radial coordinate. (This will be illustrated shortly in Figs. 1 and 2.)

The width of the 45 degree valley in the $r_1 - r_2$ plane is governed mainly by the n-n force range. Hence, for an angular force integral not too small, the cross width of the wave packet should be independent of separation distance, and the energy tied up in the zero point energy of the cross mode should be constant.

Thus, the angular integral should directly give the

barrier lowering. Let us do the angular integral using the Cartesian approximation. We take the coordinates of the first neutron about the center of the nucleus in which the neutrons originate as x_1, y_1, and z_1, where the z-axis is along the line of centers and ℓ is an average angular momentum of the nucleon orbitals involved. Neutron 2 has coordinates x_2, y_2, and z_2. Let R equal the nuclear radius from which shell model orbitals are assigned. Take

$$\psi(x_1,y_1,x_2,y_2) = N \exp[-(\ell^2/2R^2)(x_1^2 + y_1^2 + x_2^2 + y_2^2)],$$

which is an approximation to a forward-angle Legendre function, with N the normalization factor. We ignore spin and antisymmetrization.

$$V(\vec{r}_1\vec{r}_2) = -V_1 \exp\left[-\frac{(x_2-x_1)^2 + (y_2-y_1)^2 + (z_2-z_1)^2}{r_0^2}\right] \tag{1}$$

$$V(z_1 z_2) = -V_1 \exp[-(z_2-z_1)^2/r_0^2] \iiiint \psi^* \exp\{-[(x_2-x_1)^2 + (y_2-y_1)^2]/r_0^2\} \, \psi \, dx_1 dx_2 dy_1 dy_2 \quad , \tag{2}$$

where V_1 is the Gaussian strength parameter for the nn force and r_0 is the range.

Changing variables to relative ξ and center of mass X, we have

$$\xi = 2^{-1/2}(x_2 - x_1) \tag{3}$$

$$X = 2^{-1/2}(x_2 + x_1) \quad , \tag{4}$$

and similarly for the y_i coordinates.

where we have chosen a normalization that makes the transformation canonical.

$$V(z_2-z_1) = -V_1 \exp[-(z_2-z_1)^2/r_0^2]$$
$$\left\{N^2 \int_{-\infty}^{\infty}\int_{-\infty}^{\infty} \exp\left[-\frac{\ell^2}{R^2}(X^2+\xi^2) - \frac{2\xi^2}{r_0^2}\right] dX d\xi\right\}^2 . \quad (5)$$

The integral is squared because it is identical in x_i and y_i variables.

Normalization N is

$$1 = N^2 \left[\int_{-\infty}^{\infty} \exp\left[\frac{-\ell^2}{R^2} x_1^2\right] dx_1\right]^4 . \quad (6)$$

$$V(z_2-z_1) = -V_1 \left(1 + \frac{2R^2}{\ell^2 r_0^2}\right)^{-1} \exp[-(z_2-z_1)^2/r_0^2] \quad (7)$$

The critical ℓ value of orbitals to allow exploitation of the pairing attraction is

$\ell_c = \sqrt{2}R/r_0$ = Nuclear radius/neutron-neutron force range

For ^{208}Pb, this is about 8, so $i_{11/2}$ and $j_{15/2}$ are needed.

Finally, we obtain

$$\Delta B = -V_1(1 + 2R^2/\ell^2 r_0^2)^{-1}$$

$$\Delta B = -13.5 \text{ MeV for } \ell = 6$$

$$\Delta B = -\ 2.3 \text{ MeV for } \ell = 2$$

with V_1 = 35 MeV, r_0 = 1.4 fm, R = 7.5 fm.

Having performed the angular integrals, we are now able to represent the potential surface in the radial

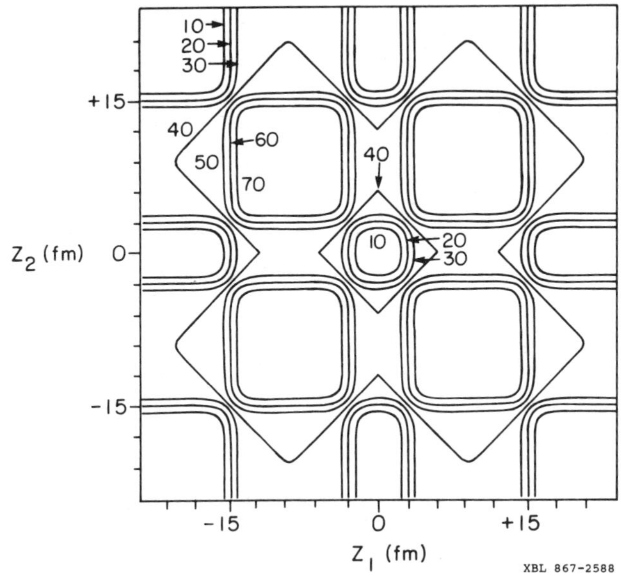

XBL 867-2588

FIGURE 1. Contour plot of the potential of a system of two neutrons in nearby Woods-Saxon rounded square wells. The abscissa represents the coordinate of neutron 1 along the center-to-center axis, and the ordinate is the corresponding value for neutron 2. Contours are at 10 MeV intervals

$$V(z_1,z_2) = -V_0 \sum_{i=1,2} \left\{ [1 + \exp((|z_i|-15)/0.6)]^{-1} - [1 + \exp((|z_i|-3)/0.6)]^{-1} \right\}$$

with V_0 = -40 MeV.

coordinates of the two neutrons. Figures 1 and 2 show contour plots of the potential for a system of two particles along the internuclear (z) dimension. Figure 1 involves two rounded square wells of the size of tin nuclei at a surface-to-surface distance of 6 fm with no pairing interaction (V_1 of Eq. (1) set to zero). Figure 2 is the same except for addition of the pairing interaction with

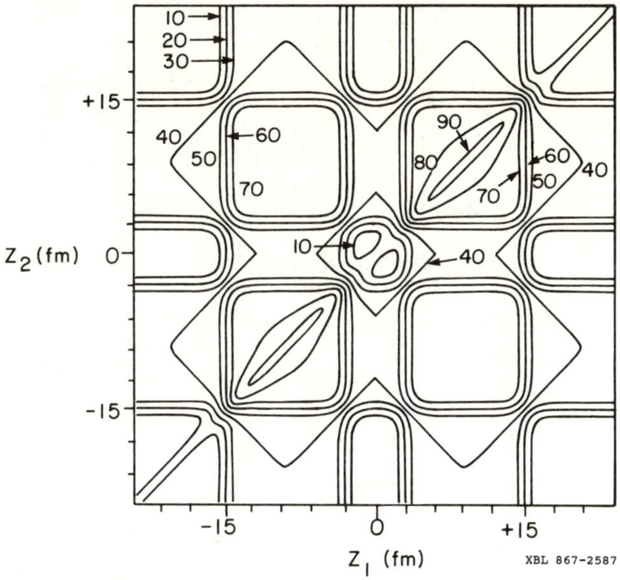

FIGURE 2. Same as Figure 1 but with a short range attractive Gaussian pairing term added (Equation 2) with V_2 ($\equiv V_1$ times the angular integral) taken as 12 MeV and the range r_0 taken as 1.4 fm.

V_2 = 12 MeV and r_0 = 1.4 fm. One sees there are four potential wells. A point in the lower left-hand well denotes the location of both neutrons in the left-hand (parent) nucleus. The upper right-hand well denotes the final state location of both neutrons in the right-hand nucleus. The other two wells represent the intermediate states of one neutron having transferred. Thus, sequential transfer is the tunneling from lower left-hand well via one of the intermediate wells to the upper right-hand well. Simultaneous transfer is the direct tunneling along the 45 degree diagonal to the upper right-hand final state. If there is no pairing potential gain along the 45 degree diagonal, then it is easy to show for finite square-well

potentials that the slope for transfer probability vs. distance of closest approach is the same for sequential or simultaneous transfer. In the latter case the barrier is twice as high, but the path length for tunneling is shorter by the square root of two, thus, giving the same WKB integral. In Figure 2 we see that the pairing interaction term has carved a valley along the 45 degree line, cutting the central plateau and thus facilitating tunneling. Of course, the initial and final states of the two neutrons in the wells are lowered also by the pairing interaction, but the configuration space in the wells is larger than that where both neutrons are in the barrier. Hence, tunneling is enhanced by the interaction.

The angular integral evaluated previously will determine the depth of the pairing-induced valley along the 45 degree line. The width of the valley is governed by the range of the attractive short-range pairing force. The most general solution would involve solving coupled-channel equations through the barrier. It is convenient to convert to relative-center-of-mass coordinates, equivalent to a rotation of axes by 45 degrees. For the purpose of rough estimates here we may use the vibrationally adiabatic approximation to the two-dimensional barrier penetration problem. That is, we may subtract the zero-point energy across the valley (relative motion) from the available pair center-of-mass motional energy for barrier penetration along the valley.

For the Pb transfer case above, we assume the $\ell = 2$ and $\ell = 6$ cases give about the same cross-valley zero-point energy. Hence the barrier for 2-neutron tunneling is reduced by 13.5 MeV - 2.3 MeV ($\approx$ -11 MeV) in the example. This lowering accounts for the slope anomaly.

NUCLEAR SQUID EFFECTS

Broglia and collaborators[7] have showed on single j-shell models that there is an oscillation of pair transfer matrix elements as one increases the cranking velocity. Nikam, Ring, and Canto[8] have recently made theoretical calculations on a more realistic system that confirm that pair transfer matrix elements might undergo sharp drops or even sign changes at modest cranking velocities near where alignment of $i_{13/2}$ orbitals sets in. The numerical example of Ref. 7 is ^{168}Hf, the neutron number of which is the same as ^{162}Dy, a target in the experimental studies on the "spin spectrometer" at Oak Ridge. Thus, it seems appropriate to search for this effect in the existing data. To gain a better understanding of the transfer process and its slope problem for deformed nuclei it is obvious that we need studies of more exclusive nature. That is, we need to be able to examine transfer to low angular momentum and low excitation states separately from other processes.

The nuclear squid prediction does not essentially alter the goals set forth for heavy-ion 2-neutron transfer in the theoretical paper of Guidry, et al.[9] a few years ago. There it was pointed out that Coulomb excitation on the inward path might pump a deformed target to rotational levels high enough that Coriolis anti-pairing (CAP) reductions in the pair transfer could be measured. Later classical orbital calculations[10] reproduced here in Figure 3 held the hope that Pb ions on rare earths might drive the spin at closest approach up to the first $i_{13/2}$ pair alignment ("band-crossing" or "backbending"). What is new in the SQUID paper is that the 2n-transfer strength might drop more sharply near the alignment than previously

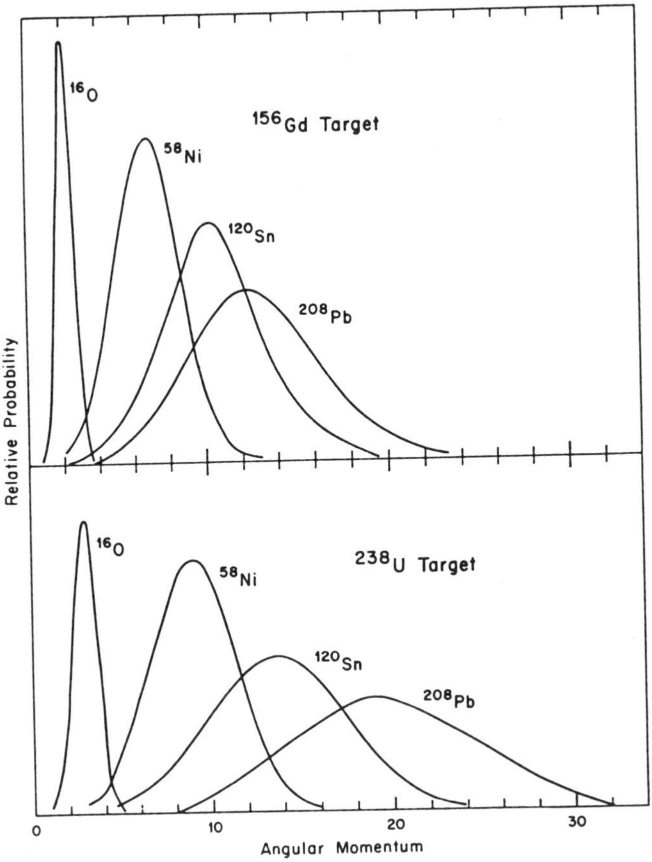

XBL 867-2589

FIGURE 3. Classical orbital calculations of the distributions of rotational angular momentum at closest approach for several heavy ion collision systems. Note that the high-spin tail of the Sn-on-Gd system extends into the rare-earth backbending (spin alignment) region of interest.

believed, and the CAP effects might be seen by careful analysis with projectiles as light as Sn, though heavier beam work is still desirable.

ACKNOWLEDGEMENTS

This work was supported in part by the Director, Office of Energy Research, Division of Nuclear Physics of the Office of High Energy and Nuclear Physics of the U.S. Department of Energy under Contract DE-AC03-76SF00098, and in part by a cooperative research grant INT-8302853 of the National Science Foundation and the Brazilian CNPq. We thank Prof. Luis Egido for helpful discussions. We thank Dr. Richard Meyer for invitations to the Dubrovnik Conference that stimulated writing this paper. Thanks are due Krysta wyatt for programming the graphics of Figures 1 and 2.

REFERENCES

1. J.O. Rasmussen, M.W. Guidry, and L.F. Canto, in Proceedings of the Niels Bohr Centennial Conference, Semiclassical descriptions of atomic and nuclear collisions, edited by J. Bang and J. DeBoer (North Holland, Amsterdam, 1985), p. 167.
2. G. Breit and M.E. Ebel, Physical Review, 103, 679 (1956).
3. W. von Oertzen, B. Gebauer, A. Gamp. H.G. Bohlen, F. Busch, and D. Schüll, Z. Phys. A, 313, 189 (1983).
4. K. Sapotta, R. Bass, V. Hartmann, H. Noll, R.E. Renfordt, and K. Stelzer, Physical Review, C31, 1297 (1985).
5. G. Himmele, H. Backe, P.A. Butler, D. Habs, V. Metag, H.J. Specht, and J.B. Wilhelmy, Nuclear Physics, A404, 401 (1983).
6. P.A. Butler, S. Juutinen, S.P. Sorensen, X.T. Liu, R.W. Kincaid, M. Carpenter, B. Cox, A. Larabee, C.R. Bingham, M.W. Guidry, C.Y. Wu, W. Kernan, A. Kacha, E. Vogt, D. Cline, C. Baktash, and I.Y. Lee, "Observation of Intrinsic Excitation in ^{160}Dy Following Heavy-Ion Transfer Reactions," ORNL preprint (unpublished), 1986. See also for 1-neutron transfer M.W. Guidry, S. Juutinen, X.T. Liu, C.R. Bingham, A.J. Larabee, L.L. Riedinger, C. Baktash, I.Y. Lee, M.L. Halbert, D. Cline, B. Kotlinski, W.J. Kerman,, T.M. Semkow, D.G. Sarantites, K. Honkanen, and M. Rajagopalan, Phys. Lett. 163B, 74 (1985).

7. R. Broglia, in Proceedings International Summer School on Theory of Nuclear Structure and Reactions, La Rabida, Huelva, Spain, June 23-July 6, 1985, Pairing phase transitions in nuclei (see Figure 7).
8. R.S. Nikam, P. Ring, and L.F. Canto, "Nuclear SQUID: Diabolic Pair Transfer in Rotating Nuclei," preprint (unpublished), 1986.
9. M.W. Guidry, T.L. Nichols, R.E. Neese, J.O. Rasmussen, L.F. Oliveira, and R. Donangelo, Nuclear Physics, A361, 274 (1981).
10. M.W. Guidry, R.W. Kincaid, and R. Donangelo, Physics Letters, 150B, 265 (1985).

BARRIER FUSION OF THE $^{64}Ni + ^{64}Ni$ SYSTEM IN A TWO-DIMENSIONAL MODEL.

DEISE A. AGRELLO AND DAVID M. BRINK
Department of Theoretical Physics, University of Oxford, 1 Keble Road, U.K.

Abstract The tunneling problem through two dimensional barriers is studied using quantum mechanical and semiclassical methods. We consider the collision of two Ni isotopes at energies around the barrier region of 100MeV. A different kind of coupling is proposed and the effect of this coupling is studied. The reflection and transmission coefficients are found in a coupled channel calculation. We also consider a semiclassical approximation, using the Esbensen model, for the transmission coefficients and find good agreement between the results.

INTRODUCTION

During the past few years a large amount of heavy ion fusion data has been collected for energies around the Coulomb barrier. For those energies the fusion cross section is wrongly predicted by standard penetrability calculations, which are two or three order of magnitude smaller than those experimentally observed.[1] This inadequacy of one-dimensional model has been taken as evidence that sub-barrier fusion reactions are strongly influenced by other intrinsic degrees of freedom of the colliding nuclei. Several candidates have been proposed in the literature. These are rotations of deformed nuclei,[2] surface vibrations[3] and nucleon transfer modes.[4]

In this work we do not propose a new candidate for a degree of freedom, but present a model calculation using a different kind of coupling to those in the literature. By

using this coupling in a two-dimensional calculation, where the degrees of freedom can be chosen to be any one of those relevant degrees of freedom, we study the effect of this coupling on the transmission coefficients. We consider the ^{64}Ni + ^{64}Ni fusion and follow the same method used by Ring et. al. in reference 5.

THE MODEL

Let us consider the following two-dimensional Hamiltonian,

$$H= \frac{-\hbar^2}{2m_x}\frac{\partial^2}{\partial_x^2} - \frac{\hbar^2}{2m_y}\frac{\partial^2}{\partial y^2} + v(x,y), \qquad (1)$$

where

$$v(x,y)=v(x) + 1/2\ C\ (y-g(x))^2 \qquad (2)$$

and g(x) and v(x) have the same form, i.e.

$$g(x)= \frac{go}{\cosh^2(x/a)}\quad ,\quad v(x)= \frac{Vo}{\cosh^2(x/a)}\ . \qquad (3)$$

The inertial tensor has been assumed to be constant and there will be no coupling between the degrees of freedom for $|x|\to\infty$. We considered the following parameters: V_o=94.403 MeV, a=3.0fm, C=0.772 MeV/fm^{-2}, $\hbar w_o$=5.214 MeV, g_o=3.219fm, m_x=m_y=32m_n and m_n=938.858 MeV/c^2 . To get these constants we adjusted the barrier to the Coulomb plus Christensen and Winther potential.[6] We introduce an energy $\hbar w_o$ characterizing the curvature of the top of the barrier for g_o=0, where $w_o=[|v''(0)|/m_x]^{1/2}$. To obtain g_o, we considered 1/2 C g_o^2=4MeV and the choice of m_y and C gives the frequency $\hbar w_y$ of 1MeV.

Figure 1.a shows the curves for the transmission coefficient versus total energy. The curve (1) is for the pure Eckart potential, curve (2) is obtained when we add

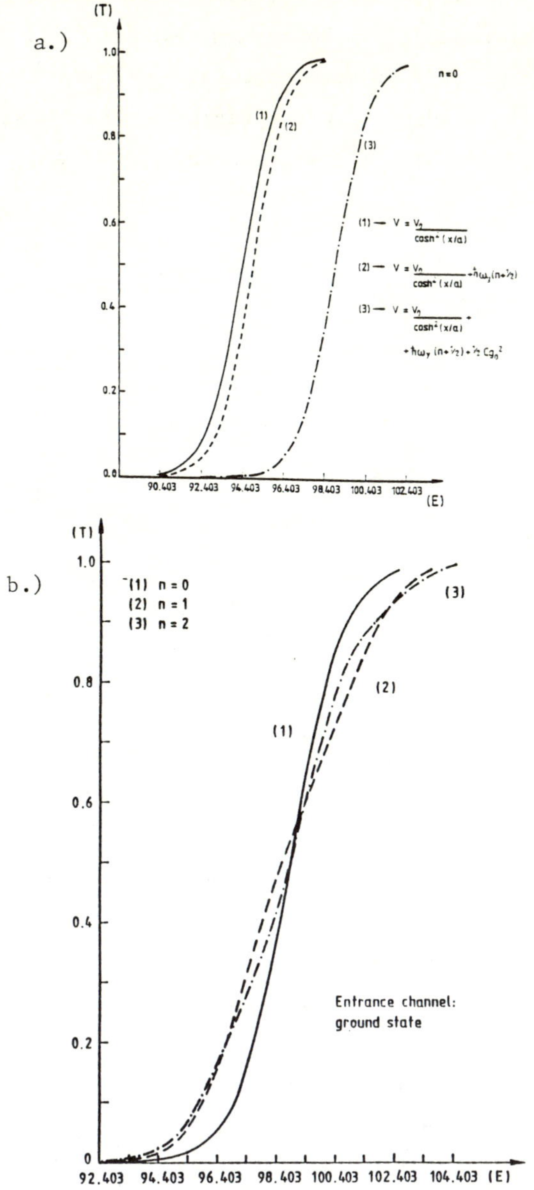

FIG. 1 - Transmission Coefficient x Total Energy (see text)

the energy $\hbar w_y$ (n+1/2) and the third one is shifted of about 4MeV from the second one and corresponds to the addition of the term $1/2C\ g(x)^2$.

Figure 1b shows the total transmission x Energy when we have up to three channels and the entrance channel is the ground state.

In the numerical calculations we considered up to n=4, i.e. five channels and have found that if the number of channels is increased further the results change only slightly. When we increase the number of channels, the biggest effect is when the number of channels changes from one to two. The effect after that is very small.

We also consider a semiclassical approximation, using the Esbensen model[7] for the transmission coefficient. Using this approximation the total fusion probability is the average

$$P_{fus}=P_o(E)= \int_{-\infty}^{\infty} p(y,E)\ (\phi_o(y))^2 dy \tag{4}$$

weighting it with the ground-state distribution of y. For p(y,E) we considered the Hill-Wheeler[8] formula,

$$P(y,E)=\{1+\exp\left[\frac{2\pi}{\hbar w_o}\ (V_B(y)-E\right]\}^{-1} \tag{5}$$

and

$$V_B(y)=V_o \quad + 1/2\ C\ (y-g(o))^2\ . \tag{6}$$

Figure 2 compares the semiclassical results obtained using the equations (4) and (5) with those found using the coupled-channels calculations considering five channels.

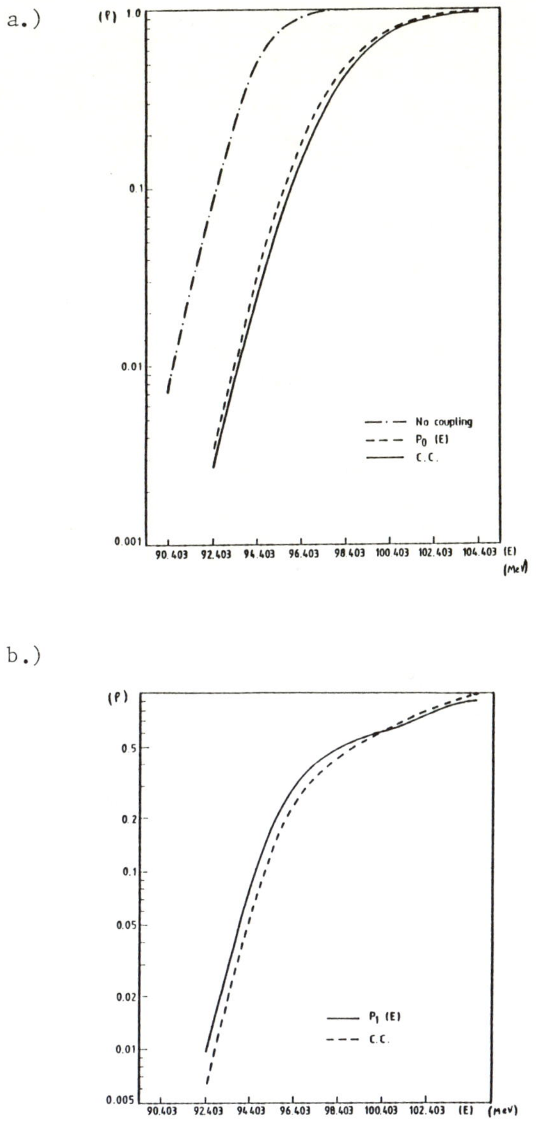

FIG.2 - Total transmission x Total Energy (see text)

Figure 2.a shows the case when the entrance channel is the ground state and figure 2.b when it is the first excited state. In this case, we use for the semiclassical calculation, the equation (4) changing ϕ_o to the first excited state ϕ_1. In figure 2.a we include the results considering the Eckart potential only, showing more clearly the effect of the coupling.

We can see that the results are in very good agreement for $\hbar w_y$=1 MeV. Calculations considering different values of the constants are now in progress.

REFERENCES

1. W. Reissdorf et al, Phys. Rev. Lett. 49, 1811, (1982). M. Beckerman et al, Phys. Rev. C25, 837 (1982).
2. R.G. Stockstad and E.E. Gross, Phys. Rev. C23, 281 (1981)
3. H. Esbensen, Nucl. Phys. A352, 147 (1984).
4. R.A. Broglia, Proc. Int. Conf. on Heavy Ions Physics and Nuclear Physics, Catania (1983).
5. R.Ring, H. Massmann and J.O. Rasmussen, Nucl. Phys. A296, 50 (1978).
6. P.R. Christensen and A. Winther, Phys. Lett. 65B, 19 (1976).
7. H. Esbensen, Jian-Qun Wu and G.F. Bertsch, Nucl. Phys. A411, 275 (1983).
8. D.L. Hill and J.A. Wheeler, Phys. Rev. 89, 1102 (1953).

THE CONSERVATION OF THE TIDAL SPIN IN NUCLEAR REACTIONS

J. GOMEZ-CAMACHO and R.C. JOHNSON
Department of Physics, University of Surrey,
Guildford GU2 5XH, U.K.

Abstract The existence of an approximate symmetry in nuclear collisions, that we call "Tidal Symmetry", is discussed. This symmetry becomes exact when a certain approximation, the "Isocentrifugal Approximation", is applied to the Hamiltonian. As a result, there is a quantum number, the "Tidal Spin", that is conserved throughout the collision. This conservation simplifies the scattering calculations and imposes strict relations for the polarization observables.

To get an idea of what is the Tidal Symmetry, let us consider the classical description of the scattering of a deformed projectile by a spherical target. The rate of change of the angular momentum of the projectile $\vec{I}$ in a coordinate system fixed in the projectile is given by

$$\left.\frac{d\vec{I}}{dt}\right|_{Int} + \vec{w} \times \vec{I} = \vec{M}\,, \qquad (1)$$

where $\vec{w}$ is the angular velocity of the projectile about the target and $\vec{M}$ the torque of the external forces about the centre of mass of the projectile. If these forces are momentum-independent, then the component of the torque along the relative coordinate $\vec{r}$ of projectile and target vanishes. This is equivalent to saying that the potential energy is not modified by rotations of the projectile around $\vec{r}$. Let us assume that the non-inertial torque $\vec{w}\times\vec{I}$ can be ignored. Then, the projection of $\vec{I}$ along $\vec{r}$ will be conserved.

In a quantum-mechanical description of the same process

the situation is similar. The equations of motion are given by

$$\left\{-\frac{\hbar^2}{2\mu}\left(\frac{1}{r}\frac{d^2}{dr^2}r\right)+\frac{\hbar^2}{2\mu r^2}L^2(\hat{r})+V(\vec{r},\varepsilon)+h(\varepsilon)-E\right\}\psi(\vec{r},\varepsilon)=0 \quad (2)$$

For momentum independent interactions the potential commutes with the operator that generates rotations of the internal coordinates around r (Tidal Symmetry). This operator (Tidal Spin Operator) is the projection of the internal spin of the projectile along $\vec{r}$. The term $L^2(\hat{r})$ of the centrifugal potential, however, does not commute with the Tidal Spin operator. If we can substitute it by a constant $\bar{L}(\bar{L}+1)$ (Isocentrifugal Approximation) then the Tidal Spin Operator will commute with the whole Hamiltonian, and its eigenvalues will be good quantum numbers.

The eigenstates of the Tidal Spin Operator can be used to construct a basis for the intrinsic and angular coordinates. This basis, the Tidal Spin basis, will be characterized by the internal state, labelled by the internal spin I, the tidal spin eigenvalue K, and the total angular momentum J and its third component M. The Tidal Spin basis is related to the usual channel basis defined in terms of the relative angular momentum L by

$$\langle ILJM | I'KJ'M' \rangle = \delta(I,I')\,\delta(J,J')\,\delta(M,M')\,(-1)^{I-K)}\langle I\ K\ J\ -K \quad L\text{\`{}}0 \rangle \quad (3)$$

Using the Tidal Spin basis and the Isocentrifugal Approximation, the eigenvalue equation for Hamiltonian (1) becomes

$$\left(-\frac{\hbar^2}{2\mu}\frac{d^2}{dr^2}+\frac{\hbar^2}{2\mu}\frac{\bar{L}(\bar{L}+1)}{r^2}+V_I(r)+e_I-E\right)f^J_{IK}(r)+\sum_{I'}V^K_{II'}(r)\,f^J_{I'K}(r)=0 \qquad (4)$$

where it is seen that different values of K are not coupled. This equation is formally equivalent to having several reduced coupled channels systems involving just spin-zero states.

When the isocentrifugal approximation is made the Hamiltonian for large values of r does not generate solutions of sufficient accuracy to easily satisfy standard scattering theory boundary conditions. To overcome this problem it is assumed that a sphere of radius R exists outside of which the coupling potential can be ignored and the centrifugal barrier is treated exactly. The isocentrifugal approximation is only used for r<R. Smooth joining of the solutions at r=R is achieved under the assumption that R is large enough for the asymptotic form of the Coulomb functions to be valid. This procedure gives the S matrix elements in the (IL)J basis

$$S^J_{IL,I'L'}(\bar{L})=\sum_K \langle IKJ-K|L0\rangle\, S^{\bar{L}K}_{II'}\, \langle I'KJ-K|L'0\rangle\,(-1)^{I-I'}\, i^{L+L'-2\bar{L}} \qquad (5)$$

$$\exp(i(\sigma^C_{\bar{L}}(\eta_i)+\sigma^C_{\bar{L}}(\eta_f)-\sigma^C_L(\eta_i)-\sigma^C_{L'}(\eta_f)))$$

where σ^C_L are the Coulomb phase shifts and $S^{\bar{L}K}_{IJ}$, is the scattering matrix associated with the reduced system (4).

The value of $\overline{L}$ is chosen so that the first order transitions induced by the non-isocentrifugal term, that is the difference between the exact centrifugal barrier and the average one, vanish. This implies that, for elastic scattering or inelastic scattering in the adiabatic limit, $\overline{L}$ is the average of the incoming angular momentum L and the outgoing one L'.

We have derived the relation (1)

$$\sum_{LL'} \{Y_L(\Omega_i)\, Y_{L'}(\Omega_f)\}_{\lambda\mu} (L0L'0|\lambda 0)\, i^{L-L'}\, \delta(L+L',2\overline{L}) = \hat{\lambda}/\sqrt{4\pi}\; Y_{\lambda\mu}(\overline{\Omega})\, P_{\overline{L}}(\cos\theta) \tag{6}$$

where $\overline{\Omega}$ is the bisecting direction of $-\Omega_i$ and Ω_o and Θ the angle between Ω_i and Ω_o. The result is correct asymptotically for large $\overline{L}$, being λ even but otherwise arbitrary, and is found to be numerically accurate for moderate values of $\overline{L}$ if sin Θ is not too small.

Using eq.(6), the transition amplitudes can be expressed as

$$A_{IM,I'M'}(\Omega_i,\Omega_f) = \sum_K A^K_{II'}(\theta)\, D^{I\,*}_{MK}(\overline{\Omega})\, D^{I'}_{M'K}(\overline{\Omega}), \tag{7}$$

where $A^K_{II'}(\theta)$ is the scattering amplitude of the reduced spin-zero coupled channels calculations. This result states that the projection of the spin along the recoil direction is conserved.

The unpolarized cross sections are obtained summing the square of the transition amplitudes with respect to all the magnetic substates. Using the orthogonality of the rotation matrices, one gets

$$\sigma_{II'}(\theta) = \sum_K \sigma^K_{II'}(\theta) \,/\, (2I+1) \tag{8}$$

where $\sigma^K_{II'}$ is the cross section obtained in the calculation

with spin zero states, for each K. The Tensor Analyzing Powers (1,2) given in the Madison coordinate system are

$$T_{kq}(\theta) = \qquad (9)$$

$$d^{k}_{q0}((\theta+\Pi)/2)\Sigma_{K}\sigma^{K}_{II'}(\theta)\ \surd(2k+1)\ (I\ K\ k\ 0\ |I\ K\ )/\Sigma_{K}\sigma^{K}_{II'}(\theta)$$

It should be pointed out that σ(K) equal to σ(-K) in this approximation, and so all the odd Tensor Analyzing Powers are zero. For k=2 the Tensor Analyzing Powers given by (9) fulfil the so-called "Shape Effect" relations. These have been obtained previously within a first order perturbation theory treatment of the coupling potential (2,3). The present treatment makes no use of perturbation theory and extends the validity of the relations to strong coupling situations.

We have applied the previous results to the reaction 23Na + 208Pb at 200 MeV. Only the effect of the reorientation of the ground state of the projectile is included. We compare the results for the elastic cross-section and the tensor analyzing powers from the exact calculation with the approximate calculation in the Tidal Spin basis (TS). The Tidal Spin basis calculations for the cross sections and the TT20 for a spin 3/2 nucleus are given in terms of cross sections for definite values of the tidal spin by

$$\sigma(TS) = \frac{\sigma(K=1/2) + \sigma(K=3/2)}{2} ; \qquad (10)$$

$$^{T}T_{20}(TS) = \frac{\sigma(K=1/2) - \sigma(K=3/2)}{2(\sigma(K=3/2) + \sigma(K=1/2)}$$

The cross sections for values of the Tidal Spin K = ½ and K = 3/2 are shown in Fig. 1. The elastic cross sections in the exact and TS calculations agree very well, as shown in

Fig 2. Although neither of them differ very much from the uncoupled cross section, the agreement is remarkable because the TS cross section is the average of two very different terms. The agreement of the exact and TS calculations for TT20 is very good also, as shown in Fig. 3.

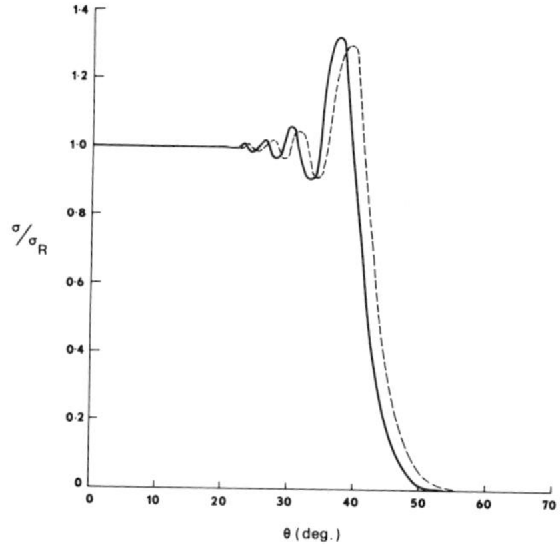

FIGURE 1. Cross sections for values of the Tidal Spin K=3/2 (dashed line) and K=½ (full line) for $^{23}Na+^{208}Pb$ at 200 MeV

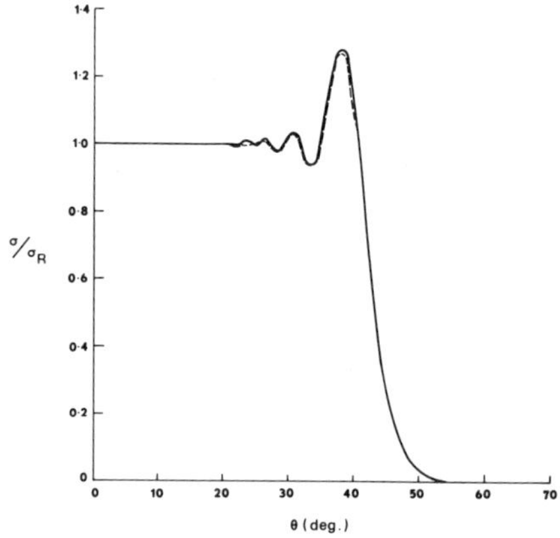

FIGURE 2. Unpolarized cross sections obtained in an exact calculation (full line) and in the Tidal Spin basis (dashed line) for the same reaction.

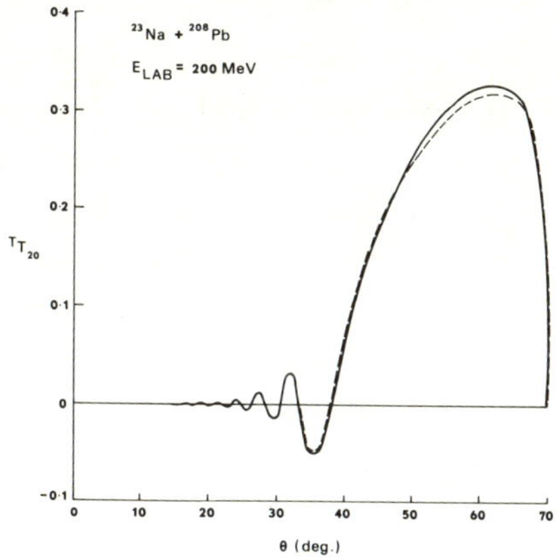

FIGURE 3. $^{T}T_{20}$ obtained in an exact calculation (full line) and in the Tidal Spin basis (dashed line) for the same reaction.

REFERENCES

1. J. Gomez-Camacho and R.C. Johnson, submitted for publication.
2. G. Tungate and D. Fick 1979, Lecture Notes in Physics, Vol. 89
3. H. Nishioka, J.A. Tostevin, R.C. Johnson and K.I. Kubo, Nucl. Phys. A415 230,1984).
4. R.C. Johnson, Nucl. Phys., A239 92 (1977)
5. D.J. Hooton and R.C. Johnson Nucl. Phys. A175 583 (1971)
6. P.Jacobs and U. Smilansky, Nucl. Phys., A438 536 (1985)
7. M.A. Nagarajan Proceedings of the Summer school on nuclear structure and nuclear reactions, La Rabida (Spain), June 1985., World Scientific Press, Singapore, 1986.

COMPARISON OF THE YIELDS OF HIGH-ENERGY GAMMA-RAYS WITH E_γ > 10 MeV MEASURED IN THE REACTIONS $^{nat}Sn + {}^{40}Ar$ AND $^{68}Zn + {}^{64}Zn$

V.V. Kamanin, A. Kugler, Yu.E. Penionzhevich,
Yu.G. Sobolev, and A.S. Fomicthev
Joint Institute for Nuclear Research, Dubna, USSR

I. David
Institute of Physics and Nuclear Technology
Bucharest, Romania

Astract: The spectra of the KX-rays of reaction produ ducts were observed in coincidence with high energy gamma-rays (E_γ > 8 MeV) in the $^{nat}Sn+{}^{40}Ar$ (E_{Ar}=300 MeV) and $^{68}Zn+{}^{64}Zn$ (E_{Zn}=290 MeV) reactions. Two groups of KX-rays can be clearly identified in the spectra which correspond to the evaporation residue of the compound nucleus and to the target-like products of the reaction. The relative yields, the position of the broad shoulders and the ratio $W=Y_{90^0}/Y_{0^0}$ certify our assumptions: the emission of high energy gamma-rays from the dinuclear system prior to the formation of the compound nucleus.

INTRODUCTION

The numerous observation of the emission of high energy neutrons and light charged particles in heavy-ion reactions renewed the interest of physicists in the study of the preequilibrium processes in reactions between complex nuclei. [1] The possibility of the emission of high-energy (E_γ>10 MeV) gamma-rays at the first stages of the heavy-ion fusion reactions was discussed in the recent papers [2-4]. The "cooling" of the evaporation spectrum of light particles (α, p, n) gated by high-energy gamma-rays as compared to the inclusive spectrum was reported both in the study of the $^{58}Ni+{}^{16}O$ (76 MeV) [2] and $^{159}Tb+{}^{12}C$ (100 MeV) [3] reactions. The enhancement

of the 1n and 2n evaporation channels as compared to the statistical theory was observed in the $^{87}Ru + ^{65}Cu \rightarrow ^{148}Dy$ reaction.[4] They did not make any definite conclusions, but it is clear that the assumption of preequilibrium emission can explain the experimental data.

At the same time there have been observed[5-8] broad "shoulders" ("bumps") in the gamma-spectra at energies $E_\gamma > 10$ MeV which were associated mainly with the decay of the giant dipole isovector resonance (GDR) built upon the high excitation states of the compound nucleus. But it is clear that the products of deep inelastic collisions,[9] can not be distinguished from the compound nuclei by the "gamma-start" method. We gated the high energy gamma-rays by a signal from the fission fragments detector in $^{192}Os+^{15}N$, $^{181}Ta + ^{22}Ne$ and $^{159}Tb + ^{40}Ar$ reactions.[10] The masses of the Bi compound nuclei formed in the reported fusion reactions differed by several units only, but the spectra of high energy gamma-rays observed in the fusion-fission channel were very different. This entry channel effect may indicate the preequilibrium emission of high-energy gamma-rays.

The present paper is devoted to a more thorough study of the role of the main reaction channels in the emission of high-energy gamma-rays in the $^{nat}Sn + ^{40}Ar$ (E_{Ar} = 300 MeV) and $^{68}Zn + ^{64}Zn$ (E_{Zn} = 290 MeV) reactions. The yields of KX-rays, which correspond to a broad range of Z, were determined as a function of the energy of the gating gamma-ray up to E_γ = 20 MeV. We report the spectrum of $E_\gamma = (4 \div 20)$ MeV gated by KX-rays of evaporation residues. The anisotropy of gamma-ray emission was determined in a wide range of E_γ for the $^{nat}Sn + ^{40}Ar$ reaction.

EXPERIMENT

The principal design of the experiment is demonstrated in fig.1. The beam of heavy ions from the U-300 cyclotron of the Laboratory of Nuclear Reactions of JINR hit targets of 15 mg/cm^2 thickness, which were backed by a Th metallic foil of 47 mg/cm^2 thickness. The KX-ray intrinsic Ge-detector was placed behind the target. The high-energy gamma-ray detectors of a Ø 75x25 mm BGO scintillator and of a Ø150x100 mm NaI(Tl) scintillator were placed perpendicularly to the beam direction and only E_γ>4 MeV were accepted.[10]

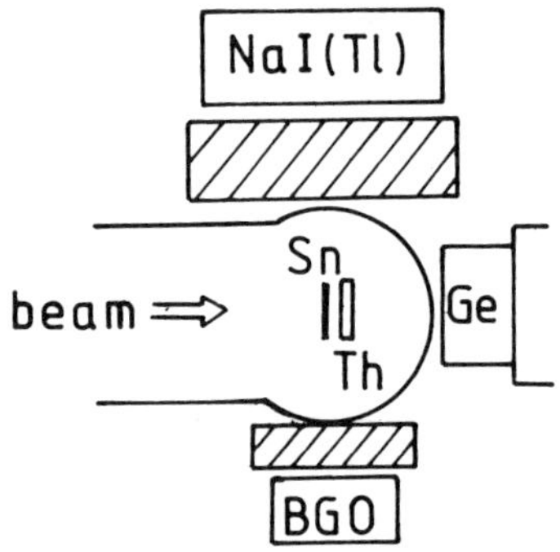

Fig.1. Experimental setup.

The setup for the angular anisotropy was: two BGO detectors were placed at angles of 90 and 0 degrees to the beam direction and the trigger of Ø 63x63 mm NaI(Tl) scintillator was placed perpendicularly both to the beam direction and to the plane determined by the two BGO.

RESULTS

The spectrum of gamma transitions in the ^{nat}Sn + ^{40}Ar reaction, which were detected in coincidence with a signal from the KX-ray detector corresponding to the broad range of energies E_{KX} = (25-400) keV, is demonstrated in fig. 2. The low-energy part of the spectrum is composed mainly of statistical transitions which cool the compound nucleus at the last stages of the evaporation cascade.[5] This region of the spectrum can be roughly approximated by the function $\exp(-E_\gamma/T_{eff})$, $T_{eff} \approx 1.1$ MeV. Above this region the

spectrum significantly deviates from the exponential fall and has the form of a broad peak convoluted with an exponentially falling function. [6] The high-energy part of the spectrum had to consist mainly of transitions due to the statistical decay of GDR's of the compound nucleus, as the authors of previous papers [5-8] had assumed. We have gated gamma transitions by KX-rays which correspond to Z not too far from the Z of the compound nucleus (see fig. 3) to determine

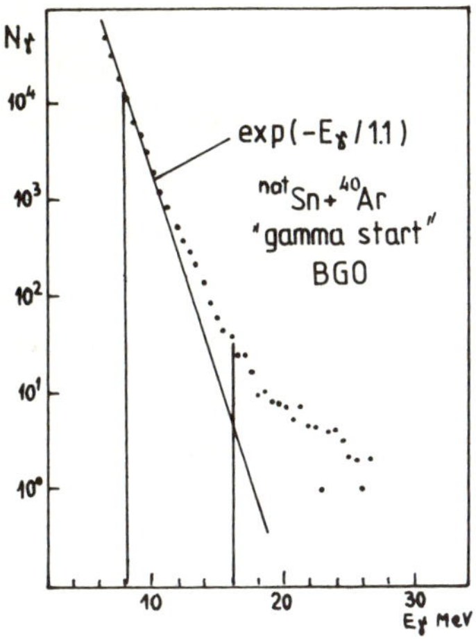

Fig.2. Spectrum of gamma transitions gated by soft gamma rays of E_{KX} = (24 - 400 keV)

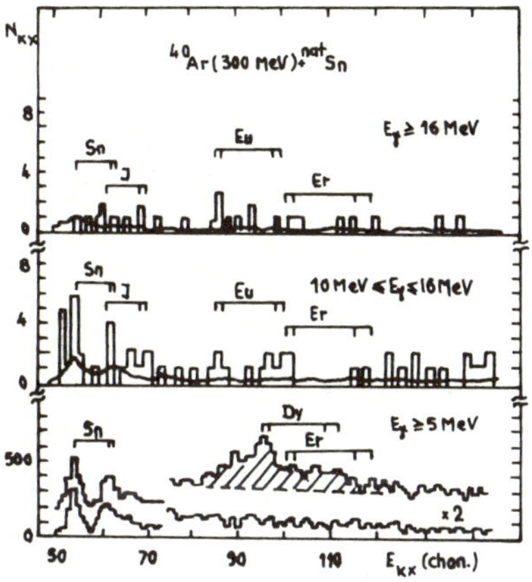

Fig.3. The KX-rays, which correspond to true and or to random coincidence with different gamma-energy ranges for $^{nat}Sn+^{40}Ar$

the spectrum of gamma-rays due to the fusion channel of the reaction. The resulting spectra of gamma transitions detected in BGO and NaI(Tl) detectors are shown in fig. 4. Because of the existence of the pronounced minimum at $E_\gamma \approx 16$ MeV in these spectra, we have divided the high energy gamma transitions into two groups, and three regions (fig. 3). The first region consists of usual

statistical transitions of E_γ < 8 MeV and the spectrum of KX-rays which were observed in coincidence with these is dominated by KX-rays of coumpound-like nuclei. There are KX-rays of both target-like and compound-like nuclei in the spectrum gated by gamma transitions of 10 MeV$\leq E_\gamma \leq$16 MeV (second region). This fact disagrees with the assumption that the main source of high-energy gamma-rays is the statistical decay of GDR's of the compound nucleus. The KX-rays of compound-like nuclei dominate again in the spectrum gated by gamma transitions of $E_\gamma \geq$16 MeV (third region). To summarize, the significant increase in the yield of gamma

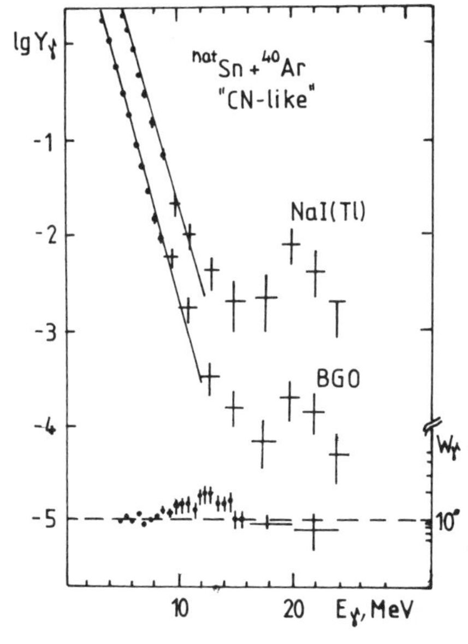

Fig.4. Spectra of γ-rays in coincidence with KX of compound-like nuclei, and W_γ in the bottom part of the fig.

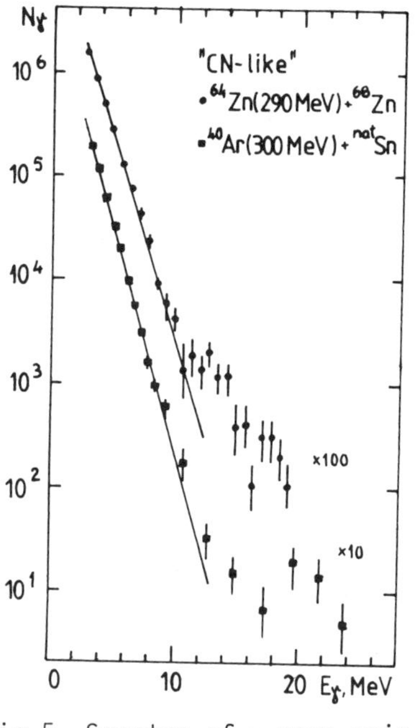

Fig.5. Spectra of γ-rays coincidence with KX of compound-like nuclei for the both reactions.

transitions above the exponentially falling tail was observed both in the second and third region of E_γ. The fusion

channel dominates in the third region of E_γ, only. The reaction channels which lead to target-like products play an important role in the emission of gamma-rays on the second region.

We have determined the ratio $W_\gamma = Y_\gamma^{90^0}/Y_\gamma^{0^0}$ of the yields of gamma transitions observed and gated by the cascade of soft gamma-rays. W_γ has to depend both on the multipolarity of gamma transitions and on the characteristics of their source. The isotropic emission of gamma-rays of E_γ=(4-8)MeV was observed in the $^{159}Tb + {}^{12}C$ reaction by the authors of paper.[11] Therefore we normalised the observed gamma spectra in the region of E_γ =(4-8) MeV (fig. 4). W_γ is significantly higher than unity in the second region, and there is no significant deviation in the third region of E_γ if one takes experimental errors into account. These facts indicate again the different character of gamma-ray emission in the second and the third region of E_γ.

It was interesting to compare the gamma spectra measured in both investigated reactions leading to the compound (residual) nucleus Yb and Er (fig. 5). An increase in the yield of gamma-rays for $E_\gamma \geq 10$ MeV in going to the reaction $^{68}Zn + {}^{64}Zn$ with a ratio $A_1/A_2 \approx 1$ has been observed. This result is in agreement with the conclusions of a previous paper.[10]

The spectra of KX-rays for the reaction $^{68}Zn + {}^{64}Zn$ measured in coincidence with the same ranges of E_γ as for the reaction $^{nat}Sn + {}^{40}Ar$ are shown in fig. 6. For $10 \text{ MeV} \leq E_\gamma \leq 16$ MeV as well as for $E_\gamma \geq 16$ MeV KX-rays are connected with the Z of the residual nucleus close to the Z of the compound nuclei.

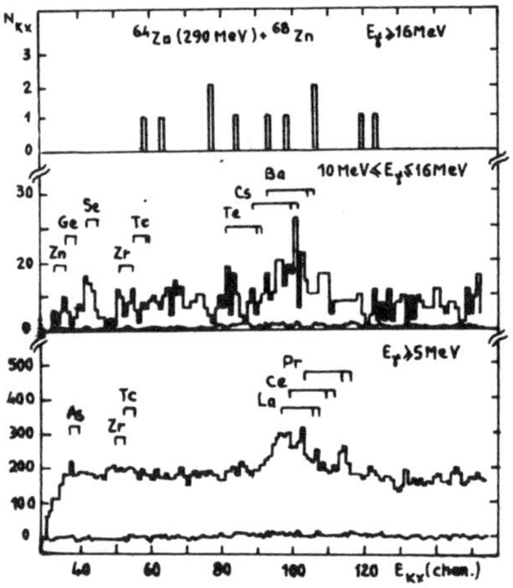

Fig.6. The same as Fig.3 for the reaction $^{68}Zn + ^{64}Zn$

CONCLUSION

Our data show that in the $^{nat}Sn + ^{40}Ar$ reaction, the main part of gamma transitions is connected with deep-inelastic collisions in the range E_γ = (10-16) MeV (in contradiction with [5-8]), the fusion channel is dominant in the range E_γ = (16-22) MeV. The increase of yield emitted perpendicularly to the beam direction in the range E_γ = (10-16) MeV and the nearly isotropic angular distribution in the range E_γ= (18-22) MeV has been observed in the $^{nat}Sn + ^{40}Ar$ reaction, too.

We have observed that the centroid of the part of the γ - spectrum exceeding the exponentially falling tail was shifted to smaller values in the $^{68}Zn + ^{64}Zn$ reaction, where the masses of projectile and target nuclei are almost equal. The yield of this part of gamma-rays increased significantly, as compared to the $^{nat}Sn + ^{40}Ar$ reaction. Finally, gamma-rays with $E_\gamma \geq 10$ MeV were mainly due to the fusion channel in the $^{68}Zn + ^{64}Zn$ reaction.

Our data indicate that the mechanism of emission of high energy gamma-rays ($E_\gamma \geq 10$ MeV) is probably more complex than it was assumed in previous papers.[5-8] The gamma-ray emission may be due to the transfer of nucleons between the two interacting nuclei (a change in the shape of the dinuclear system).

Although the data have been obtained in experiments

with thick targets and the sums of the masses of projectile and target nuclei differ slightly between the two studied reactions, the observed differences in the data can be due to the values of mass-asymmetry in the entrance channel of the studied reactions. It should be noted that the dependence of the yield of high-energy gamma-rays on the projectile energy, and on the mass of the compound nucleus A ∿ 150-200 is weak.[10]

We would like to express our sincere thanks to Acad. G.N.Flerov and Prof. Yu.Ts.Oganessian for their interest in this work and for helpful discussions.

REFERENCES

1. Penionzhkevich Yu.E. et al. "Particles and Nuclei", Moscow, Energoatomizdat, v.17, 2, 1986, p.165
2. Lazzarini A. et al. Phys.Rev.Lett., 53, 1974, p.1045
3. Arctaedius Th. et al. Phys.Lett., 158B, 1985, p.205
4. Cabot C. et al. Z.Phys., A322, 1985, p.393
5. Suover K.A. Jour.Phys., 45, 1985, p.C4-337
6. Newton J.O. et al. Phys.Rev.Lett., 36, 1981, p.1383
7. Gaardhoje J.J. et al. Phys.Lett., 139B, 1984, p.273
8. Haas B. et al. Phys.Lett., 120B, 1983, p.79
9. Volkov V.V. "Particles and Nuclei", Moscow, Energoatomizdat, v.6, 1975, p.1040.
10. Kamanin V.V. et al. JINR Rapid Communications, 8-85, Dubna, 1985, p.18
11. Bardgholtz Ch. et al. J.Phys.G: Nucl.Phys., 10, 1984, p. L275.

DIABATIC RECOIL IN CENTRAL HEAVY ION COLLISIONS

PIOTR ROZMEJ +) and WOLFGANG NÖRENBERG*)
Gesellschaft für Schwerionenforschung (GSI)
D-6100 Darmstadt, F.R. Germany
and*) Institut für Kernphysik, TH Darmstadt
D-6100 Darmstadt, F.R. Germany

Abstract The central collisions of $^{208}Pb + {}^{124}Sn$ at bombarding energies between 0.1 and 5 MeV/u above the Coulomb barrier are chosen to study the effect of nuclear elastoplasticity. The result show that for energies < 2 MeV/u a considerable amount of elastic response survives such that with ^{208}Pb as the projectile the recoiling target-(^{124}Sn-) like nuclei have larger velocities than the beam.

For finite collective velocities where the adiabatic approximation for the single-particle motion is no longer valid, a qualitatively new feature of collective nuclear motion has been predicted : elastoplasticity [1-4]. This dynamical behaviour results from a coherent coupling between collective and intrinsic degrees of freedom and subsequent equilibration by residual two-body collisions. Whithin a non-markovian transport-theoretical approach[3], the elastic response is described by scaling of diabatic single-particle wave functions[5] according to the collective deformation, while two-body dissipation is accounted for by a relaxation ansatz. This dissipative diabatic dynamics (DDD) ascribes elastoplasticity to nuclei and establishes a link between time-dependent Hartree-Fock and markovian transport theories of nucleus-nucleus collisions[6]. Isoscalar giant quadrupole vibrations of nuclei and mass diffusion in nucleus-nucleus collisions can be considered as well-established

+) on leave from the University MCS Lublin, Poland.

examples for the elastic and plastic limits of elastoplasticity. First numerical studies of dissipative diabatic dynamics have been performed for central collisions of $^{40}Ca+^{40}Ca$ and $^{90}Zr+^{90}Zr$ [7]. In the following we like to draw attention to an observable effect of nuclear elastoplasticity: the quasi-elastic recoil in nucleus-nucleus collisions.

In the present paper we limit our investigation to one collective degree of freedom - the elongation ζ. This parameter denotes the distance between the centers of the potentials in a two-center shell model[5] and for not very compact configurations roughly coincides with the distance between the centers of masses.

The collective motion of the colliding nuclei can be described in the framework of DDD[4] by the non-markovian equation

$$\frac{d}{dt}(B_{\zeta\zeta}\dot{\zeta}) - \frac{1}{2}\frac{dB_{\zeta\zeta}}{d\zeta}\dot{\zeta}^2 + \int_{t_0}^{t} C_{\zeta\zeta}\ exp\{\int_{t'}^{t} \tau_{intr}^{-1}(\theta)d\theta\}\ \dot{\zeta}(t')dt' = \bar{F} \quad (1)$$

The ingredients of this equation are calculated as follows. The mass parameter $B_{\zeta\zeta}$ is calculated from irrotational flow. It has been shown[5] that this is identical with the cranking mass parameter for the scaling diabatic states α. The equilibrium force F is approximated by the derivative of the adiabatic potential energy V_{adiab} which is calculated macroscopically[8] as the sum of the surface and Coulomb energies (without shell corrections). The stiffness coefficient $C_{\zeta\zeta}$ is calculated microscopically from

$$C_{\zeta\zeta} = -\sum_{\alpha}\{\frac{d(\epsilon_\alpha - \mu)}{d\zeta}\}^2 \frac{d\bar{n}_\alpha}{d(\epsilon_\alpha - \mu)} \quad (2)$$

where ε_α denote the diabatic single-particle energies. The equilibrium distribution $\bar{n}_\alpha$ of the occupation probabilities is determined at a given value of ζ by a Fermi function with chemical potential μ and temperature T. The temperature is calculated via $T=\sqrt{8\varepsilon^*/\text{MeV}}$ MeV from the total excitation energy ε^* given by

$$\varepsilon^* = \frac{E_{CM} - V_{adiab}(t) - E_{kin}(t)}{A} \quad (3)$$

where E_{CM}, E_{kin} and A denote the bombarding energy in the center-of-mass system, the collective kinetic energy and the total nucleon number, respectively. The intrinsic equilibration time

τ_{intr} has been estimated by Bertsch[9] within a Fermi-gas model as

$$\tau_{intr} \approx \frac{t^*}{\varepsilon^*}\, 10^{-23} s \cdot MeV \tag{4}$$

with $t^* = 20$.

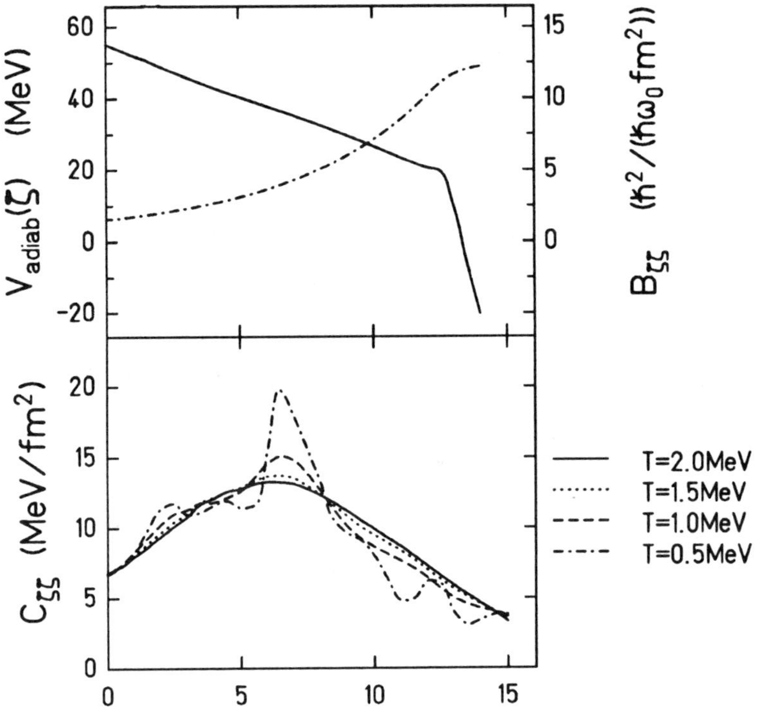

FIGURE 1. Top: the adiabatic potential (solid line) and irrotational mass parameter (dot-dashed line) as functions of the elongation coordinate ζ for $^{208}Pb + ^{124}Sn$. Bottom: the stiffness parameter calculated for different values of the nuclear temperature. The solid line represents the results for T = 2, 2.5 and 3 MeV (discrepancies within linewidth).

The numerical results for the adiabatic potential, mass and stiffness parameters in the central collision of $^{208}Pb + ^{124}Sn$ are shown in Figure 1. The system is chosen to maximize the effect of the elastic response. Indeed, the adiabatic potential has only

a very shallow pocket (less than 1 MeV) just inside the contact distance and rises almost linearly with decreasing ζ. Therefore, the elastic response cannot be obscured by the adiabatic potential. The mass parameter shows a smooth dependence on the distance and approaches the reduced mass value for large ζ. The stiffness parameter exhibits shell effects for low temperatures. These effects are smoothed out with increasing temperature and vanish practically for $T \geq 2$ MeV.

We have studied the dynamical development of $^{208}Pb + ^{124}Sn$ as function of the incident energy

$$\varepsilon'_{in} = \frac{E_{CM} - V_{Coul}}{A_{red}} \qquad (5)$$

between 0.1 and 5 MeV/u. In order to show the influence of the intrinsic equilibration time we have performed the calculations with $t^* = 10$ and $t^* = 40$ in addition to $t^* = 20$.

Figure 2 shows the dynamical development of the $^{208}Pb + ^{124}Sn$ system for three values of the incident kinetic energy (ε'_{in} = 1, 2 and 3 MeV/u) and $t^* = 20$. The trajectory $\zeta(t)$ and the intrinsic equilibration time $\tau_{intr}(t)$ are displayed in the upper parts of the figure and the kinetic, potential and excitation energies (i.e. $E_{kin}(t)$, $V_{adiab}(t)$ and $E^*(t) = A\varepsilon^*(t)$) in the lower part. The total energy of the system is denoted by the thin dotted straight line (the total energy of the system is conserved). For $\varepsilon'_{in} = 1$ and 2 MeV/u we get a remarkable elastic or quasi-elastic behaviour of the system. In these cases the kinetic energies at the exit point are 50 MeV and 30 MeV, respectively, corresponding to $\approx 65\%$ and $\approx 20\%$ of the initial energies. The interaction times (i.e. the times system stays inside the Coulomb barrier) are almost the same ($\sim 90 \times 10^{-23}$ s). The collisions proceed almost in the same way apart from the fact that for higher incident energy the interpenetration of the colliding nuclei is deeper, while the excitation energy is higher. Since τ_{intr} is inversely proportional to ε^*, it becomes smaller at higher bombarding energies. The qualitative behaviour of the trajectory changes drastically as one goes to $\varepsilon'_{in} = 3$ MeV/u. In this case the interaction time increases to more then three times the values obtained for $\varepsilon'_{in} = 1$ and 2 MeV/u. The reason is that the system is reflected back once by the elastoplastic force which is built up on the way towards reseparation. For times $t \geq 100 \times 10^{-23}$ s practically all kinetic energy is dissipated and the nuclear temperature $T \geq 2$ MeV is stabilized while the system creeps towards the Coulomb barrier.

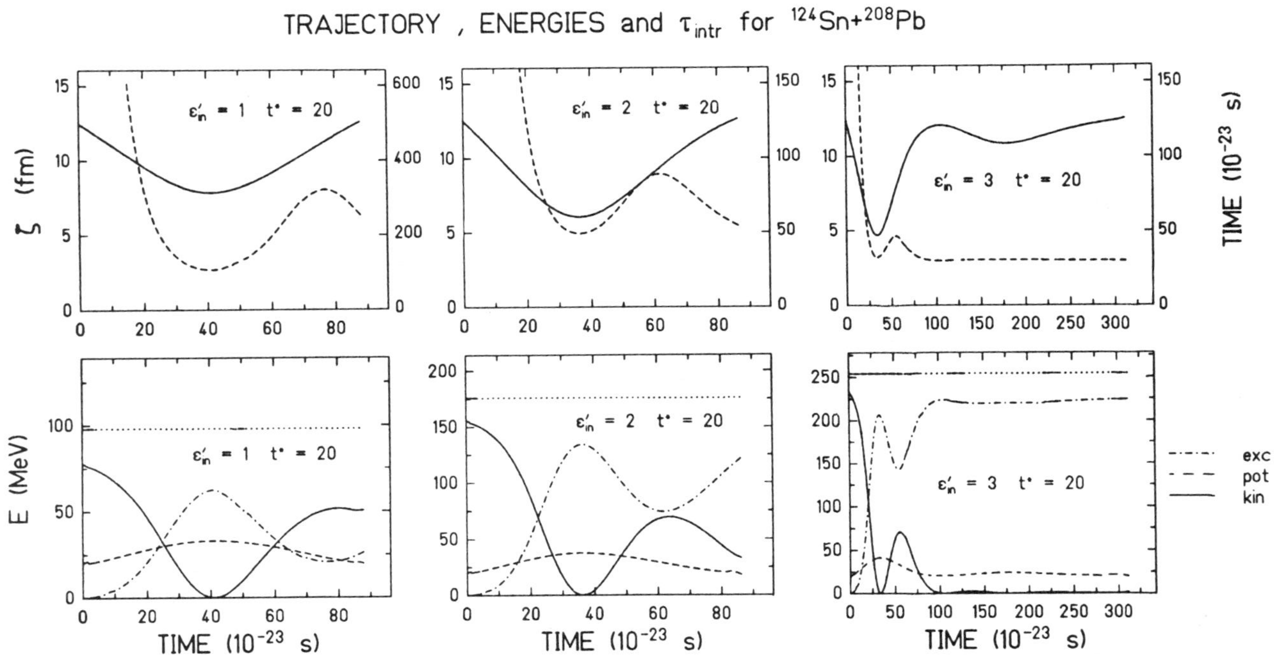

FIGURE 2. Quantities describing the time evolution of the system for three values of incident energy above the Coulomb barrier, $\varepsilon'_{in} = 1$, 2 and 3 MeV/u ($t^*=20$). Upper part: elongation ζ (solid line) and intrinsic equilibration time τ_{intr} (dashed line) as functions of time. Lower part: kinetic (solid line), potential (dashed line) and excitation (dot-dashed line) energies as functions of time. The thin dotted line denotes the total energy of the system.

The results obtained for different values of the t* parameter are summarized in Figure 3 where the kinetic energies at the exit point (Coulomb barrier) and the interaction times are shown as functions of incident kinetic energy for three different values of t*.

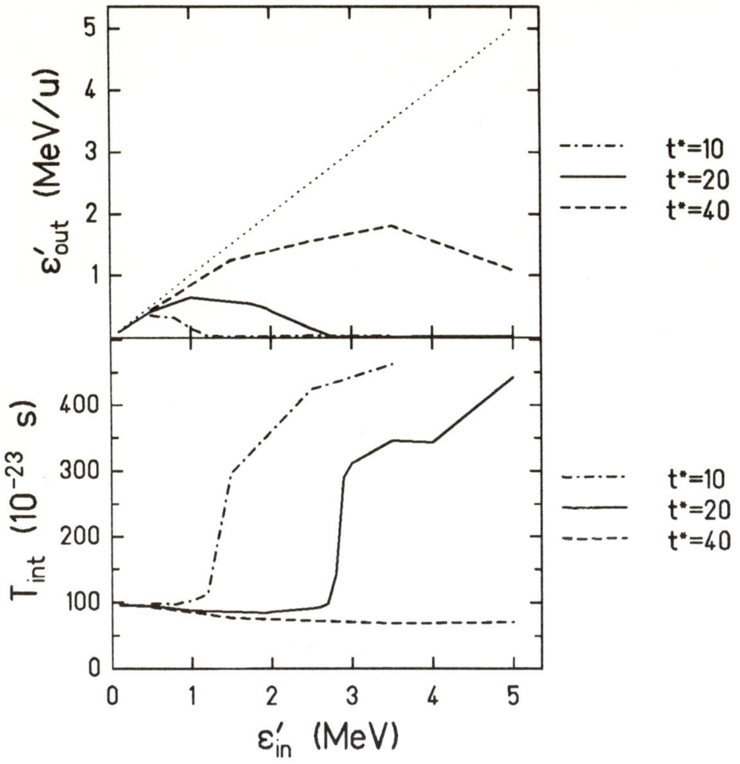

FIGURE 3. Kinetic energy at the exit point (top) and interaction time (bottom) as functions of the initial kinetic energy per nucleon above the Coulomb barrier for three different values of t*. The straight dotted line represents the initial kinetic energy (in CM system).

For the cases t* = 20 and t* = 10 (the latter is close to the frictional limit) we have found a sudden change of the interaction time as function of the initial kinetic energy. There exists a particular initial energy (the value depending on t*) beyond which the interaction time becomes roughly three times larger and the quasi-elastic behaviour of the system switches to a frictional one where the kinetic energy at the exit point becomes very small.

From the present investigation we can draw the following conclusions which are specific for the elastoplastic property tested in nucleus-nucleus collisions.

1. For energies up to roughly 2...3 MeV/u above the Coulomb barrier one can expect a quasi-elastic behaviour in central collisions of heavy systems. Therefore, in an experiment ^{208}Pb on ^{124}Sn with inverse kinematics one would observe the recoiling ^{124}Sn-like nuclei with velocities larger than the beam velocity. The critical bombarding energy, however, depends strongly on the intrinsic equilibration time (parameter t^*) which is not sufficiently well known.
2. Beyond the critical bombarding energy where the exit kinetic energy vanishes, the interaction times are more than three times larger than below.

We note in closing that our studies cannot give quantitative predictions because they are restricted to one collective degree of freedom. However, we expect that the qualitative results of the quasi-elastic recoil due to elastoplasticity remain valid also when further essential collective degrees of freedom are included.

REFERENCES

1. W. Nörenberg, Phys. Lett. 104B, 107 (1981).
2. W. Cassing and W. Nörenberg, Nucl. Phys. A401, 467 (1983).
3. S. Ayik and W. Nörenberg, Z. Phys. A309, 121 (1982).
4. W. Nörenberg, Nucl. Phys. A428, 177 (1984); W. Nörenberg, in: Theory of Nuclear Structure and Reactions, edited by M. Lozano and G. Madurga (World Scientific, Singapore, 1985), pp. 492-530.
5. A. Lukasiak, W. Cassing and W. Nörenberg, Nucl. Phys. A426, 181 (1984).
6. H.A. Weidenmüller, Progr. Part. Nucl. Phys. 3, 49 (1980); W. Nörenberg and H.A. Weidenmüller, Introduction to the Theory of Heavy-Ion Collisions (Springer, Heidelberg, 1980); A. Gobbi and W. Nörenberg, in Heavy-Ion Collisions, edited by R. Bock, (North-Holland, Amsterdam, 1980), Vol. 2, pp. 127-274;
7. A. Lukasiak and W. Nörenberg, Preprint GSI - 86, in preparation.
8. P. Möller and J.R. Nix, Nucl. Phys. A272, 502 (1976) ; A 281, 354 (1977).
9. G.F. Bertsch, Z. Phys. A289, 103 (1978).

STUDY OF ANOMALONS USING NUCLEAR CHEMISTRY TECHNIQUES

K. Aleklett(1), R. Brandt(2), M. Bronikowski(3), V.S. Butzev(4), G. Dersch(2), G. Feige(2), E.M. Friedlander(5), E.Ganssauge(6), G.Haase(2), J.Herrmann(3), D.C.Hoffman(5), B. Judek(7), D. Lee(5), E.J. Langrock(8), P. McGaughey(5), P. Kozma(4),B.A. Kulakov(4), W. Loveland(9), F. Pille(8), N.T. Porile(3), W. Schulz(6), G.T. Seaborg(5)

(1)Studsvik, Sweden, (2)Kernchemie, Marburg, F.R.G., (3)Chemistry, Purdue, Ind., USA, (4)JINR, Dubna, USSR, (5)LBL, Berkeley, Cal.,USA, (6)Physik, Marburg, F.R.G., (7)NRC, Ottawa, Canada, (8)TH Leipzig, Leipzig, German Dem. Rep., (9)Chemistry, Oregon State, USA.

ABSTRACT

Projectile fragments formed in the interaction of relativistic heavy ions within a 1cm thick Cu disk interact with a second Cu disk placed downstream. The relative yields of radioactive products in the two disks are measured by radioactivation techniques off-line using conventional Ge(Li)-detectors. Of particular interest are deep spallation products like ^{22}Na, ^{24}Na and ^{28}Mg, which are formed predominantly by high-energy particles ($E > 1$ GeV). The cross section ratio R_0 of "downstream/ upstream" Cu disks being in close contact (d=0cm) was studied for

4GeV alphas (SATURNE,Saclay), 24GeV protons (PS, CERN), 25.1GeV ^{12}C (BEVALAC,LBL), 36GeV ^{40}Ar (BEVALAC,LBL), 48GeV ^{4}He (PS,CERN), 54GeV ^{12}C (Synchrophasotron, JINR), and 72GeV ^{40}Ar (BEVALAC,LBL).

The ratio R_0 (d=0cm) scales with the total energy of the projectile; such an increase is hard to interpret - at present - with conventional models.

Furthermore, the variation of R_d with increasing distance d between the two Cu disks was studied for some of the above-mentioned systems. For total energies below $E < 40$ GeV no unusual behavior of R with d could be found. The experimental results for 72 GeV ^{40}Ar demand either (1) decay in flight ($t_{1/2} \simeq 10^{-10}$s) of some projectile fragments having large cross-sections or (2) production of secondaries with unexpectedly large transverse momenta (see G. Dersch et al., Phys. Rev. Lett. 55, 1176-1180 (1985)).

I. INTRODUCTION

The formation of projectile fragments with anomalously short mean free paths in relativistic heavy ion collisions has attracted considerable interest in recent years[1]. This anomalously short mean free path of secondary fragments could, so far, not been explained within the framework of conventional nuclear physics. This effect itself has not been definitely established experimentally[2-12]. Recently, a large number of different experiments have been presented at a conference at the GSI, Gesellschaft für Schwerionenforschung, Darmstadt, F.R. of Germany, and summarized by Heinrich et al.[13]. The results from experiments using the nuclear emulsion technique are at present controversial. The same holds for experiments using CR-39 solid state track detectors. The results from large counter experiments using thin Cerenkov detectors for the determination of Z from projectile fragments with

$8 < Z < Z_{Primary}$ were all negative. Despite the large statistical significance of their results, they can apply only to projectile fragments whose interactions lead to $\Delta Z > 0$ (i.e. at least a unit of charge change) moving within a narrow forward cone of appr. 3°. Results from bubble chamber experiments in Dubna, USSR, show some strikingly positive results[4,12]. In particular, Gasparian et al.[4] could show that for interactions of 4.2 AGeV/c ^{12}C in a propan bubble chamber one could observe a very strong "anomalon" effect for such primary interaction, where C remained C, or alternatively expressed for reactions with $\Delta Z=0$. In reactions with $\Delta Z > 0$, no such anomalously short mean free path could be observed. This was shown particularly convincing for reactions with $\Delta Z=1$. Thus, one can consider the question: Do anomalons exist? as being open from an experimental point of view. Consequently, it is important that as many different techniques as possible be brought to bear on this problem. We report on results of experiments in which the possible formation, interaction, and decay of anomalous projectile fragments has been investigated by means of activation techniques.

This radiochemical experimental technique is complementary to the other techniques used so far, as we study the target-like fragments of the interactions. Some specific radioactive target-like fragments are produced in the interaction and then identified by their well-known gamma activity. A typical example of such a radioactive nucleus is ^{24}Na ($T_{1/2}$=15h), which can easily be identified by its 1368 keV gamma-line. Given a sufficiently high beam intensity of primary heavy ions, we can measure the ^{24}Na-activity quite accurately ($\pm$ 1%). In this way, one

obtains results for the production of some target-like fragments with a very high degree of accuracy. The experimental technique and results are described in Section II. It is outside the scope of this conference contribution with its limited space available to present the quantitative interpretation of our experimental results. This will be done in forthcoming publications of this collaboration. However, we will present as the last figure our preliminary result that the "effect" observed with our experimental technique scales with the total energy of the incoming heavy ion.

II. EXPERIMENTS

We carried out the irradiations at several accelerators giving relativistic heavy ions, such as described in the preceeding abstract. The principle of our experiment is illustrated in Fig.1. Two 1.0 cm thick copper disks of circular shape (0=8 cm) were irradiated by an appropriate beam (Fig.1). Typically, appr. 10^{12} ions were obtained over a period of (2-4) hours. The beam was well focused ($\emptyset \simeq 1$ cm), as can be seen on an autoradiographic picture of an irradiated Cu-block (Fig.1b). Actually, the beam was centered (> 99%) within a central area of r=2 cm, as measured via the induced ^{22}Na-activity. These autoradiographic pictures showed no significant difference for Cu-disks numbered 1-4 (Fig.1a). For a given combination of heavy ion and its energy, pairs of Cu-disks were irradiated together in a touching-close configuration (d=0 cm), and most of the times also in 10 cm distance, and 20 cm distance. In the touching-close configuration

and in the "20 cm" configuration, the downstream Cu-disks were surrounded by a 1.0 cm thick Cu guard ring ($Ø_{out}$=14 cm, $Ø_{in}$=8 cm) (Fig.1a). Both Cu-disks in one particular configuration serve as targets for the primary beam as well as for secondaries of beam interactions interacting within the same block in which they were produced (Ar-1 and Ar-2 on Fig.1c). However, secondaries produced in the first disk (in this context "target") and interacting in the second disk ("detector", Fig.1c, Ar-3) may give an enhanced nuclide production in the "detector" disk, if such a projectile fragment has an enhanced, i.e. "anomalously" large interaction cross section.

After the completion of the irradiation one had to wait approximately 12-24 hours in order to let short-lived activity decay. Afterwards, the radionuclides present in the irradiated Cu-disks and guard rings were assayed by off-line gamma ray spectroscopy. The guard rings were cut into sections and reassembled to simulate the disk geometry. Measurements were made with Ge(Li) detectors (Δ E= 1.8 keV). The analysis of the energy spectra was based on standard radiochemical procedures[15]. Counting was carried out in the participating laboratories for several months after each irradiation.

Experimentally, we determined the activity ratio R_d for specific nuclides within any pair of Cu-disks, d being the distance betwen the Cu-disks. Due to the fact that each pair of Cu-disks was irradiated with the same particle beam at one time and afterwards asseyed for its gamma activity in a fixed position at the same Ge(Li) gamma detector, this measured activity ratio for a specific nuclide can be determined to a high degree of accuracy. All uncertainties due to particle fluxes, counting

efficiencies, uncertainties in the decay scheme for a specific radioactive nuclide, etc. cancel when we are only interested in activity ratios R. Essentially, the only experimental uncertainty in this ratio is given by the counting statistics; i.e. the number of counts in a specific gamma line as observed with the Ge(Li) gamma detector. As the number of counts is typically $\gtrsim 10^4$ events, our activity ratio R can be determined within $\pm$ 1%. Such an accuracy is comparable only to the accuracy of large counter experiments or bubble chamber experiments, as mentioned in the introduction.

Figures 2 and 3 show the dependence of R_d on product mass number for two different separations of the disks for the two ^{40}Ar energies. The dependence of R_d on A is a reflection of the energy spectrum and angular distribution of the PF (projectile fragment) inducing secondary reactions in the disks. The results for the low energy of 36 GeV ^{40}Ar show that when the two disks are in contact (R_0), the PF most likely to strike the detector disk lead to the formation (by target fragmentation) of products with $A \simeq 52$ and substantial yields are seen for all products with $A > 40$. The products with $A < 30$ (^7Be, ^{22}Na, ^{24}Na, ^{28}Mg) are formed only in high-deposition-energy target-fragmentation events. They cannot be projectile fragments because these are much too energetic to stop in the copper disks.

The R_0 values for the light fragments, e.g. 1.17 ± 0.01 for ^{24}Na, can be understood as follows: The interaction of 36 GeV ^{40}Ar with Cu leads to one heavy projectile fragment plus a substancial number of highly energetic nucleons, on average. All of these secondaries can produce ^{24}Na from copper. However, they are more likely to interact in the

downstream disk, thereby leading to $R_0 > 1$. Quantitative results will be published soon[16].

When the disks are moved to 20cm apart, the detector disk samples a different subset of the PF created in the target disk, i.e., the more strongly forward-focused and thus higher-energetic fragments. As a result, the PF most likely to reach the second disk now lead to the formation of products with $A \simeq 45$ and the formation of heavier fragments is less likely. The fragments with $A < 30$ are produced with the same yields regardless of disk separation because they are only produced by highly forward-focused, energetic projectile fragments. The activities found in the guard rings support this interpretation. Thus, the guard ring shows essentially no activity of $A < 30$ products ($< 1\%$ of the disk activity for ^{24}Na) when the disks are in contact and very little activity (4% of the disk activity for ^{24}Na) when the disks are separated by 20cm.

For the high energy of 72 GeV ^{40}Ar bombardment, the dependence of R_0 on A for $A > 40$ is similar to that obtained with 36 GeV ^{40}Ar except for the shift of the most probable product mass number to a slightly lower value ($A \simeq 48$). The values of R_{20} (for $A > 40$) are also similar to those obtained at 0.9 AGeV. However, the results for products with $A < 30$ are significantly different from those obtained with 36 GeV ^{40}Ar. The values of R_0 are substantially higher, ranging from 1.2 to 1.6. Furthermore, the ratios decrease with separation between the disks, with the values of R_{20} being 1.1 to 1.2. However, the activities of these products in the guard rings remain low ($< 1\%$ of the disk activity of ^{24}Na at d=0cm and 7% at d=20cm). Due to reasons of limited space

the reader is referred to Ref.14 for a detailed discussion of these results. However, as ^{24}Na is produced only by high-energy projectiles and as this nuclide can be determined very accurately, the results for this nuclide are shown in Fig.4 separately. Here we observe two surprising results:

1. a strong increase in R_0 when the Ar energy increases by $(28\pm2)\%$, i.e. (1.50/1.17);
2. a strong decrease in R_d with d increasing from 0cm to 20cm by $(12\pm2)\%$, i.e. ((1.50-1.34)/1.34).

A more detailed interpretation of these results have already been published[14]. A forthcoming publication of this collaboration will elucidate in more detail the quantitative arguments[16]. Here it is enough to summarize our answer: At present we are unable to interpret our results on the basis of conventional nuclear physics. It is conceiveable, that the reasons for this unability are the same as have been pointed out in other experiments, described in Ref.1 and Ref.4.

At last, we want to present some preliminary results. Recently, we obtained an irradiation of 13 Cu-disks (1cm) with 54 GeV ^{12}C at the Synchrophasotron at Dubna. Here we present preliminary results for ^{24}Na measurements in those Cu-disks normalized to the first once.

Fig.5 shows the increase in R_i for this stack of Cu-disks, together with some rather crude theoretical estimation what could be expected using conventional assumptions (details will be published later). We again observe much stronger increase of R_i with increasing depth of the Cu-block arrangement, as can be explained with

present models.

The summary of all our results for R_0-values, i.e. the activity ratio of the downstream second Cu-disk as compared to the first Cu-disk obtained for ^{24}Na activities is finally shown in Fig.6. It includes also results of two recent measurements with 4 GeV ^{4}He at SATURNE in Saclay as well as the just-mentioned results for 54 GeV ^{12}C at Dubna; results of both experiments being preliminary. Fig.6 shows that R_0 scales with the total energy of the incoming heavy ion. This holds in detail also for the three pairs of heavy ions used so far, be it ^{4}He, ^{12}C or ^{40}Ar. The experimental results are taken from Ref.15, also. For three points (36 GeV ^{40}Ar, 54 GeV ^{12}C, and 72 GeV ^{40}Ar) we tried a theoretical estimation of R_0: we were again unable to account for the steep slope of the increase in experimental values for R_0. Obviously, this work is going on to be continued.

REFERENCES

1. E. Friedlander et al., Phys. Rev. Lett., 45, 1084 (1980); Phys. Rev. C27, 1489 (1983), and references therein.
2. M.L. Tincknell et al., Phys. Rev. Lett., 51, 1948 (1983)
3. W Heinrich et al., Phys. Rev. Lett., 52, 1401 (1984)
4. A.P. Gasparian et al., Z. Phys. A, Atoms and Nuclei, 320, 459-465 (1985)
5. P.L. Jain, G. Das, Phys. Rev. Lett., 48, 305 (1982)
6. P.L. Jain et al., Phys. Rev. Lett., 52, 2213 (1984)
7. M.M. Aggarwal et al., Phys. Lett., 112B, 31 (1982)
8. H.A. Gustafsson et al., Phys. Rev. Lett., 51, 363 (1983)
9. J.D. Stevenson et al., Phys. Rev. Lett., 52, 515 (1984)
10. T.J.M. Symons et al., Phys. Rev. Lett., 52, 982 (1984)
11. Y.Golutvin et al., JINR Rapid Communications, 5-84, p.8-13 (1984), Joint Institute for Nuclear Research, Dubna, USSR, as well as I.Veres, JINR Rapid Communication, 9-85 (1985)43

12. Bano et al., GSI-85-63 (Reprint), Gesellschaft für Schwerionenforschung, Darmstadt, FRG, submitted for publication
13. "Status on Anomalon Experiments", Shop 2, in "Proc. 7th High Energy Heavy Ion Study, GSI, Darmstadt, 8.-12.10.1984, GSI-85-10 Report, ed. R. Bock et al.
14. G. Dersch et al., Phys. Rev. Lett., 55, 1176-1179 (1985)
15. G. Dersch, Dissertation, FB Physikalische Chemie, Philipps- Universität, Marburg, unpublished (1986)
16. Detailed forthcoming publication of the collaboration working with relativistic Ar-ions, to be submitted to Phys.Rev.

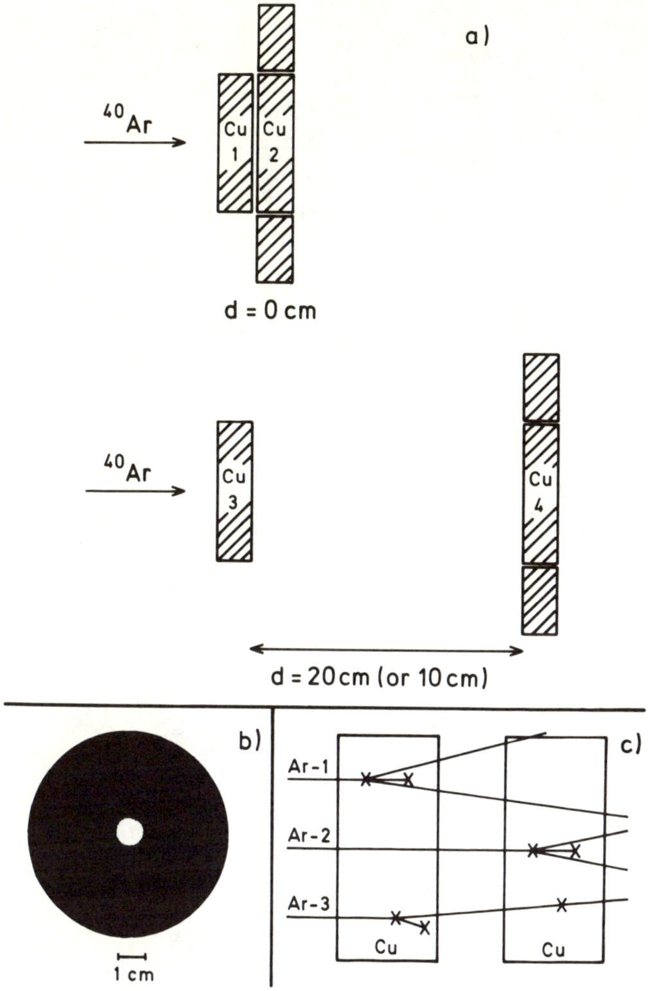

Fig.1: (a) Schematic representation of the target set-up using Cu disks and their surrounding guard rings. (b) Autoradiographic picture of a Cu disk after the irradiation. (c) Schematic representation of different reaction paths in the two Cu disks, details are given in the text.

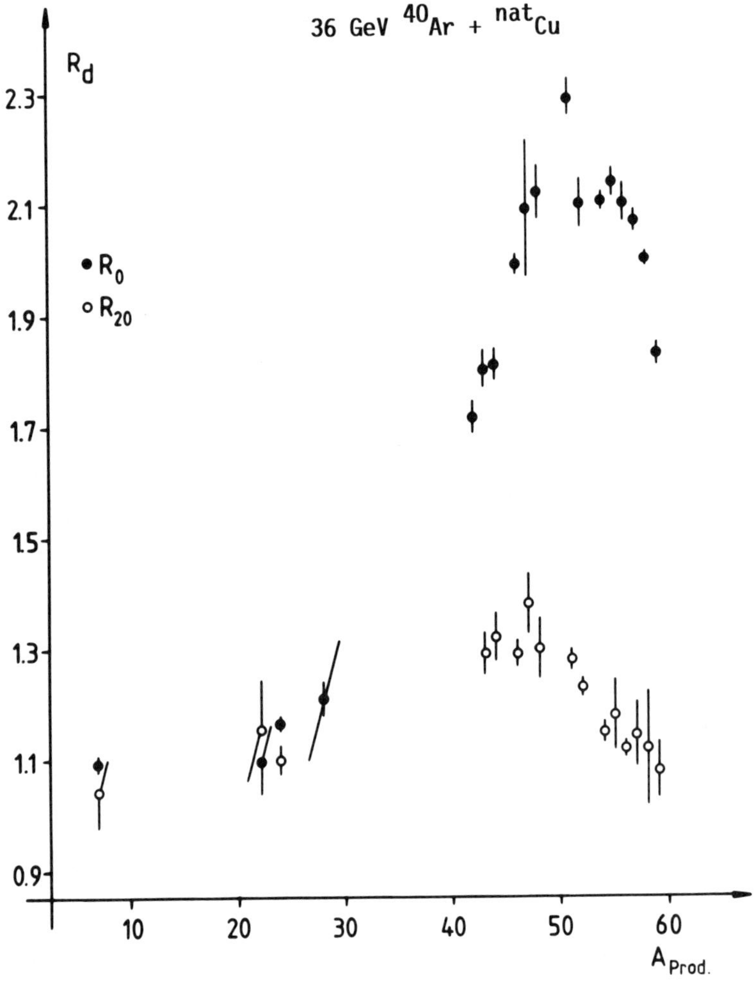

Fig.2: Activity ratios in the two Cu disks for 36 GeV ^{40}Ar in their dependence of the product mass A_{prod} for two geometric configurations with d=0cm separation between the two Cu disks (R_0) and for d=20cm, resp. (R_{20}).

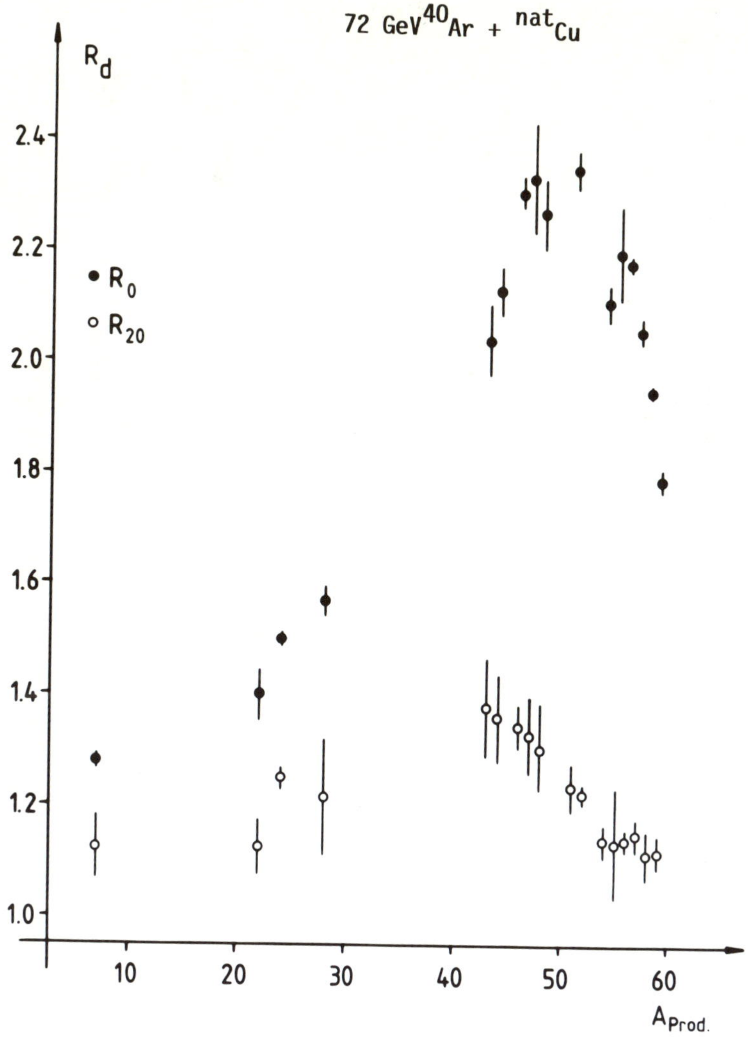

Fig.3: Activity ratios in the two Cu disks for 72 GeV ^{40}Ar in their dependence of the product mass A_{Prod} for the two geometric configurations with d=0cm separation between the two Cu disks (R_0) and for d=20cm, resp. (R_{20}).

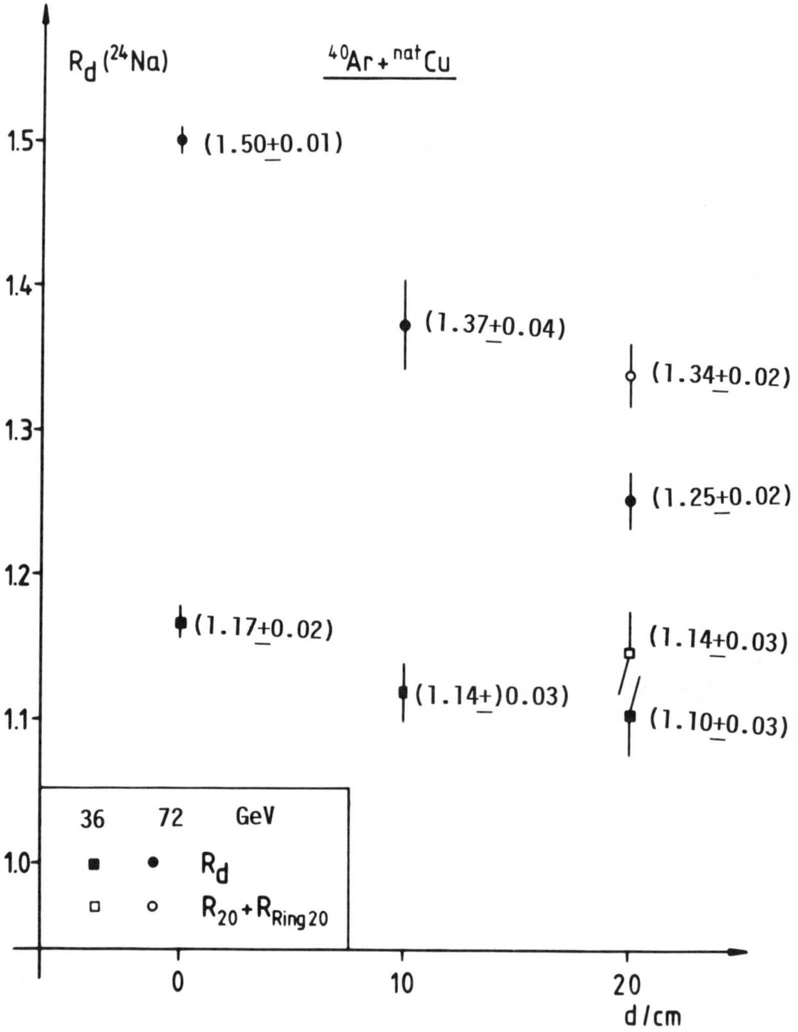

Fig.4: Activity ratios R_d for ^{24}Na produced in Cu disks with relativistic Ar ions. d is the distance between a pair of Cu disks.

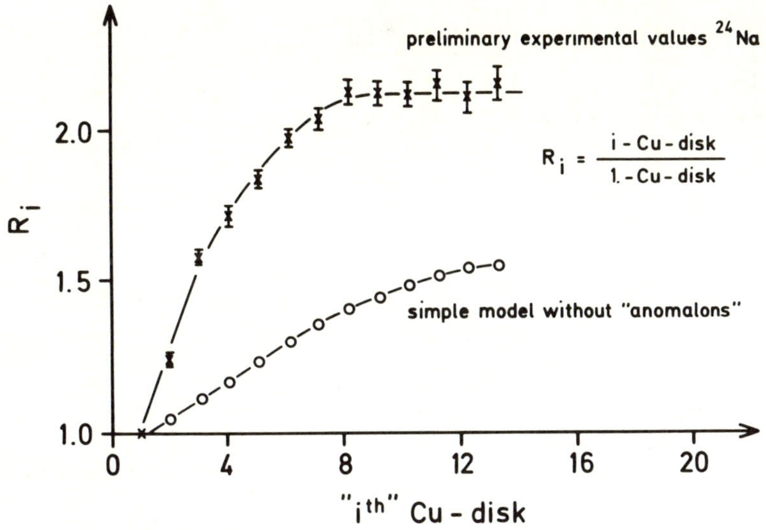

Fig.5: Preliminary results of an irradiation of 13 Cu disks with 54GeV ^{12}C at Dubna. ^{24}Na-results are given for all Cu disks, normalized to the 1. Cu disks. It is interesting to note, that this ^{24}Na-activity in the second disk (1.237 ± 0.016) is significantly higher than at the lower bombarding energy with 25 GeV ^{12}C at LBL (1.139 ± 0.009).

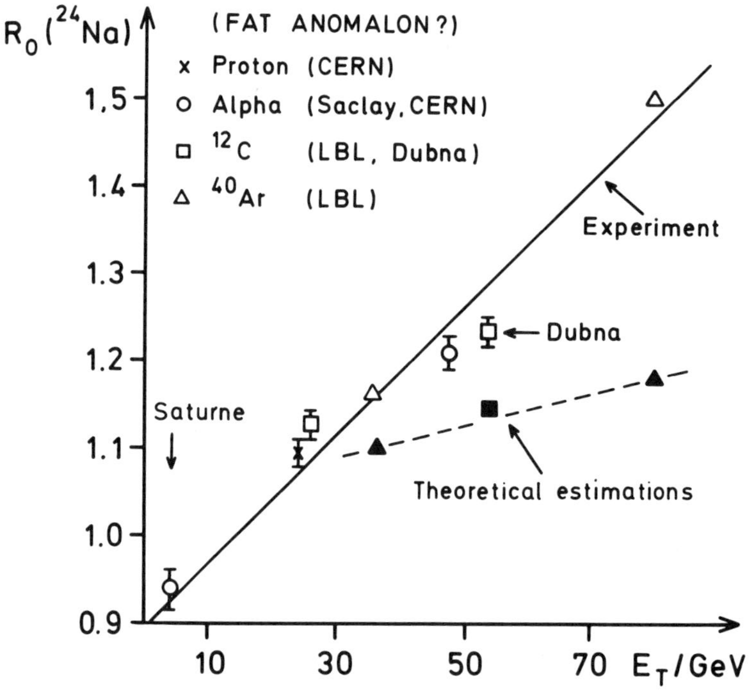

Fig.6: Increase of R_0 (^{24}Na) with the total kinetic heavy ion energy. R_0 is the ratio of ^{24}Na-activity in the "downstream" to "upstream" Cu block. The straight line through the experimental points (open symbols) is drawn in order to guide the eye. The same is true for the dashed line through the "theoretical estimations" (closed symbols). The "Dubna", the "Saclay" experimental data are preliminary, as well as the theoretical estimations.

DYNAMICAL BEHAVIOUR AND COLLECTIVE EFFECTS IN NULCEAR FISSION

AHMED OSMAN AND A.M. ABDEL-LATIF
Physics Department, Faculty of Science,
Cairo University, Cairo, Egypt.

The theory of nuclear fission is reconsidered. The dynamical behaviour as well as the collective effects for the fission process are investigated by developing the nulcear fission theory. The Strutinsky formalism and the cranking model are followed and used in calculating the components of the collective Hamiltonian. In this way, a set of collective wave functions is generated[1] from the collective Hamiltonian. This obtained set of collective wave functions is used in calculating the mass distribution of the fissioning nuclei. Theoretical expressions are derived to relate the single particle potential of the asymmetric two center shell with the collective coordinates of the fission theory. The variations of the coordinates of the centers of the two product nuclei, resulting from the fission process, with the various collective parameters are investigated. The dependence of the different oscillator frequencies on the different parameters are studied. A theoretical expression is obtained for the potential energy of a deformed nucleus as a function of the shape collective parameters.

In the present work, numerical calculations have been carried out for the dependence of the potential energy of the fissioning deformed nucleus, as being two asymmetric spheroids, on the collective shape parameters. Also, the

mass parameters of the kinetic energy operator in the fission process are calculated using the cranking model. These mass parameters are calculated for the two center oscillator model as dependent on the collective coordinates defining the shape of the deformed nucleus. Mass distributions and yields for spontaneous fission of the uranium isotopes are investigated.[2] The yields of the ^{236}U fissioning nucleus with the products of ^{95}Kr and ^{141}Ba and of the ^{238}U fissioning nucleus to the products ^{95}Kr and ^{143}Ba are calculated[3] numerically. The present obtained theoretically calculated yields are in agreement with the experimental measurements.

REFERENCES

1. J. Maruhn, and W. Greiner, Phys. Rev. C13, 2404 (1976).
2. A. Osman, Renewable Energy Sources: International Progress, Elsevier Science Publishers B.V. Amsterdam, The Netherlands, (1984), p. 301.
3. A. Osman, and A.M. Abdel-Latif, to be published (1986).

COULOMB FORCES IN THREE CHARGED PARTICLES INTERACTING VIA STRIPPING NUCLEAR REACTIONS

AHMED OSMAN
International Centre for Theoretical Physics, Trieste, Italy;
and Physics Department, Faculty of Science, Cairo University, Cairo, Egypt.

Abstract Coulomb forces are examined for three charged interating particles following the Faddeev formalism. Application of stripping nuclear reactions is studied for ^{6}Li projectile incident on ^{12}C target with alpha particle transfer. Differential cross-sections are calculated. The results are found to be improved by about 11.7% due to the inclusion of the Coulomb forces.

Coulomb force is one of the most interesting problems in the three-body problems, since it is of quite different nature than that of separable potentials. Different approaches[1-3] have been suggested, in which it is necessary to know the two-body T matrix off-the-energy-shell. One of the approaches[1] is that by approximating the Coulomb Green's functions in momentum space, while the other approach is an improved version of that approximation using the Yamaguchi potential.

In the present work, the Coulomb forces are studied in the three-body problem and included with application to direct nuclear reactions. The system is considered as a three charged interacting particles system. We considered the ^{6}Li induced stripping reaction on ^{12}C target with alpha particle transfer. The ^{6}Li projectile is a cluster composition of a deuteron and an alpha particle. The nuclear two-body interactions are taken as nonlocal separable potentials.[4,5]

The Coulomb forces are considered to act for distances much larger than the ranges of the other interactions. Thus the Coulomb Green's functions are defined by approximating the Coulomb wave functions in momentum space. Modified Faddeev equations are obtained resulting as a set of coupled integral equations. These obtained set of coupled integral equations are managable and suitable for computation.

The ^{6}Li projectile is taken as bound state of an alpha particle and a deuteron with binding energy 1.47 MeV. Then we study the nuclear stripping reaction ^{12}C(^{6}Li,d)^{16}O at ^{6}Li incident energy 42 MeV and leaving the residual nucleus ^{16}O in its ground state. Numerical calculations of the differential cross-sections are performed using the obtained integral equations which include the Coulomb forces between the interacting particles. The obtained calculations of the angular distributions are compared with the experimental data of reference 6, which are shown in Figure 1. The calculations based on the present work which include the Coulomb forces are shown by the solid curve. For the purpose of comparison, we also carried out three-body calculations for the same reaction using a previous model[7,8] and are shown by the dashed curve. The agreement between the present calculations including the Coulomb forces and between the experimental data is good as shown from Figure 1. Fitting the theoretically calculated angular distributions to the experimental data give values for the spectroscopic factors as 0.8916 for calculations due to the present model including the Coulomb forces, and 0.7834 due to previous model. Thus the effect of including the Coulomb forces in the three-body problem of the present application of ^{6}Li stripping nuclear reaction on ^{12}C target with alpha particle transfer, is improving the results with a percentage of

11.725%. This percentage of effect is not small and thus, Coulomb forces are very important and must be included in the three-body calculations.

Thus, the Coulomb forces as had been investigated[3-5] before in the case of three bound alpha particles are shown in the present work that it affect the transfer reactions with a large percentage. This investigation for the Coulomb forces in both the bound case and the transfer reactions case are performed by using three-body calculations following the Faddeev formulations. This ensures the accuracy and the large effect of the Coulomb forces in the nuclear calculations and especially following three-body problem methods. Thus, we can conclude that the Coulomb forces are important and should be taken into account in the three-body calculations.

ACKNOWLEDGEMENT

I am very thankful and grategul to Professor Abdus Salam, Professor Paolo Budini and Professor Luciano Fonda as well as to the International Atomic Energy Agency and UNESCO for their hospitality at the International Centre for Theoretical Physics, Trieste, Italy. Thanks are also due to the Centro di Calcolo dell'Università di Trieste for the use of the facilities.

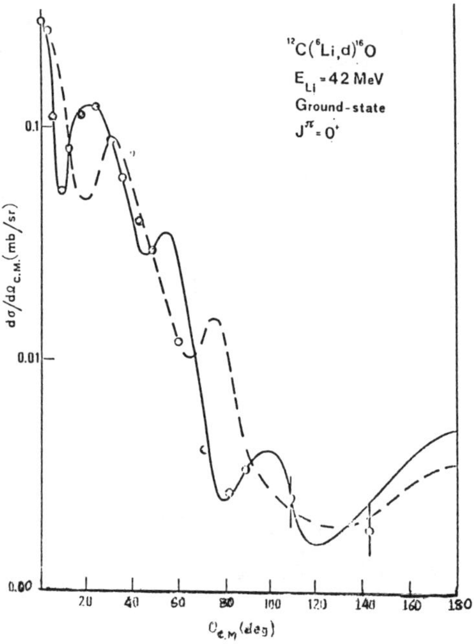

FIGURE 1. The angular distributions of the ^{6}Li stripping reaction $^{12}C(^{6}Li,d)^{16}O$ at ^{6}Li incident energy of 42 MeV leaving the ^{16}O nucleons in its ground-state. The solid curve is our present calculations. The dashed curve is calculated according to our previous model introduced in references 7 and 8. The experimental data are taken from reference 6.

REFERENCES

1. L. Schulman, Phys. Rev. 156, 1129 (1967).
2. J.V. Noble, Phys. Rev. 161, 945 (1967).
3. A. Osman, Phys. Rev. C4, 302 (1971).
4. A. Osman, Phys. Rev. C17, 341 (1978).
5. A. Osman, Phys. Rev. C19, 1127 (1979).
6. F.D. Becchetti, J. Jänecke, and C.E. Thorn, Nucl. Phys. A305, 313 (1978).
7. A. Osman, Phys. Lett. 34B, 478 (1971).
8. A. Osman, Part. Nucl. 3, 28 (1972).

Chapter 6

STUDIES AT HIGH ENERGY

Moderator:

H. MARKHUM
Institut für Kernphysik
Technische Universitat Wien
A-1020 Wien, Austria

NUCLEAR MATTER AND QUARK MATTER IN LATTICE QCD

MANFRIED FABER, HARALD MARKUM[+]
Institut für Kernphysik, Technische Universität Wien
A-1020 Wien, Austria

Abstract Energies of multi-quark systems both in the confined and deconfined phase of static QCD are calculated. In the confinement regime the free energy per quark is found to be saturated with increasing density due to color singlet formation. High above the deconfinement transition the free energy per quark is nearly independent of the density of the practically non-interacting quarks.

INTRODUCTION

According to our present understanding of the big bang the early universe has been in a state of quark matter up to a few microseconds. Afterwards quarks were confined and nuclear matter was built. To create and study such a quark-gluon plasma in the laboratory is one of the great challenges of current experimental physics. First investigations are expected to start in the summer of this year by collisions of relativistic heavy ions at Brookhaven National Laboratory and at CERN. From the theoretical point of view a phase transition between two states of strongly interacting systems is suggested by various models.[1] The most fundamental approach is the framework of lattice QCD which predicts a first order phase transition of the gluon field at critical temperatures between 150 and 200 MeV.[2] The static approximation of QCD allows for the calculation of the free energy of many-quark systems at various

[+] supported by Kulturamt der Stadt Wien

temperatures. This seems to us as a useful tool to study the behavior of quark matter. A short description of the underlying theory and a discussion of our results are the aim of this article.

THEORY

The free energy $F(\vec{x}_1,\dots\vec{x}_N)$ of N static quarks in a gluonic heat bath can be expressed as a Feynman path integral over Wilson-Polyakov loops $L(\vec{x}_1)\dots L(\vec{x}_N)$ corresponding to the world lines of the quark sources[3]

$$e^{-\frac{1}{T}F(\vec{x}_1,\dots,\vec{x}_N)} = \frac{\int \prod_{x\mu} dU_{x\mu} L(\vec{x}_1)\dots L(\vec{x}_N) e^{-S(U)}}{\int \prod_{x\mu} dU_{x\mu} e^{-S(U)}} \qquad (1)$$

In this formulation a discretization of the space-time continuum is assumed and the gauge field variables $U_{x\mu}$ live on the links of a lattice with N_x points in the space directions, N_t points in Euclidean time, and with lattice spacing a. The Wilson-Polyakov loop

$$L(\vec{x}) = \frac{1}{3}\,\mathrm{tr}\prod_{t=1}^{N_t} U_{\vec{x}t,\mu=0} \qquad (2)$$

represents the propagator of a very heavy quark which can only move in time direction. For the action of the gluon field we take the standard Wilson form

$$S(U) = \beta \sum_{x,\mu<\nu} \left(1 - \frac{1}{3}\,\mathrm{Re\,tr}\, U^{+}_{x\nu} U^{+}_{x+\hat{\nu},\mu} U_{x+\hat{\mu},\nu} U_{x\mu}\right) \qquad (3)$$

with the coupling denoted by $\beta = 6/g^2$. The temperature $T = 1/(N_t a)$ of the system is given by its inverse temporal extension. Thus, the temperature can be tuned by choosing an appropriate number of lattice points N_t in time direction or by varying the lattice spacing $a(\beta)$ which is proportional to the coupling β via the renormalization group equation. We have

investigated thermodynamical systems with three different temperatures. The spatial volume is always chosen of size 16x16x16. For the first system we take $\beta = 5.60$ and $N_t = 6$ corresponding to a temperature T = 108 MeV which lies clearly in the confinement region. Next we try $\beta = 5.85$ and $N_t = 6$ yielding T = 143 MeV which is situated close at the phase transition. Using $\beta = 5.85$ and $N_t = 3$ doubles the temperature, T = 286 MeV, to a value high in the deconfinement region. We simulate quark matter by distributing static quarks at random over a volume of size 12x12x12 cut out of the full cube in order to minimize surface effects.[4]

RESULTS

In the figure we present the behavior of the free energy per quark as a function of the number of quarks in the volume.

In the confinement regime (T = 108 MeV) quark interact through the confining forces leading to a potential of a Coulomb plus linear form. The Coulombic term comes from one-gluon exchange at short distances. The linear part stems from a string of gluons being created to locally neutralize the color of quark clusters. This singlet formation becomes easier with increasing density of the quark matter resulting in a saturation of the free energy per quark.

At the phase transition (T = 143 MeV) the gluon string breaks down and only Coulomb forces survive. As a consequence of the bounded Coulomb potential the energy content of the system is lowered for low quark densities. For high densities we find again a saturation of the energy per quark due to singlet formation.

High above the deconfinement transition (T = 268 MeV) the quarks become more and more screened by a cloud of gluons. The corresponding flat screened Coulombic potential represents

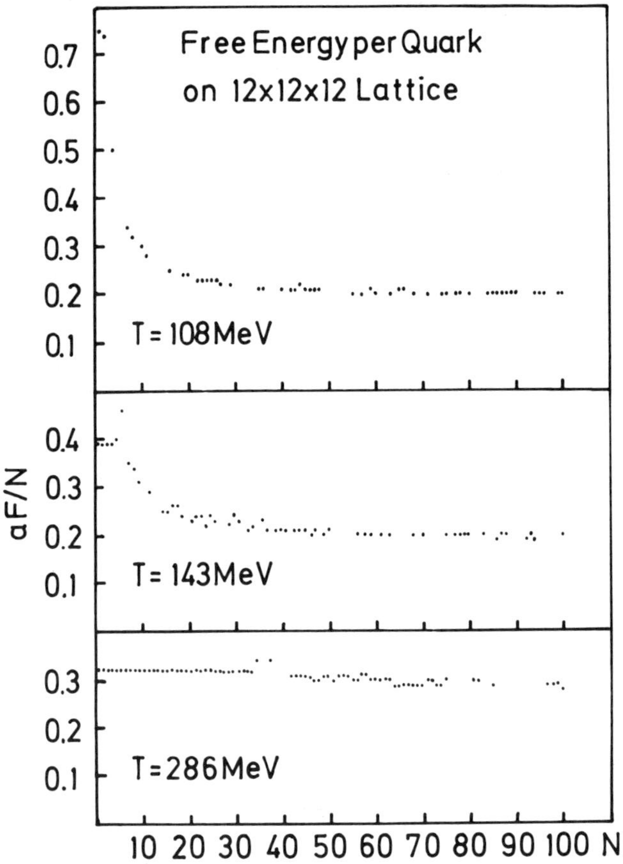

FIGURE 1. Free energy per quark in dimensionless units as a function of the number of static quarks in a volume of size 12x12x12. The thermodynamics of the quark-gluon system is studied in the confinement region (T = 108 MeV), at the phase transition temperature (T = 143 MeV), and in the deconfinement region (T = 286 MeV). For the missing data points it was impossible to resolve a signal.

practically free quarks. This leads to a constant energy contribution per quark added to the system demonstrating non-interacting quark matter at extreme temperatures.

Having presented our results we should make a remark on the length of the lattice spacing $a(\beta)$. The choice of $\beta = 5.60$ corresponds to $a = 0.30$ fm and $\beta = 5.85$ yields $a = 0.23$ fm. In order to compare the calculations for T = 108 MeV and T = 143 MeV one should consider a sublattice of 10^3 in the first case and also scale to physical energy units. Doing this does not change the consequences drawn above.

Let us conclude with a criticism of our investigations. Because static quarks have no kinetic energy we compute only the potential energy of the multi-quark system. An expected increase of the energy per quark due to repulsive forces at high densities can thus not be obtained. A consideration of the kinetic energy necessitates an extension of the static quark propagator of Polyakov's theory to the dynamic quark propagator of full QCD.

REFERENCES

1. B. Müller, The Physics of the Quark-Gluon Plasma, Lecture Notes in Physics No. 225 (Springer, Berlin, 1985).
2. H. Satz, Ann.Rev.Nucl.Part.Sci. 35, 245 (1985).
3. L.D. McLerran, B. Svetitsky, Phys.Rev. D24, 450 (1981). H. Markum et al., Phys.Rev. D31, 2029 (1985).
4. H. Markum et al., in Proceedings of the 16th Mikolajki Summer School on Nuclear Physics, edited by Z. Wilhelmi and M.Kicinska-Habior (Nuclear Physics Laboratory, Warshaw, 1985) p.143.

QUARK EFFECTS IN NUCLEAR STRUCTURE

Zoltán Árvay
Institute of Nuclear Research of the Hungarian Academy of Sciences, Debrecen, Hungary

We first present a hybrid-model, which takes into account the quark and meson degrees of freedom in the description of nuclear structure of a system with j^2 configuration coupled to I=0 total angular momentum. As an application of the model the radius of the nucleon described as a 3-quark cluster has been calculated to be 0.44 fm in agreement with the little-bag model. It was determined by evaluating a residual interaction matrix element constructed on the base of QCD assumptions and relating it to the value extracted from experimental nuclear data.

1. INTRODUCTION

Because of the fact that the nucleus consists of number A nucleons and the nucleon itself is built from three quarks, the atomic nucleus can be considered as a system of number A quark-triplets. Short range forces act between the 3-quark clusters in a nucleus and the quark structure of nucleons can play an essential role in the description of nuclear forces.

At low energies the nuclear constituents may be divided into two groups:

(i) nucleons of the inert, closed core and

(ii) nucleons of the active, valence shell.

The core nucleons occupy shell-model orbits up to the Fermi level of the core and the valence nucleons fill states of the next open shell. The shell-model potential already describes a large amount of nuclear forces. The residual interaction between valence nucleons, which accounts for the

part of interaction not included in the average central field consists of a direct and an indirect part.The indirect part takes place due to the interaction by means of nuclear core polarisation effects (e.g. phonon exchange). Free nucleons don't interact via such type of force, it can occur only between nucleons bounded in a nucleus. The direct part originates from the free nucleon-nucleon interaction. Quark-antiquark pair exchange descending from the meson clouds of nucleons can explain the medium range attractive component of the direct force[13,14]. The short range (<0.5 fm) repulsive core of this potential arises from quark-gluon exchange process between the nucleon 3-quark clusters[1,2,3,5,11].

The properties of the low-lying nuclear states are determined by the residual interaction of valence nucleons. Taking into account quark structure of nucleons in the description of the residual interaction one can determine Quantum Chromodynamical (QCD) parameters from nuclear structure data or vice versa. At present there is an interesting experimental evidence, the so called EMC effect[4] concerning the above mentioned possibility. It was experimentally observed that the quark structure of a nucleon bounded in a nucleus significantly differs from the quark distribution in a free nucleon. The EMC phenomenon is interpreted by different authors[16,17] as a result of six-quark cluster or nucleon size effects and on the base of purely nucleonic degrees of freedom[18]. The EMC effect can be regard as the first empirical fact that demonstrates the role of quark constituents of nucleons already in the ground state of the nucleus.

In this paper we take a simple model inspired by Quantum Chromodynamics to deal with a nucleus, explicitly introducing quark structure of nucleons and, in part, of nuclear forces, during the evaluation of a Cohen-Kurath matrix element of the residual interaction. Comparing the calculated value with the experimental one extracted from nuclear spectros-

copic data, we have determined the radius of the nucleon, treated as a three-quark cluster.

2.QUARK COMPONENTS OF NUCLEAR FORCES

We next consider a nucleus with two extra nucleons outside a closed-shell core. In this case only the two valence nucleons contribute to the low energy excitation spectrum. In the framework of the applied model based on the underlying Quantum Chromodynamics the two nucleons are treated as a six-quark system. The quark-gluon interaction in described by the next fundamental lagrangian

$$L_{int} = \frac{1}{2}\, g\, \bar{q}\, \lambda^{(i)} \gamma^{\mu} q G_{\mu}^{(i)} \qquad (1)$$

Where g is the coupling constant, q and $G_{\mu}^{(i)}$ denote the quark and gluon fields respectively, γ^{μ} are the Dirac matrices and $\lambda^{(i)}$ are the generators of the color SU(3) group. From the above lagrangian one can gain[10], the quark-quark interaction potential of the general form

$$V_{qq} = -\lambda_i \lambda_j [v(r_{ij}) + \sigma_i\, \sigma_j\, u(r_{ij})], \qquad (2)$$

where $v(r_{ij})$ is to describe color confinement, $u(r_{ij})$ characterises the radial dependence of the one-gluon exchange potential and σ_i, σ_j are the quark spins. Following Liberman[1]) the color electric and color-magnetic parts of the quark-quark interaction are given by

$$v(r_{ij}) = a\, r_{ij}^2, \quad u(r_{ij}) = \frac{1}{6}\, \frac{\pi}{m^2}\, \alpha_s\, \nabla^2\, v(r_{ij}) \qquad (3)$$

where r_{ij} is the distance between quarks, a is the confinement strength and $\alpha_s = g^2/4\pi$.

The quark-quark interaction potential given in (2,3) satisfies the color confinement and assymptotic freedom requirements of QCD. The six-quark wave function has been constructed from the totally antisymmetric three-quark states

$$\Psi_6 = A \Psi_3 \Psi_3 \qquad (4)$$

with the antisymmetrizer

$$A = (1 - \sum_{i=1}^{3} \sum_{j=4}^{6} P_{ij}) (1-P), \tag{5}$$

which makes Ψ_6 totally antisymmetric under the exchange of quarks between the nucleons.

Than the effective nucleon-nucleon potential is defined as the difference between the expectation values of the energy of the six-quark cluster at finite and infinite separation

$$V_{NN} = \langle \Psi_6 | \Psi_6 \rangle^{-1} \langle \Psi_6 | 2H_3 + \sum_{i=1}^{3} \sum_{j=4}^{6} V_{ij} | \Psi_6 \rangle - 2 \langle \Psi_3 | H_3 | \Psi_3 \rangle \tag{6}$$

Here H_3, Ψ_3 stands for the internal Hamilton operator and wave function of the quark-triplet, respectively. The six quarks don't fuse into one bag in this description but form identity saving 3-quark clusters at a distance R between their centres. Since the main contribution to the NN potential originates from the term describing the interaction between the 3-quark clusters, in a first approximation we can characterise the NN interaction by the next expression following Holinde[5])

$$V'_{NN} = \langle \Psi_6 | \Psi_6 \rangle^{-1} \langle \Psi_6 | [\sum_{i=1}^{3} \sum_{j=4}^{6} V_{ij}] (-9P_{36}) (1-P) | \Psi_6 \rangle \tag{7}$$

where P_{36} and $P = P_{14} P_{25} P_{36}$ are permutation operators. When the (7) expression is applied we neglect some corrective terms. This quark-gluon exchange potential quite well describes the short range repulsive core of the effective nucleon-nucleon interaction. Meson (quark-antiquark pair) exchange gives rise to the attractive part of the NN interaction. A potential of the Gaussian form has been introduced for taking into account of the meson exchange effect[13,14]). As a result the direct part of the NN force can be written in the next form

$$V_{NN}^{dir} = V'_{NN} + V_0 \exp(-r^2/2r_0^{m2}) \tag{8}$$

The one-particle quark wave function is taken as

$$\psi = \psi^{X} \psi^{SF} \psi^{C} \tag{9}$$

with the s-state orbital part of

$$\psi^{X} = \left(\frac{\nu}{\pi} \right)^{3/4} \exp\left\{ - \frac{1}{2} \nu (\vec{R}_1 - \vec{R}_a)^2 \right\} \tag{10}$$

and with ψ^{SF} and ψ^{C} that are the spin-flavour wave function of SU(4) symmetry and the color-octet SU(3) wave function, respectively[8,2]. In the description of the quark flavour states we are restricted to up(u) and down (d) states. The nucleons are characterised by colorless wave functions so the antisymmetric three-quark wave function is given by

$$\Psi_3 = \Psi_3^{X}([3])\, \Psi_3^{ST}([3])\, \Psi_3^{C}([1^3]) \tag{11}$$

where Ψ_3 and Ψ_3^{ST} are the symmetric space and spin-isospin 3-quark wave functions, respectively and Ψ_3^{C} is an anti-symmetric color-singlet wave function. The [3] and [1^3] prescriptions denote the appropriate Young symmetries. The spins and isospins of the nucleons are coupled to total S and T.

3. THE RADIUS OF THE NUCLEON QUARK-TRIPLET

For the determination of the nucleon radius the interactions of a nucleon in a nucleus give a good opportunity. Features of the nucleon-nucleon interaction can be derived from nuclear excitation spectra caused by itself the interaction. The finite size of nucleon can be considered in a very natural way, supposing quark structure for it.

The low energy excitation spectrum of a nucleus containing two nucleons outside the closed core is described by the next formula

$$\langle \psi_{IM}(jj) | H | \psi_{IM}(jj) \rangle = 2\,\varepsilon_j + E_I(jj \mid jj) \tag{12}$$

where

$$E_I(jj \mid jj) = \langle \psi_{IM}(jj) \,|\, V_{12} \,|\, \psi_{IM}(jj) \rangle \tag{13}$$

is the matrix element of the V_{12} residual interaction. ε_j is the one-nucleon energy relative to the core and $\psi_{IM}(jj)$ is the wave function of the two nucleon system when both 3-quark clusters are restricted to one orbit with angular momentum j and are coupled to angular momentum I. Placing the two nucleons into a harmonic oscillator shell-model potential the corresponding one particle orbitals are given by oscillator wave functions

$$\Phi_{nl}(r) = R_{nl}(r)\, Y_{lm}(\vartheta,\varphi) \tag{14}$$

where R_{nl} is the oscillator radial part and Y_{lm} is the spherical harmonic.
The effective residual interaction is

$$V_{12} = V_{NN}^{dir.} + V_{NN}^{indir.} \tag{15}$$

and further the indirect part will be neglected (weak coupling, hard inert core).

For the description of effective residual interaction between two nucleons as an approximation one can use only the color-magnetic part of the underlying quark-quark interaction. This idea is supported by the contrary behaviour of the two parts of the quark-quark interaction, because in the $R \to 0$ limit the color-magnetic part gives the main component of the interaction whereas the color-electric part vanishes. As a direct part we adopt the above gained quark-gluon-meson exchange hybrid potential (9) excluding the color-electric part

$$V_{12} = V'^{M}_{NN} + V_{meson} \tag{16}$$

At the evaluation of the V'^{M}_{NN} expression the orto-normalizing factor has been approximated by the value $1/\sqrt{20}$ in the $R \to 0$ limit taken from work of Harvey[2]) for the case of Young symmetry [6] of the six-quark space wave function. Thus the matrix element of the above residual interaction can be expressed in terms of the Talmi integrals. In case of oscillator orbitals the interaction energy is given by

$$E_{IT}(j_1j_2;j_1j_2) = \sum_{\substack{p\lambda S\\ nlNL}} \gamma_{\lambda S1} C^{nl}_{NL} B_{nlp} I_p \tag{17}$$

where $\gamma_{\lambda S1}$ are the antisymmetrised LS→jj transformation coefficients, C^{nl}_{NL} are the coefficients of the Brody-Moshinsky transformation from the laboratory system to the center of mass system, B_{nlp} are coefficients taken from [15], I_p are the Talmi integrals, I, T are the total angular momentum and total isospin, λ denotes the resulting orbital angular momentum, S stands for the total spin, nl and NL are the main and orbital quantum numbers, respectively of the relative and the center of mass motion in the center of mass system.

Cohen and Kurath[6] have determined the value of the matrix element (17) for $p_{1/2}$ orbitals from experimental data and find $E^{exp}_{01}(\frac{1}{2}\,\frac{1}{2}\,;\,\frac{1}{2}\,\frac{1}{2})=-0.26$ MeV. Evaluating the theoretical matrix element formula (17) the strength $V_o= -20$ MeV and range $r^m_o= 1.5$ fm consistent with[7] has been used for the attractive meson-exchange part of the NN interaction. The QCD parameters for the evaluation of the repulsive core potential were a=10.9 MeV fm^{-2} and b=18.6 MeV ($b=\pi\alpha_s/6m^2$)[11].

Finally, equating the theoretical value of the matrix element with the experimental data, we have gained ς_N=0.44 fm for the radius of the nucleon assumed to be a 3-quark cluster.

Y. He et al.[12] have investigated the rms radius of quarks in nucleons and find that it ranges from 0.3 to 0.5 fm. Their result was obtained in the framework of a hybrid quark model of the NN spin-orbit forces with one-gluon-exchange and one-meson-exchange quark-quark interaction and fitting it to the Paris-potential. Our result is in good agreement with the work of Y. He et al. and also is in consistence with the little bag model value for the bag radius[9].

REFERENCES

1. Liberman, D.A.: Phys. Rev. D 16 1542 (1977)
2. Harvey, M.: Nucl. Phys. A 352, 301, 326 (1981)
3. Árvay, Z.: Proceedings of the X. European Symposium on

the Dynamics of Few-Body Systems, pp. 4-6, Balatonfüred, Hungary, 1985.

4. Aubert, J.J. et al.: Phys. Lett. B123, 275 (1983)
5. Holinde, K.: Nucl. Phys. A 415, 477 (1984)
6. Cohen, S., Kurath, D.: Nucl. Phys. 73, 1 (1965)
7. Auger, G., Lagrange, J.M., Pantrat, M., Vanhorenbeeck,J.: Nucl. Phys. A 426, 109 (1984)
8. Feynman, R.P., Kislinger, M., and Ravndal, F.: Phys. Rev. A 2706 (1971)
9. Brown, G.E., Rho, M.: Phys. Lett. B 82, 177 (1979)
10. de Rujula, A., Georgi, H. and Glashow, S.L.: Phys. Rev. D12, 147 (1975)
11. Toki, H.: Z. Phys. A 294 173 (1980)
12. He, Y., Wang, F. and Wong, C.W.: Nucl.Phys.A 448 652 (1986)
13. Faessler, A., Fernandez, F., Lübeck, G., Shimizu, K.: Phys. Lett.B112, 201 (1982)
14. Yamauchi, Y., Yamamoto, R., Wakamatsu, M.: Nucl. Phys. A443 628 (1985)
15. Brody, T.A., Moshinsky, M.: Tables of transformation brackets. 2nd Edn., New York, London, Paris: Gordon and Breach 1967.
16. Faissner, H., Kim, B.R.: Phys. Lett. B130 321 (1983)
17. Jaffe, R.L., Close, F.E., Roberts, R.G., Ross, G.G.: Phys. Lett. B134 449 (1983)
18. Birbrair, B.L., Gridnev, A.B., Zhalov, M.B., Levin, E.M., Starodubski, V.E.: Phys. Lett. B166 119 (1986)

THE STRUCTURE OF A 3-QUARK SYSTEM IN THE HYPERSPHERICAL-FUNCTIONS METHOD

K.V.SHITIKOVA

Institute of Nuclear Physics, Moscow State University, USSR

Within the hyperspherical functions method [1] properties were studied of a 3-quark system. The analytical expression for effective potential have been obtained. The excitation energies and the density distribution have been found for the ground and monopole-excited states of N-particle. The N - Δ -mass difference, the r.m.s. nucleon radius, the elastic eN scattering formfactor are studied.

I. INTRODUCTION

The nonrelativistic potential model with the harmonic oscillator functions [2-4] was successfully developed and made it possible to obtain some important results concerning the masses and the decay properties of light baryon systems. However, the parameters of the quark-quark potential used

$$v_{ij} = (\lambda_i \lambda_j)\left[f(r_{ij}) + (\vec{\sigma}_i \vec{\sigma}_j)\, g(r_{ij})\right],$$
$$f(r) = A e^{-r^2/\alpha^2} + B r^2 + C + K\delta(r),$$
$$g(r) = \frac{2}{3} K e^{-r^2/\Lambda^2} . \qquad (1)$$

are highly senstive just to the value of the oscillator radius. We apply the hyperspherical functions method where the oscillator radius as a free parameter is replaced by the collective variable ρ, so that we get eventually a self-adjusted quark system whose q-q interaction parameters determine the dimensions of the system.

2. CALCULATED METHOD

In the hyperspherical functions method the wave function of an n-quark system is sought to be of the form of expansion on K-harmonic polynomials [5]:

$$|nK[f]\varepsilon LST>: \Psi(1,2\ldots n)=\rho^{-\frac{1}{2}(3n-4)}\sum_{K\gamma} \chi_{K\gamma}(\rho)|nK\gamma> \quad (2)$$

where $\gamma=[f]\,\mathcal{E}_c\,LST, \quad \int \chi^2_{K\gamma}(\rho)\,d\rho=1.$

The Hamiltonian is

$$H=-\frac{\hbar^2}{2m}\frac{1}{\rho^{3n-4}}\frac{d}{d\rho}\left(\rho^{3n-4}\frac{d}{d\rho}\right)-\frac{\hbar^2}{2m}\frac{1}{\rho^2}\Delta_\theta+V(\rho) \quad (3)$$

and the set of equations to find the eigenfunctions and the eigenvalues E takes the form:

$$\left\{\frac{d^2}{d\rho^2}-\frac{L(L+1)}{\rho^2}-\frac{2m}{\hbar^2}\left(E+W^{K\gamma}_{K\gamma}(\rho)\right)\right\}\chi_{K\gamma}(\rho)= \quad (4)$$

$$=\frac{2m}{\hbar^2}\sum_{K'\gamma'\neq K\gamma} W^{K'\gamma'}_{K\gamma}(\rho)\,\chi_{K'\gamma'}(\rho),$$

where $L=K+\frac{3A-6}{2}$ and $W^{K\gamma}_{K\gamma}(\rho)$ is the effective potential. We obtain the analytical expressions of the effective potential $W(\rho)$ in terms of the hyperspherical functions method.

$V(r)$	$-ar$	$-a'r^2$	b/r	$æ\delta(\bar{r})$	$V_1 e^{-r^2/\alpha^2}$

$W(\rho)$	$-0.96\,a\rho$	$-a'\rho^2$	$1.2\,b/\rho$	$\frac{\sqrt{2}}{\sqrt{\pi}}$	$\frac{\varkappa}{\rho^3}$	$\frac{2V_1\alpha^2}{\rho^2}e^{-\rho^2/\alpha^2}I_1(\rho^2/\alpha^2)$

where $I_1(\rho^2/\alpha^2)$ is the modified Bessel function. An important feature of the δ-like spin-spin interaction is that its contribution to the effective potential acts throughout the domain of the collective variable ρ .

3. RESULTS AND DISCUSSIONS

The calculation results are shown in Tables I and 2 and in Fig.1–3 .

Table I

$b=\alpha$ fm	A MeV	B MeV·fm^2	C MeV	K MeV·fm^3	Λ fm
0.8	3810.0	–12.5	–479.8	–911.1	0
0.8	3810.0	–12.5	–1715	0	0
0.8	3810.0	–12.5	–1618	–73	0.2
0.8	3810.0	–12.5	–1618	–73	0.2

Table 2

M_N MeV	M_Δ MeV	E_0 MeV	E_1 MeV	r_0 fm	r_1 fm
1086	1086	0	1858	0.39	0.60
938.4	–	147.6	1673	0.39	0.60
	1228	142.3	1869	0.42	0.63

We renormalized the constant C of the Harvey potential [1] (the upper line of Table I) in such a way as to reproduce the nucleon mass, while the parameter K of this quark–quark potential was so varied that the N–Δ mass difference ($M_\Delta - M_N$ = 293 MeV) could be described. Then we arrive at a new set of parameters reported in the

Table I. In Fig.1 drawn are effective potentials $W_n(\rho)$, $W_\Delta(\rho)$ and solutions corresponding to the ground and monopole-excited states with excitation energies $E_N^* = 1.54$ GeV and $E_\Delta^* = 1.75$ GeV. In contrast to the results of calculating with the harmonic oscillator functions where the spin-spin interaction enters additively, the inclusion of spin-spin interaction in the hyperspherical functions method results in variations of the core width and of the well depth (the core width gets narrower, and the effective potential depth increases, for the nucleon). As a result, the energy of the monopole-excited state proves to be lower for N (~ 1.7 GeV) than for Δ (~ 1.9 GeV), while the radius for N (r_N =0.39 fm) is smaller than for Δ (r_Δ =0.42 fm). Besides that, the calculations have shown that the dimensions of the system in its excited state vary substantially (by a factor of ~ 1.6), namely, $r_N^* = 0.6$ fm and $r_\Delta^* = 0.64$ fm.

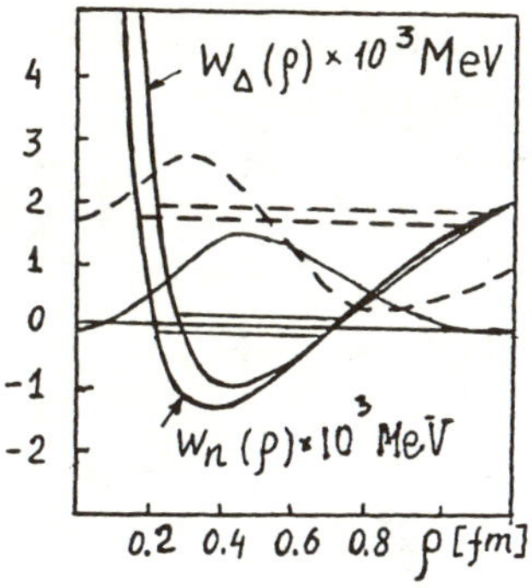

Fig.1. Effective potentials $W_n(\rho)$, $W_\Delta(\rho)$ and solutions correspoding to the ground and monopole-excited states.

4. ELASTIC eN SCATTERING IN THE THREE-QUARK SYSTEM

The radial functions $X_i(\rho)$ may be used to find the density distribution of the ground and excited states in the three-quark system

$$n_{ij}(r) = \mathcal{N} \int_r^\infty \frac{(\rho^2 - r^2)^{\frac{5n-16}{2}}}{\rho^{5n-13}} \chi_i(\rho)\chi_j(\rho)\, d\rho. \tag{5}$$

Fig.2 presents the results of calculating the nucleon density distribution in ground state (the solid line) and in monopole-excited state and the transitional density (the dash-dotted line). The densities obtained were used to make nonrelativistic calculations of the form-factor of the eN scattering in the three-quark system. Fig.3 shows the non-relativistic formfactor $F_{NR}(q^2)$ (the solid line). It is seen that a satisfactory agreement with the phenomenological dipolar formfactor

$$F_\Delta(q^2) = (1 + q^2/.71)^{-2} \tag{6}$$

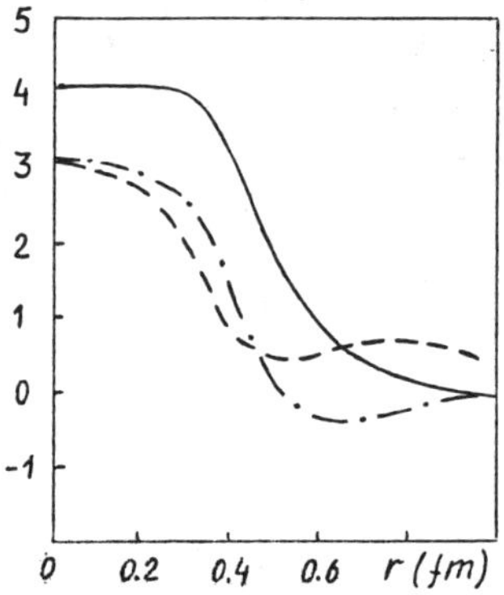

Fig.2. The density distribution for the nucleon in the ground-state (the solid line) and monopole-excited state (the dotted line) and the transitional density (the dash-dotted line)

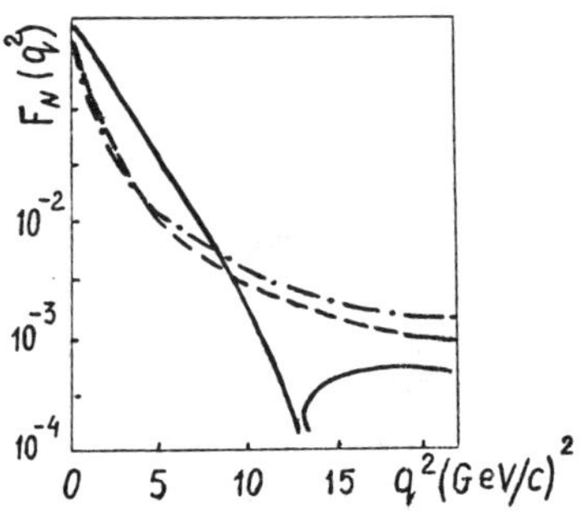

Fig.3. The elastic formfactor of the nucleon; the nonrelativistic and relativistic calculation results (the solid and dotted lines, respectively); the dipolar formfactor (the dash-dotted line).

is obtained only at $q^2 \ll M_N^2$.

In view of the above, we used the formfactor relativisation method proposed in the work [6] where allowance was made essentially for the effect of the Lorentzian transformation of the longitudinal distances when going from the Breit system to laboratory system. In this case the relativistic formfactor may be inferred in a sufficiently simple way from the nonrelativistic one:

$$F_R(q^2) = (1 + q^2/4M_A^2)\, F_{NR}\left(q^2/(1+q^2/4M_A^2)\right) \tag{7}$$

$$M_A^2 = \sum_i \left(m_i^{eff}\right)^2 = n m_q^2$$

Here, m_q is the quark mass which was selected in our calculations (see Table 2) to be m_q=1.086/3 = 0.362 GeV, whence M_A^2 = 0.393 GeV^2 . The allowance for the relativistic effects gives rise to an increased r.m.s. radius of the n-quark system:

$$\langle r^2 \rangle_R = \langle r^2 \rangle_{NR} + \frac{3(n-1)}{2M_A^2} \tag{8}$$

For example, r_R = 0.66 fm for the nucleon, which agrees with the experimental nucleon radius 0.78 fm.

It is seen that the inclusion of relativistic effects makes it possible to describe the experimental data.

5. CONCLUSION

Thus, the following effects appear when the three-quark system is solved in terms of the hyperspherical functions method:

renormalization of the parameters C and K compared with those proposed by Harvey in the harmonic oscillator model;

nonadditivity of the spin-spin interaction contribution resulting in variations of the core width and well depth for the effective potential.

It is of pleasure for the author to thank my collaborator V.V.Burov.

REFERENCES

1. Burov V.V., Shitikova K.V. Proc. VIII Intern.Seminar High Energy Physics Problems, Dubna, 33,34 (1986).
2. Harvey M. Nucl. Phys., A352, 326 (1981); Nucl. Phys. A424, 419 (1984).
3. Ohta S., Oka M., Arima A., Yazaki K. Phys.Lett., 119B, 35 (1982).
4. Oka M., Yazaki K. Nucl. Phys. A402, 477 (1983).
5. Simonov Yu.A. Sov.Nucl.Phys., 3, 630 (1966). Smirnov Yu.F., Shitikova K.V. Particles and Nuclei, 8, 847 (1977); Shitikova K.V. Nucl. Phys. A231, 365 (1979).
6. Stanley D.P., Robson D. Phys.Rev., 26D, 223 (1982).

ISOBAR DEGREES OF FREEDOM AT FINITE TEMPERATURE IN A MEAN FIELD CALCULATION

R.M. QUICK AND H.G. MILLER
Theoretical Physics Division, NRIMS, CSIR, P O Box 395, Pretoria 0001, South Africa.

J.P. VARY
Physics Dept., Iowa State Univ., Ames, IA50011, U S A

W. FABIAN
Frankfurt, West Germany

Abstract Finite temperature Hartree Fock (FTHF) calculations have been performed in ^4He in a quasi particle basis containing both nucleons and Δ(1232) isobars. For temperatures below 10 MeV there is a neglibibly small admixture of the isobar states.

The importance of the Δ(1232) resonance in pion condensation and the formation of giant Gamow Teller states is well known. However previous calculations of these phenomena have always assumed there to be no Δ admixture in the uncorrelated nuclear ground state. At zero temperature one would probably not expect to find Δ's in the nuclear ground state obtained in a Hartree Fock mean field calculation. Although several of the NΔ matrix elements are large as a result of the larger basic coupling constants involved they do not entirely compensate for the mass difference between the nucleon and delta states. At higher temperatures, however, one will ultimately find Δ contributions when the thermal energy becomes sufficient to excite the internal degrees of freedom. A crucial question is at what temperature this occurs in a realistic calculation.

In the present finite temperature Hartree Fock calculation the quasi particles may contain a mixture of nucleon

and Δ particles. In ^4He the calculation has been carried out in a model space containing 200 single particle levels. These comprise 20 proton, 20 neutron, 40 Δ^{++}, 40 Δ^{+}, 40 Δ^{0} and 40 Δ^{-} states in the 0s, 0p, 1s-0d shells. Isospin symmetry has been assumed and the quasi-particle states are taken to have good $T_z = \pm\frac{1}{2}$. For ^4He then the problem may be effectively reduced to a 2 particle problem in a basis space consisting of 60 single particle states. The Hamiltonian for the system is

$$H = (T_{rel} + V_{eff})_N + \Delta M + T_{lab} + V_{N\Delta} + V_{\Delta\Delta}$$

where ΔM is zero for the nucleon states and the Δ-nucleon mass difference for the Δ sector. In the nucleon sector the Hamiltonian contains the relative kinetic energy operator and the effective two body potential described in reference 1. In the Δ sector the laboratory kinetic energy operator has been used for simplicity. The transition potentials $V_{N\Delta}$ and $V_{\Delta\Delta}$ are the bare interaction arising from one pion and one rho exchange [2]. In order to avoid the singularities in these potentials a dipole regularization factor, dependent on the regularization parameter has been used.

In principle the transition potentials should be renormalized to account for the truncation introduced by the model space. However, rather than renormalizing, we have performed the calculation for a range of values of Λ, the regularization parameter. Calculations in the deuteron show that the percentage admixture of isobars in the deuteron is roughly 1% for $\Lambda = 5fm^{-1}$ [2]. Increasing Λ makes the potentials more attractive and thus increases the isobar admixture probabilities. The present calculations were performed for Λ = 5, 10 and 20 fm^{-1}. A more consistent procedure would involve calculating the ground state of the deuteron in the

model space and adjusting Λ, and ultimately the coupling constants, to reproduce the isobar admixture obtained in standard calculations.

In the present work no Δ admixtures were obtained in the HF ground state for Λ = 5, 10 and 20fm^{-1} for temperatures ranging from 0 to 10 MeV, the temperature at which a deformed to spherical phase transition occurs. This suggests that Δ degrees of freedom will have little effect on shape transitions at normal densities within the mean field approximation. However, it should be noted that the double Δ component is important because of its strong coupling to the nucleon sector. Therefore the Hartree Fock Bogolyubov method and its finite temperature extension may provide a better description. Future work will study the effect of increasing pressure on the rate of the Δ's.

This work was supported in part by the U.S. Department of Energy under Contract No. DE-AC02-82ER40068, Division of High Energy and Nuclear Physics.

REFERENCES

1. G. Bozzolo and J.P. Vary, Phys Rev C31 (1985) 1909.
2. H. Arenhövel, Z. Phys A275 (1975) 189.

SEARCH FOR DYNAMICAL SYMMETRY IN THE $^{12}C+^{12}C$ RESONANCES

J. Cseh* and J. Suhonen**

*Institute of Nuclear Research, Debrecen, Pf. 51, Hungary-4001

**Department of Physics, University of Jyväskylä, SF 40100 Jyväskylä, Finland

The energy spectrum of the $^{12}C+^{12}C$ resonances is described within the dipole (vibron) and the quadrupole (IBM-1) algebraic models. The Hamiltonians corresponding to each dynamical symmetries and to the general cases of the models are used. Based on the present experimental data no choice can be made between the dipole and quadrupole descriptions, but in both cases the best fit is quite close to the dynamical symmetry corresponding to a soft vibrator.

As it is well known the reactions of some light heavy ion systems show resonances. Many experimental and theoretical investigations concluded that this fact is due to some collective behaviour of the compound nuclei. For a long time this collectivity has been considered to be the same kind as the one appearing in the low-energy region of nuclei i.e. to be of quadrupole type. In 1981, however, Iachello suggested an interpretation in terms of dipole collectivity[1]) . An algebraic description of the dipole motion can be given by means of the vibron model [2]). One of its special cases called O(4) dynamical symmetry was used by Erb and Bromley [3]) for the description of the $^{12}C+^{12}C$ re-

sonances. They obtained good agreement between the experimental and model spectra. Investigating the question of multipolarity Cindro and Greiner applied a geometrical quadrupole model to the same system [4]. The goodness of their description is similar that of ref. [3].

The interacting boson approximation offers a convenient way for a uniform description of collective spectra with different multipolarities and dynamical symmetries. Both in the dipole and in the quadrupole models the spectra are generated by taking into account a finite number of bosons which are in two-body interactions with each other. In the dipole case the bosons can occupy single particle levels with $\ell=0$ or $\ell=1$ angular momentum, the general Hamiltonian has U(4) group structure, and there are two dynamical symmetries labeled by the $U(4)\rightarrow U(3)\rightarrow O(3)$ and $U(4)\rightarrow O(4)\rightarrow O(3)$ group chains. In the quadrupole case $\ell=0$ and $\ell=2$ single particle states are taken into account, the general Hamiltonian has U(6) group structure and there are three dynamical symmetris corresponding to the group chains: $U(6)\rightarrow U(5)\rightarrow O(5)\rightarrow O(3)$; $U(6)\rightarrow SU(3)\rightarrow O(3)$; $U(6)\rightarrow O(6)\rightarrow O(5)\rightarrow O(3)$ [5].

We have performed calculations [6,7] within these models for the $^{12}C+^{12}C$ system with twofold aim. On one hand we wanted to investigate the question of multipolarity in a slightly different way than in ref. [4]. On the other hand we wanted to compare the agreement between the experimental spectrum and those corresponding to all possible dynamical symmetries of these two models and to explore what kind of preference can be found for these symmetries.

The experimental states selected by Erb and Bromley [3] have been used. We took the spin-parity values for the uncertain levels as suggested in ref. [3], but in the fitting procedure these states had only 50 % of the weight given to the exactly known resonances. In the first step the Hamiltonians of the dynamical symmetries were applied. The experimental spectrum consists of 38 resonances and they are

arranged into multiplets according to the relative energy differences, as in ref. [3]), also keeping in mind the relations among the quantum numbers of the dynamical symmetries [2,5]). For the total number of bosons we have chosen the smallest value that gave enough bands. This number turned out to be N=20 and N=8 for the dipole and quadrupolc cases, respectively. The parameters of the dynamical symmetric Hamiltonians were fitted by a least square procedure, and the results were used as initial values for the general Hamiltonians. The latter ones were those of the codes ROTVIB [8]) and PHINT [9]) equipped also with least square fitting. The results are summarized in Table I. The agreement between the experimental spectrum and the model one corresponding to the last row of Table I is shown in Fig. 1.

Table I. Goodness if fits (χ^2) obtained with different descriptions

Model	Dyn. symm.	$\chi^2\,(\mathrm{MeV}^2)^a$
U(4)	U(3)	2.07
U(4)	O(4)	1.89
U(4)	No	1.57
U(6)	U(5)	2.02
U(6)	SU(3)	3.70
U(6)	O(6)	3.83
U(6)	No	1.47

a $\chi^2 = \sum_{i=1}^{N} W_i (E_i^{EXP} - E_i^{THEOR})^2$

When we deal with situations intermediate between the limiting cases of models it is useful to display the positions of the system on the phase diagrams of the models. In the dipole case this diagram is one dimensional while for the quadrupole model it is two dimensional. We have

found it convenient to measure the distances from the ending points by quantities which take values between 0 and 1.

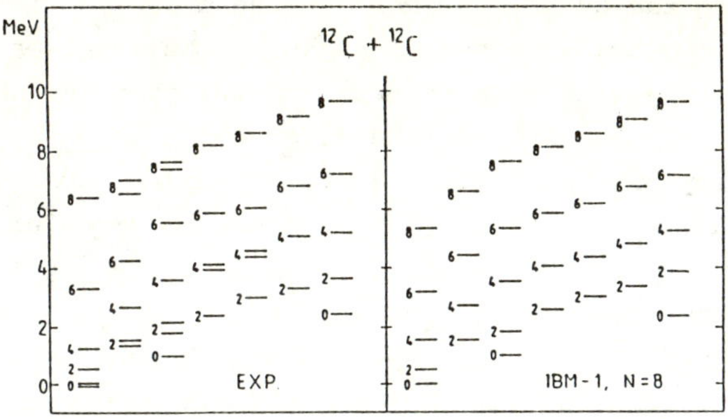

Fig. 1. Energies of experimental and model states measured from the lowest lying resonance. Only those theoretical levels are shown which have experimental correspondents.

These are defined by the parameters of the Hamiltonians in the following may. For the dipole model:

$$H_d = \bar{\alpha} C_{2O4} + \bar{\beta} C_{2O3} + \bar{\gamma} C_{1U3} + \bar{\delta} C_{2U3} \tag{1}$$

$$m \equiv \frac{\bar{\alpha}^2}{\bar{\alpha}^2 + \bar{\gamma}^2 + \bar{\delta}^2} , \tag{2}$$

and for the quadrupole model:

$$H_q = \varepsilon''' C_{1U5} + \alpha C_{2U5} + \beta C_{2O5} + \gamma C_{2O3} + \delta C_{2SU3} + \zeta C_{2O6} \tag{3}$$

$$m_1 \equiv \frac{\varepsilon'''^2 + \alpha^2}{\varepsilon'''^2 + \alpha^2 + \delta^2 + \zeta^2} , \quad m_2 \equiv \frac{\delta^2}{\delta^2 + \zeta^2} \tag{4}$$

Here m is the distance on the one dimensional phase diagram of the U(4) model from the U(3) ending point, towards the O(4) end. Similarly: m_1 and m_2 determine the position within the triangle of the phase diagram of the U(6) model as fol-

lows: m_1 is the distance of the phase space point from the O(6)-SU(3) edge towards the U(5) corner, while m_2 gives the distance from the U(5)-O(6) edge along a line which is parallel to the O(6)-SU(3) edge.

With the parameters of our best fit we obtain m=0.003 or m_1=0.95 and m_2=1.00. These values correspond to points quite close to the U(3) and U(5) corners (Fig. 2).

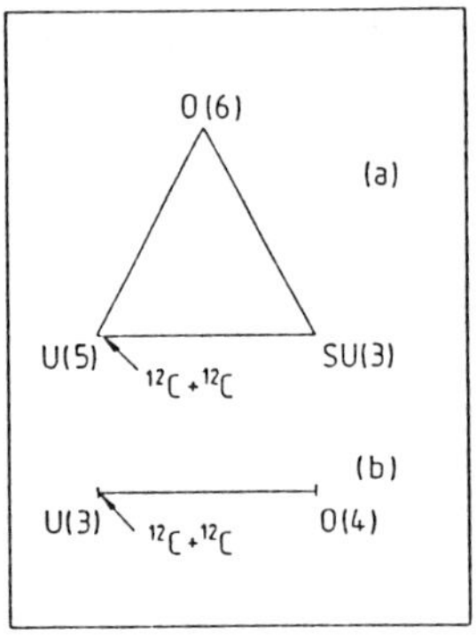

Fig. 2. Positions of the $^{12}C+{}^{12}C$ system on the phase diagrams of the (a) quadrupole and (b) dipole model.

As for the question of multipolarity, our calculation support the conclusion of Cindro and Greiner: based on the present experimental data no choice can be made between the dipole or quadrupole nature of these resonances. On the other hand it is remarkable that in both the models the best fit is very close to the exact dynamical symmetry corresponding to the soft vibrator [U(5) limit in the IBM-1 and U(3) limit in the vibron model]. This would indicate that the deformation is dynamic rather than static. These conclusions are drawn, of corse, on the basis of the present simple approximation. More elaborate ones may be used for a better understanding of these resonances among which the one that takes into account both the dipole and the quadrupole degrees of freedom is a straightforward candidate.

Helpful discussions with Professors F. Iachello, P.O. Lipas and Dr. H.J. Daley during this wcrk are gratefully acknowledged.

1) Iachello, F., Phys. Rev. C23, 2778 (1981).
2) Iachello, F. and Levine, R.D., J. Chem. Phys. 77, 3046 (1982).
3) Erb, K.A. and Bromley, D.A., Phys. Rev. C23, 2781 (1981).
4) Cindro, N. and Greiner, W., J. Phys. G9, L175 (1983).
5) Arima, A. and Iachello, F., Adv. Nucl. Phys. 12, 139 (1984).
6) Cseh, J., Phys. Rev. C31, 692 (1985).
7) Cseh, J. and Suhonen, J., Phys. Rev. C33, 1553 (1986).
8) van Roosmalen, O.S., Computer Code ROTVIB, Univ. Groningen, 1981.
9) Scholten, O., Computer Code PHINT, Univ. Groningen, 1975.

BREMSSTRAHLUNG IN THE πP-SYSTEM

PAUL FUCHS
Oregon State University, Corvallis, Oregon 97331, USA

A. STETZ
Oregon State University, Corvallis, Oregon 97331, USA

Abstract We review the present status of bremsstrahlung in the πp-system and outline the possibilities of a new experiment.

This work is supported by the National Science Foundation

Elastic Scattering in the πp-system is dominated by the Δ-resonance at energies up to a few hundred MeV. Fig. 1a. shows the corresponding

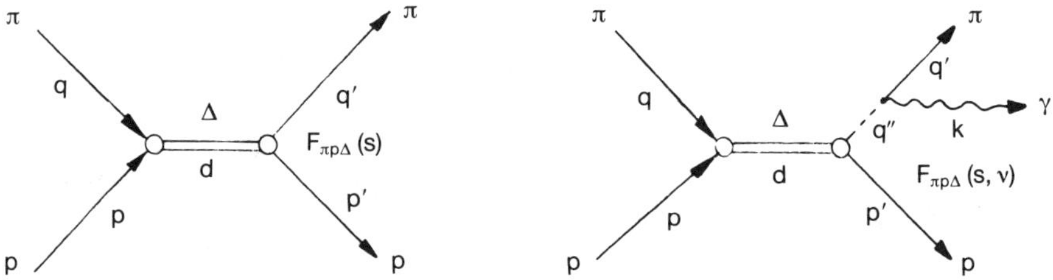

Fig. 1a Fig. 1b

Feynman diagram. The vertex function $F_{\pi p\Delta}$ is only dependent on the invariant mass $\sqrt{s}$ of the system. The term "bremsstrahlung" describes a system, where a photon appears in the final state. One such process is pictured in Fig. 1.b. for the πp-system with a photon k as the third

particle. The characteristic feature of this diagram is that the pion, which is going to radiate away a photon, is not "on its mass shell", meaning that its four-momentum squared is not equal to its rest mass squared. If we define the parameter ν as

$$\nu = q''^2 - m^2 = (q' + k)^2 - m^2 = 2q' \cdot k \qquad (1)$$

(q", q', k are the four-momenta of the intermediate, pion outgoing pion and photon, respectively; p',p describe the proton and m is the rest mass of the pion). We see that its value is, indeed, not zero. Since the additional freedom in the scattering process (due to the pion going off its mass shell) is conveniently described by ν, this parameter also enters into the vertex function $F_{\pi p\Delta}(s,\nu)$.

The five independent coordinates for the bremsstrahlung process can be chosen to be the polar and azimuthal angles for the pion and photon and the momentum of the photon. Let me thus establish a notation for the cross-sections of interest:

$$\sigma_B = \frac{d^5\sigma}{d\Omega_\pi \, d\Omega_k \, dk}$$ Bremsstrahlung cross-section

$$\sigma_E = \frac{d\sigma}{d\Omega}$$ Cross-section for the corresponding elastic scattering process.

In 1958, Low[1)] showed that the cross-section σ_B can be expanded in powers of k as

$$\sigma_B = \frac{\sigma_{-1}}{k} + \sigma_0 + \sigma_1 k + \ldots \qquad (2)$$

where σ_{-1}, σ_0, ... are constants, which are determined in principle by elastic scattering data. Thus for low photon

momentum k, a measurement of σ_B would not give any information on the off-shell behavior of the vertex function $F_{\pi p\Delta}$ of Fig. 1.b. Various authors[2,3,4] tried to incorporate off-shell effects of the $\pi p\Delta$- as well as of the electromagnetic vertex-functions of the pion and nucleon to get estimates of the contributions of higher order terms. Their predictions all show an increase of the cross-section for photon momenta above 60 MeV as illustrated by fig. 2a, b. These figures also show data, measured by Nefkens et al.[5] in the 1970's, which follow a simple 1/k dependence in contradiction to those predictions. So far, there are two models, which can fit the data well. These are the so-called External Emission Dominance Model (EED)[6] and a calculation by Liou and Nutt [7], which is based on the proper expansion of σ_B in the sense of Low's theorem, retaining only the first two terms.

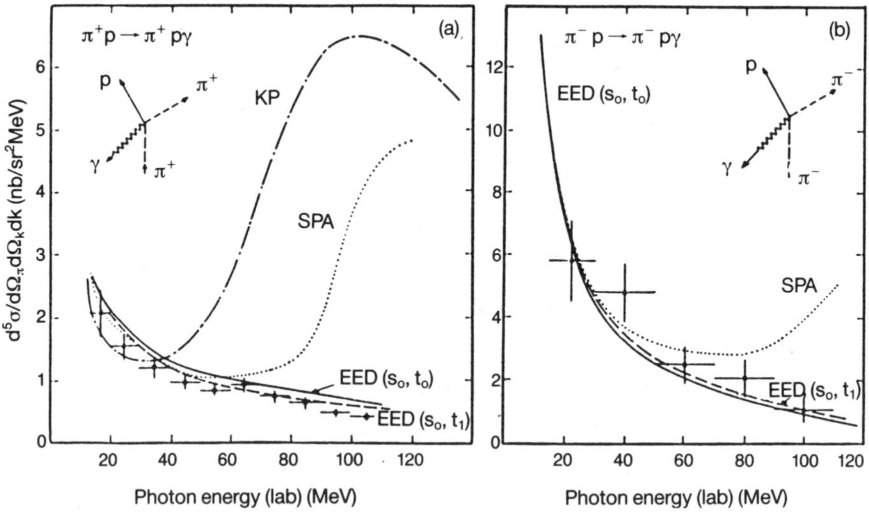

Fig. 2: Differential cross section in the lab system for $\pi^\pm p \to \pi^\pm p\gamma$ in the backward photon geometry of Ref. 2 for 298–MeV incident $\pi^\pm$. KP is the model calculation of Kondratyuk et.al. (Ref. 3) using $\mu_\Delta = +2\mu_p$; SPA is the soft-photon approximation, namely the first two terms in a Low-type expansion (Ref. 2); EED is the prediction using external emission dominance, the definitions of (s_o, t_o), (s_o, t_1) are given in the text. The data points are from Ref. 2.

The EED-calculation assumes that the photon is radiated only from external legs (like fig. 1.b.), neglecting all structure-dependent contributions (like internal radiation due to the magnetic moment of the delta). This leads to a term of the form $\sigma_B = \sigma_{-1}(k)/k$, where σ_{-1}, however, is now a function of k.

It is important to note that both models fail to explain why higher order terms cancel or are negligible, and are thus only justified by how well they fit the data.

The data that have been obtained by the Nefkens group[5)] and on which the model calculations have been based are restricted to one certain geometry, which is coplanar (all particle trajectories lie in one plane) and which corresponds to just one value of the polar angle between incoming and outgoing pion. Furthermore, the measured cross-sections are those for an unpolarized target. Our motivation for a bremsstrahlung experiment is strongly connected to these observations. We think that it is necessary to extend the set of available data to other, namely non-coplanar, geometries, as well as to measurements of asymmetry parameters on a polarized proton target, which will add a new kind of information to the problem. These new data will allow us to determine whether the above models really hold over the whole range of possible geometries, or whether they are only applicable for the special setup of coplanar scattering. The latter would imply that more sophisticated models would be necessary, whose construction would have to be based on our measurements.

In order to show that a new choice of geometries might allow an answer to this question, let us discuss the EED-model in a little more detail:

If we evaluate the four diagrams, which correspond to external emission of a photon (compare fig. 1.b., the internal structure is unimportant for this model), the cross-section σ_B [ref. 5] can be written as:

$$\sigma_B = \frac{e^2}{4\pi^2} A^{\mu} A_{\mu} \sigma_E (s,t)$$

$$A_{\mu} = \left(\pm \frac{q'_{\mu}}{q'\cdot k} \mp \frac{q_{\mu}}{q\cdot k} + \frac{p'_{\mu}}{p'\cdot k} - \frac{p_{\mu}}{p\cdot k}\right) \text{ for } \pi^{\pm}$$

where s and t are usually taken to be

$$s = \frac{1}{2} [(p+q)^2 + (p'+q')^2]$$

$$t = \frac{1}{2} [(q'-q)^2 + (p'-p)^2]$$

However, this choice of s,t is not determined unambiguously. Since one only retains the first term of the Low expansion in σ_B, other definitions of s and t are also possible. We therefore take as a first hint that geometries, which yield values of σ_B or a_B (asymmetry parameter), that are especially sensitive to the choice of s and t, will constitute a good test. Under the resulting geometries, the non-coplanar ones are to be preferred, since their off-shell effects are expected to be more important. (A coplanar geometry puts some serious constraints on the data, like the vanishing of the a_B's in some planes). These ideas are pictured in figs. 3 and 4, where we show calculations based on EED for some cases (s_0, t_0 are taken as above, $t_1 = (q'-q)^2$; a_B is calculated for the proton

polarization taken to be perpendicular to the scattering plane defined in the k=0 limit).

We are presently working on the design of the experiment to be run at the meson facility at TRIUMF, Vancouver, British Columbia, Canada. The corresponding proposal will be submitted to the evaluation committee this winter ('86), where details of the experiment will be worked out.

1) F. E. Low, Phys. Rev. 110, 974 (1958).
2) D. Sober et al., Phys. Rev. D 11, 1017 (1975).
3) L. A. Kondratyuk, L. A. Ponomarev, Sov. J. Nucl. Phys. 7, 82 (1968).
4) E. M. Nyman, Nucl. Phys. 154 A, 97 (1970).
5) B. M. K. Nefkens et al., Phys. Rev. D 18, 3911 (1978).
6) B. M. K. Nefkens, D. I. Sober, Phys. Rev. D 14, 2434 (1976).
7) M. K. Liou, W. T. Nutt, Phys. Rev. D 16, 2176 (1977).

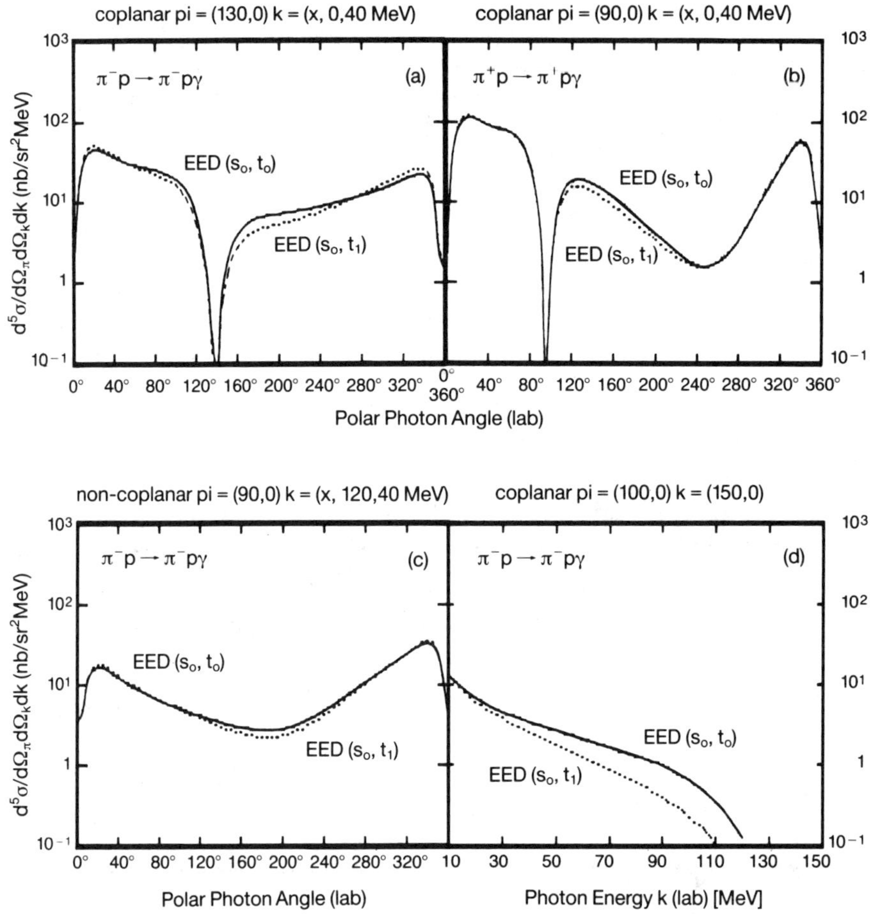

Fig. 3: Bremsstrahlung Cross-Section for different geometries and choices of s and t (see text) (EED = External Emission Dominance Model [6]). The directions of the particles are shown in brackets: (polar, azimuthal angle).

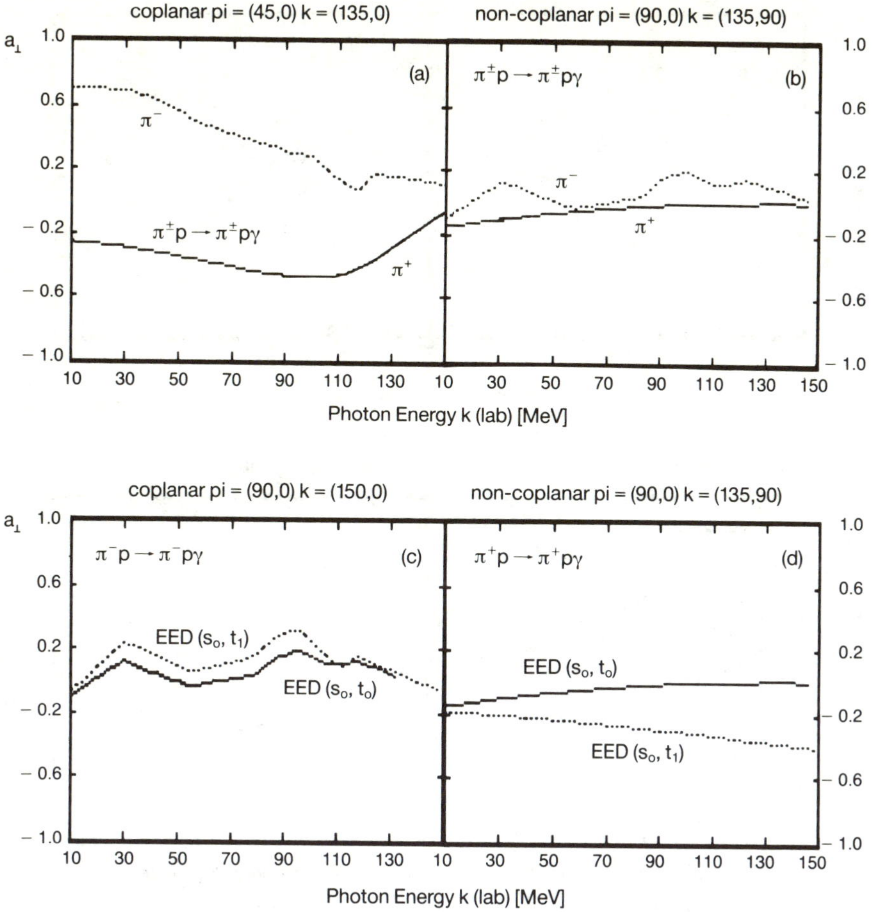

Fig. 4: Asymmetry Parameter normal to the scattering plane (defined for $k \to 0$) for different geometries and choices of s and t (see text), the directions of the particles are shown in brackets: (polar, azimuthal angle).

Chapter 7

THEORETICAL STUDIES

Moderator:

M. SAMBATARO
I.N.F.N.- Sezione di Catania
Corso Italia 57
95129 Catania, Italy

MEAN FIELD THEORY OF NUCLEAR SPIN

F. Dönau

Zentralinstitut für Kernforschung, Rossendorf
8051 Dresden, PSF 19

1. Motivation

The Cranking concept is rather fruitful in describing structures in rotating nuclei. In present applications of this theory the Cranking axis is ad hoc identified with a selected principal axis of the deformed nucleus. This restriction is certainly justified when studying the Yrast states in axially deformed even-even nuclei.

However, one cannot expect that this limitation makes sense in the general case. We are going to show that the Cranking axis freely chosen defines an instantaneous direction of the nuclear spin vector and this relative orientation must be considered as a particular collective degree of freedom in deformed systems. The study of the spin orientation opens in fact a new field of applications as, for instance, the calculation of electromagnetic transition amplitudes for complex quasiparticle configurations.

2. Mean field theory of nuclear spin

The starting point is the general Cranking Hamiltonian

$$h' = h_{def} - \vec{\omega}\cdot\vec{I} \tag{1}$$

where the part h_{def} implies the deformed single particle potential parametrized in terms of some deformation parameters $\varepsilon = (\beta, \gamma, \varepsilon_4, \ldots)$ as well as pair field properties specified by the gap parameter Δ and the Fermi energy. In what follows we refer to the intrinsic coordinate system (x,y,z) which is determined by the principal axes of the deformed potential. Hence the frequency vector $\vec{\omega}$ is given by the tripel (ω_x, ω_y, ω_z) and the Cranking term in eq. (1) reads explicitly

$$\vec{\omega}\cdot\vec{I} = \omega_x I_x + \omega_y I_y + \omega_z I_z \tag{2}$$

involving, in general, all three angular momentum (a.m.) components I_i (i = x,y,z) when comparing with the usual Cranking approach. The Cranking Hamiltonian (1) is used to form the energetically lowest determinantal-like state

$$|\vec{\omega}> = \phi(\omega_x, \omega_y, \omega_z) \tag{3}$$

which describes an ensemble of independent quasiparticles in a deformed potential that rotates in direction of the ω-vector with the frequency $|\vec{\omega}|$.
The so-defined Cranking state (3) is uniquely related to a spin vector

$$<\vec{I}> = <\vec{\omega}|\vec{I}|\vec{\omega}> \equiv (<I_x>,<I_y>,<I_z>) \tag{4}$$

calculated as the expectation value of the a.m. components.

The spin vector (4) provides a relevant information on the relative orientation of the lab space reference. Since the a.m. operator $\vec{I}$ commutes with the total Hamiltonian H of the system the spin vector (4) is a constant of motion, i.e. it is still-standing in the lab system. Therefore, it is possible to fix the lab Z-axis, for instance, along the spin vector. In this way the relation (4) serves as a link between the intrinsic and the lab system.

The mean field approach is sketched by the following steps:

(i) Construct the family of Cranking determinants $|I(\vec{\omega})>$ that belong to a given spin value I, i.e. find all states $|\vec{\omega}>$ the spin vectors (4) of which have the same length I by requiring

$$<\vec{I}>^2 = <I_x>^2 + <I_y>^2 + <I_z>^2 = I^2. \tag{5}$$

This constraint defines a two-dimensional submanifold of states $|\vec{\omega}\rangle$ which "covers" the a.m. sphere $\langle\vec{I}\rangle^2 = I^2$ being the natural condition for conserving the a.m. value (fig. 1)

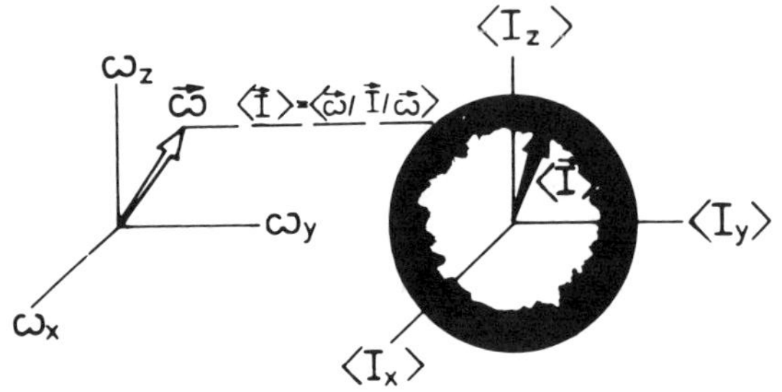

Fig. 1
Mapping of a Cranking state $|\vec{\omega}\rangle$ onto the angular momentum sphere.

(ii) Evaluate the true price for such a state $|I(\vec{\omega})\rangle$ by taking the energy expectation value

$$E_I(\vec{\omega}) = E_I(\langle\vec{I}\rangle) = \langle I(\vec{\omega})|\, H \,|I(\vec{\omega})\rangle \tag{6}$$

where H is the many body Hamiltonian of the nuclear system under study and the argument $\vec{\omega}$ is replaced by using the relation (4).

Consequently the mean field solution is found by searching the minimal points of the energy surface (6), i.e.

$$E_I(\vec{\omega}) \longrightarrow \mathit{Minimum} \tag{7}$$

supplies the optimal orientation $\langle\vec{I}\rangle_0$ within the set of Cranking solutions (the search for the minimum (7) may include also shape parameters to be optimized).

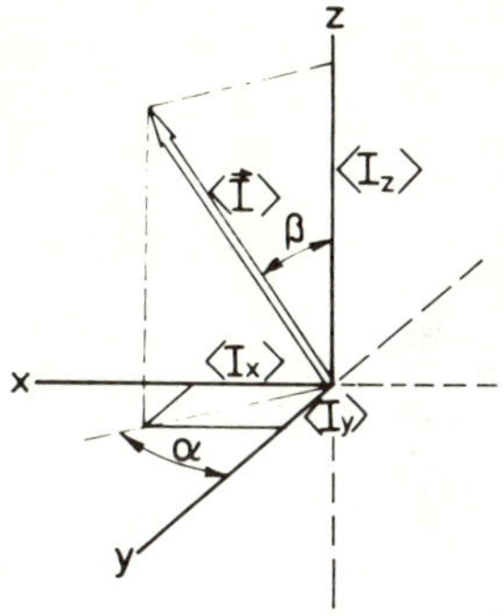

Fig. 2
Definition of the Euler angles (α,β)

For later purpose the direction of the spin vector $\langle \vec{I} \rangle$ pointing to the a.m. sphere surface is parametrized by the Euler angles (α,β) as displayed in fig. 2:

$$\sin\alpha = \frac{\langle I_x \rangle}{\sqrt{I^2 - \langle I_z \rangle^2}} \quad , \cos\beta = \frac{\langle I_z \rangle}{I} . \qquad (8)$$

Accordingly we will apply also the notation $|I(\alpha,\beta)\rangle$ for the Cranking states $|I(\vec{\omega})\rangle$.

3. Symmetries

All symmetry operations S which transform a given intrinsic coordinate system into an equivalent one leave the deformed part h_{def} of the Cranking Hamiltonian invariant.
The group of these operations is given by

(i) all reflexions of the intrinsic axes ($x \rightarrow -x$, $y \rightarrow -y$, $z \rightarrow -z$) and

(ii) any rotation about the symmetry (z) axis in the axial symmetric case.

Hence, by operating with S onto a Cranking determinant $|I(\vec{\omega})\rangle$ one obtains another Cranking state $S|I(\vec{\omega})\rangle$ with a changed spin orientation

$$\langle \vec{I}_S \rangle \equiv \langle S^+ \vec{I} S \rangle \neq \langle \vec{I} \rangle \qquad (9)$$

but the value $\langle \vec{I}_S \rangle^2 = \langle \vec{I} \rangle^2$ remains constant because all operations (i) and (ii) conserve the length of the spin vector.

The symmetry operation allows one to shift the spin vector along the surface of the a.m. sphere. On the other hand the

points with the same weight (the irreps of the symmetry groups are one-dimensional), i.e.

$$\phi_I \propto \sum_S e^{i\varphi_s} S|I(\vec{\omega})\rangle_0 \tag{11}$$

where $\varphi_s = 0, \pm\pi$ denotes possible phase angles.
In the axial case, for example, the symmetry operation $S = e^{i\alpha I_z}$ rotates an instantaneous optimal spin orientation $\langle\vec{I}\rangle_0$ about the symmetry (z) axis. The corresponding continuous superposition ($\alpha = 0,\ldots,2\pi$, $\varphi_s = 0$) can be interpreted as quantum counterpart of the classical precession. Note that the energy (10) is independent on the angle α but the state $\exp(i\alpha I_z)|I(\vec{\omega})\rangle$ depends on it.

4. Application to the rotor plus particle system

We illustrate the method for the axial rotor plus particle system which is a simple but non-trivial case with respect to the a.m. conservation. This system has, in addition, the advantage to be exactly solvable. Instead of a real many-body Hamiltonian we use the axial rotor Hamiltonian

$$H_{rot} = \frac{1}{2\mathcal{J}_0}(R_x^2 + R_y^2) + h_{def} \tag{12}$$

where R_i (i=x,y,z) denotes the rotor a.m. components and $\mathcal{J}_0$ is the moment of inertia parameter. The deformed part h_{def} is specified by a quasiparticle (qp) Hamiltonian

$$h_{def} = \sum_{k=-j}^{j} E_k \alpha_k^+ \alpha_k = \sum_{k=-j}^{j} \sqrt{[\varkappa(k^2 - k_F^2)]^2 + \Delta^2}\, \alpha_k^+ \alpha_k \tag{13}$$

of a high-j intruder shell where $\varkappa$ characterizes the deformation, k_F denotes the Fermi level and Δ is the pairing gap. Accordingly the operators α_k^+ (α_k) create (destruct) a quasiparticle on the orbital k.
The a.m. conservation concerns the total spin

$$\vec{I} = \vec{R} + \vec{j} \tag{14}$$

symmetry group leaves the full Hamiltonian invariant too and because of $[H,S] = 0$ follows

$$E_I(\langle \vec{I}_s \rangle) = \langle S^+ H S \rangle = \langle H \rangle = E_I(\langle \vec{I} \rangle) \tag{10}$$

i.e. the energy at the points $\langle \vec{I}_S \rangle$ and $\langle \vec{I} \rangle$ is equal.

In particular, the following symmetries result:

(i) In triaxially deformed systems one observes a cristal-like (8 symmetry points) symmetry (see fig. 4) that is equivalent to the threefold signature quantum numbers described by Bohr and Mottelson /BOH 75/.

(ii) In axially deformed systems exists a double-cone symmetry as illustrated in fig. 3.

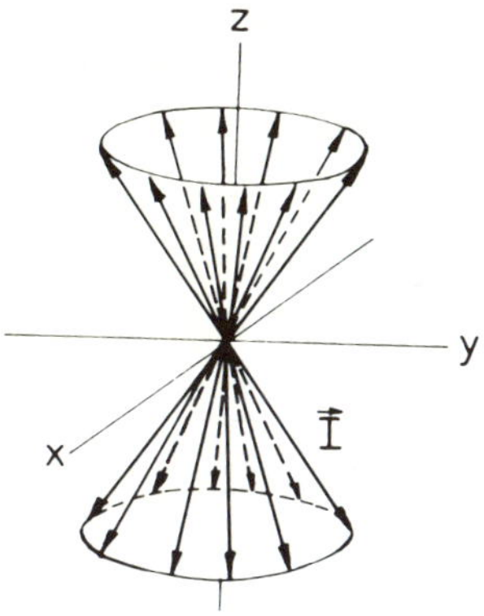

Fig. 3

Double-cone symmetry of an axially deformed system

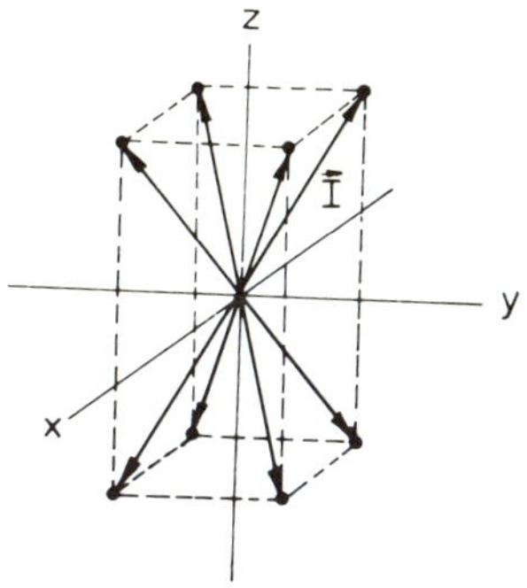

Fig. 4

8-Point symmetry of triaxially deformed systems

The consequences of the symmetries (i) and (ii) are obvious. Suppose an optimal direction $\langle \vec{I} \rangle_0$ is found. The symmetries demand that the nuclear state must be a superposition of all symmetry

where the qp contribution $\vec{j}$ acts by definition only onto the quasiparticles outside the rotor. The same restriction holds for the Cranking Hamiltonian

$$h' = h_{def} - \vec{\omega}\cdot\vec{j}\,. \tag{15}$$

The mean field approximation of the rotor a.m. means

$$R_x = \mathcal{J}_0\omega_x \quad , \quad R_y = \mathcal{J}_0\omega_y \quad , \quad R_z = 0 \tag{16}$$

leaving out here possible quantal fluctuations (which may be added afterwards (cf. sec. 5)). Because of the axial symmetry we can restrict ourselves to $\omega_y = 0$ that corresponds to the plane $\langle I_y\rangle = 0$. Therefore the a.m. sphere is determined by the condition (cf. eq. (5))

$$(\mathcal{J}_0\omega_x + \langle j_x\rangle)^2 + \langle j_z\rangle^2 = I^2 \tag{17}$$

for the Cranking state $|I(\omega_x,\omega_z)\rangle$. The energy $E_I(\omega_x,\omega_z)$ gets

$$E_I(\omega_x,\omega_z) = E_I(\langle I_z\rangle) = e'(\omega_x,\omega_z) + \omega_x\langle j_x\rangle + \omega_z\langle j_z\rangle + \tfrac{1}{2}\mathcal{J}_0\omega_x^2 \tag{18}$$

where $e'(\omega_x,\omega_z)$ is the Cranking energy.
The energy E_I is explicitly calculated for a system composed of one $j = 13/2$ quasiparticle coupled to an axial rotor.
The parameters $\Delta = 1$ MeV, $\varkappa = 0.2$ MeV, $\mathcal{J}_0 = 60$ MeV characterize a typical deformed odd-mass nucleus. In fig. 5,6 the energy curves E_I ($\langle I_z\rangle$) for two distinct positions of the Fermi level ($k_F = 7/2$, $k_F = 1/2$) are displayed. The double-cone symmetry of axial symmetry implies $E_I(\langle I_z\rangle) = E_I(-\langle I_z\rangle)$ and consequently the point $\langle I_z\rangle = 0$, $\langle I_x\rangle = I$ is always an extreme point of the energy surface but not necessarily a minimum. At low spin values one recognizes that in the case $k_F = 7/2$ the optimal state belongs to a finite component $\langle I_z\rangle_0$ which is close to the k_F-value of the

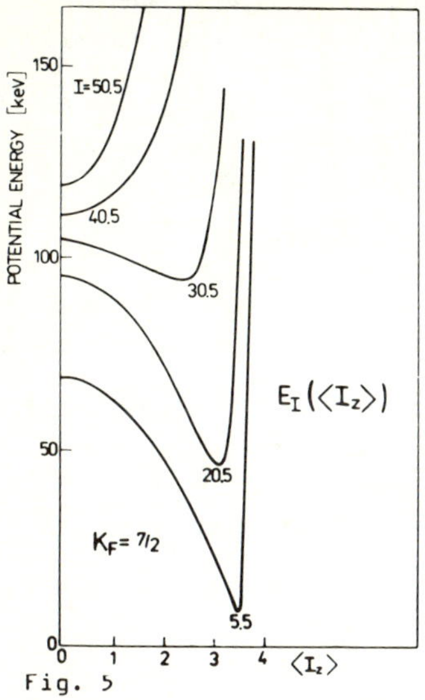

Fig. 5

Energy curve $E_I(\langle I_z\rangle)$ for an axial rotor ($\mathcal{J}_0$ =60 MeV) plus j = 13/2 quasiparticle (k_F = 7/2, Δ = 1 MeV) system

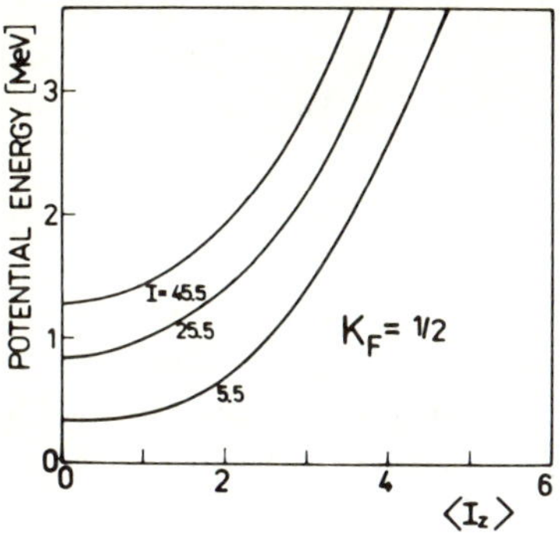

Fig. 6

The same as in fig. 5 for k_F = 1/2

odd quasiparticle. The optimal spin direction found there outside the x-axis is by no means identical with the axis of the collective rotation as assumed in common Cranking approaches. One notices the turning of the optimal spin orientation (minimum) to the rotation-aligned regime ($\langle I_z\rangle_0$ =0) for high spin values as expected for the increasing Coriolis force.
In the case k_F = 1/2 the optimal state is rotation-aligned already from the beginning with the optimal point at $\langle I_z\rangle_0$ = 0.

Note that all equivalent optimal points must be superimposed according to the symmetry requirement. Thus, the fully aligned state, for example, described by the point ($\langle I_x\rangle$ =I, $\langle I_z\rangle$= 0) is only an "instantaneous" component of the complete state (11) which implies the quantal precession about the symmetry axis.

5. Beyond the mean field solution - Requantization of the spin orientation

The optimal state characterized as a minimum in the energy surface (6) corresponds to a stationary point in which an analogous classical system would stay in rest. The quantal system does oscillate around the classical equilibrium. To account for these quantal fluctuations we apply a generator coordinate ansatz for the nuclear states, i.e.

$$\Psi_I = \int d\alpha \, d\cos\beta \; \varphi_I(\alpha\beta) \, |I(\alpha\beta)\rangle \tag{19}$$

superimposing the Cranking determinants $|I(\alpha\beta)\rangle$ with respect to the various spin orientation parametrized by the Euler angles $(\alpha\beta)$ (cf. eq. (8)). In eq. (19) the quantity $\varphi_I(\alpha\beta)$ denotes the weight function which can be determined, in principle, by the techniques of the generator coordinate method /see, for instance, RIN 80/. For the rotor problem investigated in sec. 4 the ansatz (19) reduces to the product form

$$\Psi_I^{rotor} = \varphi_I(\alpha\beta) \, |I(\alpha\beta)\rangle \tag{20}$$

since in this case the Euler angles are in fact additional dynamical variables.

We may derive also a collective Hamiltonian for the weight function $\varphi_I(\alpha\beta)$ by using intuitive arguments. The energy $E_I(\alpha\beta) = \langle I(\alpha\beta)|H|I(\alpha\beta)\rangle$ enters there as a potential part measuring the amount of energy to bring the spin vector into the static position $I(\alpha\beta)$. According to fig. 2 one can express the mass parameters belonging to the angle variables $(\alpha\beta)$ in terms of the principal moments of inertia $\mathcal{J}_i$ $(i=x,y,z)$. In this way one arrives at the following top-like kinetic energy term

$$T(\alpha\beta) = -\frac{\hbar^2}{2}\left[\frac{1}{\mathcal{J}_x\cos^2\alpha + \mathcal{J}_y\sin^2\alpha}\frac{\partial^2}{\partial\beta^2} + \frac{1}{\mathcal{J}_z}\frac{\partial^2}{\partial\alpha^2}\right]. \tag{21}$$

The complete collective Hamiltonian the eigensolutions of which provide the desired weight function $\varphi_I(\alpha\beta)$ is given by

$$h(\alpha\beta) = T(\alpha\beta) + E_I(\alpha\beta) + E_I^{zero}(\alpha\beta) \tag{22}$$

where $E_I^{zero}(\alpha\beta)$ is a zero-point correction term that compensates for the kinetic energy contribution in the static limit, i.e.

$$E_I^{zero}(\alpha\beta) = - \langle I(\alpha\beta)|T(\alpha\beta)|I(\alpha\beta)\rangle. \tag{23}$$

In axially symmetric systems ($\mathcal{J}_x = \mathcal{J}_y$, $\mathcal{J}_z = 0$) the α-dependence in eq. (22) drops out. The rotor function (20) reads in this special case as

$$\Psi_I^{rotor}(\alpha\beta) = \varphi_I(\beta)\, e^{-i\alpha I_z} |I(\alpha=0,\beta)\rangle \tag{24}$$

where the determinantal state $|I(\alpha\beta)\rangle$ has been expressed as a rotation of the one at $\alpha = 0$ (i.e. $\omega_y = 0$). The exponential factor means essentially a K-projection of the states. The symmetry $h(\beta) = h(\pi - \beta)$ leads to the signature "parity" $\pi_z = \pm 1$ of the weight function which is equivalent to the usual signature classification. The eigensolutions of the collective Hamiltonian satisfy automatically the symmetry requirements.

6. Calculation of transition matrix elements

The possibility to calculate electromagnetic transition amplitudes is a crucial point for a detailed analysis of experimental data and for predicting measurable effects. Since electromagnetic radiation is observed in the lab frame of reference but all previous considerations are based entirely on the intrinsic coordinates we need to transform the multipole radiation operators to the intrinsic system. Choosing the lab Z-axis in direction of the spin vector $\langle\vec{I}\rangle$ (4) this transformation can be written in terms of the Euler angles $(\alpha\beta)$ (cf. eq. (8)) as

$$(\mathcal{M}\lambda)^{lab}_{\mu} = \sum \mathcal{D}^{\lambda}_{\mu\varkappa}(\gamma,\beta,\alpha)\ (\mathcal{M}\lambda)_{\varkappa} \tag{25}$$

where $(\mathcal{M}\lambda)_{\varkappa}$ are the intrinsic multipole components and $D^{\lambda}_{\mu\varkappa}$ mean the Wigner D-functions. The Euler angle γ specifies the lab X- and Y-axis and it can be fixed to $\gamma = 0$. Since our choice of the lab Z-axis means to consider the states with the maximal lab z-component $M = I$ the component μ in eq. (25) is determined by the selection rule $\mu = I' - I$ (i.e. by the spin difference ΔI) for the transition $I \rightarrow I'$.

Now, the transition amplitudes between the states (19) represented as superposition of Cranking determinants $|I(\alpha\beta)\rangle$ with weight functions $\varphi_I(\alpha\beta)$ can be formed. Employing the transformation (25) one obtains

$$\begin{aligned} &\langle I'M'=I' \,|\, (\mathcal{M}\lambda)^{lab}_{\mu=I'-I} \,|\, I M=I \rangle \\ &= \int d\alpha\, d\alpha'\, d\cos\beta\, d\cos\beta'\ \ \varphi^{*}_{I'}(\alpha'\beta')\,\varphi_I(\alpha\beta)\,\times \\ &\quad \times \sum_{\varkappa} \mathcal{D}^{\lambda}_{\mu\varkappa}(0,\beta,\alpha)\,\langle I'(\alpha'\beta')|(\mathcal{M}\lambda)_{\varkappa}\,|\,I(\alpha\beta)\rangle . \end{aligned} \tag{26}$$

This transition matrix element contains the geometrical dependence which is due to the fact the orientation of the radiating source in lab space is relevant for the observed strength. The microscopic structure of this source (i.e. how do the individual quasiparticle contribute) is sticking in the intrinsic matrix element $\langle I'(\alpha'\beta')|(\mathcal{M}\lambda)_{\varkappa}|I(\alpha\beta)\rangle$. It is worth mentioning that the above combination of geometrical factors with individual quasiparticle contributions may lead to coherence effects in complex qp configurations. Thus, the enhancement of M1 transitions observed in three quasiparticle states can be interpreted as an indication of

favored geometrical coupling schemes in which the individual magnetic moments add up constructively /FUN 83/.

A rigorous simplification of the transition matrix element is achieved when the states can be approximated by optimal determinants, i.e. if the potential energy surface shows up well-pronounced minima. In this case the nuclear state is approximated as

$$\phi_I \approx P_S \,|I(\alpha\beta)_0\rangle \tag{27}$$

where $(\alpha\beta)_0$ denotes a minimal point of the energy surface $E_I(\alpha\beta)$ and P_S means the symmetry projection (cf. eq. (11)).

The transition matrix element reduces in this case to

$$\langle I'M'=I' \,|\, (\mathcal{M}\lambda)^{lab}_{\Delta I=I'-I} \,|\, IM=I\rangle \approx$$
$$\approx \sum_{\varkappa} \mathcal{D}^{\lambda}_{\Delta I\varkappa}(0,\beta_0,\alpha_0)\langle I'(\alpha\beta)_0 \,|\, P_S'(\mathcal{M}\lambda)_{\varkappa} P_S \,|\, I(\alpha\beta)_0\rangle \tag{28}$$

involving only the relevant spin orientation $(\alpha\beta)_0$. The form (28) of the matrix element resembles strongly the radiation of a classical multipole that rotates with the frequency $\omega_{rot} = \Delta E/\Delta I$ about the lab Z-axis while its intrinsic orientation is determined by $(\alpha\beta)_0$.

The transition matrix elements of a rotor plus particle system gets identical with the local approximation $(\alpha\beta) = (\alpha'\beta')$ of the general expression (26).

7. Summary

The spin orientation appears naturally as a new collective degree of freedom in deformed systems when insisting in the body fixed frame of reference. General Cranking determinants are taken as suitable states for preparing both the spin value and its orientation in order to satisfy angular momentum conservation on mean field level. In this way the geometrical aspects of a rotating

deformed nucleus can be treated on a microscopic basis. This enables one to calculate straightforwardly electromagnetic transition amplitudes where the knowledge of the spin orientiation relative to the intrinsic system is a necessary ingredient.

The theory is promising with concern to all geometrical problems as, for instance, configurations with finite K-values including their decay properties or mixing to other bands. Furthermore, it supplies the theoretical framework for investigations of quantal fluctuations of the spin orientation like signature oscillations or geometrical changes when the system passes band crossings (backbending). It should be also possible to obtain more detailed information on the spin orientation in rotating triaxial systems.

References

/BOH 75/ Aa. Bohr and B.R. Mottelson, Nuclear Structure, Vol. 2 (Benjamin, London, 1975)

/FUN 83/ L. Funke et al., Phys. Lett. 120 B (1983) 301

/RIN 80/ P. Ring and P. Schuck, The Nuclear Many-body Problem (Springer, NY, 1980) p. 420

THE CRANKING APPROACH TO THE COLLECTIVE PAIRING HAMILTONIAN FOR ASYMMETRIC SYSTEMS*

J.DOBACZEWSKI AND S.G.ROHOZIŃSKI
Institute of Theoretical Physics, Warsaw University, Hoża 69, 00-681 Warsaw, Poland.

Abstract The collective pairing Hamiltonian is derived within the adiabatic cranking approach for asymmetric systems. The laboratory gauge angle ϕ and the energy gap Δ are used as the collective variables. The Hamiltonian is expressed in terms of four gap-dependent functions: the potential, the moment of inertia for rotations in the gauge space, the mass parameter and the Fermi energy, and contains a term linear in the particle number excess.

INTRODUCTION

It is well known since many years that the pairing residual forces play an important role in nuclei. In particular, the pairing forces affect strongly the parameters of the collective quadrupole motion like the deformation potential and, especially, the moments of inertia and the vibrational masses. However, their influence on the collective motion is most likely not restricted to static effects and a dynamic coupling between the quadrupole and the pairing vibrations is expected to be essential[1].

In order to treat both modes on an equal footing in the frame of a collective model the collective pairing Hamilto-

*Supported in part by the Ministry of Science and High Education under the contract CPBP01.09.

nian has to be constructed. Such a Hamiltonian was originally derived by Bés et al.[2] within the cranking approach for symmetric systems, i.e. for systems with the single-particle levels symmetric with respect to the Fermi level. Such an assumption is in an obvious contradiction with the nuclear shell model and leads to the collective pairing Hamiltonian quadratic in the particle number excess. For asymmetric systems we expect that such a Hamiltonian contains also odd powers of the particle number excess operator. It might, however, be a purely nonadiabatic effect and thus impossible to explain within the cranking or the adiabatic time-dependent approach.

Recently Góźdź et al.[3] have constructed a collective pairing Hamiltonian using the generator coordinate method. Their Hamiltonian contains the first power of the particle number excess and may thus account for an asymmetry between the systems with the same number of pairs added or removed.

The aim of the present note is to derive the collective pairing Hamiltonian for asymmetric systems within the adiabatic cranking approach and to show that this Hamiltonian is able to explain the effect of asymmetry in question as well.

THE ADIABATIC CRANKING APPROXIMATION FOR THE PAIRING HAMILTONIAN

Let the nuclear system be described by the pairing Hamiltonian,

$$H = \sum_{\nu} e_{\nu} a^{+}_{\nu} a_{\nu} - \frac{1}{4} G \sum_{\nu\mu} s^{*}_{\nu} s_{\mu} a^{+}_{\nu} a^{+}_{\bar{\nu}} a_{\bar{\mu}} a_{\mu} \quad , \tag{1}$$

where indices ν and $\bar{\nu}$ label mutually time-reversed single-particle states and s_{ν} are the phase factors associated with the time reversal, $T^{+} a^{+}_{\nu} T = s^{*}_{\bar{\nu}} a^{+}_{\bar{\nu}}$. The time evolution of the system in the Time-Dependent Hartree-Fock-

-Bogolyubov approximation[4,5] is governed by the following equation of motion

$$i\hbar\dot{\mathcal{R}} = [\mathcal{W},\mathcal{R}] \quad , \qquad \mathcal{R}^2 = \mathcal{R} \quad , \qquad \text{(2ab)}$$

where $\mathcal{R}$ and $\mathcal{W}$ are the generalized density matrix and the self-consistent Hamiltonian, respectively. Since $\mathcal{R}$ and $\mathcal{W}$ do not commute with the generalized particle number operator $\mathcal{N}$,

$$\mathcal{N} = \begin{pmatrix} 1 & 0 \\ 0 & -1 \end{pmatrix} \quad , \qquad (3)$$

we can construct the matrices $\mathcal{R}'$ and $\mathcal{W}'$ rotated in the gauge space by the angle ϕ,

$$\mathcal{R}' = e^{i\phi\mathcal{N}}\mathcal{R}e^{-i\phi\mathcal{N}} \quad , \quad \mathcal{W}' = e^{i\phi\mathcal{N}}\mathcal{W}e^{-i\phi\mathcal{N}} \quad . \qquad (4)$$

Assuming the time-dependent gauge angle we can study the quasiparticle system in an intrinsic frame of reference which rotates in the gauge space. The equation of motion (2a) transformed to the rotating frame reads

$$i\hbar\dot{\mathcal{R}}' = [\mathcal{W}' - \hbar\dot{\phi}\mathcal{N} \, , \, \mathcal{R}'] \quad , \qquad (5)$$

and the stationary problem, $\dot{\mathcal{R}}'_0 = 0$, in the intrinsic **rotating** frame,

$$[\mathcal{W}'_0 - \lambda\mathcal{N} \, , \, \mathcal{R}'_0] = 0 \quad , \qquad (6)$$

corresponds to the BCS approximation with the time-independent rotation velocity equal to the Fermi energy λ, $\hbar\dot{\phi}_0 = \lambda$.

We do not report here the standard formulae for the stationary solution $\mathcal{R}'_0$ and the associated self-consistent Hamiltonian $\mathcal{W}'_0$, which can, for instance, be found in ref.[4]. However, one should note that, due to the fact that the gauge-space rotation generator $\mathcal{N}$, eq.(3), is a time-even operator, the stationary solution in the rotating frame can be chosen to be time-even too. This is a main difference with respect to the spatial rotations where the corresponding generator - the angular momentum - is time-odd.

Looking for the adiabatic solution of eq.(5) we assume that $\mathcal{R}_0'$ varies in time due to a slow time dependence of Δ and small deviations of the angular velocity $\dot{\phi}$ from $\dot{\phi}_0$. The family of states $\mathcal{R}_0'(\Delta)$ results from the stationary equation (6) by treating the gap parameter appearing in $\mathcal{W}_0'$ as an independent variable. In order to obtain the stationary solutions with the fixed particle number N_0 the Fermi energy λ for an asymmetric system has to depend on the pairing gap Δ.

The first order adiabatic correction $\mathcal{R}_1'$ to the stationary density matrix $\mathcal{R}_0'$ can be found from eq.(5) by keeping the first-order terms only,

$$i\hbar\dot{\mathcal{R}}_0' = [\mathcal{W}_0' - \lambda\mathcal{N}, \mathcal{R}_1'] - \hbar\dot{\phi}'[\mathcal{N}, \mathcal{R}_0'] \quad , \tag{7}$$

where

$$\dot{\phi}' = \dot{\phi} - \dot{\phi}_0 = \dot{\phi} - \lambda/\hbar \quad . \tag{8}$$

The first order adiabatic correction $\mathcal{W}_1'$ to the self-consistent Hamiltonian vanishes for the pairing interaction and thus the corresponding term is absent in eq.(7). Let us note that the second term on the right-hand side of eq.(7) is time-even, and thus the first order adiabatic correction $\mathcal{R}_1'$ will contain an unusual time-even component.

Within the common basis of $\mathcal{W}_0' - \lambda\mathcal{N}$ and $\mathcal{R}_0'$ such that

$$\begin{aligned} &\mathcal{R}_0'|\mathcal{A}_\nu\rangle = 0 \quad , \qquad \mathcal{R}_0'|\mathcal{B}_\nu\rangle = |\mathcal{B}_\nu\rangle \quad , \\ &(\mathcal{W}_0' - \lambda\mathcal{N})|\mathcal{A}_\nu\rangle = E_\nu|\mathcal{A}_\nu\rangle \quad , \\ &(\mathcal{W}_0' - \lambda\mathcal{N})|\mathcal{B}_\nu\rangle = -E_\nu|\mathcal{B}_\nu\rangle \quad , \end{aligned} \tag{9}$$

where E_ν are the usual quasiparticle energies, the non-vanishing matrix elements of $\mathcal{R}_1'$ read

$$\begin{aligned} \langle\mathcal{A}_\nu|\mathcal{R}_1'|\mathcal{B}_\mu\rangle &= \langle\mathcal{B}_\mu|\mathcal{R}_1'|\mathcal{A}_\nu\rangle^* = \\ &= -\frac{s_\nu\hbar\delta_{\nu\bar{\mu}}}{2E_\nu^2}\left[\dot{\phi}'\Delta + i\dot{\Delta}\,\frac{e_\nu - \lambda + \Delta\, d\lambda/d\Delta}{2E_\nu}\right] \quad . \end{aligned} \tag{10}$$

These matrix elements of the second order correction $\mathcal{R}_2'$, depending only on velocities $\dot{\phi}'$ and $\dot{\Delta}$ can be expressed[6] by the matrix elements of $\mathcal{R}_1'$, as to fulfil the projectivity condition (2b), and look

$$\langle \mathcal{A}_\nu | \mathcal{R}_2' | \mathcal{A}_\mu \rangle = - \langle \mathcal{B}_\nu | \mathcal{R}_2' | \mathcal{B}_\mu \rangle = \tag{11}$$

$$= \frac{\hbar^2 \delta_{\nu\mu}}{4E_\nu^4} \left[\dot{\phi}'^2 \Delta^2 + \dot{\Delta}^2 \frac{(e_\nu - \lambda + \Delta d\lambda/d\Delta)^2}{4E_\nu^2} \right] .$$

The remaining matrix elements of $\mathcal{R}_2'$ depend on the accelerations $\ddot{\phi}'$ and $\ddot{\Delta}$ as well.

THE COLLECTIVE HAMILTONIAN

It is well known that the mean value of the pairing Hamiltonian (1) within the BCS wavefunction given by $\mathcal{R}_0'$ is equal to

$$V(\Delta) = 2 \sum_{\nu>0} e_\nu v_\nu^2 - \tfrac{1}{4} G \Delta^2 \left[\sum_{\nu>0} E_\nu^{-1} \right]^2 , \tag{12}$$

where v_ν^2 are the usual BCS occupation numbers. One refers to the collective pairing Hamiltonian as the mean value of the microscopic pairing Hamiltonian calculated up to the second order adiabatic corrections to $\mathcal{R}_0'$. Since the mean values are gauge-invariant, they can be determined in either reference frame. Therefore, the collective Hamiltonian reads

$$H_{coll} = V(\Delta) + \tfrac{1}{2}\mathrm{Tr}(\mathcal{R}_1'(\mathcal{W}_0' - \lambda\mathcal{N})) + \tfrac{1}{2}\mathrm{Tr}(\mathcal{R}_2'(\mathcal{W}_0' - \lambda\mathcal{N})) +$$
$$+ \tfrac{\lambda}{2}\mathrm{Tr}(\mathcal{R}_1'\mathcal{N}) + \tfrac{\lambda}{2}\mathrm{Tr}(\mathcal{R}_2'\mathcal{N}) \quad . \tag{13}$$

The second term on the right-hand side of eq.(13) vanishes because the matrix $\mathcal{R}_1'$ has vanishing diagonal matrix elements in the basis of eq.(9). Using equations (10) and (11) and neglecting the second order contribution to the particle number, which depends on the accelerations, one obtains

$$H_{coll} = \tfrac{1}{2}B(\Delta)\dot{\Delta}^2 + \tfrac{1}{2}J(\Delta)\dot{\phi}'^2 + \lambda(\Delta)J(\Delta)\dot{\phi}' + V(\Delta) \quad , \tag{14}$$

where

$$B(\Delta) = \frac{1}{4}\hbar^2 \sum_{\nu>0} (e_\nu - \lambda + \Delta d\lambda/d\Delta)^2/E_\nu^5 \ , \ J(\Delta) = \hbar^2\Delta^2 \sum_{\nu>0} E_\nu^{-3} . \quad (15)$$

Substituting eq.(8) to eq.(14) we obtain

$$H_{coll} = \frac{1}{2}B(\Delta)\dot{\Delta}^2 + \frac{1}{2}J(\Delta)\dot{\phi}^2 - \frac{1}{2}J(\Delta)\lambda^2/\hbar^2 + V(\Delta) \quad , \quad (16)$$

which means that the collective Hamiltonian is quadratic in the velocity $\dot{\phi}$ corresponding to the nonadiabatic collective variable ϕ.

In order to know how to quantize the collective Hamiltonian we express it in terms of the particle number excess,

$$N - N_0 = \frac{1}{2}\mathrm{Tr}(\mathcal{R}_1' \mathcal{N}) = \hbar^{-1}J(\Delta)(\dot{\phi} - \lambda/\hbar) \quad , \quad (17)$$

which is the canonical momentum adjoint to the variable ϕ, and thus corresponds to the quantal operator $-i\partial/\partial\phi$. From eqs.(16) and (17) we have

$$H_{coll} = \frac{1}{2}B(\Delta)\dot{\Delta}^2 + \frac{\hbar^2}{2J}(N - N_0)^2 + \lambda(N - N_0) + V(\Delta) \ . \quad (18)$$

Eq.(17) tells us that the momentum is not proportional to the corresponding velocity and therefore the usual Pauli prescription[7] may not be simply applied to the Hamiltonian of eq.(16). Instead, the quantized Hamiltonian, as determined from eq.(18), reads

$$\hat{H}_{coll} = -\frac{1}{2}\frac{\hbar^2}{\sqrt{JB}}\frac{\partial}{\partial\Delta}\sqrt{\frac{J}{B}}\frac{\partial}{\partial\Delta} - \frac{\hbar^2}{2J}\frac{\partial^2}{\partial\phi^2} - i\lambda\frac{\partial}{\partial\phi} + V \quad , \quad (19)$$

which differs in form from that of ref.[2] by the term with the first derivative $\partial/\partial\phi$. In the case of the symmetric system, when λ is a constant, this term gives nothing but a shift of the entire spectra of neighbouring even nuclei.

FINAL REMARKS

We have derived the collective pairing Hamiltonian for

asymmetric systems. In comparison to that for symmetric systems it contains the term which is proportional to the particle number excess with the gap-dependent Fermi energy as a coefficient. To realize the geometrical picture of the problem one should remember that the collective motion in the gauge space is adiabatic with respect to a frame of reference which already rotates around the laboratory frame with the angular velocity $\dot{\phi}_0 = \lambda/\hbar$ determined by the condition for the mean particle number. Therefore, the Ruthian, and not the Hamiltonian, governs the adiabatic collective motion and the gauge angle in the rotating reference frame is the adiabatic collective variable. On the other hand, the Hamiltonian is quadratic in the velocity of the laboratory gauge angle which is not an adiabatic variable. The Pauli prescription may not be applied for the quantization of the Hamiltonian since the momentum adjoint to ϕ is not proportional to $\dot{\phi}$. Analogies of the pairing collective motion to the quadrupole motion are thus limited.

The authors appreciate helpful discussions with M.Brack, A.Góźdź, K.Pomorski and W.Stępień-Rudzka.

REFERENCES

1. S.G.Rohoziński, J.Dobaczewski, B.Nerlo-Pomorska, K.Pomorski and J.Srebrny, Nucl.Phys. A292 (1977) 66.
2. D.R.Bés, R.A.Broglia, R.P.J.Perazzo and K.Kumar, Nucl.Phys. A143 (1979) 1.
3. A.Góźdź, K.Pomorski, M.Brack and E.Werner, Nucl.Phys. A442 (1985) 50.
4. M.Baranger and K.Kumar, Nucl.Phys. A122 (1968) 241.
5. P.Ring and P.Schuck, The Nuclear Many-Body Problem (Springer-Verlag, New York, 1980)
6. M.Baranger and M.Veneroni, Ann.Phys.(N.Y.) 114 (1978) 123
7. W.Pauli, Handbuch der Physik, Vol 24/1 (Springer, Berlin, 1933) p.120

EXACT TREATMENT OF THE PAIRING FORCE IN LARGE SHELL MODEL SPACES

DENIS KARL SUNKO
Department of Theoretical Physics, University of Zagreb, 41000 Zagreb, Yugoslavia

Abstract A generating function method is described which gives the exact spin distribution of a multilevel configuration with prescribed seniority for each level. The method is equivalent to an exact diagonalization of the pairing Hamiltonian with no interaction between the levels. Possible applications are discussed.

INTRODUCTION

There are essentially two methods to deal with large discrete configuration spaces. One is statistical, which smoothes out the discreteness to extract the salient features of the problem analytically. The level-density expression of Bethe[1] is a classic example of this approach. Another is combinatorial, which uses precisely the discrete nature of the space to investigate it with finite algorithms. A natural counterpart to Bethe's result is the work of Hillman and Grover[2] on shell model level densities, which also showed the limitations of the combinatorial method: the "odometer" (exhaustion) algorithms become prohibitively inefficient at higher excitations, and the residual pairing force had to be taken into account by the BCS approximation, an analytical approach.

This is a preliminary report on a method that promises to extend the applicability of the combinatorial approach. Basically it amounts to replacing certain exhaustion algorithms by much more efficient ones. It is also possible to include the pairing force in a natural way. The most important deficiency is that no interactions between the levels are at present taken into account.

GENERATING FUNCTIONS

When n identical particles of angular momentum j are coupled, the possible resulting angular momenta J are restricted by symmetry requirements on the total wave function. The usual method to deal with the problem was proposed by Condon and Shortley[3]. One lists (exhausts) all the possible m-scheme states, and the J-multiplicities are deduced by subtracting the numbers of these with successive total projection M.

Another way is to use a generating function. It turns out that the generating function for the m-scheme multiplicities is[4]

$$G(n,j) = q^{-n(r-n)/2} \begin{bmatrix} r \\ n \end{bmatrix}_q \qquad (1)$$

where

$r = 2j + 1$ (fermions)

$r = 2j + n$ (bosons)

and the square brackets denote a Gaussian polynomial in q. It is a generalization of the ordinary binomial coefficient to which it reduces for q = 1. An efficient recursion algorithm is avail-

able to calculate the coefficients in (1). The method is easily extended to the case of many levels[5].

In order to treat the pairing force, the generating function of a configuration with given seniority must be found. It is given by

$$V(s=n,j) = G(n,j) - G(n-2,j) \tag{2}$$

since any angular momentum state containing pairs coupled to zero must also be present in the (n-2)-particle configuration. The generating function (2) is extended to the multilevel case in the same way as before.

DYNAMICS

The pairing Hamiltonian is here[6]

$$H = \sum_i n_i \varepsilon_i + G \sum_i S_+^{(i)} S_-^{(i)} \tag{3}$$

where n_i are the number operators of the single-particle levels with energy ε_i. Notice that (3) does not exchange zero-coupled pairs among the levels: configurations are not mixed.

One may generate by odometer all possible distributions of uncoupled nucleons in a given set of levels, and all arrangements of zero-coupled pairs for each of them. The energy of each such configuration is given by a simple expression[7]. The essential point of the method is that for each configuration with prescribed seniority structure, a generating function like V(s) gives the exact spin distribution. The calculation time is limited by the number of configurations, and not the number of states.

An example is a valence-shell calculation for ^{137}Ce. It has 8 protons and 3 neutron holes distributed over five levels. The complete spectrum was calculated using Kisslinger-Sorensen energies[8] and G = 23/137. It contains 2.68 x 10^9 states distributed over 72451 levels, with 9010 distinct spin distributions. The calculation time was 2.9 minutes on a UNIVAC 1100.

Part of the results is shown in the figure. It can be seen that configurations predicted by the shell model to be lowest in energy for a given distribution of unpaired particles are actually yrast only for J ≥ 17/2. At lower excitations, configurations containing promoted zero coupled pairs

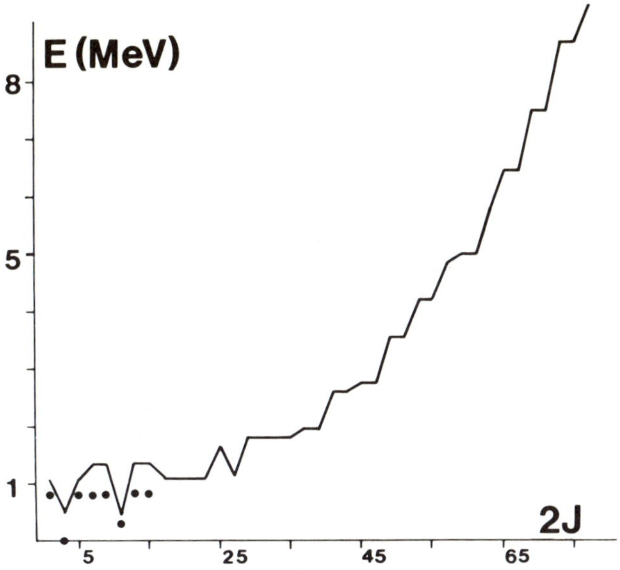

FIGURE 1. Shell-model ground state band for ^{137}Ce (full line). Points show yrast states that do not belong to it.

are lower: the shell model ground state is unstable for J up to 15/2. Thus a superfluid effect is observed.

DISCUSSION

Two possible areas of application come to mind. One is an extension of the combinatorial calculation of shell model level densities. Another is an exact investigation of the superfluid phase transition as a function of both the excitation energy and the number of particles in a major shell.

Although the Hamiltonian (3) correlates zero-coupled pairs only within levels and not among them, these correlations are capable of destabilizing the shell-model ground state and thus modelling the essential feature of superfluidity.

At the moment of writing, the method easily handles valence-shell problems. Improvements in the configuration generator need to be made before several major shells can be treated.

On the theoretical side, one should include some interaction between the levels. A neutron-proton interaction is presently being considered which also contains particle-hole correlations to an extent. It is hoped that investigating the interplay of these with the pairing correlations by an exact method suited for finite systems will lead to interesting results.

REFERENCES

1. A. Bohr and B.R. Mottelson, Nuclear Structure,

Vol.1 (W.A. Benjamin, New York, 1969), p. 288 and references therein.
2. M. Hillman and J. Grover, Phys. Rev.,185, 1303 (1969).
3. A. de-Shalit and I.Talmi, Nuclear Shell Theory (Academic Press, New York, 1963) pp. 253-255.
4. D.K. Sunko and D. Svrtan, Phys. Rev,, C31, 1929 (1985).
5. D.K. Sunko, Phys. Rev. C33, 1811 (1986)
6. P. Ring and P. Schuck, The Nuclear Many Body Problem (Springer Verlag, New York, 1980) p. 449.
7. Ref. 6, p. 223.
8. L.A. Kisslinger and R.A. Sorensen, Rev. Mod. Phys., 35, 853 (1963)

ON THE COHERENT-STATE REPRESENTATION FOR THE HOMOGENEOUS SPACES OF $Sp(6,\mathbb{R})$ +

Z. PAPADOPOLOS*, P. KRAMER, W. SCHWEIZER
Institut für Theoretische Physik der Universität
Tübingen, D-7400 Tübingen, FRG

Abstract This contribution is complementary to [1] in which we studied analytic coherent states for the general open-shell problem. We also introduce unitary coherent states. Using a coordinate transformation for the system of $n \equiv A-1$ particles we show explicitly, using the techniques in Bargmann[2] space, how to determine the finite action of $Sp(6,\mathbb{C})$ on the analytic (and unitary) coherent state. The matrix element of $Sp(6,\mathbb{C})$ between any two coherent states determines the corresponding integral kernel for finite symplectic transformations.

Def. 1: The complex symplectic group in three dimensions is defined

$$Sp(6,\mathbb{C}) = \{{}^c g \,|\, {}^{tc}g K {}^c g = K \;, \quad K := \begin{pmatrix} 0 & \mathbb{I}_3 \\ -\mathbb{I}_3 & 0 \end{pmatrix}\}.$$

In the matrix block form the element ${}^c g$ can be written ${}^c g = \begin{pmatrix} \lambda & \mu \\ \nu & \rho \end{pmatrix}$, where 3×3 complex matrices λ,μ,ν,ρ are such that ${}^t({}^t\lambda\nu) = {}^t\lambda\nu$, ${}^t({}^t\mu\rho) = {}^t\mu\rho$ and ${}^t\lambda\rho - {}^t\nu\mu = \mathbb{I}_3$.

Now we introduce a chart for $Sp(6,\mathbb{C})$. We demand that there exists λ^{-1} and we introduce the new parametrization $A := \nu\lambda^{-1}$ and $B := -\lambda^{-1}\mu$. We also demand that there exists

+Work supported by the Deutsche Forschungsgemeinschaft
*Permanent adress: Institute of Physics, Belgrad, Jugoslavia

$(\mathbb{I}_3 - B_1A_2)^{-1}$ where A_2 and B_1 are A's and B's that can be attached with any two ${}^c g_1$ and ${}^c g_2$. The new parametrization actually defines a restricted set of matrices. This subset we denote by G with $g \in G$

$$g \equiv g(A, {}^t\lambda^{-1}, B) := \begin{bmatrix} \lambda & -B \\ A & {}^t\lambda^{-1} - A\lambda B \end{bmatrix} = \begin{bmatrix} \mathbb{I} & 0 \\ A & \mathbb{I} \end{bmatrix} \begin{bmatrix} \lambda & 0 \\ 0 & {}^t\lambda^{-1} \end{bmatrix} \begin{bmatrix} \mathbb{I} & -B \\ 0 & \mathbb{I} \end{bmatrix}$$

where ${}^tA = A$ and ${}^tB = B$, A and B are complex matrices. $\mathbb{I}$ stands from now on for $\mathbb{I}_3$. The multiplication properties are:

i) $g_1 g_2 = g$

$$\lambda = \lambda_1(\mathbb{I} - B_1A_2)\lambda_2$$

$$A = A_1 + {}^t\lambda_1^{-1} A_2 (\mathbb{I} - B_1A_2)^{-1} \lambda_1^{-1}$$

$$B = B_2 + \lambda_2^{-1} (\mathbb{I} - B_1A_2)^{-1} B_1 {}^t\lambda_2^{-1}$$

ii) $g^{-1} = -K {}^t g K$

iii) $e := g(0, \mathbb{I}, 0)$

All elements of the complex form of $Sp(6,\mathbb{R})$ belong to G (compare also [3]).

<u>Def. 2:</u> Define the kernel in Bargmann space

$$T_g(x,\tilde{x}) = \nu_g({}^t\lambda^{-1}) \exp\tfrac{1}{2}\mathrm{tr}({}^t\tilde{x}A\tilde{x}) \exp \mathrm{tr}({}^t\tilde{x}\,{}^t\lambda^{-1}\bar{x}) \exp\tfrac{1}{2}\mathrm{tr}({}^t\bar{x}B\bar{x})$$

where x and $\tilde{x}$ are the $3\times n$ complex matrices and $\nu_g = (\det {}^t\lambda^{-1})^{n/2}$.

<u>Prop. 3:</u> The kernels $T_g, \forall g \in G$ under multiplication obey $T_{g_1} T_{g_2} = T_{g_1 g_2}$ under the additional condition $\mathbb{I} - \frac{1}{4}(A_2 + \bar{B}_1)(\bar{A}_2 + B_1) > 0$. Proof: One uses the results from Itzykson[4].

<u>Def. 4:</u> Let us introduce the coordinate transformation

$$\{x_{is}, \bar{x}_{is} \mid i=1,2,3 \ s=1,2,\ldots,n\} \overset{\phi}{\to} \{z'_{is} \mid i=1,2,3 \ s=1,2,3;$$

$z''_{is} \mid i=1,2,3 \;\; s=4,5,6; \;\; x'''_{is} \mid i=1,2,3 \;\; s=7,8,\ldots,n, \;\; c.c.^{*}\}$

represented in matrix form

$$\begin{pmatrix} x \\ \bar{x} \\ 0 \end{pmatrix}_{n \times n} \begin{pmatrix} R^{-1} & 0 \\ 0 & \mathbb{I}_{n-6} \end{pmatrix}_{n \times n} = \begin{pmatrix} z' & z'' & x''' \\ \overline{z''} & \overline{z'} & \overline{x'''} \\ 0 & 0 & 0 \end{pmatrix}_{n \times n}$$

where $R = \sqrt{\tfrac{1}{2}} \begin{bmatrix} \mathbb{I} & i\mathbb{I} \\ \mathbb{I} & -i\mathbb{I} \end{bmatrix}$

Prop. 5: The kernel from definition 2 under the coordinate transformation (c.t.) takes the form

$$T_g(z'z''x''' \,,\, \tilde{z}',\tilde{z}''\tilde{x}''') = \nu_g \exp E$$

$$E := \mathrm{tr}({}^t\tilde{z}'' A \tilde{z}') + \tfrac{1}{2}\mathrm{tr}({}^t\tilde{x}''' A \tilde{x}''') + \mathrm{tr}({}^t\tilde{z}'' \, {}^t\lambda^{-1} \bar{z}'') + \mathrm{tr}({}^t\tilde{z}' \, {}^t\lambda^{-1} \bar{z}')$$
$$+ \mathrm{tr}({}^t\tilde{x}''' \, {}^t\lambda^{-1} \overline{x'''}) + \mathrm{tr}({}^t\bar{z}' B \overline{z''}) + \tfrac{1}{2}\mathrm{tr}({}^t\overline{x'''} B \overline{x'''})$$

---

* c.c. means complex conjugate.

and the measure in Bargmann space separates under ϕ

$$d\mu(x) = d\mu(z')d\mu(z'')d\mu(x''')$$

where all $d\mu$'s are Bargmann measures

$$d\mu(z)=\pi^{-k}\exp-(z\cdot\bar{z})\prod_{i=1}^{k} d\,\mathrm{Re}(z_i)\, d\,\mathrm{Im}(z_i).$$

Prop. 6: Let us introduce the extremal or "reference" state denoted by $|extr\rangle$ which takes the following form in Bargmann space

$$\langle x|extr\rangle \equiv f_h(x) \equiv f'_h(z')=(\Delta^1_1)^{h_1-h_2}(\Delta^{12}_{12})^{h_2-h_3}(\Delta^{123}_{123})^{h_3}(z'),$$

Δ's are the subdeterminants of the matrix z' written as z'^s_i. This state is the minimal state w.r.t. $Sp(6,\mathbb{R})$ and the maximal w.r.t. $U(3)$. The $h:\{h_1h_2h_3\}$ is the label for irr of $Sp(6,\mathbb{R})$.

Proof: We make use of the c.t. ϕ and take the definition of the generators for $Sp(6,\mathbb{R})$ from [5], proposition 2.2.

Prop. 7: We determine the finite action of G on the extremal or "reference" state in Bargmann space using the c.t. ϕ and proposition 5 to obtain

$$\langle extr|T_g(A,{}^t\lambda^{-1},B)|x\rangle =$$

$$=\nu_g\exp\{tr({}^t\overline{z''}B\overline{z'})\}\exp\{\tfrac{1}{2}tr({}^t\overline{x'''}B\overline{x'''})\}\overline{f'_h({}^t\overline{\lambda}^{-1}z')}$$

ν_g is as in definition 2.

Prop. 8: We determine the expectation value of T_g in the extremal state using the c.t. ϕ

$$\langle extr|T_g|extr\rangle=(\Delta^1_1)^{h_1-h_2}(\Delta^{12}_{12})^{h_2-h_3}(\Delta^{123}_{123})^{w_3}({}^t\lambda^{-1}),\ w_3:=h_3+\tfrac{n}{2}.$$

Prop. 9: The adjoint kernel $T^+_g(x,\tilde{x})$ to the kernel $T_g(x,\tilde{x})$ is the kernel of $g^{\S}$, $T_{g^{\S}}(x,\tilde{x})$, where $g^{\S}=Mg^+M$,

$M=\begin{vmatrix} \mathbb{I} & 0 \\ 0 & -\mathbb{I} \end{vmatrix}$, $g^{\S}\epsilon G$. All the results in the remaining part of the paper follow from the propositions and definitions stated till now.

Corollary 10:

$$\| \langle extr | T_g(A, {}^t\lambda^{-1}, B) | x \rangle \|^2 = (\Delta_1^1)^{h_1-h_2} (\Delta_{12}^{12})^{h_2-h_3} (\Delta_{123}^{123})^{w_3} (\;\;),$$

where $(\;\;) := ({}^t\lambda^{-1} (\mathbb{I} - B^{+}B)^{-1} ({}^t\lambda^{-1}))$.

It is valid for $(\mathbb{I} - B^{+}B) > 0$.

Proof: Using c.t. ϕ, proposition 7 and the results from Bargmann[2] from Appendix. The ket $|x\rangle$ reminds us that the integration is over $\mathbb{C}^{3n}, x\epsilon\mathbb{C}^{3n}$.

Def. 11: For the elements g_c of the coset $U(1)\times U(1)\times U(1)\backslash Sp(6,\mathbb{R})$, which are the group elements, we introduce the parametrization as in [1], relations (A.7,10,11):

$$g_c = y(k,B)\, c\,(k,B),$$

where

$$y(k,B)=\begin{bmatrix} {}^t\ell^{-1} & 0 \\ 0 & \ell \end{bmatrix}\begin{bmatrix} \mathbb{I} & 0 \\ A' & \mathbb{I} \end{bmatrix} \text{ and } c(k,B)=\begin{bmatrix} {}^tk^{-1} & 0 \\ 0 & k \end{bmatrix}\begin{bmatrix} \mathbb{I} & -B \\ 0 & \mathbb{I} \end{bmatrix}$$

and A' and B are complex 3×3 symmetric matrices, k is a strictly upper triangular complex and ℓ is lower triangular complex with strictly positive diagonal elements. Note that $c(k,B)$ is the analytic parametrization, but is not the group element.

Corollary 12:

$$\langle extr | T(y) | x \rangle = \kappa(y) \langle extr | x \rangle, \text{ where } \| \, | extr \rangle \| = 1$$

and

$$\kappa(y) := (\Delta_1^1)^{h_1-h_2} (\Delta_{12}^{12})^{h_2-h_3} (\Delta_{123}^{123})^{w_3} (\ell) > 0$$

and y and ℓ are as in definition 11.

Proof: We use propositions 7, 10 and definition 11.

Corollary 13:

$\langle extr|T(y)T(c)|x\rangle = \kappa(y)\langle extr|T(c)|x\rangle$,

and the analytic coherent state $\langle extr|T(c)|x\rangle$ we denote by

$\langle kB|x\rangle = \exp(\overline{z''}\cdot B\overline{z'})\exp(\frac{1}{2}\overline{x'''}\cdot B\overline{x'''})\overline{f'_h(\overline{k}z')}$.

Proof: From corollary 12 and proposition 7.

Corollary 14:

(1) $\|\langle extr|T(c)|x\rangle\|^2 = \langle kB|kB\rangle = (\Delta_1^1)^{h_1-h_2}(\Delta_{12}^{12})^{h_2\,h_3}(\Delta_{123}^{123})^{w_3}(\ )$

where $(\ ) := (k(\mathrm{I\!I}-B^+B)^{-1}k^+)$

(2) $\|\langle extr|T(y)|x\rangle\|^2 = [\kappa(y)]^2$

(3) $\|\langle extr|T(yc)|x\rangle\|^2 = 1$, what means that $T(yc) \equiv T_{g_c} \equiv U(g_c)$.

U stands for the unitary operator.

Proof: Follows from corollary 10 and definition 11.

Prop. 15: The relation between the unitary and analytic coherent states is

$\langle extr|U(g_c)|x\rangle = |\langle kB|kB\rangle^{-\frac{1}{2}}|\langle extr|T(c)|x\rangle$

$\kappa^2(y) = \langle kB|kB\rangle^{-1}$.

Proof: It follows from corollary 13 and 14(3).

Prop. 16: The finite action of G on the analytic function $F(c) \equiv \langle kB|F\rangle$ is given by

$T(\tilde{g})F(c) = \kappa({}^t\lambda'^{-1})F(c')$

where κ is as in corollary 12. The notation is introduced as follows

$c(kB)\tilde{g}(\tilde{A}, {}^t\tilde{\lambda}^{-1}, \tilde{B}) = g'(A', {}^t\lambda'^{-1}, B')$

where ${}^t\lambda'^{-1} = k(\mathrm{I\!I}-\tilde{A}B)^{-1}\,{}^t\tilde{\lambda}^{-1}$, $c' = c'(k'b')$ with $B' =$

$=\tilde{B}+\tilde{\lambda}^{-1}B(\mathbb{I}-\tilde{A}B)^{-1}\,{}^t\tilde{\lambda}^{-1}$ and k' determined explicitly from ${}^t\lambda'^{-1}$, and $A'=k\tilde{A}(\mathbb{I}-B\tilde{A})^{-1}\,{}^tk$. The proposition is valid under the conditions that there exists $(\mathbb{I}-\tilde{A}B)^{-1}$ and $\mathbb{I}-\frac{1}{4}(\tilde{A}+\overline{B})\overline{(\tilde{A}+\overline{B})}>0$.

Proof: One determines $\langle kB|T(\tilde{g})|x\rangle$ using propositions 3 and 7, corollary 13 and the multiplication properties of 6.

Prop. 17: The matrix elements of the kernels of G in the analytic coherent states are given

$$\langle kB|T(\tilde{g})|\hat{k}\hat{B}\rangle=(\Delta_1^1)^{h_1-h_2}(\Delta_{12}^{12})^{h_2-h_3}(\Delta_{123}^{123})^{w_3}\{\ \}$$

$$\{\ \}:=\{k(\mathbb{I}-\tilde{A}B)^{-1}\,{}^t\tilde{\lambda}^{-1}[\mathbb{I}-\hat{B}^+(\tilde{B}+\tilde{\lambda}^{-1}B(\mathbb{I}-\tilde{A}B)^{-1}\,{}^t\tilde{\lambda}^{-1})]^{-1}\hat{k}^+\}$$

under the conditions that there exists

$(\mathbb{I}-\tilde{A}B)^{-1}$ and $\mathbb{I}-\frac{1}{4}(\tilde{A}+\overline{B})\overline{(\tilde{A}+\overline{B})}>0$

and that there exists

$[\mathbb{I}-\hat{B}^+(\tilde{B}+\tilde{\lambda}^{-1}B(\mathbb{I}-\tilde{A}B)^{-1}\,{}^t\tilde{\lambda}^{-1})]^{-1}$ and

$\mathbb{I}-\frac{1}{4}[\ \ ]\overline{[\ \ ]}>0$ where

$[\ \ ]:=[\mathbb{I}-\hat{B}^+(\tilde{B}+\tilde{\lambda}^{-1}B(\mathbb{I}-\tilde{A}B)^{-1}\,{}^t\tilde{\lambda}^{-1})]$.

In the special case g=e we obtain the overlap

$$\langle kB|\hat{k}\hat{B}\rangle=(\Delta_1^1)^{h_1-h_2}(\Delta_{12}^{12})^{h_2-h_3}(\Delta_{123}^{123})^{w_3}(k(\mathbb{I}-\hat{B}^+B)^{-1}\hat{k}^+)$$

under the corresponding conditions.

Proof: Follows from the multiplication properties of G, propositions 3, 8, 9 and definition 11.

REFERENCES

1. P. Kramer and Z. Papadopolos, J. Phys. A 19, 1083 (1986).
2. V. Bargmann, Analytic Methods in Mathematical Physics, ed. by R.P. Gilbert and R.G. Newton (Gordon & Breach, New York, 1968), p. 27.
3. M. Brunet and P. Kramer, Rep. Math. Phys. 17, 205 (1980).
4. C. Itzykson, Commun. Math. Phys. 4, 92 (1967).
5. P. Kramer, Ann. Phys. 141, 269 (1982).

A MICROSCOPIC STUDY OF NUCLEON ELASTIC SCATTERING FROM COMPLEX NUCLEI

C. I. CALLE
Sweet Briar College, Sweet Briar, VA 24595, U.S.A.

Abstract Using a microscopic theoretical approach, a completely off-shell t-matrix, obtained from a Reid hard-core nucleon-nucleon potential, is used to construct an energy-dependent, first-order optical potential for elastic scattering. In this microscopic treatment, use is made of finite nuclei, including complete off-shell effects, insuring that the solutions include all orders of multiple scattering for the ground state. Calculations were performed for the elastic scattering of protons on ^{4}He at several low and intermediate energies. The cross-sections reproduced the general experimental angular distributions at these energies and specifically showed the energy dependence of the back angle rise.

I. INTRODUCTION

As is well known, the behavior of an incident nucleon as it interacts with a complex nucleus at low energies can be represented by use of a complex, energy dependent, nonlocal, optical potential. Although a phenomenological approach at obtaining this potential has given good agreement with experimental data, a microscopic theoretical approach is more desirable, but immensely more difficult to achieve.

Watson[1] and later KMT[2] gave microscopic basis to the calculation of the optical potential. In the KMT approximation, the calculation of the optical potential makes use of a factorized approximation which leads to a local t-

matrix. However, the microscopic multiple scattering series is equivalent to the use of an energy-dependent optical potential which is nonlocal. Thus, an off-shell, nonlocal t-matrix should be used in the wave equation.

In this paper we present an off-shell t-matrix obtained from a Reid hard-core nucleon-nucleon potential. This t-matrix is used to construct an energy dependent, first-order, optical potential for elastic scattering. In this microscopic treatment use is made of finite nuclei, including off-shell effects. This insures that the solutions of the integro-differential equations include all orders of multiple scattering for the ground state of the target nucleus.

II. FORMALISM

We shall present here a derivation neglecting spin and isospin. The complete derivation with spin and isospin is presented in reference 3.

Neglecting spin and isospin, the total wavefunction is

$$\Psi(\vec{r}_P) = 4\pi \sum_{\substack{LL' \\ JM}} \frac{g^{L'}_{JL}(k_P r_P)}{k_p r_p} Y^{*}_{LM_L}(\hat{k}_P) Y_{L'M'_L}(\hat{r}_P), \tag{1}$$

where the subindex p refers to the incoming projectile. The integro-differential equation we must solve is

$$\left[\frac{d^2}{dr^2_P} - \frac{L(L+1)}{r^2_P} - \frac{2\mu}{\hbar^2} V_C(r_P) + k_P\right] g^{L}_{JL_i}(k_P r_P)$$
$$= \frac{2\mu}{\hbar^2} r_P \sum_{L'} \int r'_P dr'_P \langle LM_L | V(\vec{r}_P, \vec{r}'_P) | L'M'_L \rangle \, g^{L}_{JL_i}(k_P r'_P), \tag{2}$$

where V_L is the Coulomb potential,

$$V(\vec{r}_p,\vec{r}_p') = \langle\vec{r}_p|V|\vec{r}_p'\rangle$$

is a nonlocal potential, and L_i is the incident angular momentum of the incoming nucleon, determined by the boundary conditions.

In the solution of our integro-differential equation, Eq. (2), the calculations can be simplified by using the plane wave expansion method.[4] In this method, the solutions $g^L_{JL_i}(k_p r_p)$ can be represented by an expansion in spherical Bessel functions

$$g^L_{JL_i}(k_p r_p) = \sum_{m=1} a_m^{JL}\, \phi^L_m(r_p), \tag{3}$$

where

$$\phi^L_m(r_p) = r_p j_L(k_m r_p),$$

and the $j_L(k_m r_p)$ are the spherical Bessel functions. The expansion (3) is a Fourier-type series known to converge.

In the KMT formulation, the nonlocal potential $V(\vec{r}_p,\vec{r}_p')$ is proportional to the t-operator

$$V(\vec{r}_p,\vec{r}_p') \propto \langle\vec{r}|\langle\Phi_0|t|\Phi_0\rangle|\vec{r}'\rangle \tag{4}$$

where $|\Phi_0\rangle$ is the state vector of the correlated ground state of the target nucleus. The target wavefunctions Φ_0 in the t-matrix element should be antisymmetrized by forming the Slater determinant

$$\Phi_0 = \frac{1}{\sqrt{A!}} \Sigma_k (-)^k\, \hat{P}_k u_1(1)\ldots u_A(A) \tag{5}$$

for the A noninteracting nucleons, where $u_n(i)$ are the single-particle harmonic oscillator wavefunctions for the ith nucleon. For the spinless case,

$$u_n(i) = u_{n_x l_x}(r_x). \tag{6}$$

To take into account the correct center-of-mass motion, the spherical Bessel functions $\phi_m^L(r_p)$ from the plane wave expansion are expanded in terms of harmonic oscillator wavefunctions $U^{LN}(r_p)$, so that a coordinate transformation of the product of these wavefunctions and the harmonic oscillator wavefunctions (6) of the target nucleons can be performed.

The transformation between the two wavefunctions is caused by the transformation of a pair of linearly independent vectors $\vec{r}_p$ and $\vec{r}_x$ into another pair of independent vectors $\vec{r}$ and $\vec{R}$. The transformation bracket is known as the Moshinsky bracket.[5] The harmonic oscillator wavefunctions of the relative coordinate $\vec{r}$ so obtained can be expanded in a series of spherical Bessel functions

$$\langle nl|\vec{r}\rangle = \sum_m d_m j_l(k_m r) \tag{7}$$

The completely off-shell matrix element

$$\langle k_m|t|k_m'\rangle \tag{8}$$

is calculated using the code TOFF developed by Koshel and Griffen[6] assuming a realistic two-body Reid hard-core potential.[7] Once this realistic, two-body, local potential with hard core is assumed, the off-shell elements of the t-matrix should be uniquely determined.

In order to simplify the calculations in the solution of our integro-differential equation, Eq. (2), we use the plane wave expansion method,[4] in which the wavefunctions can be represented as an expansion in spherical Bessel func-

tions, allowing the solutions of the inhomogeneous differential equation to be replaced by the product of matrices. The expansion introduces no real approximations since one can keep as many plane waves as needed to obtain a good representation of the scattering wavefunctions. This is due to the fact that the expansion method is uniformly convergent.[8] Thus, only a subset of expansion coefficients need be retained, not the wavefunction at every coordinate point. As the orbital angular momentum of the state increases, fewer plane wave states are needed.

III. RESULTS AND DISCUSSION

Figures 1 and 2 show the results of our calculations for the p-^{4}He differential cross sections at 200 and 85 MeV. The data are taken from G. A. Moss, *et al.*,[9] and Comparat, *et al.*[10] These cross sections reproduce the general experimental angular distributions at these energies and specifically show the energy dependence of the back angle rise, present in the 85 MeV cross-sections and absent at 200 MeV. If we compare Figs. 1 and 2, it is apparent that our results for the differential cross sections are too weak at 200 MeV and too strong at 85 MeV.

The t-matrix used in this calculation includes multiple scattering. However, this multiple scattering was only for the ground state and did not include excitation. It is likely that this multiple scattering with excitation could make the quantitative agreement, only fair in the present calculation, somewhat better.

Most of the difficulty with the numerical calculations came from the expansions of the Bessel functions into the harmonic oscillator wavefunctions required by our method.

Thus, special numerical techniques were used to avoid inaccuracies. The expansion of the solution in plane waves, Eq. 3, was of greater sensitivity to inaccuracies. The expansion would not converge unless a different set of expansion coefficients was used with each partial wave, in accordance with previous successful calculations using the same method.

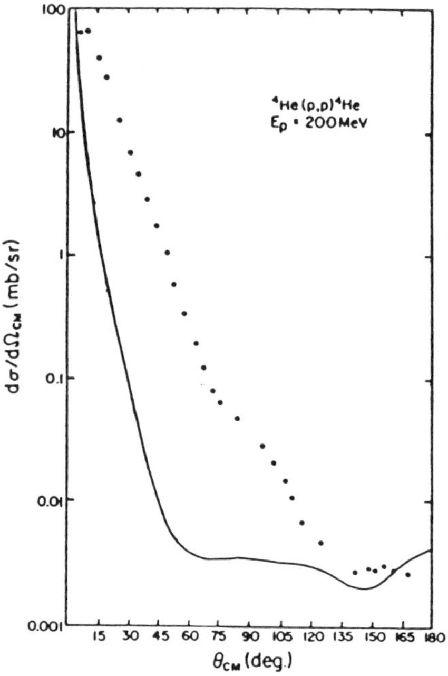

FIGURE 1 Differential cross section for E_p=200 MeV. The experimental data points are denoted by dots. Data from Ref. 9.

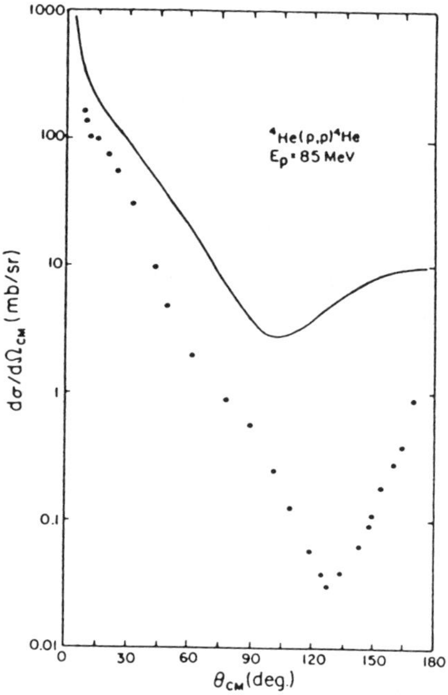

FIGURE 2 Differential cross section for E_p= 85 MeV. The experimental data points are denoted by dots. Data from Ref. 10.

IV. CONCLUSIONS

Using a microscopic theoretical approach, a completely off-shell t-matrix obtained from a realistic nucleon-nucleon potential has been used to construct an energy-dependent, first-order optical potential for elastic scattering. The important aspect of this work is that it was derived from first principles, with no free parameters to fit.

This theoretical approach has allowed us to obtain elastic scattering cross-sections that reproduce the p-^{4}He backward peaking and its energy dependence, flat at 200 MeV and rising at 85 MeV. We believe this is an important advance. Since the nucleon-nucleon interaction used was empirically determined, exchange effects, which we believe are important, were inherently included.

As we mentioned earlier, we did not include multiple scattering with excitation. This was due to limitations of the computer used. It is likely that these virtual nuclear excitations present an important contribution to this interaction, and their inclusion should yield results with better quantitative agreement.

REFERENCES

1. K. M. Watson, Phys. Rev. 89, 575 (1953).
2. A. K. Kerman, H. McManus, and R. M. Thaler, Ann. Phys. 8, 551 (1959).
3. C. I. Calle, Ph.D. thesis, Ohio University, 1984.
4. D. Robson and R. D. Koshel, Phys. Rev. C6, 1125 (1972).
5. M. Moshinsky, Nucl. Phys. 13, 104 (1959).
6. R. D. Koshel and P. J. Griffen, Bull. Am. Phys. Soc. 20, 67 (1975).
7. R. V. Reid, Ann. Phys. 50, 411 (1968).
8. P. J. Griffen, P. Nagel, and R. D. Koshel, J. Math. Phys. 15, 1913 (1974).
9. G. A. Moss, *et al*., Phys. Rev. C21, 1932 (1980).
10. L. G. Votta, *et al*., Phys. Rev. C10, 520 (1974).

SPATIAL SYMMETRIES OF PROBABILITY EXTREMES RELATED TO LEGENDRE FUNCTIONS

GERASSIMOS S. ANAGNOSTATOS
National Research Center of Natural Sciences "Demokritos"
Aghia Paraskevi-Attiki, Athens, Greece 15310

Abstract Some hitherto unknown features of the angular part of the Independent Particle Model wave functions are presented and discussed in relation to previous similar findings. Specifically, angles θ which correspond to the extremes of probability related to the Associated Legendre Functions $P_\ell^m(\cos\theta)$ have been found to be identical to characteristic angles of the regular and quasi-regular polyhedra. Cases for $\ell \leq 4$ are reported here.

In recent years many hitherto unknown features of the angular part of the Independent Particle Model (IPM) wave functions have been found. These features are identically related to characteristic angles of the regular (R) and quasi-regular (Q-R) polyhedra.[1-4] It is interesting to briefly review the most prominent of them in order for the reader to have an overall picture of all these related and very interesting features.

In Refs. 1 and 2 the orbital angular momentum (OAM) quantization of direction for $\ell = 0 \div 6$ is discussed in every detail. Specifically, it has been found that there are characteristic axes of the R and Q-R polyhedra (e.g., some of their axes of symmetry) which, with respect to a common axis taken as quantization axis, form angles θ identically equal to the angles

$$\theta_\ell^m = \cos^{-1} \frac{m_\ell}{\sqrt{\ell(\ell+1)}} \quad . \tag{1}$$

That is, these angles are identical to the angles formed by

the OAM vectors($\underline{\ell}$) and their projections on the quantization axis(m_ℓ). In Ref. 3 one sees that, besides the OAM, the intrinsic spin($\underline{s}$) and its projection(m_s) is identically presented by the symmetry of the same polyhedra. Moreover, the vector addition of the representation vectors for OAM($\underline{\ell}$) and for spin($\underline{s}$) leads identically to total angular momentum vectors($\underline{j}$) which are quantized and have quantized projections on the quantization axis. That is, the resultant vectors identically follow the relation

$$\underline{j} = \underline{\ell} + \underline{s} \ . \tag{2}$$

In addition, all pairs of $\underline{j}$ (e.g., $\underline{j}_1$ and $\underline{j}_2$)geometrically derived as above, identically follow the relation

$$\underline{J} = \underline{j}_1 + \underline{j}_2 \ . \tag{3}$$

In Ref. 4 a complete symmetry description of the IPM is given, based on the $\underline{\ell}$, $\underline{s}$, $\underline{j}$ vectors discussed earlier. Fig. 1 could be considered as a summary representation of all previously found features, taking $\ell = 1$ as an example. Isospin ($\underline{\tau}$) is also presented there.

In the present work the spatial symmetries of probability extremes related to associate Legendre functions are presented and related to the previous features. As will become apparent, such extremes are identically connected to the symmetries of the R and Q-R polyhedra. Cases for $\ell \leq 4$ only are reported here.

The R and Q-R polyhedra are completely specified by their three characteristic angles ϕ, χ, and ψ. That is, the angle ϕ equals $\widehat{031}$; χ, $\widehat{032}$; and ψ, $\widehat{132}$; where 0 labels a vertex, 1 the middle of an edge, 2 the center of a face, and 3 the center of the polyhedron itself. All these angles for the R polyhedra are given in Table I of Ref. 5. The same angles numerically stand for the Q-R polyhedra, as one can

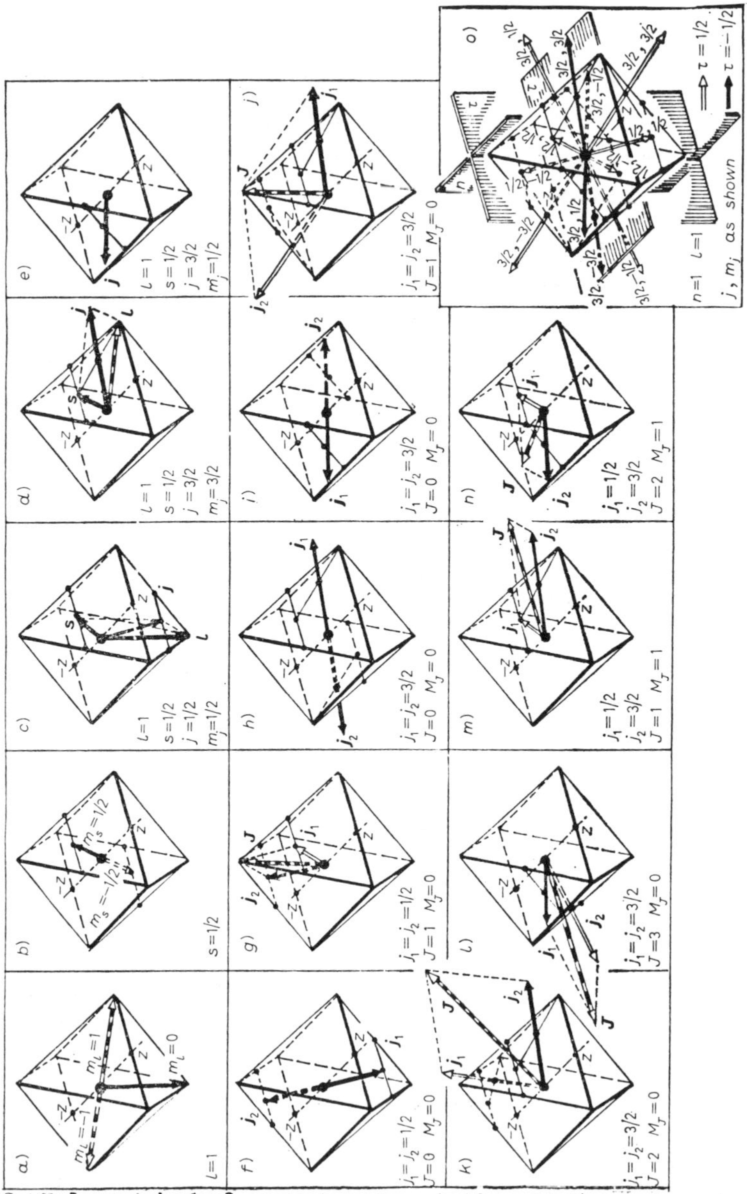

FIGURE 1. Octahedral symmetry description of the independent particle model for the 1p-shells. Each mark • stands for the middle point of the corresponding straight-line segment, while ● stands for the centre of the octahedron. Dark arrows stand for protons and open ones for neutrons, or the opposite.

easily verify by taking the appropriate relative orientation between R and Q-R polyhedra as shown in Figure 2 of Ref. 1. It has been found that the probability density function $(P_2^0)^2$ takes a minimum value when $\theta = 54°44'08''$, which is identical to the characteristic angle χ of an octahedron or, equally, of a hexahedron. The same value of θ also gives maximum value to the p.d.f. $(P_3^{\pm 2})^2$. The p.d.f. $(P_3^{\pm 1})^2$ has a minimum at $\theta = 63°26'06''$ which is identical to twice the angle ϕ of an icosahedron or, equally, the angle ψ of a dodecahedron. The angle θ which makes the p.d.f. $(P_1^{\pm 1})^2$ maximum is $90°$ and is identical to twice the angle ϕ of an octahedron or, equally, the angle ψ of an hexahedron. An equal angle gives an extreme value to each of the following p.d.f.: $(P_1^0)^2$, $(P_2^{\pm 2})^2$, $(P_2^{\pm 1})^2$, $(P_3^{\pm 3})^2$, $(P_3^{\pm 2})^2$, $(P_3^0)^2$, $(P_4^{\pm 4})^2$, $(P_4^{\pm 3})^2$ and $(P_4^{\pm 1})^2$.

It is interesting to compare these angles θ of probability extreme for the functions $P_\ell^m(\cos\theta)$ with the angles θ_ℓ^m, θ_ℓ^j, θ_j^{mj} discussed earlier in this paper and the cited references. Specifically, one can notice the following identities: The angle θ for $(P_2^0)^2$ or $(P_3^{\pm 2})^2$ to θ_3^2, for $(P_3^0)^2$ or $(P_3^{\pm 1})^2$ to θ_4^2, for $(P_4^0)^2$ or $(P_4^{\pm 1})^2$ or $(P_4^{\pm 3})^2$ to $\theta_3^{7/2}$, and for $(P_4^{\pm 2})$ to $\theta_{7/2}^{3/2}$.

It is also interesting to observe that while the symmetries associated with the quantization of direction of angular momenta (angles θ_α^β) refer to rotation axes, round which particles with definite quantum numbers α and β (e.g., ℓ and m) could rotate no matter what their instantaneous positions may be, the symmetries associated with the probability extremes of associate Legendre functions refer to positions of particles with definite ℓ and m values. Thus, both axes of rotation and probability extreme positions

are related to R and Q-R polyhedra and could be used to create a mental image of an instantaneous picture of a nucleus repeated periodically.

At this stage, one should correlate these findings with the isomorphic shell model[6], which considers that the most probable positions of the nucleons in a nucleus are highly correlated and, as a consequence, lie at the vertices of nested R and Q-R polyhedra, each of which represents an instantaneous form of a particular nuclear shell periodically appearing.

Specifically, in the framework of the isomorphic shell model, the main assumption of the simple shell model--that the nucleons in a nucleus move relatively freely in a central potential created by all nucleons--is understood as a result of a dynamic equilibrium in the following sense. Each nucleon in a nucleus is in a dynamic equilibrium with all other nucleons and, as a consequence, its motion may be described independently of the motions of the other nucleons. Given that equilibrium of particles on a sphere (e.g., the average sphere assumed by the nucleons of a particular nuclear shell, i.e., $\langle r_n^2 \rangle = \frac{\hbar}{m\omega}(n+3/2)$ for an harmonic oscillator central potential[7]) is permissible only for a certain number of particles in the specific arrangement of an equilibrium[8] (R,Q-R) polyhedron, this conclusion directly relates the IPM with the R and Q-R polyhedra. This relation is completely satisfied if one simultaneously considers all nested polyhedra presenting the successive nuclear shells, as shown in Fig. 2 (see Ref. 6). From this figure it is apparent that the known magic numbers 2, 8, 20, 28, 40, 50, 82, and 126 coincide with the cumulative numbers of vertices written inside brackets at each block of the figure. That is, according to the isomorphic shell model, each time an equilibrium

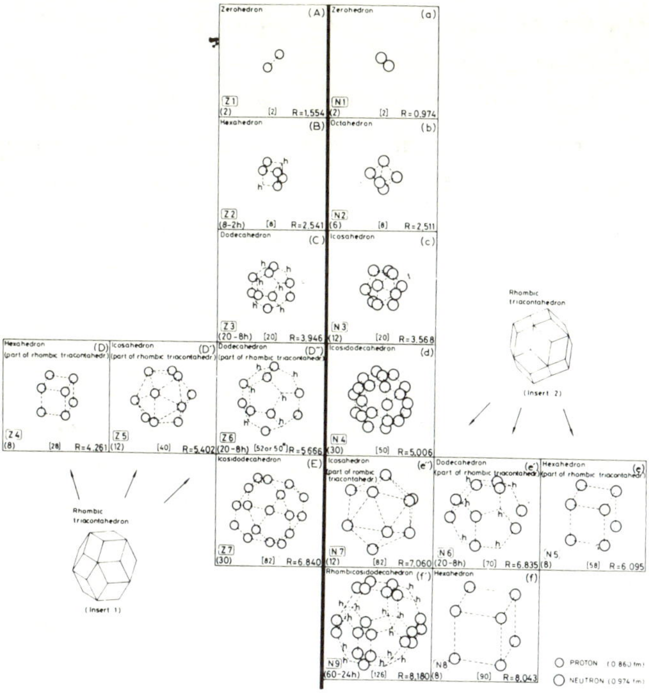

FIG. 2 Instantaneous dynamic average forms of nuclear shells from ^{4}He to ^{208}Pb presented by equilibrium polyhedra. Polyhedra lying on the left of the heavy central vertical line stand for proton shells, while those on the right stand for neutron shells, with a correspondence between them such that next to a proton (neutron) polyhedron (in each of the levels of the figure, which are labelled A–F for protons and a–f for neutrons) a neutron (proton) polyhedron exists that has the same rotational group (i.e., it is its reciprocal in geometrical language). The angular vertex distribution of the rhombic triacontahedron in insert 1 is identical to the angular vertex distributions of the three proton polyhedra hexadedron (cube), icosahedron, and dodecahedron (Figures 1D, 1D′, and 1D″, respectively) taken together [where vertices labeled h (holes) are excluded]. The same is true for insert 2 (which differs from insert 1 only in orientation) and the neutron polyhedra in Figures 1e, 1e′, and 1e″. Also, the angular vertex distribution of the imaginary dodecahedron made of the centers of all triangular faces of the rhomb-icosi-dodeca-hedron in Figure 1f′ is analyzed into the angular distribution of dodecahedral vertices occupied by sets of three neutrons (Figure 1f′) and the angular distribution of dodecahedral vertices (see letters h in Figure 1f′) forming an hexahedron (cube; Figure 1f). Even similar polyhedra in the figure (with the exception of the cubes) have different angular vertex distributions, since they have different orientations. The occupied polyhedral vertices stand for most probable positions of nucleons presented by hard spheres of definite sizes ($r_n = 0.974$ fm and $r_p = 0.860$ fm). All polyhedra of the figure of the figure are considered superimposed with a common center and relative orientations as shown, which bring the "hard spheres" of each proton (neutron) shell in contact with the "hard spheres" of the previous and of the next proton or neutron shell. For the derivation of the cumulative proton number $Z = 50^*$ see Ref. 6.

polyhedron employed by the model is completed, the cumulative number of vertices of all previous polyhedra and this particular one is equal to a magic number. That is, the model[6] broadens the basis of the independent particle model explanation of magic numbers.

REFERENCES

1. G. S. Anagnostatos, Lett. Nuovo Cimento, 22, 507 (1978).
2. G. S. Anagnostatos, J. Yapitzakis, and A. Kyritsis, Lett. Nuovo Cimento, 32, 332 (1981).
3. G. S. Anagnostatos, Lett. Nuovo Cimento, 28, 573 (1980).
4. G. S. Anagnostatos, Lett. Nuovo Cimento, 29, 188 (1980).
5. H. S. M. Coxeter, Regular Polytopes, 3rd edition (Macmillan, New York, 1973), p. 292.
6. G. S. Anagnostatos, Int. J. Theor. Phys., 24, 579 (1985).
7. W. F. Hornyak, Nuclear Structure (Academic Press, New York, 1975), p. 240.
8. J. Leech, Math. Gazette, 41, 81 (1957).

NEW SOLUTIONS TO SOME STATE LABELLING PROBLEMS OF INTEREST IN NUCLEAR STRUCTURE

C. QUESNE
Physique Théorique et Mathématique CP 229, Université Libre de Bruxelles, Bd. du Triomphe, B 1050 Brussels, Belgium.

Abstract New solutions to some state labelling problems associated with non-canonical chains of groups are presented. They all use complementarity properties generalizing the relation between seniority and quasi-spin. As an example, the case of U(3) x U(3) $\supset$ U(3), occurring in the construction of SU(3) Wigner coefficients, is reviewed.

INTRODUCTION

In many applications of group theory to nuclear structure, there occur non-canonical chains of groups, i.e., sequences G$\supset$H wherein H does not provide quantum numbers enough to completely specify the states transforming under an irreducible representation (irrep) of G. They give rise to so-called state labelling problems.

Over the past few years, new solutions have been proposed to some of them, associated with the chains U(n)$\supset$O(n),[1,2] U(n)$\supset$USp(n) (n even),[3] and U(n) x U(n)$\supset$U(n).[4,5] All such solutions use complementarity properties, generalizing the relation between seniority and quasi-spin.[6]

The purpose of the present communication is to review their underlying ideas. First the use of complementarity properties is illustrated by the (trivial) construction of SU(2) Wigner coefficients. Then the non-trivial case of

SU(3) Wigner coefficients is discussed. Finally the analysis is extended to the three above-mentioned chains.

SU(2) WIGNER COEFFICIENTS

The SU(2) Wigner coefficients $\langle jm,j'm \mid J\ M \rangle$ are the transformation coefficients between the states classified according to the two group chains

$$\underset{j}{SU(2)} \times \underset{j'}{SU(2)} \supset \underset{m}{U(1)} \times \underset{m'}{U(1)} , \tag{1}$$

$$\underset{j}{SU(2)} \times \underset{j'}{SU(2)} \supset \underset{J}{SU(2)} \supset \underset{M}{U(1)} , \tag{2}$$

where below each group we have indicated the quantum numbers specifying its irreps. The generators of the first SU(2) group in the product SU(2) x SU(2) are realized in terms of one pair of boson creation operators η_{1s}, s=1,2, and of the corresponding annihilation operators ξ_{1s}, s=1,2, as follows:

$$j_+ = \eta_{11}\xi_{12},\ j_- = \eta_{12}\xi_{11},\ j_0 = \frac{1}{2}(\eta_{11}\xi_{11} - \eta_{12}\xi_{12}). \tag{3}$$

For the generators of the second SU(2) group, a different realization in terms of another pair of boson creation operators η_{2s}, s=1,2, and of the corresponding annihilation operators ξ_{2s}, is chosen, namely

$$j'_+ = -\eta_{22}\xi_{21},\ j'_- = -\eta_{21}\xi_{22},\ j'_0 = \frac{1}{2}(\eta_{22}\xi_{22} - \eta_{21}\xi_{21}). \tag{4}$$

The generators of the SU(2) subgroup are then $\underline{J} = \underline{j} + \underline{j}'$.

The coupled states $|jj'\ J\ M\rangle$ are most easily built by considering the operators[5]

$$K_+ = \sum_s \eta_{1s}\eta_{2s},\ K_- = \sum_s \xi_{1s}\xi_{2s},$$
$$K_0 = \frac{1}{2}\sum_s(\eta_{1s}\xi_{1s} + \eta_{2s}\xi_{2s} + 1),\ G_1 = \sum_s (\eta_{1s}\xi_{1s} - \eta_{2s}\xi_{2s}). \tag{5}$$

The three former satisfy SU(1,1) commutation relations and commute with the latter. All four operators commute with the total angular momentum operators J_+, J_-, J_0. Hence we may consider the group chain

$$\begin{array}{ccccc} \varrho = 2(j-j') & & J & & K=J+1 \\ U(1) & \times & SU(2) & \times & SU(1,1) \\ & & \cup & & \cup \\ & & U(1) & & U(1) \\ & & M & & Q=j+j'+1 \end{array} \quad . \tag{6}$$

The eigenvalues of $\mathfrak{G}_1$, K_0, and $\underline{K}^2 = - K_+K_- + K_0(K_0-1)$ are denoted by ϱ, Q, and K(K-1), respectively. They are related to j,j', and J as shown in (6). The coupled states $|jj'JM\rangle$ therefore coincide with the states $|\varrho JMQ\rangle$, classified according to (6). All such states corresponding to given values of ϱ,J,M, and increasing values of Q=K,K+1,..., or j+j'=J,J+1,..., belong to a single infinite-dimensional irrep of SU(1,1), and may be obtained from the stretched state, for which $\bar{j}+\bar{j}'=J$, by successive applications of K_+ :

$$|jj'JM\rangle \propto (K_+)^{j+j'-J}\left|\frac{1}{2}(j-j'+J)\frac{1}{2}(j'-j+J)JM\right\rangle \quad . \tag{7}$$

Here, K_+,K_-, and K_0 play the same role with respect to total angular momentum as quasi-spin with respect to seniority in nuclear structure.

By considering matrix elements of K_+ or K_- between a coupled state and an uncoupled one, one easily obtains two recursion relations for the SU(2) Wigner coefficients, wherein J and M are kept fixed, j and j' are decreased or increased by 1/2. Such recursion relations differ from the standard ones, wherein j,j', and J are kept fixed. By iterating one of the relations, one can express $\langle jm,j'm'|JM\rangle$ in terms of stretched Wigner coefficients $\left\langle\frac{1}{2}(j-j'+J)\bar{m},\frac{1}{2}(j'-j+J)\bar{m}'\right|JM\right\rangle$, which can then be determined from the other rela-

tion. The result is Racah's formula.

For the purpose of extending the present approach to SU(3), it is convenient to incorporate $\mathcal{G}_1$ in SU(2) and SU(1,1), thereby obtaining U(2) and U(1,1). This leads to the following chains and irreps :

$$\begin{array}{ccccccc} U(2) & \times & U(2) & \supset & U(1) & \times & U(1)\,, \\ [h\ 0] & & [0,-h'] & & q & & -q' \end{array} \qquad (8)$$

$$\begin{array}{ccccccc} U(2) & \times & U(2) & \supset & U(2) & \supset & U(1)\,, \\ [h\ 0] & & [0,-h'] & & [k,-k'] & & \mu \end{array} \qquad (9)$$

$$\begin{array}{ccccc} U(1,1) & \supset & U(1) & \times & U(1)\,, \\ [k;k'] & & h & & h' \end{array} \qquad (10)$$

where $h=2j$, $h'=2j'$, $q=j+m$, $q'=j'-m'$, $k=j-j'+J$, $k'=j'-j+J$, and $\mu=j-j'+M$. Note that the U(2) irreps $[0,-h']$ and $[k,-k']$ have a negative label $-h'$ or $-k'$, and are therefore mixed irreps.[7]

SU(3) WIGNER COEFFICIENTS

For SU(3), the generalization of (8),(9) and (10) reads[5]

$$\begin{array}{ccccccccccc} U(3) & \times & U(3) & \supset & U(2) & \times & U(2) & \supset & U(1) & \times & U(1)\,, \\ h & & h'^{*} & & q & & q'^{*} & & r & & r'^{*} \end{array} \qquad (11)$$

$$\begin{array}{ccccccccc} U(3) & \times & U(3) & \supset & U(3) & \supset & U(2) & \supset & U(1)\,, \\ h & & h'^{*} & & \varkappa & & \lambda & & \mu \end{array} \qquad (12)$$

$$\begin{array}{ccccccccc} U(2,2) & \supset & U(2) & \times & U(2) & \supset & U(1) & \times & U(1)\,, \\ [k;k'] & & h & & h' & & h_1 & & h'_1 \end{array} \qquad (13)$$

where $h=[h_1h_2]$, $h'=[h'_1h'_2]$, $h'^{*}=[0,-h'_2,-h'_1]$, $\varkappa = [k_1,k_2-k'_2,-k'_1]$, $[k;k'] = [k_1k_2;k'_1k'_2]$, and either $k_2=0$ or $k'_2=0$. The generators of the groups in (11), (12), and (13)

can be realized in terms of four triplets of boson creation operators η_{is}, i=1,2,3,4, s=1,2,3, and of the corresponding annihilation operators ξ_{is}. For instance, those of U(2,2) are

$$D^{+}_{ij} = \sum_{s} \eta_{is}\eta_{js},\ D_{ij} = \sum_{s} \xi_{is}\xi_{js},\ i=1,2,\ j=3,4, \tag{14}$$

$$E_{ij} = \sum_{s} (\eta_{is}\xi_{js} + \frac{1}{2}\delta_{ij}),\quad i,j=1,2 \text{ or } i,j=3,4, \tag{15}$$

where the latter generate the U(2) x U(2) subgroup.

All the states corresponding to a given U(3) irrep $\varkappa$, and given irreps of its subgroups, belong to a single U(2,2) irrep $[k;k']$,labelled by the same quantum numbers as the U(3) one. The coupled states being equivalent to states classified according to (13) can be constructed by applying a polynomial in the D^{+}_{ij} operators to the stretched states, corresponding to h=k and $h'^{*}= k'^{*}$.

In contrast with the SU(2) case, in the SU(3) one there is a state labelling problem since a U(3) irrep $\varkappa$ may occur more than once in the product $h\mathrm{x}h'^{*}$: there is one missing label in the chain (12). The proposed coupled state construction method provides us with a solution to this problem. The stretched states are characterized by an irrep kxk' of U(2)x U(2) in (13).On the other hand, the polynomials in D^{+}_{ij} can be specified by an irrep gxg of U(2) x U(2), where g= $[g_1 g_2]$ is the same for both U(2) groups. Thus to obtain the U(3) coupled states, one just couples kxk' and gxg to hxh' in U(2) x U(2) as follows :

$$\left|(h\ h'^{*})g\ \varkappa\lambda\mu\right\rangle = \left[P^{g\mathrm{x}g}(D^{+}_{ij})\ \mathrm{x}\ \left|(kk'^{*})\varkappa\lambda\mu\right\rangle\right]^{h\mathrm{x}h'}_{h_1\mathrm{x}h'_1}, \tag{16}$$

where the irrep g solves the state labelling problem. Since $g_1+g_2=h_1+h_2-k_1-k_2=h'_1+h'_2-k'_1-k'_2$, g contains only one independent label, as it should be.

The irrep g provides an elegant solution to the state labelling problem because it reflects in a direct way the operation of Littlewood-Richardson rule,[8] as reformulated by King[7] to deal with mixed irreps. King's rule indeed states that the multiplicity of κ in $h \times h'^{*}$ is given by

$$m_{hh'^{*}\kappa} = \sum_{g} m_{kgh} \times m_{k'gh'} , \qquad (17)$$

where m_{kgh} and $m_{k'gh'}$ are either equal to 1 or 0. The states (16) are however not orthogonal with respect to g since the latter cannot be associated directly to eigenvalues of Hermitian operators.

Consider for instance the product $[h_1h_2] \times [21]$, containing $[h_1+1,h_2+1,1]$ with multiplicity 2 whenever $h_1 > h_2 > 0$. In terms of mixed irreps, it becomes $[h_1h_2] \times [0,-1,-2]$, containing two times $[h_1-1,h_2-1,-1]$. Hence $h = [h_1h_2]$, $h' = [21]$, $k=[h_1-1,h_2-1]$, and $k' = [1]$. Since for $h_1 > h_2 > 0$, $m_{[h_1-1,h_2-1]\, g\, [h_1h_2]}$ and $m_{[1] g [21]}$ are equal to one if $g = [20]$ or $[11]$, and zero otherwise, from King's rule one gets $m_{[h_1h_2]\,[0,-1,-2]\,[h_1-1,h_2-1,-1]} = 2$, as it should be. Moreover, for the additional label in (16), one obtains $g = [20]$, $[11]$.

SU(3) reduced Wigner coefficients are the overlaps $\langle hq,h'^{*}q'^{*},\lambda | (hh'^{*}) g \kappa \lambda \rangle$ with maximal μ, and contain g as additional label. They satisfy two recursion relations different from the standard ones. In the former, κ and λ are indeed kept fixed, while g is decreased to $\bar{g} = [g_1,g_2-1]$ or $[g_1-1,g_2]$. By iterating such relations, it is possible to express any reduced Wigner coefficient in terms of stretched ones $\langle k\,\bar{q},k'^{*}\bar{q}'^{*},\lambda \mid (kk'^{*}) \kappa \lambda \rangle$, for which $g = [00]$. The latter are then easily calculated.

GENERALIZATION

The cases of U(n) x U(n)$\supset$ U(n), U(n)$\supset$ O(n), and U(n)$\supset$ USp(n) (n even) can be treated in the same way as U(3) x U(3)$\supset$ U(3). The analogs of chain (13) are respectively U(p,q)$\supset$ U(p) x U(q), Sp(2d,R)$\supset$ U(d), and $SO^*(2d)\supset$ U(d), where p,q, and d are the number of rows of the considered U(n) irreps.[1-4] The obtained solutions to the state labelling problems are very elegant from a group theoretical viewpoint because they directly reflect how the branching rules operate. In addition, they are quite useful for practical calculations because complementarity leads to many simplifications, the state non-orthonormality being the price to be paid for such nice properties. However, at the end, orthonormality can be restored by standard procedures, among which coherent state methods.[5,9]

REFERENCES

1. J. Deenen and C. Quesne, J. Phys. A: Math. Gen., 16, 2095 (1983).
2. C. Quesne, J. Phys. A: Math. Gen., 17, 777, 791 (1984).
3. C. Quesne, J. Phys. A: Math. Gen., 18, 2675 (1985).
4. C. Quesne, J. Phys. A: Math. Gen., 19 (1986) in press.
5. C. Quesne, to be published.
6. M. Moshinsky and C. Quesne, J. Math. Phys., 11, 1631 (1970).
7. R.C. King, J. Math. Phys., 11, 280 (1970).
8. D.E. Littlewood and A.R. Richardson, Phil. Trans. Roy. Soc. London, A 233 , 99 (1934).
9. R. Le Blanc and D.J. Rowe, J. Phys. A: Math. Gen., 18, 1891 (1985).

A NEW METHOD TO SOLVE THE SCHRÖDINGER EQUATION FOR ATOMIC NUCLEI BOUND STATES

GINTAUTAS P. KAMUNTAVIČIUS

Institute of Physics, K.Požėlos 54, 232600 Vilnius, USSR

Abstract Properties of solutions of the simplest functional-differential equations equivalent to multiparticle Schrödinger equation in the case of atomic nuclei bound states are concerned.

For any system of identical particles with a pair interaction Hamiltonian may be represented as a sum of two-particle operators:

$$H = \sum_{i,j=1(i<j)}^{A} h_{i,j} , \tag{1}$$

$$h_{i,j} = - \frac{\hbar^2}{2mA} (\nabla_i - \nabla_j)^2 + V(\vec{r}_i - \vec{r}_j, \sigma_i \tau_i \sigma_j \tau_j). \tag{2}$$

Each of these operators is called a reduced two-particle Hamiltonian (RH2). When a complete system of its eigenfunctions

$$h_{A-1,A} \varphi_i (A-1,A) = \varepsilon_i \varphi_i (A-1,A) \tag{3}$$

is used, the energy of any bound state α may be represented in a form

$$E_\alpha = \binom{A}{2} \sum_i \varepsilon_i W_{i,\alpha} , \tag{4}$$

where $W_{i,\alpha}$ values possess the properties characteristic for probabilities of different two-particle states i in a definite state of all the system:

$$W_{i,\alpha} \geqslant 0; \quad \sum_i W_{i,\alpha} = 1. \qquad (5)$$

They are defined as components of a reduced two-particle density matrix of the system in a basis of eigenfunctions of RH2.

Eq. (4) is the simplest one among familiar expressions for energy of a multiparticle system, and $W_{i,\alpha}$ seem far simpler than the corresponding wave-function. A problem of reasonable approximations for their direct calculations, named a problem of N-representability[1], attracted the attention of a great number of scientists for a long time.

From the results set forth below it follows that solution of a problem of N-representability in the form suggested by Prof. A. Coleman is hardly possible at all. However, we managed to obtain rather a simple expression for $W_{i,\alpha}$. Namely this result shows that one cannot determine W values for any fixed α irrespective of other possible states of the system. This task may be solved by using the functional-differential equations of a multiparticle quantum system theory, defined in[2]. These equations emerge as a result of a special simplification of multiparticle Schrödinger equation (SE). Only one (RH2) operator is present in place of Hamiltonian in equations of such a type. Moreover, we

managed to prove that each solution of these equations corresponds uniquely to a solution of initial SE[3]. From here, final results of[2,3] can be represented in the formalism, which follows.

First of all, a complete system of eigenfunctions for RH2 must be constructed. This operator may be expressed in a form:

$$h_{A-1,A} = -\frac{\hbar^2}{mA}\Delta_{\vec{\varrho}} + V(\sqrt{2}\vec{\varrho}, \sigma_{A-1}\tau_{A-1}\sigma_A\tau_A) \quad (6)$$

$$\vec{\varrho} = (\vec{r}_{A-1} - \vec{r}_A)/\sqrt{2}\,. \quad (7)$$

It describes only a relative motion of nucleons, therefore determination of its eigenfunctions , like in analogous problem for deutron, offers no difficulties. This operator has a very important property - the effective reduced mass of two "nucleons" is directly proportional to A. However, RH2 is only a formal structure,not describing any real susystem in fact. Nevertheless, it possesses very interesting properties in the case of realistic NN potentials. The approximate method for probabilities values $W_{i,\alpha}$ determination[6] allowed a qualitative description of many of known light nuclei properties, such as bound states quantum numbers, mags location and poor stability of some nuclei[7].

Furtheron, for exact solution of SE a complete basis of functions depending on all the variables of the system[3,8] , must be constructed of RH2 eigenfunctions. Each function in this basis possesses only a minimal antisymmetry, i.e.

does not require a complete antisymmetrization. Thus, construction of this basis offers no difficulties. Then matrix of a certain projector X in this basis is calculated. This projector is defined as a special sum of one-particle coordinates permutations operators[3]. Assume, matrix of this operator becomes a projection after a certain number n of basis functions is considered. Hence, it may be expressed in a form:

$$X_{(nxn)} = F_{(nxr)} \cdot F^{+}_{(rxn)} , \qquad (8)$$

where r - rank of matrix X, equal to its trace. Matrix F is constructed of X eigenvectors corresponding to unity eigenvalues, as columns. They are orthnormalized, therefore :

$$F^{+}_{(rxn)} \cdot F_{(nxr)} = 1 . \qquad (9)$$

For calculations of nucleus spectrum it is sufficient to diagonalize one r by r matrix, i.e. to define:

$$G^{+}_{(rxr)} \left\{ F^{+}_{(rxn)} \cdot \varepsilon_{(nxn)} \cdot F_{(nxr)} \right\} G_{(rxr)} = = \binom{A}{2}^{-1} E_{(rxr)} . \qquad (10)$$

Here ε - a diagonal matrix on eigenvalues, corresponding to basic functions of the problem[3].

When the last expression is rewritten in a form of Eq. (4), we clearly see, that

$$W_{i,\alpha} = \left| (FG)_{i,\alpha} \right|^{2} . \qquad (11)$$

Here i = 1,2, ...,n specifies basic states ,

and $\alpha = 1,2, \ldots ,r$ - various bound states of the system.
The wave-function of α state (possessing E_α energy) in such a basis is the α -th column of matrix (FG).
Possibly, in the problems of practical interests X becomes a projection only after infinite number ($n = \infty$) of basis states is considered. The developed method of successive approximation to exact solutions allows a quantitative control of the obtained functions proximity to exact ones, meanwhile providing a uniform approach of energies to exact values from below.

REFERENCES

1. A. J. Coleman, Rev. Mod. Phys., 35, 668 (1963).
2. G. P. Kamuntavičius, Few-Body Systems, 1, 3, (1986).
3. G. P. Kamuntavičius, Izvestija AN SSSR, ser. fiz. (in Russian), 51, 1 (1987) (in print).
4. A. Bernotas, G. P. Kamuntavičius, Soviet Physics-Collection, 21, N 5, 1 (1981).
5. A. Bernotas, G. P. Kamuntavičius, Soviet Physics-Collection, 21, N 6, 1 (1981).
6. G. P. Kamuntavičius, Sov. Journ. Nucl. Phys., 34, 661 (1981).
7. A. Bernotas, G. P. Kamuntavičius, in Microscopic Calculations of Light Nuclei, edited by A. Gorbatov (Kalinin, USSR, 1981), pp. 15-19.
8. G. P. Kamuntavičius, Soviet Physics-Collection, 26, N 2, 1 (1986).

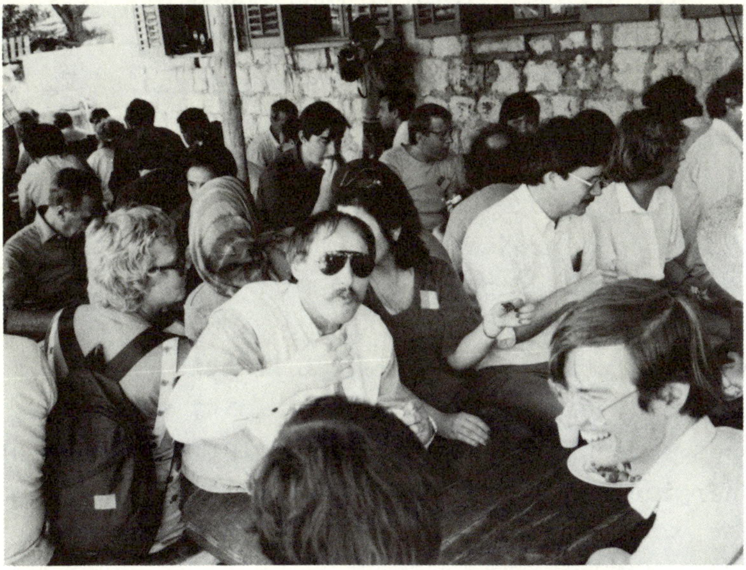

Chapter 8

EXPERIMENTAL TECHNIQUES

Moderator:

D.M. MOLTZ
Nuclear Science Division
Lawrence Berkeley Laboratory
University of California
Berkeley, California 94720, U.S.A.

PHOTON SCATTERING TECHNIQUES FOR THE INVESTIGATION OF LOW LYING COLLECTIVE STATES*

RALF-DIETER HEIL+

Institut für Kernphysik, Justus-Liebig-Universität Giessen, Leihgesterner Weg 217, D-6300 Giessen, Germany

Abstract: The experimental techniques for the investigation of low energy collective states by nuclear resonance fluorescence (NRF) performed at the bremsstrahlung facilities of the Stuttgart Dynamitron and the Giessen Linac by a Giessen-Cologne-Stuttgart-collaboration are described. Details of the experimental set-up and the analysis are discussed.

INTRODUCTION

The recently discovered mixed symmetry rotational M1 states[1] represent one example of various predicted new collective modes[2] that can be studied by the photon scattering technique (nuclear resonance fluorescence NRF). The results obtained so far in our NRF investigation of these low lying collective states in heavy deformed nuclei were presented by U.Kneissl in his contribution to the NSRS conference[3].

* Supported by the Deutsche Forschungsgemeinschaft

+ The experiments have been performed in collaboration with U.E.P.Berg, C.Bläsing, A.Jung, U.Kneissl, H.H.Pitz, U.Seemann, R.Stock, F.J.Urban (Giessen), C.Wesselborg, P.v.Brentano, K.O.Zell (Köln), B.Fischer, H.Hollick, D.Kollewe (Stuttgart)

Since the NRF method is described in a couple of review articles (see U.E.P.Berg[4] for references) we will only recall some of its main features:

- due to the low momentum transfer selectively dipole and to a lesser extend quadrupole transitions are excited.
- excitation energies, spins, parities, decay widths and mixing ratios of multipole transitions can be obtained completely model-independent with the high precision of γ-ray spectroscopy.

Using bremsstrahlung produced by electrons from high

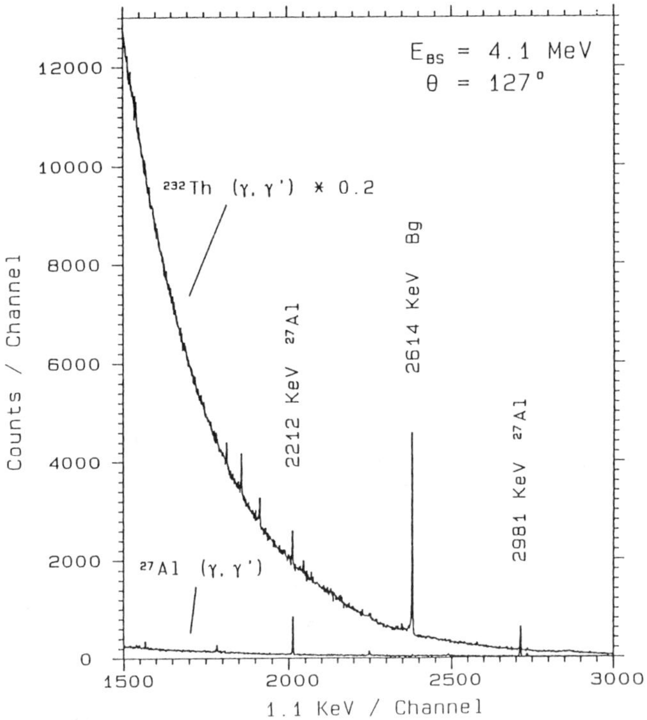

Fig.1. γ-spectra of bremsstrahlung scattered on ^{27}Al and a mixed Th-Al-target.

current electron accelerators one benefits from a very intense incoming photon beam ($\sim 10^9 \gamma$/MeV.s) and a simultaneous excitation of all states by its continuous energy spectrum. But on the other hand one has to cope with an enormous background of non-resonant scattered photons increasing exponentially with the atomic number Z of the nucleus that is to be investigated as illustrated in fig.1.

THE STUTTGART BREMSSTRAHLUNG FACILITY

We will only describe the bremsstrahlung facility at the Stuttgart Dynamitron accelerator shown in fig.2. The Giessen linearly polarized bremsstrahlung facility is explained elsewhere[4].

The Stuttgart Dynamitron provides a cw-electron beam up to 4.3 MeV in energy and 4 mA in current. The electrons are bent by 120° and focused on the radiator by two quadrupole dubletts and 2 pairs of steering coils. The Au-Cu-radiator is thick enough ($\sim$ 3g/cm^2) to stop 4.5 MeV electrons completely. The bremsstrahlung produced passes a lead-carbon beam-hardener and hits after very careful collimation the NRF target located in the neighbouring experimental hall. The target is placed in vacuum preventing small angle Compton scattered γ-quanta from the surrounding air from reaching the germanium detectors (countrate without NRF target $\sim$ 20 Hz for 100 mA of 3 MeV electrons on the radiator).

The three germanium detectors (Ge(Li)or Ge(HP)) placed at 90° or 95°, 127° and 150° are carefully shielded with a lead and an interior copper shield. Increasing the 90° scattering angle to a slight

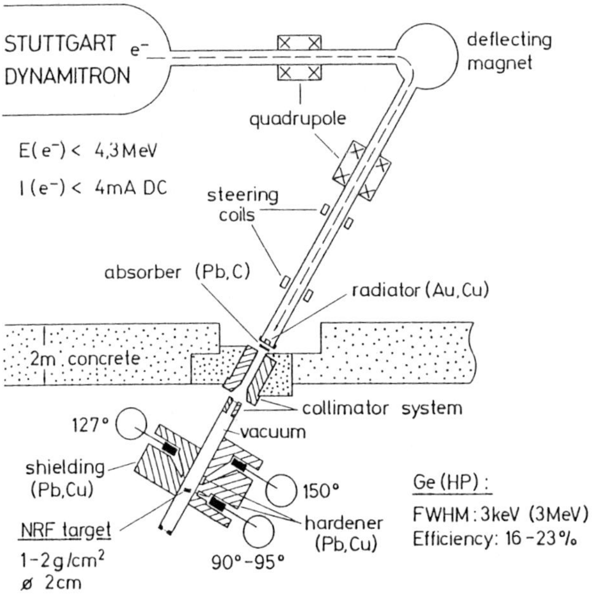

Fig.2. The bremsstrahlung facility at the Stuttgart Dynamitron accelerator (not to scale).

backward angle of 95° improved the peak-to-background ratio significantly. Lead and copper filters in front of the crystals attenuate the resonant radiation slightly but absorb the majority of the degraded low energy γ-quanta. The filter thickness is optimized according to the energy region of interest.

DATA ANALYSIS

The decay width for transitions to excited states (Γ_f) or to the ground state ($\Gamma_f=\Gamma_0$) are determined by the peak areas observed in the γ-spectra for thin targets according to:

$$N_d(E_R) = N_\gamma(E_R).e_n.e_d.g.W(\Theta).n_s.\lambda\!\!\!^-{}^2.\pi^2.\Gamma_0.\frac{\Gamma_f}{\Gamma}$$

The number of detected γ-quanta N_d resonantly scattered at a level with energy $\lambda\!\!\!^{-}=\frac{\hbar\cdot c}{E_R}$ depends experimentaly on the detector efficiency e_d, the absorption e_n of the filters, the number n_s of scattering nuclei and the number of γ-quanta N_γ incident on the scatterer. The product $e_d \cdot e_n$ is determined by means of a ^{56}Co source in place of the target emitting γ-rays of well known intensities and energies between 800 and 3600 KeV. These lines serve for the energy calibration too. The number of target nuclei n_s is known from the target thickness. Since only dipole and quadrupole transitions are excited in photon scattering the measurement of the intensities of the scattered radiation at only two angles (Θ= 90° and 127°) is sufficient to determine the multipolarity of the elastically scattered photons by a spin J_o=0 target. The intensity ratio $W(\Theta = 90°)/W(\Theta = 127°)$ is 0.736 or 2.206 for pure dipole or quadrupole radiation. Fig.3 shows the results for the ground state transitions observed in ^{160}Gd. Obviously all these states have spin J_e=1 and thus the statistical factor $g=2J_e+1/2J_o+1$ amounts to 3.

Besides the elastic transitions to the J_o=0 ground state inelastic scattering only to the first excited 2^+ level is observed in the investigated nuclei. An upper limit of $\Gamma_r=(5\pm5)\%$ for the branching to the γ-and β-band compared to the ground state decay width Γ_o can be given[5]. The only quantity still unknown is $N_\gamma(E_R)$ the number of γ-quanta incident on the scatterer having energy E_R. Since an absolute

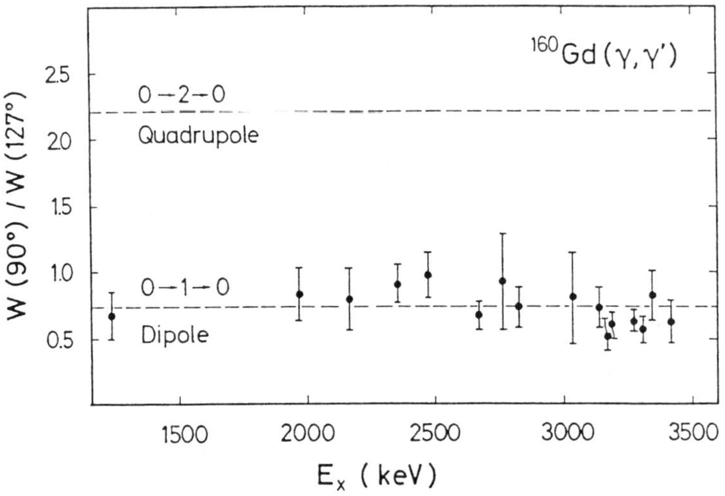

Fig.3. Intensity ratio of the ground state transitions ($J_o \rightarrow J_e \rightarrow J_o$) observed in ^{160}Gd (γ,γ') under scattering angles of 90° and 127°.

measurement of the photon flux is very difficult we calibrated it by counting the number of γ-quanta resonantly scattered on levels of known widths. For this flux calibration mixed targets were used, where the NRF targets were sandwiched between aluminum disks of the same diameter. The lifetimes and decay properties of the three aluminum levels below 4 MeV are well known[6]. Between this calibration points and the bremsstrahlung endpoint energy a linear extrapolation of the shape of the bremsstrahlung spectrum is assumed.

The sensitivity level obtained in the Stuttgart experiments, where each run took 1-3 days of beam time, ranges from $\Gamma_o<2$ meV at 1 MeV to $\Gamma_o<10$ meV at 3.5 MeV. This corresponds to e.g.B(M1)↑-values

of < 0,5 μ_k^2 and 0,06 μ_k^2, respectively. An overall absolute accuracy of about 10-20% for the strong transitions is obtained.

PARITY DETERMINATION IN NRF

Parities of excited states can be determined model-independently in NRF experiments by either using linearly polarized photons in the entrance channel[4] or by measuring the polarization of the scattered γ-rays[7]. The applicability of the Giessen polarized bremsstrahlung facility for studying this low energy states (< 3 MeV) is limited because of experimental reasons. Only for the strong 3070 KeV level in ^{156}Gd the positive parity could be confirmed[8]. Therefore we examine the possibility of using a Compton polarimeter similar to the one of Bass et al.[9] for the polarization measurements in our future NRF studies. Such a very efficient device will give an improved sensitivity compared to the results of Metzger[10] and will allow us to distinguish between M1 and E1 excitations.

REFERENCES

1. D.Bohle et al, Phys.Lett.137B (1984) 27
2. F.Iachello, Phys.Lett.160B (1985) 1
3. U.Kneissl, Proc.Nuclear Structure, Reactions and Symmetries, Dubrovnik 1986, to be published*
4. U.E.P.Berg, Journ. de Phys. C4 (1984) 360
5. U.E.P.Berg, Phys.Lett.149B (1984) 59
6. P.M.Endt and C.van der Leun, Nucl.Phys.A310(1978)1
7. L.W.Fagg and S.S.Hanna, Rev.Mod.Phys.31 (1959) 711
8. U.E.P.Berg, Habilitation Thesis, Giessen 1986, unpublished
9. R.Bass et al, Nucl.Instr.Methods 163 (1979) 377
10. F.R.Metzger,Phys.Rev. C 14 (1976) 543

*See page 362-367 in "Nuclear Structure, Reactions, and Symmetries" R.A. Meyer and V. Paar, Eds., World Sci. Pub. Co. Inc., Singapore, (1986).

NUCLEAR LEVEL-WIDTH MEASUREMENT BY γ-RAY RESONANCE SCATTERING

H. Tagziria*, W.D. Hamilton*, F. Hoyler+ and S.J. Robinson+

* Physics Division, University of Sussex, Brighton BN1 9QH U.K.

+ Institute von Laue-Langevin, 38042 Grenoble Cedex, France.

Abstract Nuclear resonance scattering using transmission and self-absorption methods have been carried out to examine the feasibility of level-width measurements of states populated by neutron capture γ-rays. The nuclear recoil folowing the primary radiation provides the necessary resonance condition. Large source strengths in the region of the absorber and scatterer can be produced by placing target material in a neutron flux of $\sim 10^{15}$n $cm^{-2}s^{-1}$ in the ILL reactor. A preliminary test was made to measure the widths of the 2.186 MeV 1^- level and the 2.073 MeV 2^+ mixed symmetry state in ^{144}Nd. It would appear to be a generally applicable method for targets having a sufficiently large cross-section and for states with mean-lives less than 10^{-14}s.

INTRODUCTION

Nuclear Resonance Fluorescence experiments can be used to measure lifetimes shorter than $\sim 10^{-10}$s as the resonance cross-section is proportional to the width of the excited state and hence inversely proportional to the mean-life. The main experimental problem to overcome is that of providing a source of radiation with sufficient intensity at the resonant energy. Such sources can be produced by neutron capture reactions, the limit being set by the source heating. The resonance condition is provided by the Doppler shift imparted to a ground state transition following the emission of a primary γ-ray. Now since the latter may have an energy substantially greater than that of the ground state transition the Doppler loss can be compensated. The form of the recoil energy distribution in the case of a

simple γ_2-γ_1 cascade, is rectangular (fig.1) and is centred about the energy $E_e = E_0 - E_0^2/Mc^2$, E_0 being the level energy. The distribuiton must have sufficient width to cover the recoil energy loss E_0^2/Mc^2 suffered by the ground

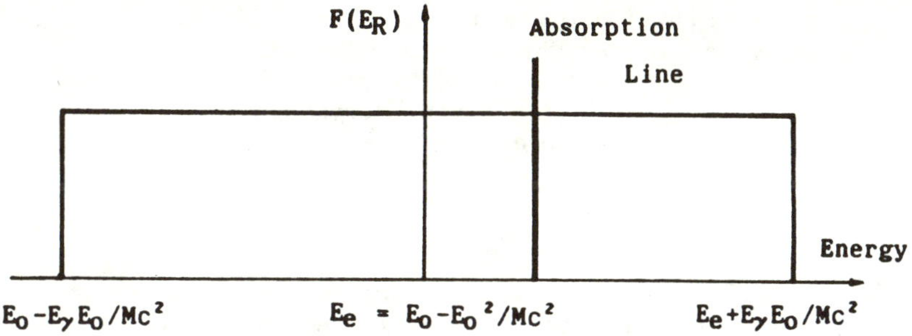

Figure 1: Energy distribution of the γ radiation emitted following a primary transition of energy E_γ. $F(E_R)$ represent the fraction of γ-rays satisfying the resonant condition. Thermal Effects are neglected.

state transition on emission and absorption. The number of γ-rays satisfying this condition will be maximum when the primary γ-ray is just greater than the energy of the ground state transition.

Thus we have a situation which is roughly equivalent to the usual Laboratory experiments using radioactive sources in which $E(\gamma_2) > E(\gamma_1)$.

APPLICATION TO ^{144}Nd

Experimental Procedure

In order to test the feasibility of level-width measurements of states populated by primary γ-rays, self-absorption and transmission methods have been applied in an attempt to measure the mean-lives of the 2.186 MeV 1^- level and the 2.073 MeV 2^+ mixed symmetry state in ^{144}Nd [1]. A high neutron flux of $\sim 10^{15}$n cm^{-2}s^{-1} occurs close to the core at the position normally used in GAMS experiments at the ILL reactor. The target was ~3g of pure natural Nd_2O_3 in which there is 23.9% ^{144}Nd and 12.3% ^{143}Nd which has $\sigma(n,\gamma) = 325$ b. The experiment was mounted ~10 m from the source and despite the small solid angle a source strength of ~10 mCi was available at the position of the absorber.

The Self-Absorption Experiment

The experimental arrangement is shown in figure 2; the absorber and scatterer being of the same material. The scatterer, together with the Ge(Li) detector (2keV resolution) form a high resolution detector as the absorber removes only those γ-rays which can be resonantly scattered by the scatterer, hence filtering the spectrum. Cerium oxide comparators were used to correct for the normal electronic attenuation which is dominated by Compton scattering. Sheets of B_4C were put on the beam line and reduced very much the background due to neutrons. A good peak-to-background ratio was obtained. The only effects competing with the resonant process are the elastic Rayleigh and Thomson scatterings which are respectively $(Z/E)^3$ and $(Z^2/A)^2$ dependant.

The comparative reduction of the resonant effect with and without the absorbers was measured.

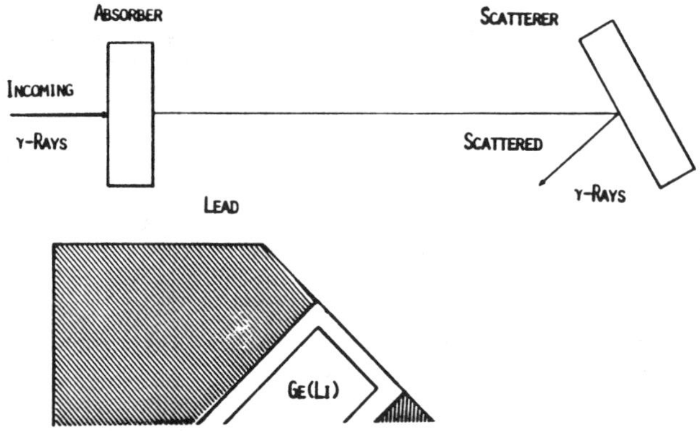

Figure 2. Experimental arrangement for the self-absorption experiment.

The Transmission Experiment

A transmission experiment was mounted at the other end of the beam line. Its geometry was similar to that of figure 2 but with the scatterer omitted. Therefore this method lacks the high resolution detection which is characteristic of the self-absorption technique although better statistics may be obtained. The detector used was a well shielded pair spectrometer which has a good peak-to-background ratio and measurements were carried out with four different thicknesses of Nd absorbers and matched Ce comparators. The

electronic contributions turned out to be rather high and this together with the high density of transitions in the region of interest in the γ-ray spectrum made the method rather inaccurate for the measurement of mean-lives $\gtrsim 10^{-14}$s in heavy nuclei.

RESULTS AND CONCLUSION

The steps involved in extracting the level width from nuclear resonance fluorescence data may be found in refs 2,3 where the relative self-absorption coefficient R is defined. The total number of counts under the 2.186 MeV peak, with and without the absorbers are summarised in Table 1 from which we obtain R = 0.87 ± 0.13.

Table 1: Areas under the 2.186 MeV peak with and without Nd (Ce) absorbers (comparators) from the self-absorption experiment.

Absorber Thickness Atom.b^{-1}	0	0.082
Counting Period (hr)	10	6
Total number of counts; Nd abs.	835 ± 28	342 ± 18
Total number of counts; Ce comp.	—	292 ± 17

Using this value of R and with $\Gamma_0/\Gamma = 0.71$, the total width of 2.186 MeV 1^- state in ^{144}Nd was found to be $\Gamma = (151 \pm 0.25) \times 10^{-2}$ eV which corresponds to a mean-life $\tau = (3.10 \pm 0.51) \times 10^{-14}$s. This is in agreement with Rice-Evans[4] who found a lower limit for the mean life to be 1.6×10^{-14}s and also lies within the interval $(1.4 \text{ to } 3.4) \times 10^{-14}$s set by Blanc[5]. Later, Metzger[6] using a bremsstrahlung source of radiation obtained $\tau = (1.5 \pm 0.2) \times 10^{-14}$s. No resonant effect was registered for the 2.073 MeV 2^+ level; both the relative intensity and

the branching ratio (~0.3) are small, and the corresponding level-width is below the limit of detectability.

It would appear from these preliminary measurements that the nuclear resonance fluorescence technique using neutron capture γ-rays is generally applicable provided the targets have a sufficiently large capture cross-section, the branching to ground state is intense and the mean-life to be measured is $\lessdot 10^{-14}$s.

REFERENCES

1. W.D. Hamilton, A. Irbäck and J.P. Elliott, Phys. Rev. Lett., 53, 2469 (1984).
2. F.R. Metzger, Progr. Nucl. Phys., 7, 54 (1959).
3. S.J. Skorka, in The Electromagentic Interactions in Nuclear Spectroscopy, edited by W.D. Hamilton (North Holland, 1975), p.283.
4. P. Rice-Evans, Proc. Roy. Soc (London), 82, 914 (1963).
5. J.P. Blanc, M. Lambert and C.F. Perdrisat, Helv. Phys. Acta, 36, 820 (1963).
6. F.R. Metzger, Phys. Rev., 187, 1700 (1969).

PHOTOEXCITATION OF $^{189}Os^m$

LASZLO LAKOSI, ISTVAN PAVLICSEK, ZSOLT NÉMETH AND ARPAD VERES
Institute of Isotopes, H-1525, Budapest, POB77, Hungary

Abstract γ-excitation of the 5.8 h half-life $^{189}Os^m$ has been observed for the first time. Both isotopic (^{60}Co) and bremsstrahlung sources of energies from the X-ray region up to 4 MeV (Linac) have been used for excitation. Contribution from the NEET effect is found to be present besides (γ,γ') inelastic resonance scattering.

INTRODUCTION

Present work is the first report of photoactivation experiments on the stable ^{189}Os nuclide. ^{189}Os was chosen for this study because its excitation by γ-rays to the isomeric state could not be detected so far, in spite of extensive studies[1], whereas it was successfully excited by 85 keV electrons[2] on the other hand. This latter achievement has been reported as a first evidence of a new mechanism for nuclear excitation called NEET (nuclear excitation by electron transition), which may occur, when an electron of inner shells is kicked off by the bombarding electron or X-ray, and the excitation energy is transferred from the orbital electron system to the nucleus.

Activation levels of 69.54, 216.68, 219.40, 275.88, 427.91, 549.91 and 599.61 keV were identified from Ref.[3].

These levels can directly be excited from the ground state and feed the 30.81 keV energy (fully converted) isomeric level of 5.8 half-life. Above 600 keV the activation levels are not known. The atomic deexcitation following a vacancy creation in the K orbit can also induce the excitation of the 69.54 keV nuclear level instead of X-ray or Auger electron emission, because both transitions have a common E2 multipolarity and nearly the same energies, which is necessary to NEET process, (according to its theory[4]). Due to electron excitation experiments, nonzero branching ratio was postulated to the metastable state[2,5].

EXPERIMENTS

Natural Os samples in powder form of 0.5-2.0 g were irradiated in the radiation field of the $2.28*10^{15}$ Bq (61 kCi) ^{60}Co γ-ray source of cage type and the 4 MeV LPR4 linear electron accelerator in the Institute of Isotopes of the Hungarian Academy of Sciences. In the first case the irradiation arrangement was similar to that of a previous investigation[6], but samples were placed on the surface of sources themselves, where the dose rate was $3.48*10^4$ Gy/h (3.48 MR/h), i.e. by 50 % higher than in the center. In the latter case the bremsstrahlung produced by a 20 μA electron beam in a 0.9 mm thick Pt converter was used. Irradiations have also been made by a 200 kV, 15 mA and a 300 kV 5 mA X-ray machine in the Hungarian National Bureau of Measurements.

The $^{189}Os^m$ activities have been detected via LX-rays (following the emission of internal conversion electrons) by a Si(Li) (30 mm^2 sensitive area, 3 mm thickness), or alternatively a pure Ge planar spectrometer (10 mm diam. of sensitive area, 7 mm thickness) connected to a 1024 channel

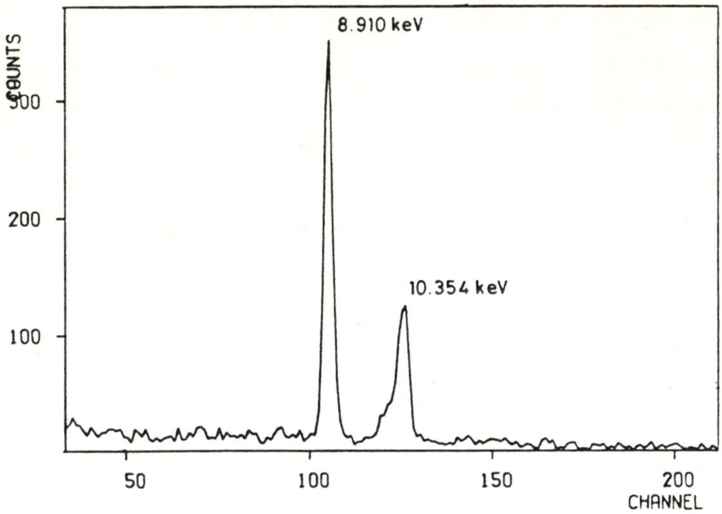

FIGURE 1. Si(Li) spectrum of $^{189}Os^{m}$ excited by LINAC bremsstrahlung.

analyzer.

A typical Os LX spectrum is shown on Figure 1. L_α and L_β lines are at 8.91 and 10.35 keV. Since L_γ line could not be seen well, the number of counts in L_α and L_β photopeaks were mostly evaluated, extrapolated to saturation and corrected for the time delay and intensity decay during the counting period.

From the thus determined intensities $I(L_\alpha)$ and $I(L_\beta)$, the photoactivation yield Y was calculated considering an infinitely thick sample because of the strong attenuation of L_α and L_β photons in Os, according to Eq. (1):

$$Y = \frac{I(L_\alpha) + I(L_\beta)}{a\varepsilon A\omega_L \frac{ce(L)}{ce} \left[\frac{L_\alpha}{L} \frac{1}{\sigma(L_\alpha)} + \frac{L_\beta}{L} \frac{1}{\sigma(L_\beta)} \right]} \qquad (1)$$

Here a = 0.161 is the natural abundance of ^{189}Os. For detection efficiency ε values of 0.4 (Si(Li)) and 0.17 (Ge) were taken for L_α and L_β photons alike. As active detector

areas A of Si(Li) and Ge, 0.3 and 0.785 cm^2 values were used, respectively. L fluorescence yield ω_L as 0.35 and ratio of L to all conversion electrons $\frac{ce(L)}{ce}$ as 0.704 were taken from Refs.[7,3], respectively.

The total microscopic absorption cross sections $\sigma(L_\alpha)$ and $\sigma(L_\beta)$ of 8.91 and 10.35 keV energy L_α and L_β photons were found as $4.60*10^{-20}$ and $2.89*10^{-20}$ cm^2/atom for Os, respectively[8]. The intensity ratios of L_α and L_β to all L photons were determined from the present measurements as 0.708 ± 0.040 and 0.2475 ± 0.040, respectively, so as $\frac{L_\alpha}{L_\beta}\frac{\sigma(L_\beta)}{\sigma(L_\alpha)}$ be equal to the intensity ratio $\frac{C(L_\alpha)}{C(L_\beta)} = 1.795 \pm 0.132$. Based on $\sigma(L_\alpha)$ and $\sigma(L_\beta)$ values above, samples of 40 mg/cm^2 or thicker were used as to be considered infinitely thick.

Experimental data are summarized in Table I. In the case of irradiations with ^{60}Co, spectra of 9 activation runs were summed up to increase sensitivity.

EVALUATION AND RESULTS

Photoactivation yields were evaluated for the various cases on the basis of Eq. (1).

In order to extract information from the measured yields, they were compared to the calculated ones, which, on the other hand, can be given as

$$Y = \sum_{i=1}^{n} \phi(E)_i \sigma_i + PB_1 \int_{73.87\ keV}^{E_0} \phi(E)\sigma_K(E)\, dE \qquad (2)$$

The first term is due to inelastic resonance excitation, whereas the second one accounts for NEET effect. The resonance fluorescence term consists of contributions from the

TABLE I Irradiaticn and measurement data

Radiaticn scurce (Energy)	Irradiaticn time	Ccunting time	Number of ccunts			Intensities $10^{-3}s^{-1}$		
	h	s	$C(L_\alpha)$	$C(L_\beta)$	$C(L_\gamma)$	$I(L_\alpha)$	$I(L_\beta)$	$I(L_\gamma)$
Linac brems-strahlung (4 MeV)	1	18672	999 (36)	557 (35)	63 (36)	651 (24)	363 (23)	41 (24)
^{60}Cc (1.33, 1.17 MeV)	17-66 (9 runs)	12000-22000 (9 runs)	1650 (113)	894 (83)		15.3 (1.1)	8.3 (0.8)	
X-ray machine 200 kV 15 mA	5.83	18860	22 (10)			3.6 (1.6)		
300 kV 5 mA	5.75	16900	15 (9)			2.5 (1.5)		

i-th activation level as the product of flux density $\phi(E)_i$ taken at the level energy and the respective integrated cross section σ_i for isomer activation:

$$\sigma_i = \frac{\lambda^2}{4} g \Gamma_{0\gamma} \frac{\Gamma_i}{\Gamma} = \frac{\lambda^2}{4} g \Gamma \frac{B_0}{\alpha+1} B_i \qquad (3)$$

Here λ is the wavelength of the γ-transition to the ground level. The statistical weight factor $g = (2I+1)/(2I_0+1)$, being I and I_0 the total angular momenta of the excited and ground levels, respectively. The radiation width of ground state transition is $\Gamma_{0\gamma} = \Gamma B_0/(\alpha+1)$, where Γ is the total level width, B_0 is the branching ratio to the ground state and α is the internal conversion coefficient. $B_i = \Gamma_i/\Gamma$ is the branching ratio to the metastable state.

Only the first activation level at 69.54 keV is involved also in the NEET effect taking place with a probability P, which appears in the second term of Eq. (2), together with the unknown branching ratio B_1 of this level.

K-shell ionization cross section σ_K is the function of the γ-ray energy. Integration starts with the K absorption edge of Os.

In the case of X-ray experiments, the flux density $\phi(E)$ was determined on the basis of a rough approximation for the number of bremsstrahlung photons of E energy emitted from the converter of atomic number Z by an impinging electron of the energy E_0 as

$$\frac{dN_E}{dE} = \frac{2kZ\,(E_0 - E)}{E} \qquad (4)$$

where $k = 7*10^{-8}\ eV^{-1}$. Calculations were performed on Z=74 (tungsten) taking into account the electron current and the anode-sample distance.

For ^{60}Co irradiation, primary photon flux ϕ_0 and Compton-scattered flux density values $\phi(E)$ at single level

TABLE II Photoexcitation yields, calculated photon fluxes and overall cross sections.

Source	Y $10^{-20}\ s^{-1}$	ϕ_0 $10^{12}\ cm^{-2}s^{-1}$	$\sigma_{expt} = \frac{\sum \phi(E)_i \sigma_i}{\phi_0}$ $10^{-32}\ cm^2$
Linac bremsstrahlung	889(93)		
^{60}Co	19.1(2.5)	0.84(0.24)	22.7(9.2)
200 kV X-rays	4.9(2.8)	3.1(0.8)	1.57(0.96)
300 kV X-rays	3.4(2.5)	3.0(0.7)	1.15(0.92)

energies were determined on the basis of $^{115}In^m$ activation experiments carried out in the same radiation field, by using the value of integrated cross section for $^{115}In^m$ excitation determined in a specific geometry[9]. When the activation levels are not fully known, as in the present case, it is usual to give the experimental overall cross section σ_{expt}, which is related to the γ-flux ϕ_0 of primary energy. Its value is the largest one ever obtained for ^{60}Co photoexcitation among 18 nuclear isomers. The ϕ_0 values given for X-rays are integrated ones from the Os K absorption edge to the endpoint energy, see Table II.

DISCUSSION

X-RAY EXCITATION

Below 200 kV endpoint energy we estimated the ratio of resonance fluorescence contribution to be $2*10^{-3}$ by determining σ_1 from Eq. (3), i.e. negligible. Consequently, the X-ray excitation yield was attributed completely to NEET, and we

obtained a value of $1.4(1.0)*10^{-11}$ for PB_1, by one order of magnitude less than $2*10^{-10}$ reported by Otozai et al[5].

In case of 300 kV X-ray excitation there are three additional activation levels between 200 and 300 keV. The effect was smaller because the electron current was one third of that of the 200 kV machine, but still is less than could be expected. The contributions of resonance and NEET processes were estimated to be comparable, in the order of 10^{-19} and $4*10^{-20}$ s^{-1}, respectively. It is not well understood, why our experiments provided a substantially lower yield of $3.4*10^{-20}$ s^{-1} in total, as shown in Table II.

^{60}CO EXCITATION

We calculated the integrated cross sections σ_i for isomer excitation by using known branching ratio and lifetime values of activation levels. Since no lifetime data of 427.91 and 549.61 keV levels were available, their contributions were estimated on the basis of multipolarities. Integrated cross sections of five levels fall in the order of magnitude of 10^{-26} cm^2eV, whereas those of 69.54 and 599.61 keV levels in the order of 10^{-30} and less than 10^{-28} cm^2eV, respectively. Contribution of the NEET is negligible.

By summing up over the seven known activation levels, we obtained an estimated yield of $5*10^{-20}$ s^{-1}. As compared to the experimental yield of $19.1*10^{-20}$ s^{-1} indicated in Table II, it means, that roughly a quarter of the photoactivation can be attributed to levels lying below 600 keV.

LINAC EXCITATION

Photoactivation by bremsstrahlung of the 20 μA Linac is about 40 times larger, than that by ^{60}Co irradiation. Since

information about higher lying levels are missing and no photon flux data are available, evaluation of the photoactivation cross section will follow.

CONCLUSION

The present work provided an evidence on the γ-ray excitation of $^{189}Os^m$. For ^{60}Co irradiation the largest photoactivation cross section has been obtained. The presence of NEET process besides resonance fluorescence excitation has been proved. Possibility of X-ray excitation has also been shown. Refined calculations and more precise measurements via detection of conversion electrons from large surface samples are in preparation to improve counting statistics in the low energy region.

In view of the numerous low-lying activation levels, measurement of the photoexcitation function would be useful by an electron accelerator of continuously variable energy in the range of 1-2 MeV, in order to assign photoactivation to individual levels and to find new details in level structure of ^{189}Os.

REFERENCES

1. Á. Veres, Atomic Energy Rev., 18, 271 (1980).
2. K. Otozai, R. Arakawa and M. Morita, Prog. Theor. Phys., 50, 1771 (1973).
3. R.B. Firestone, Nucl. Data Sheets, 34, 537 (1981).
4. M. Morita, Prog. Theor. Phys., 49, 1574 (1973).
5. K. Otozai, R. Arakawa and T. Saito, Nucl. Phys., A297, 97 (1978).
6. K.Yoshihara, Zs. Németh, L. Lakosi, I. Pavlicsek and Á. Veres, Phys. Rev., C33, 728 (1986).
7. R. Woldseth, X-Ray Energy Spectrometry (KEVEX Corp., 1973).
8. Wm. J. Veigele, Atomic Data Tables, 5, 51 (1973).
9. L. Lakosi, M. Csürös and Á. Veres, Nucl. Instr. Meth., 114, 13 (1974).

A NEW POLARIMETER FOR BETA RAYS TO STUDY MANIFEST LEFT-RIGHT SYMMETRY IN ELECTROWEAK INTERACTIONS

V.A. WICHERS, D. ATKINSON, T. HAGEMAN, J. van KLINKEN
A. STOLK* and H.W. WILSCHUT

Kernfysisch Versneller Instituut
and Instituut voor Theoretische Natuurkunde
University of Groningen, The Netherlands

During and after "the fall of parity", three decades ago, nuclear-physics played a central role. Parity violation and broken symmetries being well verified, a decade of phenomenology was followed by a period with full attention for unification theories, leading to the actual CERN-finding of massive W and Z bosons. Fig.1, mainly borrowed from cosmology[1)], illustrates our motivation to consider another contribution via nuclear β^+ decay. The figure shows the well-known desert between electroweak and grand unification, extending over 15 orders of magnitude in energy. On our low-energy side of the desert, minimal $SU(2)_L*U(1)$ theories with left-handed (V-A) currents describe all experimental findings until now. On the high-energy side, however, it seems not possible to explain the long lifetime ($>10^{32}$ y) of the proton with minimal SU(5). In a way, this underlines the urgence of probing "the edge of the desert" as acurately as possible; e.g. by searching for a signal from right-handed (V+A) currents. These can follow from a $SU(2)_L*SU(2)_R*U(1)$ Lagrangian with manifest symmetry. In such a model, parity

is spontaneously broken. The mass eigenstates are linear combinations of unmixed chiral states

$$W_L = W_L \cos\zeta - W_R \sin\zeta$$
$$W_R = W_L \sin\zeta + W_R \cos\zeta$$

with mixing angle ζ and finite masses M_L (= 83 GeV) and M_R (being infinite in V-A). This model leads to a possibly measurable difference in longitudinal polarization P_F and P_G for Fermi and for Gamow-Teller positron decay, respectively. We perceived a practicable method to observe the ratio $R=P_F/P_G$ with an order of magnitude better accuracy than as known before.

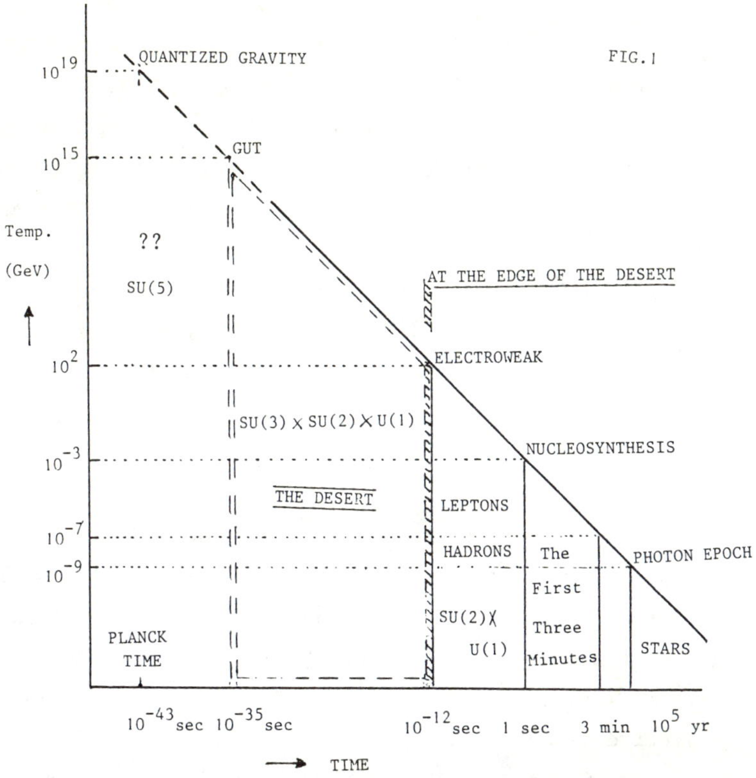

At this moment on-line measurements proceed with a new positron polarimeter at the cyclotron of the Free University in Amsterdam. The cyclotron produces periodically two β^+ activities, ^{26m}AL(6.4 s, F) and ^{30}P(2.5 m, GT), which are ideal for intercomparison. The instrument implies Bhabha scattering of the positrons by atomic electrons in a cylindrical scattering foil. This foil is made of soft-magnetic CoFe-alloy, of which up to 7 % of the atomic electrons can be polarised with a small external field. After careful magnetic annealing the foil possessed a remanence of 80%. The the new polarimeter is carried out in fourfold, and added as basic improvements to a prototype polarimeter[2)] twice the polarization sensitivity and five times higher event rates.

Fig.2 shows the arrangement of the four polarimeters, two on both sides of a system for transport of targets (T) from the (p,n) activation site (center) towards two measuring sites with the polarimeters I+II and III+IV, respectively. The targets are mounted on two rotors which are rotated over 180o within 0.25 s by powerful stepping motors M. During Fermi runs with ^{26}Mg(p,n)^{26m}Al the cycle time was 2*2 s; during the Gamow-Teller runs with ^{30}Si (p.n)^{30}P this time was adapted to 53*2 s. Both runs are alternated every hour. Accumulation of statistics proceeds several weeks with aid of computer control. Each polarimeter is made with strict axial symmetry. The magnitization of the scattering foils S inside yokes Y is reversed periodically. Positrons from the activated targets (^{26}Mg either ^{30}Si) are guided by special four-gap Mini-Orange magnets towards S. Upon Bhabha scattering into forward angles e^++e^- pairs can be pairwise detected in coincidence by two of five plastic scintillators; one on axis and four corresponding with

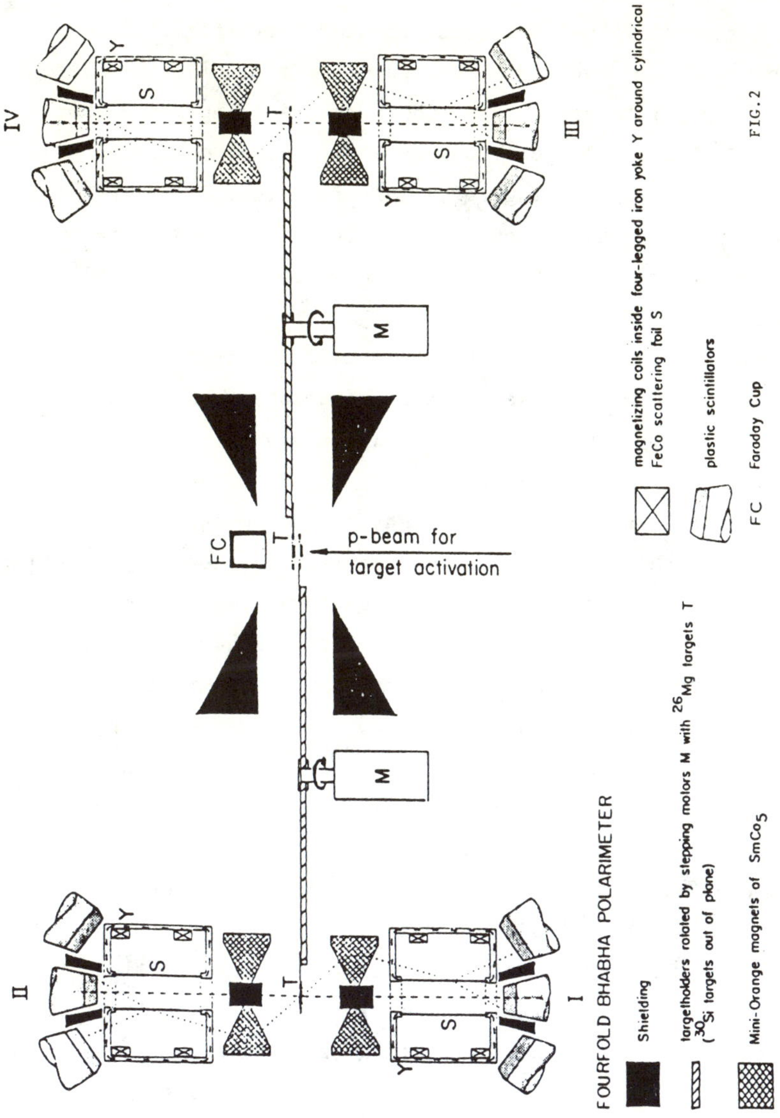
FOURFOLD BHABHA POLARIMETER
I
II
III
IV
S
Y
T
M
FC
p-beam for
target activation
Shielding
targetholders rotated by stepping motors M with 26Mg targets T
(30Si targets out of plane)
Mini-Orange magnets of SmCo5
magnetizing coils inside four-legged iron yoke Y around cylindrical
FeCo scattering foil S
plastic scintillators
FC Faraday Cup

FIG.2

the four Mini-Orange gaps.

The coincidence rate is about 5 kHz at singles rates of 300 kHz per detector. The amount of data needed for $3*10^{-3}$ accuracy in P_F/P_G is about 10 Gb, depending on the polarization sensitivity. This volume of data made on-line data analyses desirable. For this purpose the events of each independent polarimeter are processed by dedicated hardware handlers. The data-acquisition system is composed of homemade and commercially obtained high-density NIM and CAMAC modules, controlled by a micro-computer.

In a first test the observed ratio R did not differ from unity to within 1 % accuracy after 84 hours running time. At this level it remains allowed to ignore recoil-order terms. Our aim is to bypass this summer a statistical accuracy of about 0.4 %.

It may be conceivable that the novel polarimeter can facilitate other experiments in the future.

* Address: Natuurkundig Laboratorium, Free University Amsterdam, the Netherlands

1. D.N. Schramm, Physics Today, April 1983, 27.
2. J. van Klinken, K. Stam, W.Z. Venema, and V.A. Wichers, Phys. Rev. Lett.50(1983)94.

WEAK INTERACTION STUDIES OF ORIENTED NUCLEI FAR FROM STABILITY

L. VANNESTE, N. SEVERIJNS, J. WOUTERS, D. VANDEPLASSCHE AND
E. VAN WALLE
Instituut voor Kern- en Stralingsfysika, Celestijnenlaan 200 D,
B-3030 Leuven, Belgium

Abstract This study focuses mainly on the possibility of systematic measurements of A_1 asymmetries of on-line separated nuclei.
We stress that a new situation with respect to weak interaction studies of nuclei (especially for the very short lived ones or the ones far from stability) has evolved recently. Several factors contribute to this new situation.

EXPERIMENTAL DEVELOPMENTS IN WEAK INTERACTION STUDIES

A. The first factor is the use of on-line facilities, which allow the study of a previously inaccessible but very large number of nuclei. It is interesting to note that despite the tremendous flurry of activity in β-decay in the first decade following parity violation, not more than 5 cases of allowed but isospin forbidden β-transition have been studied with nuclear orientation in order to extract A_1. Isospin-allowed decays amount even to only 3 cases.

Note that the A_1 parameter of β-decay:

$$A_1(\beta^{\pm}) = \frac{\sum_{JJ'} (-1)^{J+J'+1} b_1 (J,J') F_1(J,J',I_f,I_i)}{\sum_J b_o (J,J)} \qquad (1)$$

for angular distribution of the form

$$W_{\beta^{\pm}}(\Theta) = 1 + A_1(\beta^{\pm}) B_1(I_i)\cos\Theta \qquad (2)$$

reduces to

$$A_1 = \frac{v/c}{1+y^2}\left\{\frac{\mp 1}{\sqrt{3I(I+1)}} + \frac{2}{\sqrt{3}} y \cos\phi\right\}$$

for $I_f = I_i = I \neq 0$.

The relevant parameters are contained in

$$y\, e^{i\phi} = \frac{C_V \langle 1 \rangle}{C_A \langle \sigma \rangle}$$

for the allowed approximation of the V-A theory. After ten years of intense activity, it was generally accepted [1] that time-reversal violation was disproven with $\phi = 180(8)°$ and

$$\frac{C_A}{C_V} = 1.175(20)$$

mainly on basis of the polarized neutron decay. In the same period also an impressive number of laboratories engaged in the study of β-γ(CP) correlation, which requires no special orientation equipment. Over the same decade some twenty cases of allowed but isospin forbidden decay were studied. Some individual cases, such as ^{46}Sc, were reported in sometimes up to 19 different papers with an asymmetry coefficient $\tilde{A}_1$ ranging from 0.33(4) to 0.06(3) [2]. $\tilde{A}_1$ is similar in expression to A_1 with however the opposite sign for y, if the β-γ(CP) correlation is described by

$$W_{\beta^{\pm}\gamma}(\Theta) = 1 - \sqrt{\frac{I+1}{3I}}\, \tau\, \tilde{A}_1\, A_1(\gamma)\cos\Theta$$

where $\tau = +1(-1)$ denotes right (left) circular polarization of the γ-radiation and $A_1(\gamma)$ describes the following γ-transition. This technique gives basically the same information as direct measurement of the β-asymmetry, but has a much inferior signal - to - background

ratio. Since however this technique seemed more readily adaptable for for different nuclei, results have normally been expressed in $\tilde{A}_1$, even for the (small) number of NO cases.
With the coming of age of NO-On-Line this restriction can be lifted now. At the KOOL-facility nuclei have been oriented down to lifetimes in the second region. Of course β-γ(CP) can be performed on-line as well, but requires a coincidence: attempts at more efficient "scattering" analysers have not led to substantial improvement. Obtaining data with this technique of good statistical quality at an on-line facility is therefore extremely difficult.

B. Working at an on-line separator facility allows not only access to very wide range of isotopes, but also β-spectra free from contaminants if the mass resolution is sufficient. For the light and medium heavy nuclei on which we concentrate them, this condition is fulfilled at the KOOL-LISOL-CYCLONE chain.

C. Progress in our understanding of implantation phenomena, esp. at low temperature, is fundamental as well. As where the older experiments of NO were nearly all performed using single crystals for adiabatic demagnetisation, recent experiments use almost exclusively implantation in ferromagnetic foils. For the second decade following parity violation practically no new results were reported.
We showed by systematic tests that off-line implanted sources lead to reliable results for β-decay studies. For this we rely on the small penetration depth at low implantation energy (~50 kV), which leads to small corrections for scattering in the source. In these cases we used elements, the implantation behaviour of which had been studied extensively (I, Te, Xe, Co).

D. The major breakthrough is however implantation at low temperatures, more in particular;

1) the demonstration of high substitutional fractions for implantations.

2) the developments of reliable and high resolution detectors for very low temperature work.

For the latter point I refer to the work of J. Wouters et al. reported in ref. [5].

The former point has come as a surprise [3], but has been repeatedly confirmed by our recent work [4] on a large series of isotopes. The general conclusion is that in conditions of sufficiently low dose and careful surface preparation, the substitutional fraction is normally higher than 90% in the two site approximation. Together with the demonstration of the possibility of high precision measurements on off-line prepared samples - for purposes of tests of P- and T-violation esp. by the Columbia and Cal Tech groups - we stress that in on-line working conditions with suitable detectors a whole new field of weak interaction studies has become possible.

STATUS OF WEAK INTERACTIONS IN NUCLEI.

Weak interactions in nuclei is a very much neglected field, esp. since the first measurements indicated led fast to the generally accepted (V-A) theory. Even for the bare nucleon (n decay), the most elementary properties have remained a matter of debate: even very recent measurements of the lifetime deviate by as much as 7% while the C_A/C_V ratio has been in constant motion from a "plateau" at 1.14 to one at 1.23 and now at 1.27. For nuclei matters are not settled at all.

A. C.V.C. and P.C.A.C.

One generally accepts that the Fermi part of an allowed transition is conserved. Hence the coupling constant C_V should not deviate from the value used for quarks and leptons. Fixing this constant has been

a long and major undertaking involving the precise determination of the ft-value of superallowed 0^+-0^+ transitions and is still subject to small revisions.

For C_A there is no equivalent procedure, although P.C.A.C. (Partially Conserved Axial Current) is as well established as C.V.C. (Conserved Vector Current). The effective axial constant C_{Ae} cannot be accepted a priori to be the same in a multinucleon system.

B. Mirror decays

Nevertheless it has been argued at several occasions that a C_{Ae} can be predicted esp. in the case of "good single-particle" systems without major reliance on wave functions. In mirror nuclei it is possible to deduce the value of C_{Ae} from the β decay and the isoscalar part μ_o of the magnetic moments. Another recent even more direct approach is possible using sum rules, and even more convincing since it does not rely on the intrinsically small μ_o [11]. The surprising result is that $C_A = 1.00(2)$ with a very high correlation coefficient for all mirror decays. How does this reconcile with the two nuclear cases for which A_1 has been measured? In the neutron the A_1 asymmetry is needed to obtain a highly accurate value of C_A/C_V: how is it possible that no drastic effect shows up esp. in the very accurately measured ^{19}Ne? The latter has been used effectively to look for effects in the promille range connected with time-reversal invariance, while a 27% deviation in C_A/C_V goes unnoticed.

Although C_A and C_V come always in combination with the respective matrix elements, it is expected - esp. since the large shell-model calculation of Wildenthal [10] for the sd-shell - that one should be able to calculate the ratio $\langle\sigma\rangle/\langle 1\rangle$ to much better accuracy than 25%.

Other factors may be uncertainties in the ft-value (compare with the neutron case). In any case the A_1 coefficient is strongly dependent on the C_A/C_V ratio: in the allowed approximation for the neutron one obtains

$$A_1 = - \frac{2\lambda(\lambda-1)}{1+3\lambda^2}.$$

In the case of ^{19}Ne the A_1-asymmetry is small (of the order 4% vs. 12% for the neutron). In ^{35}Ar it is much larger (22%) but the extracted A_1 has raised controversy. Indeed, when combined with the ft-value the value for C_V is anomalous (about 4% too large). This has led to the suggestion that the Cabibbo angle may vanish in nuclei for high magnetic fields (~10^{16} gauss). The Cabibbo angle - more exactly a quark mixing angle - turns out to have the normal value in a supposed high field nucleus (^{24}Al) according to a recent measurement [6]. Is it possible that the field is just large enough to induce a phase transition by which the quark mixing angle disappears?

From the J→J (J=4) analog transition in the β^+ decay of $^{24}Al \to ^{24}Mg$, one can derive the f matrix element if the GT one is known. Relying on the extracted life-time of $t_{1/2} = 2.053 \pm 0.004$ sec., and on the assumption that GT is vanishingly small, a G_V is extracted in good agreement with the value from ^{19}Ne and the superallowed (0^+)-(0^+) transitions.

From this, the authors of ref. [6] claim that the ^{35}Ar result is due to an experimental error. Considering the energy put into checking all parts of the ^{35}Ar puzzle, this is puzzling. The most suspected value is the one of the asymmetry parameter $A_1 = 0.22(3)$. Indeed it was remeasured by the same method with results varying from 0.16(4) to 0.33(6), but in order to obtain agreement a value of 0.43 is required. We are not convinced that the results obtained on ^{24}Al allow to reject the results of this AB measurement. The problem is that only one method (AB) gives access to A_1 information of mirror decays. That method is restricted to two cases, both - not by coincidence - noble gases. In terms of nuclear structure both are less than ideal. The normal behaving one (^{19}Ne) gives no problems at first glance but has a small A_1: the one with a large A_1 (^{35}Ar) leads to problems.

The recent A_1 value from the neutron also leads to disagreement with Cabibbo theory, in as far that the derived C_A/C_V disagrees with the

result from equally recent semileptonic hyperon decays.
Obviously there is an urgent need for additional measurements, preferentially with

- a different technique
- on mirror nuclei with "good" nuclear structure, i.e. "closed shell ± one" nuclei.

The problems concern both anomalies in C_V and C_A, as extracted from neutron decay and "nuclear" mirror transitions.

THE DETERMINATION OF THE AXIAL VECTOR CURRENT CONSTANT

We first describe how one can obtain largely model independent infoımation on the axial vector constant C_A in nuclear matter, using two different approaches. From these we show how with a new experimental approach it is possible to proceed to systematic and accurate measurements of C_{Ae}, hopefully thereby also resolving the remaining problems in connection with C_V and Cabibbo (quark mixing) angle.

A. Combination of magnetic moment and ft values

Wilkinson devised the first strategy by relying on combination of the isocsalar moemnt μ_o and ft in order to extract the M_{GT} matrix element without needing "perfect" nuclear wave function. He pointed out that according to the
Sachs theorem, mesonic exchange effects are absent from the isoscalar magnetic moment.
The difficulty remains that - in order to combine with the ft-value and hence extract C_{Ae} - we need $<\sigma_3\tau_3> = M_{GT}$, rather than $<\tau_3>$.
If one can accept that $<(1\pm\tau_3)\sigma_3>$ is small the problem disappears. Indeed, if an odd-mass nucleus can be described in jj coupling as a single configuration, this matrix element vanishes for closed-shell plus (or minus) - one nuclei. Also in LS coupling with single confi-

guration lowest-seniority states it vanishes. Raman et al. [7] have refined the shell-model estimates of Wilkinson for the difference between $<\sigma_3>$ and $<\sigma_3\tau_3>$ by explicitely taking into account core polarization corrections to second order with intermediate-state summation.

Of an initial 19 cases of isobaric mirror pairs with known magnetic moments, six are rejected because of unsufficient experimental accuracy and another six because of theoretical uncertainties in the calculation of $<\sigma_3(1\pm\tau_3)>$. We investigated the remaining seven.

Besides ^{41}Sc, in which large errors are involved, both the "A_1 cases" ^{19}Ne and ^{35}Ar stand out as surprises. ^{19}Ne is disregarded in all fits because its data are out of systematics. The tritium case is probably not satisfying the conditions of a real "multi-nucleon" system. For closed-shell minus one nuclei such as 15 and 39 the core polarisation correction is most difficult to estimate and may rise to 20%. The only accurate value remains with ^{17}F, the lightest double closed-shell nucleus plus one nucleon. This is in fact the one isotope pointed out as the outstanding "good" case by Wilkinson in his earliest paper describing the way to extract C_A in nuclei. We will return to this later.

One last remark: when using the full body of all 19 mirror transitions with known magnetic moments, one obtains a C_{Ae} with a value of constantly around 1.0 with a small scatter. It is strange that the more fundamental theories, using PCAC to obtain the complete one-pion exchange two-body contribution to the GT matrix element, seem less successful, even for closed-shell cases (see refs [8] and [9]). Therefore, we repeated calculations with recent shell model estimates [10]. The attempt at separating configuration mixing from pionic contributions by Raman, in evaluating $<\sigma_3(1\pm\tau_3)>$ may suffer from similar problems, as noted before. Coupled to the intrinsic smallness of μ_0, any attempt at determining C_{Ae} will remain extremely sensitive to the model dependent values of the above parameter.

Fits according to different approximations lead to

$1.00 < g_A/g_V < 1.60$ (Raman et al.)

$0.90 < g_A/g_V < 1.05$ (this work).

B. Sum-rule approach

A really model-independent way of analyzing the data on magnetic moments and β-decay of mirror-nuclei was introduced by Buck and Perez [11] by using sum-rules. One of these strongly suggests that all odd-rank multipole operators in the variables of the nucleons with even number have vanishing expectation values in the ground state. Instead of combining μ_p and μ_N into an isoscalar and isovector part, the trick used is to rewrite μ_p and μ_N such that the contributions from only odd-nucleon terms are retained for further discussion.

$$\mu_p = g_p L_o + G_p S_o + g_n L_e + G_n S_e$$

$$(g_p = 1,\ G_p = 5.586,\ g_n = 0,\ G_n = -3.826)$$

Their analysis leads to

$$\frac{C_A}{C_V} = 1.00(2)$$

with again a very high correlation coefficient for all accurately known mirror transitions (lifetime and magnetic moment). It is difficult to believe that this is a mere coincidence.

Recently theoretical calculations have been performed of the Gamow-Teller matrix elements for sd-shell nuclei [12] based on complete $(0d_{5/2}, 1s_{1/2}, 0d_{3/2})$ - space shell model wave functions. Reasonable agreement is obtained with experimental data from ft-values alone. Small deviations between the theoretical and experimental M_{GT} induce however large variations in $\frac{C_A}{C_V}$ when using ft alone! This is the difference with the procedure using the magnetic moment (but not the isoscalar part alone): for ^{17}F there is a very reasonable agreement between experimental and theoretical value of M_{GT}. Using ft alone nevertheless leads to an unreasonable assumption regarding C_{Ae}/C_V, while combined with the value of y, one obtains 1.12(2). A direct

determination of y by measurement of A_1 might be a way of solving the apparent discrepancy between different approaches.

C. Direct asymmetry measurement in mirror decays

Unlike the case of isospin-forbidden decays where the β-γ(CP) method can be used, the g.s. to g.s. (often close to 100%) transitions restrict the techniques to Nuclear Orientation. Up-to-now this has been performed on the neutron by scattering, and on ^{19}Ne and ^{35}Ar using Atomic Beams. We present here the first result using low temperature orientation on-line with isotope separator and accelerator, for the "best" case, first pointed out by Wilkinson more than a decade ago but never performed. At the same time we show that the technique can easily be extended to all other mirror decays. For details and results on the measurement of the large A_1 of ^{17}F we refer to a forthcoming paper by N. Severijns et al.

We lend credence to our first on-line A_1 result, since the extracted value corresponds to the theoretically maximum, just as we expected from our experience with off-line implanted long lived isotopes [14,15], and from the implantation behaviour of on-line implanted activities [3,4].

Unfortunately, in the case of ^{17}F, the value of A_1 is insensitive to variations in y (y = F/GT), exactly because of the maximum in the $A_1(y)$ curve. Hence at this point we conclude only that also in on-line conditions A_1 can be measured to an accuracy of a few percent. However, ^{17}F is not the ideal case Wilkinson supposed in spite of the large A_1.

From the theoretical curves for other spins we can devise a better strategy for measuring C_{ae} since we have also a fair idea of the range of magnitude of A_1 in advance and the sensitivity to y. The doubly magic cases ^{15}O, ^{17}F, ^{39}Ca, ^{41}Sc are not all equally "ideal" also because these cases have to be judged in function of hyperfine field and lifetime.

We are in the process of optimising production rates for these and other mirror cases, and believe we can extend the number of measured

A_1's from the previous two - for nearly two decades - to about twenty, and build up a systematics of accurate A_1. From there sensitive tests of C_{Ae} in multinucleon systems become possible.
We believe that the "circumstantial" evidence for a renormalisation of the axial current in nuclei has grown so large, that this effort is largely justified.

ACKNOWLEDGEMENT

We thank the separator and the cyclotron crew of Cyclone for their unrelenting support, and appreciate the encouragement from many "weak" colleagues for continuing this new endeovour.

REFERENCES

1. H. Schopper, Weak Interactions and Nuclear Beta Decay, (North-Holland Publ. Co 1966)
2. H. Behrens, KFK-report 1249 (1970) and S. Raman, T.A. Walkiewicz, H. Behrens, At. Data & Nuclear Tables 16 (1975) 451-494
3. C. Nuytten, D. Vandeplassche, E. van Walle and L. Vanneste, Phys. Lett. 92A (1982) 139
4. Proc. of the Int. Conf. on Nuclear Orientation and Nuclei far from Stability, Eds. B.I. Deutch & L. Vanneste, North-Holland Publ. Co (1985)
5. J. Wouters et al., Phys. Rev. Lett. 56 (1986) 901
6. E.G. Adelberger et al., Phys. Rev. Lett. 55 (1985) 2129
7. S. Raman, C.A. Houser, T.A. Walkiewicz and I.S. Towner, NDT 21, 567 (1978)
8. A. Barroso, R.J. Blin-Stoyle, Nucl. Phys. A251, 446 (1975)
9. H. Khanna, G. Tower, J. Lee, Nucl. Phys. A (1978) 521
10. Wildenthal, Progress in Particle and Nuclear Physics, Vol.11, ed. D.H. Wilkinson (Pergamon Press, Oxford, 1984), p. 5
11. G. Buck & J. Perez, Phys. Rev. Lett. 50, 1975 (1983)
12. B.A. Brown & B.H. Wildenthal, Nucl. Data Tables 33, 347 (1985)
    Phys. Rev. C28, 2397 (1983)
13. N. Severijns et al., to be published
14. R. Geerts, Ph. D. Thesis, Leuven (1977)
    unpublished, Phys. Rev. C20 (1979) 1171
    Phys. Rev. C21 (1980) 439
15. R. Geerts, Contr. Paper. Int. Conference on Hyp. Interactions, Madison (1977)

MEASUREMENTS FOLLOWING THERMAL NEUTRON CAPTURE BY NATURAL KRYPTON AND XENON GAS TARGETS

S.A. Hamada*, W.D. Hamilton*, B. More*† and F. Hoyler†

* Physics Division, University of Sussex, Brighton BN1 9QH, U.K.

† Institut Laue-Langevin, 38042 Grenoble, France.

Abstract The directional correlation of γ-ray transitions in ^{84}Kr, ^{130}Xe and ^{132}Xe have been measured following thermal neutron capture by natural krypton and xenon pressurised gas targets at the Institut Laue-Langevin. Gamma-ray singles spectra were measured up to 5.5 MeV. The decay schemes were developed on the basis of coincidence measurements. Correlation data were obtained for transition in the range 0.2 to 2.5 MeV. The data allow spin-parity assignments to be made to most levels lying below 3.5 MeV and multipole mixing ratios to be evaluated for the more intense transitions. The results are fitted to IBM-1[1), IBM-2[2) and are also compared with the prediction of the Dynamic Deformation Model[3). Mixed-stymmetry states in ^{84}Kr and ^{132}Xe are discussed.

[RADIOACTIVITY ^{83}Kr, ^{129}Xe, ^{131}Xe (n,γ) ^{84}Kr, ^{130}Xe, ^{132}Xe; natural gas targets, on-line directional correlation measured: E_γ, I_γ, $\gamma\gamma(\theta)$; deduced I, π, δ]

INTRODUCTION

Thermal neutron capture in natural krypton is dominated by the production of ^{84}Kr in a highly excited state at 10.5 MeV

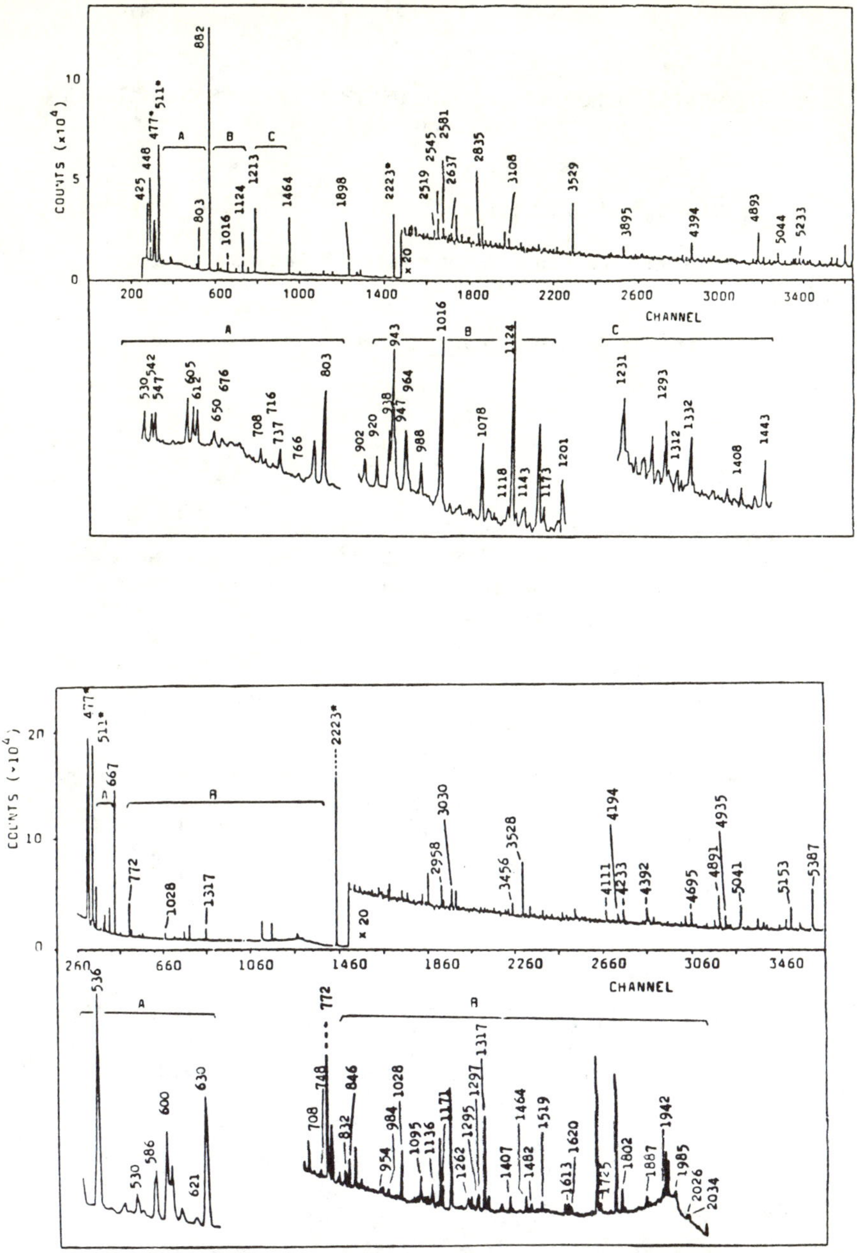

Figure 1: Singles γ-ray spectra neutron capture by krypton (above) and xenon (below). The low parts shows expansions of the labeled regions. Transition energies are in keV.

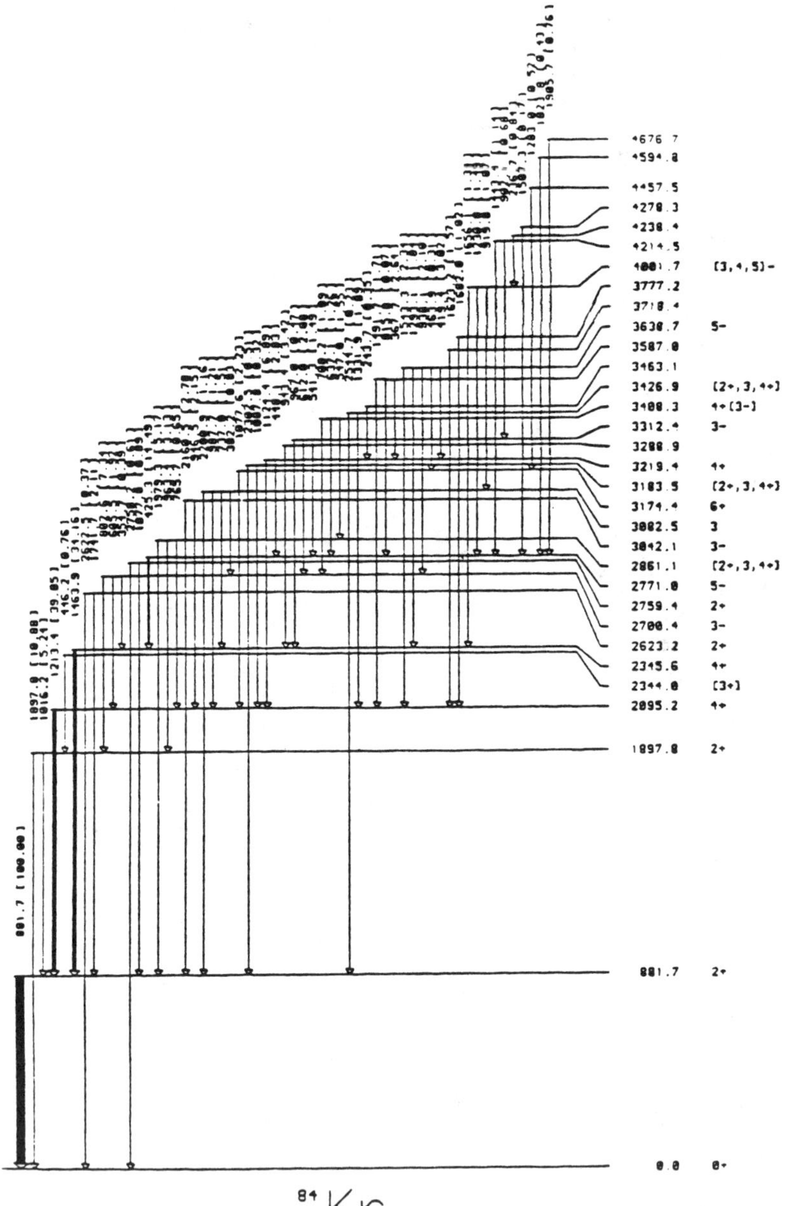

^{84}Kr

Figure 2a: The decay scheme of ^{84}Kr.

with spin values of 4^+ or 5^+. This gives the possibility of observing transition from the capture states to a high spin and high energy levels. Such levels cannot be reached by β-decay. Measurements on krypton have been made[4,5] to identified primary γ-rays and to study the multipolarity of γ-ray transitions.

In a xenon target neutron capture is dominated by the production of $^{130,132}Xe$, in which the capture states are 0^+, 1^+ in ^{130}Xe and 1^+, 2^+ in ^{132}Xe, and low lying low spin states are populated strongly by primary γ-ray transitions. Several studies have been made on $^{130,132}Xe$[6,7,8,9] to build the decay schemes from n-capture reactions and to evaluate the multipole mixing ratios for γ-ray transitions.

RESULTS

Two singles spectra wave measured for both targets; one in the range 0.0, to 2.5 MeV and the other in the range 2.5 to 5.5 MeV. Energies and intensities calibrations, were made with ^{152}Eu 10) and

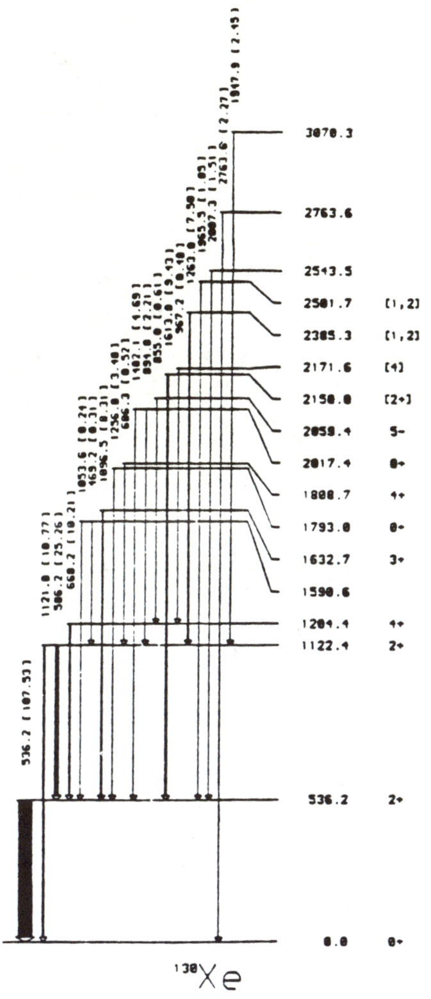

Figure 2b: The decay scheme of ^{130}Xe.

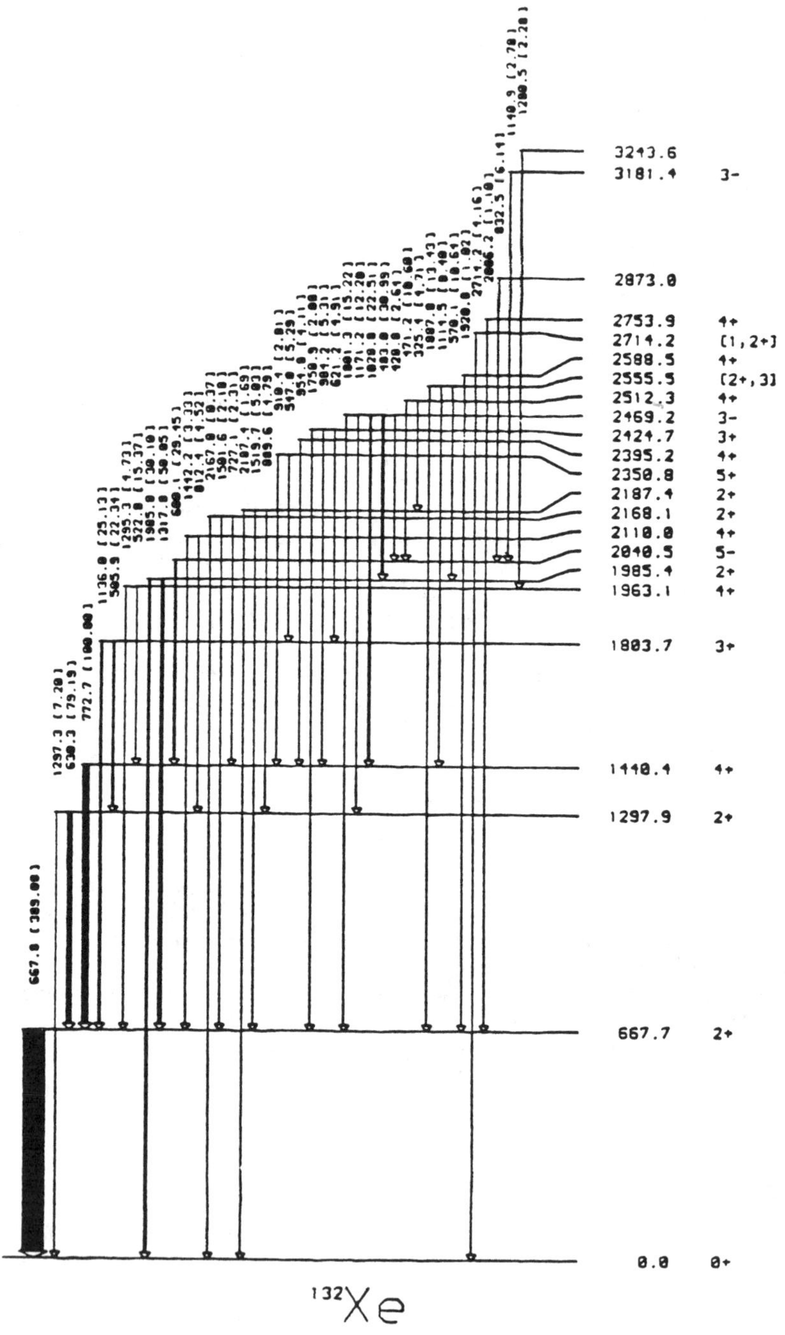

Figure 2c: The decay scheme of ^{132}Xe.

^{56}Co[11] sources for the low and high energy region respectively. Figure 1 shows the singles spectra. Twenty coincidence gates were set on the total coincidence spectra, from each target, and only transition positively identified as coming from ^{84}Kr and $^{130,132}Xe$ are included in the decay schemes which are shown in Figs. 3a, 3b and 3c. Directional correlation data were obtained and the mixing ratios for transitions in ^{84}Kr, $^{130,132}Xe$ were evaluated. Further details about the experimental procedure and results can be found in ref.[12].

DISCUSSION

The level structure of ^{84}Kr shows vibrational characteristics and the energy ratio of the two-phonon to one-phonon states is close to that of a vibrator. Calculated[13] energies of 3^- states are 2.89 and 3.87 MeV, and the correlation results indicate a 3^- assignment for the 2.700 MeV level. This 3^- state may be considered to be the first member of the octopole band and the 5^- state at 2.771 MeV, the first excited state in this band. This agrees with the systematic of the 5^- states, in neighbouring nuclei. We found the next 3^- level at 3.042 MeV which is much lower than predicted. The energy systematic of the 3_1^+ states in even krypton isotopes suggests that the new level which we proposed at 2.344 MeV may be the 3_1^+ in ^{84}Kr and this fits well with the IBM-1 calculation.

In the xenon experiment, new levels with spin-parity assignments and transitions are proposed. The energy ratio of the 2_2^+ to the 2_1^+ are close to that expected for the simple vibrational model. The SU(5) limit of the IBM-1 give a good description of the levels and suggests that the 0_2^+ states occur at about 1.0 and 1.3 MeV in ^{130}Xe and ^{132}Xe which so far have not been observed. The DDM energy level prediction for these nuclei show a vibrational character although the potential energy surfaces indicate that they are γ-soft. IBM-2 calculations and the decay properties of the 2_3^+ states in ^{84}Kr and ^{132}Xe are associated with a mixed-symmetry description.

REFERENCES

1. Arima, A. and Iachello, F., Ann. Phys. 99, 253 (1976).
Ann. Rev. Nucl. Part. 31, 74 (1981)
2. Arima, A., Otsuka, T., Iachello, F. and Talmi, I., Phys. Lett. 66B 205 (1977).
3. Kumar, K., Prog. Part. Nucl. Phys. A233 9 (1983).
4. Mattsson, G.G., Arnell, S.E. and Jonson, L., Phys. Scripta Vol 5, 58 (1972).
5. Saxena, R.A., Jahnel, L,C. and Zawislak, R.C., Phys. Rev. C21 1531 (1980).
6. Groshev, L.V., Govor, L.I., Demidouand, A.M. and Rakimov, A.S., Sov. J. Nucl. Phys. 13 647 (1971).
7. Gelletly, W., Kane, W.R. and Mackenzie, P.R.,
Phys. Rev. C9 2363 (1974).
Phys. Rev. C3 1678 (1971).
8. Girit, G., Hamilton, W.D. and Michelakakis, E., J. Phys. G6 1025 (1980).
9. Sooch, S.S., Kauv, R., Sinch, N. and Trehan, P.N., J. Phys. Soc. of Japan No.1 1419 (1983).
10. Debertin, K., Nucl. Inst. Meth. 158 479 (1979).
11. Katou, T., Nucl. Inst. Meth. 124 257 (1975).
12. Hamada, S.A., Hamilton, W.D., More, B. and Hoyler, F., to be published in J. of Phys. G (1986).
13. Veje, C.J., Mat. Fys. Medd. Dan. Vid. Selsk. 35 Vol.1 (1966).

(D,p) REACTIONS AND THE COMBINATION OF PIXE - PIGME TECHNIQUES FOR THE ANALYSIS OF BIOMASS SAMPLES

H.G. FAKHOURI* AND N.A. DYSON**
*University of Jordan, Amman, Jordan
**University of Birmingham, Birmingham, England.

Abstract

To control environmental pollution, a biodegradation technique involving biomass is used to degrade the effluent from certain industrial wastes before disposal. (D,p) reactions were used to study the degradation process and the extent of removal of carbon and nitrogen. The PIXE - PIGME technique was used for the simultaneous study of the compositions of the biomass, the feed and the treated material. Results show that 54–58% of the carbon and 45-50% of the nitrogen is removed from the wastes and that the biomass is concentrating Cl, Ti, Ni and Fe.

INTRODUCTION

Toxic metal pollutants are often introduced into the environment in the form of industrial wastes. To control the environmental pollution, these wastes should be treated to reduce their metal content. Different biological techniques have been used to degrade industrial wastes including the development of "digesters" in which a

biomass is allowed to grow from suitable seed material. Some interest is attached to the liquid waste from the manufacture of propellants and explosives. This contains aromatic compounds which could be toxic and so should be degraded before disposal.

The method used to treat synthetic compounds (in the Department of Chemical Engineering, University of Birmingham) is biodegradation[1]. An activated sludge system was devised and set up to acclimatise to phenolics. The apparatus consists essentially of a vessel containing the biomass through which the synthetic waste is passed under aerobic conditions. During the treatment, this effluent is mixed and aerated with the activated sludge. In these circumstances the retention time (contact period between the effluent and the biomass) is sufficient to allow the biodegradation of the chemical present, i.e., the substances are metabolised into energy and cellular material.

The primary degrading organisms in activated sludge plants are bacteria, and they derive their nourishment from the raw input to the plant. They are able to synthesise their enzymes, metabolic requirements and energy from the chemicals present in the influent. Although many micro-organisms can degrade a wide range of organic compounds, the enzymes are often produced only when particular compounds are present. In mixed systems such as the activated sludge system, only a few of the organisms may have the ability required to degrade a particular compound. Since it takes time to adapt systems to biodegrade a particular compound, it was decided to set up two identical plants, one of which could be used as a

control and additionally, as a source of inoculum should the experimental plant fail. The second plant was therefore set up and was adapted to phenolic waste. Once this experimental plant was established, its feed was "doctored" with additions of the industrial waste requiring processing, starting with a concentration of 0.5% and increasing gradually up to 95%.

To find out how toxic the phenolic waste is to the treatment process and to know which compounds are present in the liquid before and after degradation, and the extent of removal of organic material, the composition of the feed and the treated material has to be determined.

ANALYSIS OF CARBON AND NITROGEN DEGRADATION BY (d,p) REACTIONS

Samples were obtained from the reactor, dried, ground to a powder using an agate ball-mill and pressed into pellets 11 mm in diameter and 1 mm in thickness. The pellets were then coated with a thin layer of silver.

The semiconductor detector used for this work is a surface barrier detector capable of measuring protons of up to 10 MeV and has a resolution of less than 45 keV. It is covered by a 25μm aluminium foil to stop scattered deuterons, low energy protons and alpha particles. The energy of the incident deuteron beam was 2.0 MeV for the measurements described here.

Fig. 1 shows the carbon peak for the feed and the treated material from the control system. 65% of the carbon is removed by the treatment process. The removal from the experimental system ("doctored" system) ranged between 54 and 58%, for additions of waste of 5% and 95%

added to the feed. The percentage removal is somewhat less (≃10%) in the experimental system than the control

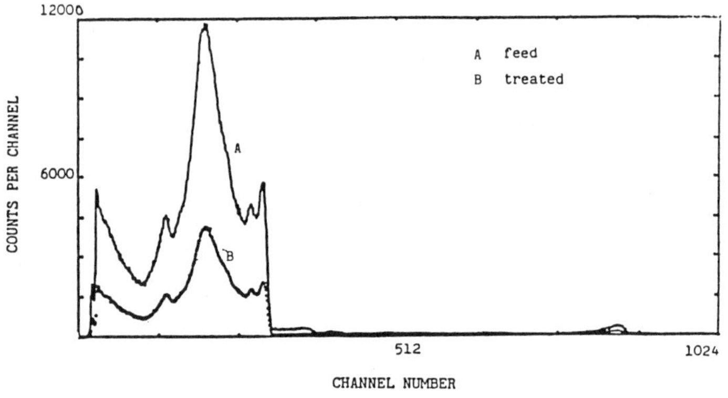

FIGURE 1. The carbon (D,P_0) peak for the feed and treated material from the control system

system. These figures were confirmed also by analysis using chemical methods.

The nitrogen P_0 peak was used to find the degradation of nitrogen. Fig. 2 shows the nitrogen content of the

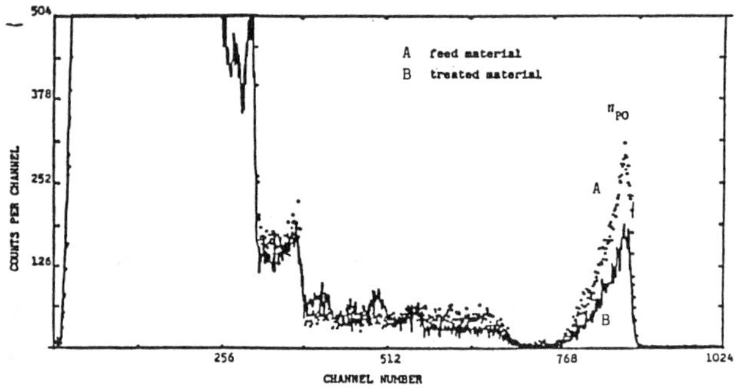

FIGURE 2. The proton spectrum from the (d,p) reactions on the experimental feed and treated material

experimental feed and treated material. 40% of the nitrogen is removed from the control feed while 45 - 50% is removed from the experimental feed for a concentration

of 5 - 95% of the industrial waste.

ELEMENTAL ANALYSIS IN THE BIOMASS USING PIXE - PIGME TECHNIQUES

To discover whether the feed has any toxic effect upon the biomass, and to know which elements are present in the feed before and after degradation, the composition of the feed and of the treated material should both be determined. To check the possibility of some elements being taken selectively by the biomass, careful monitoring of the biomass is necessary.

Since the experimental arrangement of PIGME corresponds to that of PIXE (both methods use the same charged particle, the same target and the same beam line) the combined use of these methods is possible. The main idea is to cover the low Z elements ($Z \leqslant 14$) using proton induced gamma ray analysis and the high Z elements ($Z > 14$) using PIXE.

Liquid samples were collected from the digester, and wet biomass samples of 20 ml volume were centrifuged. After decanting the supernatant solution, the solid is dried, shaken and pelletised. Samples were mounted on a 12-target holder placed in the target chamber. The experimental arrangement is described elsewhere[2]. A proton beam of 2.5 MeV and a current of 10 - 30 nA were used. The gamma rays were detected in a 76 mm × 76 mm NaI crystal placed at 90° to the direction of the incident beam at a distance of 35 mm from the samples. X-ray and γ-ray data were recorded simultaneously using the multiplexing facility on the Hewlett-Packard ADC's.

RESULTS AND DISCUSSION

X-ray and gamma-ray spectra of all the samples were obtained. Figs. 3 and 4 are the X-ray spectra of the feed and the output from the control system. It is clear that

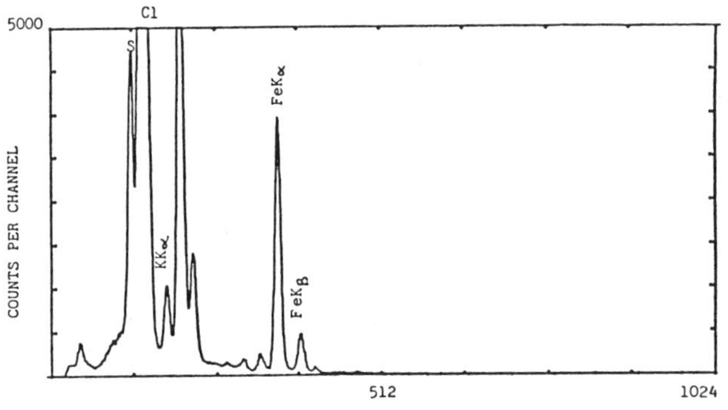

FIGURE 3. X-Ray spectrum of the control feed

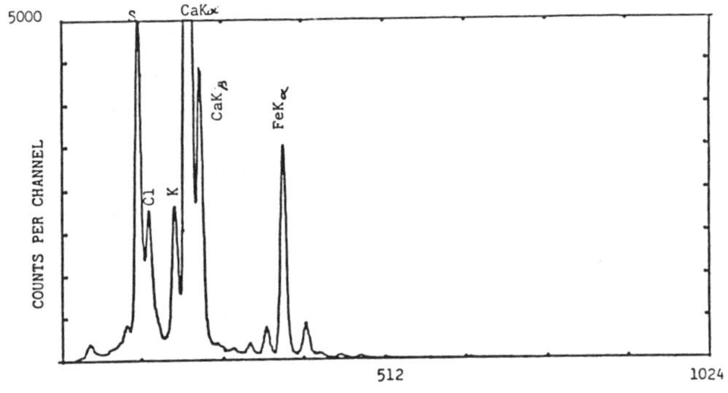

FIGURE 4. X-ray spectrum of the effluent (output) from the control system

the chlorine in the feed is degraded, and as the chlorine, nitrogen and carbon were degraded the concentration of calcium increased.

An example of a γ-ray spectrum obtained by the PIGME technique is given in Fig. 5 which illustrates spectra of the output material from the control system. Al, Mg, Na, N and F are the main gamma-rays observed.

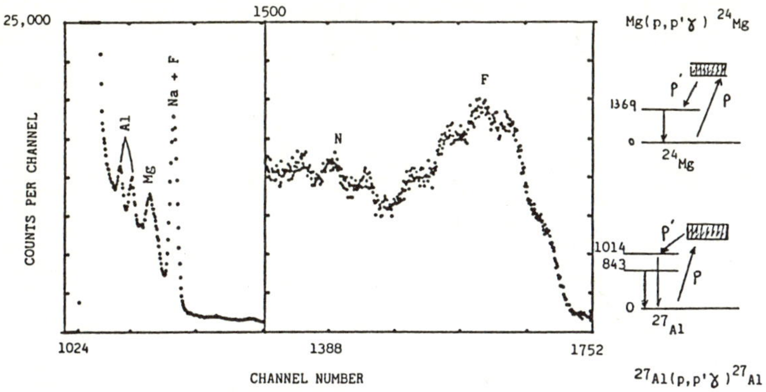

FIGURE 5. Gamma-ray spectrum of the output from the control system

To summarise, the system is reasonably effective for treatment of aromatic compounds, the amount of carbon and nitrogen is reduced by the process, leaving an increasing proportion of, e.g., calcium in the output. It is therefore, a cheap and efficient biological method of treatment for degradation. The efficiency of the system can be improved by the removal of some of the chlorine in the input to the system before treatment.

REFERENCES

1. M. Dosanjh, Report No. 2, Chemical Engineering Department, University of Birmingham.
2. H.G. Fakhouri, Ph.D Thesis, Birmingham University, 1984.

Chapter 9

CONFERENCE ACTIVITIES

Master of Ceremonies:

S. BRANT

Prirodoslovno-matematicki Fakultet

Zavod za Teorijsku Fiziku

Marulicev trg 19/I

41000 Zagreb, Yugoslavia

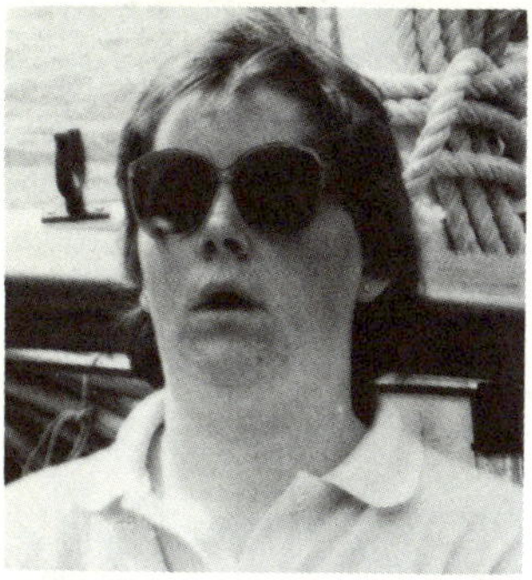

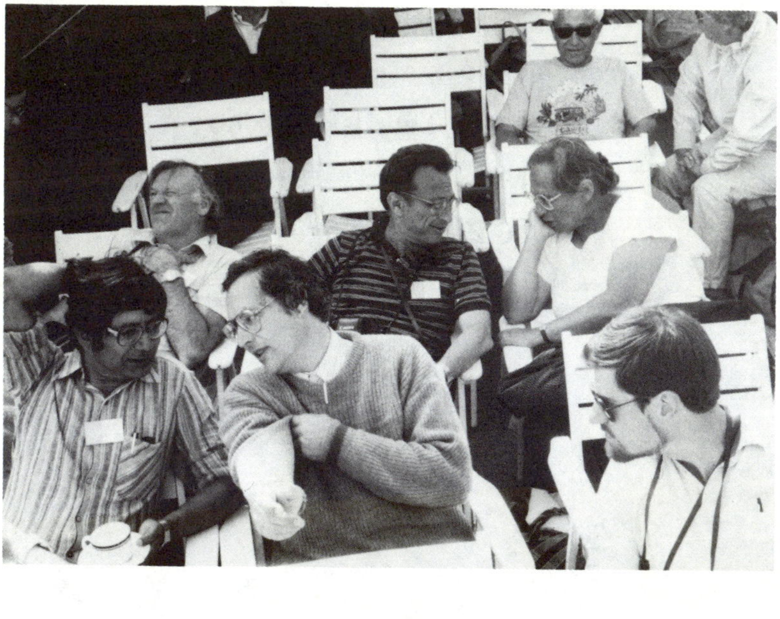

SELECTIONS OF SACRED MUSIC FROM THE BAROQUE BY CROATIAN, FRENCH, ITALIAN, GERMAN, AND ENGLISH COMPOSERS

JEAN WELLS STRUBLE
Alamo
California

Abstract: The SANS-NSRS conference concert consisted of sacred music spanning the Baroque period. The music of the little known Croatian composers Ivan Lukacic and Vincenz Jelic was featured. In addition compositions by the French composers Louis-Nicolas Clerambault and Froncois Couperin, the English composers Hemry Purcell and Jeremiah Clarke, the Italian compower Francesco Durante and the German composer George Fredich Handel were performed. The program notes for the concert are reproduced in this article with the program given as an appendix.

Louis-Nicolas Clerambault, an organist at St. Sulpice in Paris, served under Louis XIV as supervisor of concerts. It was, however, as a composer of French cantatas, not as an organist, that he surpassed all others. The Grand Plein Jeu which is being performed tonight is the opening piece of the first suite from his Premier Livre d'Orgue. This work was published in 1710 and consisted of two suites. The term, plein jeu designates what stops the organist should use, specifically a full chorus of principal stops with mixtures added.[1] Clerambault's organ music exploits the color of the French Baroque organ and reflects the characteristic

French love of ornament and decorated melodic line which contrasted drastically with the smooth, luxuriant lines of the middle and late Italian music of the Baroque which we will hear later in the program.

The next group of pieces were written by **Croatian composers** of the early Baroque. Little is known today about the musical situation in Croatia in the Baroque and very few musical works are extant.[2] Continual political turmoil, wars, and interference from Hungary, Austria, Venice and Turkey did not allow the development of aristocratic courts in Croatia as sources of patronage for musicians as existed in Italy and elsewhere. Thus musicians in Dalmatia were only employed in the the service of the church and many composers sought their fortune in more lucrative European centers. Among the most famous musicians who left Croatia were Georg Slaktonia, Jacob Handl and **Vincenz Jelich**. Georg Slaktonia from Laibach went to Vienna as the director of Emperor Maximillian's chapel. Later Jacob Handl went to Vienna and then to Bohemia. Vincenz Jelich went to Austria. In 1612 a theater was opened in Hvar and operas were performed by visiting companies from Italy and Germany but apparently there were no works by local musicians. After the Baroque, however, musical societies were organized and the musical situation improved.

Ivan Lukacic, a Croatian organist and composer, was born in Sikenik in 1587 and died in Split in 1648. At the age of ten he entered the Franciscan order. Subsequently he went to Italy where he studied theology and music, receiving a master's degree in music from Rome in 1615. In 1618 he returned to Sibenik but he soon went to Split were he became a prior of the Franciscan monastery and also director of music at the Split cathedral. Only one collection of his music survives (it was found in a library in Berlin in the middle of this century). It contains twenty-seven Latin motets for one to five voices with the accompaniment of <u>basso</u>

continuo. Since originally the collection had been published in Venice in 1620, undoubtedly Lukacic's music was performed in Italy.

Sicut cedrus by Lukacic represents a piece in the new monodic style of the Baroque. This was a time when a new style was introduced to music. As a reaction to to the tradition of polyphony where the text was unintelligible since different voices were singing different texts at the same time, monody was born. Composers began setting the text expressively for solo voice with just a simple accompaniment of an instrument to provide the harmonic background which allowed the text to be heard and understood. This monodic style gave rise to a whole new genre: opera. In this composition when the text and music are repeated, elaborate figurations are added. The harmony contributes to the dramatic character of the piece. Lukacic set a contrasting section in the middle in a completely different style - - syllabic and chordal. This section is repeated at the end and the song closes with a dramatic expressive slow ending.

Although **Vincenz Jelic** (later changed to Jelich) was born in Croatia but he lived most of his life in Austria; in one prominent music book of today he is even referred to as an Austrian composer. Even though he is unfamiliar to concert audiences today, his name appeared in numerous dictionaries and music histories not only in his lifetime but even through the nineteenth century. When we was twelve, Jelich became a choir boy in the chapel of Archduke Ferdinand at Graz. He attended a Jesuit University from 1610-1615 and then he went to Zabern in Alsace where he spent the rest of his life. There he was employed at the court of Leopold. In Zabern he took holy orders and, in addition to being an instrumentalist at the court, he was vicar and then canon at the church of St. Marie. Since no records after 1636 mention Jelic, it is assumed that he died in an epidemic after one of the battles of the

Thirty Years War. Three collections of Jelich's music were published in Strasbourg during his lifetime.

The Trumpet Voluntary has become a favorite of audiences today. However it is neither a voluntary, a trumpet piece, nor is it by Purcell. The term voluntary designated an improvisatory piece and later it was applied to pieces that were played as preludes or postludes. This piece is a march. It was originally published in 1700 in A Choice Collection of Ayres for Harpsichord or Spinet with the title The Prince of Denmark's March by **Jeremiah Clarke**. Jeremiah Clarke lived an interesting life. He was about 10 years younger than Purcell. He had lucrative jobs but unfortunately was unlucky in love. The story is told that one night when riding home from a friend's house, he was depressed. He dismounted from his horse to toss a coin to decide whether or not he should end his life. Instead of landing with a face showing, the coin stood on end in the clay. Not heeding this omen, he rode home and shot himself anyway.

The piece itself, the Trumpet Voluntary also has an interesting history. It sees to have disappeared until 1870 when it was published by the organist William Spark with Purcell's name on it. Soon afterwards Sir Henry Wood made an arrangement as a voluntary for trumpet and published it with the designation Purcell-Wood. At the same time it was also published under the true composer's name although today the composition is usually connected with Purcell.

Francois Couperin belonged to a dynasty of musicians, as did Bach. The Couperin dynasty dominated the French musical scene for about 200 years.[3] Couperin became known as Le Grand because of his superiority as an organist. He stated his philosophy of music: "I prefer that which touches me to that which surprises me." Known principally for his harpsichord pieces with programmatic titles such as The Cuckoo, The Hen and The

Nightingale, he also introduced the trio sonata into France and his important work, L'art de Toucher de Clavecin is a valuable primer for playing the harpsichord. He wrote two masses for the organ, Messe solemnelle a l'usage des Paroisses (Mass for the parishes) and Messe pour les Convents (Mass for the convents), which were published when he was only 21 years old. The organ masses consist of short pieces for the Mass with the organ playing between veres which are sung by the chior. This demonstrates that the place of the organ in the French church did not receive the prominence and prestige that it did in the German churches of the time. The Mass of the parishes was intended to be used only on principal feast days and the Mass of the convents was intended for more modest occasions. The Title, Offertoire sur les Grand Jeaux indicates that the composer wanted the organist to use the reeds on the organ when he played the piece.

Vergin tutta amor is a simple sorrowful prayer to the Virgin Mary asking for mercy. **Francesco Durante**, little known today, was a Neapolitan composer who was born a year before Handel. He was famous especially as a composer of church music and a teacher of the famous opera composers, Paisello, Piccini and Pergolesi, who dominated the opera scene in Italy at the end of the eighteenth century. This short song illustrates his expressive lyrical style which is established by the peaceful use of 12/8 meter. The urgency of the plea is made more poignant by repetitions of the text and passages with chromatic harmony.

In the Baroque, trumpet players were more highly regarded and highly paid than most other musicians.[4] In no other period were they held with more esteem and affection and understanding. Trumpets were used in operas, cantatas, military scenes, and, of course, for all special occasions. The most famous center for trumpet players and composition was in Bologna. In England, the trumpet was frequently imitated in compositions for harpsichord

and organ where the trumpet stop was used. **Henry Purcell** was the English musical genius of the Baroque and the last great English composer until the twentieth century. Even though he only lived a short time, his output was prodigious; he wrote vocal music of all kinds: odes for chorus and orchestra, anthems, cantatas, solo songs, chamber sonatas, keyboard music, one opera and incidental music for plays. This sonata is in three movements (fast, slow, fast) with the trumpet silent in the second movement when the top line was meant to be highly varied and ornamented. In the third movement, Purcell introduces a fanfare section which is repeated. The movement is unusual because when the main theme reappears in the second section, it is inverted.

The Messiah needs no introduction to concert audiences. It is perhaps the most famous and favorite oratorio that has ever been written.[5] **Handel** wrote the Messiah in just twenty-four days. Handel had abandoned operas in the 1730's and turned his attention to oratorios because opera had lost its popularity in England. Several versions of the Messiah exist; Handel altered the music for each performance, sometimes changing the voice parts. For example, he would change a soprano solo into a solo for tenor or he might change the key. The first performance was given in Dublin in April 1742. The first rehearsal met with instant success, so in order to accommodate more people in the hall, the ladies were asked not to wear hoopskirts and the men were asked to leave their swords at home. However when the Messiah was first performed in London, it received mixed reviews. Many objections were raised because Handel performed a sacred piece in a theater and he used theatrical singers in a biblical work. The Messiah eventually came to be a tradition in London; it was performed every year in the Lenten season as the last oratorio just before Easter. The Messiah also was the last oratorio that Handel conducted before he died.

The aria, Wie lieblich ist der Boten Schritt (How beautiful are the feet of them) is from Part II. For the performance in Dublin, Handel changed the soprano solo to a duet for soprano and alto followed by a chorus which sang the chorale, Break forth into joy. However, the original soprano solo was restored for the London performance in 1744. Handel wrote it as a siciliano (characterized by 12/8 meter and dotted rhythm) which was typical for pastoral scenes and produces a tranquil affect.

The text Erwach, froliche, O Tochter von Zion (Rejoice Greatly Oh Daughter of Zion) comes from Zecharaiah. It portrays the joy that the long-awaited saviour is coming. Notice the change in mood in the middle section where the text describes the fact that the saviour will teach peace. This is created by the employment of smooth lines and slower stepwize descending lines. These descending lines also graphically depict Christ descending to earth. In the first version, this aria was written with a dance rhythm but in a later version Handel changed to a simple meter which is what we commonly hear today.

REFERENCES

1. For a description of French organs at the time, see F. Douglass, "The Language of the Classical French Organ", Yale University Press, New Haven, CN, p.1 (1969).
2. See Josip Andreis, "Music of Croatia", trans. by Vladimir Ivir, Institute of Musicology, Zagreb, Yugoslavia (1974).
3. See W. Mellers, "Francois Couperin and the French Classical Tradition", Dover Publications, New Yourk (1968).
4. See D. Smithers, "The Music and History of the Baroque Trumpet before 1721" J.M. Dent, London (1973).
5. See J. Larsen, "Handel's Messiah" 2nd ed., W.W. Norton, N.Y. (1972).

APPENDIX

SANS-NSRS CONFERENCE CONCERT

DOMINICAN MONASTERY

Meduvratima od Ploca

"Old City" of Dubrovnik

2100 Saturday 7 June 1986

Performers

Milica Maravic, Soprano

Miroslav Zivkovic, Baritone

Zivko Kocev, Trumpet

Hrvojka Mihanovic, Organ

Program

L. Clerambault	Grand Plein Jeu (Organ solo)
I. Lukacic	Sicut cedrus (Soprano solo)
V. Jelic	Cantate Domino (Baritone solo)
	Viri sancti (Soprano solo)
	Quae est ista (duet)
H. Purcell	Trumpet Voluntary (Trumpet & Organ)
	---Intermission---
F. Couprin	Offertoire sur les Grands Jeux (Organ)
F. Durante	Vergin tutto amor (Baritone solo)
G.F. Handel	Dank Sei Dir Herr (Baritone solo)
H. Purcell	Sonata (Trumpet & Organ)
	Pomposo, Adagio, Presto
G.F. Handel	Selections from the Messiah (Soprano)
	Wie lieblich ist der Boten Schritt
	Erwach, froliche, O Tochter von Zion

AUTHOR INDEX

*** ALAGA PRIZE AWARDEE (see Announcement on page xix)**